Introduction to Solid-State NMR Spectroscopy

Introduction to Solid-State NMR Spectroscopy

Melinda J. Duer
Department of Chemistry
University of Cambridge

Blackwell
Science

Editorial offices:
Blackwell Publishing Ltd, 9600 Garsington Road, Oxford OX4 2DQ, UK
Tel: +44 (0)1865 776868
Blackwell Publishing Inc., 350 Main Street, Malden, MA 02148-5020, USA
Tel: +781 388 8250
Blackwell Publishing Asia Pty Ltd, 550 Swanston Street, Carlton, Victoria 3053, Australia
Tel: +61 (0)3 8359 1011

First published 2004 by Blackwell Publishing Ltd

Library of Congress Cataloging-in-Publication Data

Duer, Melinda J.
Introduction to solid-state NMR spectroscopy / Melinda J. Duer.
p. cm.
Includes bibliographical references and index.
ISBN 1-4051-0914-9 (acid-free paper)
1. Nuclear magnetic resonance spectroscopy. 2. Solid state chemistry. I. Title.

QD96.N8D84 2004
543′.66–dc22

2003063027

ISBN 1-4051-0914-9

A catalogue record for this title is available from the British Library

Set in Sabon
by SNP Best-set Typesetter Ltd., Hong Kong

For further information on Blackwell Publishing, visit our website:
www.blackwellpublishing.com

To my Father, John Duer
and
Ron Trevithick
who both inspired me

Contents

Preface

The days when nuclear magnetic resonance (NMR) was a technique applicable only to solution-state samples have long since gone. In the last thirty years, NMR spectroscopists have striven to make solid-state NMR a truly useful method, applicable to a very wide range of samples. In this, they have succeeded admirably, and today NMR spectroscopists have at their disposal a huge battery of solid-state NMR experiments that allow features of molecular structure and molecular dynamics to be determined.

Spectral resolution used to be a problem for solid-state NMR. It is no longer. The resolution routinely obtainable in solid-state NMR spectroscopy is now largely limited by the homogeneity (or lack of it!) of the samples being investigated, and not by the solid-state NMR technique. Consequently, the kind of resolution typically seen in solution-state NMR spectra is now realistically achievable for the solid state.

But solid-state NMR can do so much more than simply repeat the experiments of its solution-state counterpart. Solid-state NMR experiments can be set up such that the anisotropic nuclear spin interactions, which vanish in solution-state NMR experiments, remain in force. Thus, the anisotropy of nuclear spin interactions, such as chemical shielding (giving rise to the chemical shift in the NMR spectrum), can be measured and, more to the point, *utilized*, by chemists.

Chemical shift anisotropy, dipole-dipole coupling and quadrupole coupling can all be used by the chemist to give *quantitative* information on molecular structure, conformation and dynamics. One of the huge advances in solid-state NMR in the last twenty years is the spectroscopist's ability to, in effect, switch on and off these anisotropic interactions (at least as far as the resulting NMR spectrum is concerned). This is what has enable us to measure accurately the strengths of these anisotropic interactions for use in chemistry.

So what kinds of problems can solid-state NMR solve? The short answer is many. The detailed answer is what prompted me to write this book. It can be used, for instance, to measure internuclear distances, quantitatively, which might enable one to determine the conformation of a molecule, or the length of a hydrogen bond, or determine a significant bond angle. Of course, diffraction techniques have been traditionally used to do this job but diffraction techniques require a crystalline lattice, on a relatively long length scale, before they can give useful results on structural problems. The excellent feature of solid-state NMR is that solid-state NMR can be used effectively even in inhomogeneous or amorphous systems.

The chemistry of today, and probably of the future, has a lot to do with heterogeneous (solid) systems. Polymers are an obvious example. Beyond simple polymers, there are polymer blends where two or more polymers are mixed on a molecular scale. New polymer materials involve the mixing of polymers with inorganic components, such as clays to improve the desired material properties. None of these systems can be usefully studied by traditional diffraction techniques. Solid-state NMR on the other hand can give huge amounts of information on such systems, ranging from features of molecular structure, to the length scale of mixing in blends, to information on the nature of the interaction between the components in organic-inorganic composites. Moreover, solid-state NMR can give very useful information on the molecular dynamics in such systems. Molecular dynamics are very important in determining material properties. For example, in molecular solids, stresses are generally dissipated by deformation or displacement (temporary or otherwise) of the molecules. The least damaging way of dissipating a stress is by a distortion of the molecular conformation, possibly with concomitant (small) displacements of surrounding molecules. This route requires molecules to have certain degrees of freedom in their molecular conformation, preferably with some mechanism for restoring the original structure. One of the few ways we can study the molecular degrees of freedom is with solid-state NMR.

Many catalytic systems consist of the active catalyst material mounted on a solid support – an intrinsically heterogeneous, solid system. Again, solid-state NMR, being a technique which probes local environments, can give information about the structure and siting of catalytic species, in situ. Similarly, it can give useful structural information on glasses, or so-called amorphous materials, and on microcrystalline materials, where crystals suitable for diffraction methods are not obtainable.

In a completely different area, biology has always dealt with heterogeneous systems. Solid proteins in particular, are receiving increasingly large amounts of interest, especially with the linking of debilitating diseases such as Alzheimer's, CJD and Type II diabetes with solid protein deposits in vital organs of the body. Proteins are notoriously difficult to crystallize, and a great deal of useful information on the structure of amyloid proteins, amongst others, is now being accumulated through solid-state NMR methods.

This book is intended to provide the necessary background for those wishing to use solid-state NMR to solve problems in chemistry, biochemistry, materials, geology and engineering. As such, it is suitable for undergraduates embarking on a specialist NMR course and graduate students, as well as potential solid-state NMR spectroscopists. I hope that it will give a useful starting point from which to embark into this very interesting and exciting branch of spectroscopy.

Melinda Duer

Acknowledgements

No book can be written in isolation, and this one is no exception. The discussions I have had with numerous colleagues over the years have all contributed to this book; they have been invaluable and long may scientists go on talking to each other.

One never comes to really understand a subject until one has had to teach it. I should acknowledge here the many students, both undergraduate and postgraduate, who by their insistent questioning, and their own striving to understand the principles of NMR, have surely deepened my own understanding.

There are several people who deserve special thanks: Dr James Keeler and Dr Sharon Ashbrook for discussion and their perceptive comments; Francesca Wood, Nick Groom, Peter Gierth, Robin Stein, Robin Orr and Dr Oleg Antzutkin for their painstaking work in proof reading and gently pointing out the more incomprehensible sentences. Finally, my husband Dr Neil Piercy has, as ever, kept my computer working and borne my bad tempers with patience and even a little humour.

The Basics of NMR 1

This chapter is primarily concerned with the basics of how to describe nuclear spin systems in NMR experiments. To this end, we first consider the classical vector model, which in many cases provides a sufficient description of an uncoupled spin system. As soon as there are interactions between the spins, such as dipolar coupling, we must use a quantum mechanical model to describe the dynamics of the spin system. We will use the *density operator* approach, which combines a quantum mechanical modelling of individual spins or sets of coupled spins with an ensemble averaging over all the spins (or sets of spins) in the sample.

The latter sections of the chapter deal with the essentials of recording Fourier transform (FT) NMR spectra. This is essential as it affects the way in which we view the spins in the sample and thus must influence our theoretical description of the spin system. Throughout, each topic is dealt with in such a way as to introduce the nomenclature which will be used in the rest of the book and to remind readers of the salient points. Those requiring a more in-depth discussion of these points are strongly recommended to read the superb book by Levitt [1].

1.1 The vector model of pulsed NMR

In the semi-classical model of NMR, only the net magnetization arising from the nuclei in the sample and its behaviour in magnetic fields is considered. It is a suitable model with which to consider the NMR properties of isolated spin-$\frac{1}{2}$ nuclei, i.e. those which are not coupled to other nuclei. This model also provides a convenient picture of the effects of radiofrequency pulses on such a system. Only a brief description is given here in order to define the terms and concepts that will be used throughout this book.

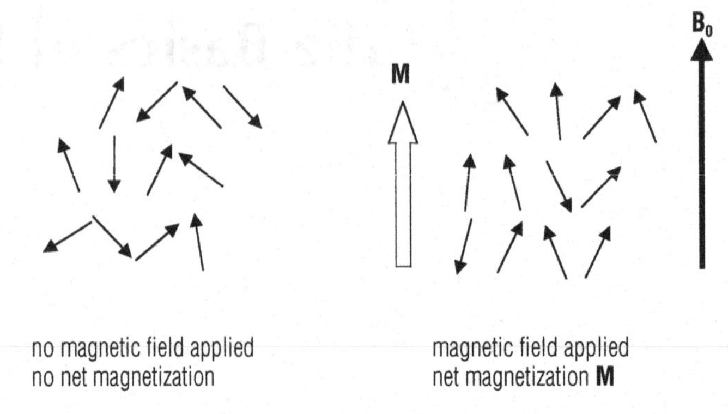

Fig. 1.1 The classical model of the formation of net nuclear magnetization in a sample. In the absence of a magnetic field, the individual nuclear magnetic moments (represented by vector arrows here) have random orientation so that there is no net magnetization. In the presence of an applied magnetic field, however, the nuclear magnetic moments are aligned preferentially with the applied field, except that thermal effects cause a distribution of orientations rather than perfect alignment. Nevertheless, there is in this case a net nuclear magnetization.

1.1.1 Nuclei in a static, uniform magnetic field

The net magnetization (which is equivalent to a bulk magnetic moment) arising from the nuclei in a sample is **M** and is the vectorial sum of all the individual magnetic moments associated with all the nuclei (Fig. 1.1):

$$\mathbf{M} = \sum_i \boldsymbol{\mu}_i \tag{1.1}$$

where $\boldsymbol{\mu}_i$ is the magnetic moment associated with the ith nucleus. In turn, each nuclear magnetic moment is related to the *nuclear spin* $\mathbf{I}_i$ of the nucleus by

$$\boldsymbol{\mu}_i = \gamma \mathbf{I}_i \tag{1.2}$$

where γ is the magnetogyric ratio, a constant for a given type of nucleus. Thus we can write the net magnetization of the sample as

$$\mathbf{M} = \gamma \mathbf{J} \tag{1.3}$$

where **J** is the net nuclear spin angular momentum of the sample giving rise to the magnetization **M**. If the nuclei are placed in a uniform magnetic field **B** as in the NMR experiment, a torque **T** is exerted on the magnetization vector:

$$\mathbf{T} = \frac{\mathrm{d}}{\mathrm{d}t}\mathbf{J} \tag{1.4}$$

In turn, the torque in this situation is given by

$$\mathbf{T} = \mathbf{M} \times \mathbf{B} \tag{1.5}$$

Combining Equations (1.3) to (1.5), we can write

$$\frac{\mathrm{d}\mathbf{M}}{\mathrm{d}t} = \gamma \mathbf{M} \times \mathbf{B} \tag{1.6}$$

which describes the motion of the magnetization vector **M** in the field **B**. It can be shown that Equation (1.6) predicts that **M** precesses about a fixed **B** at a constant rate $\omega = \gamma B$.

In NMR, the applied magnetic field is generally labelled $\mathbf{B}_0$ and is taken to be along z of the laboratory frame of reference, i.e. $\mathbf{B} = (0, 0, B_0)$ in the above equations. The frequency with which the magnetization precesses about this field is defined as ω_0, the *Larmor frequency*:

$$\omega_0 = -\gamma B_0 \tag{1.7}$$

1.1.2 The effect of rf pulses

An electromagnetic wave, such as a radiofrequency (rf) wave, has associated with it an oscillating magnetic field, and it is this field which interacts with the nuclei in addition to the static field in the NMR experiment. The rf wave is arranged in the NMR experiment so that its magnetic field oscillates along a direction perpendicular to z and the $\mathbf{B}_0$ field. Such an oscillating field can be thought of as a vector which can be written as the sum of two components rotating about $\mathbf{B}_0$ in opposite directions. The frequencies of these two components can be written as $\pm\omega_{rf}$, where ω_{rf} is the frequency of the rf pulse. Furthermore, it can be shown that only the component which rotates in the same sense as the precession of the magnetization vector **M** about $\mathbf{B}_0$ has any significant effect on **M**; we will henceforth label this component $\mathbf{B}_1(t)$. The effect of this field is most easily seen by transforming the whole problem into a rotating frame of reference which rotates at frequency ω_{rf} around $\mathbf{B}_0$; in this frame $\mathbf{B}_1$ appears static, i.e. its time dependence is removed.

We can see what happens to the $\mathbf{B}_0$ field in this frame by examining the effect of a similar rotating frame in the absence of an rf pulse, i.e. the case of the static, uniform magnetic field considered previously. We concluded that in the presence of a field $\mathbf{B}_0$ the magnetization vector **M** would precess around $\mathbf{B}_0$ at frequency ω_0. If the pulse is *on resonance*, i.e. $\omega_0 = \omega_{rf}$, then the magnetization vector appears stationary in the rotating frame. In effect, then, the $\mathbf{B}_0$ field is removed in this frame; the effective static field parallel to z is zero and hence the magnetization **M** is stationary. So, in the presence

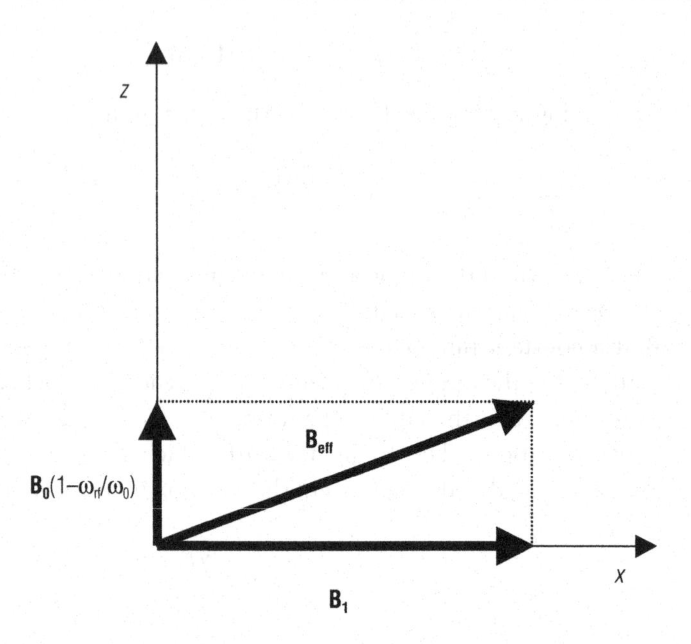

Fig. 1.2 The magnetic fields present in the rotating frame of reference. The rotating frame rotates about the laboratory *z*-axis at the frequency of the rf pulse, ω_{rf}. The rf irradiation is applied such that its oscillating magnetic field is along the laboratory *x*-axis – in common parlance, we say that the pulse is applied along *x*. The field due to the pulse appears static in the rotating frame, and the static field $\mathbf{B_0}$ appears to be reduced by a factor of ω_{rf}/ω_0, where ω_0 is the Larmor frequency, $\omega_0 = -\gamma B_0$. The net effective field in the rotating frame is the vectorial sum of the components along *x* and *z*, $\mathbf{B_{eff}}$. It is this field that the nuclear spin magnetization precesses around.

of a pulse, the only field remaining in the rotating frame is the $\mathbf{B_1}$ field. As in the case of the magnetization experiencing the static field $\mathbf{B_0}$ in the laboratory frame, the result of this interaction is that the magnetization vector **M** precesses about the resultant field, which is now $\mathbf{B_1}$, at frequency $-\gamma B_1$. We define this *nutation frequency* $-\gamma B_1$ as ω_1.

The direction of the magnetic field due to the rf pulse can be anywhere in the *xy* plane of the rotating frame. The *phase* of a pulse, ϕ_{rf}, is defined as the angle $\mathbf{B_1}$ makes to the *x*-axis in the rotating frame. The pulse does not have to be applied on resonance; indeed there will be many, many cases in solid-state NMR experiments when the pulse will be off resonance at least for part of the total spectrum available. In a frame rotating at ω_{rf} about $\mathbf{B_0}$, in the absence of a pulse, the Larmor precession frequency is reduced from ω_0 to $\omega_0 - \omega_{rf}$ about $\mathbf{B_0}$. We can infer from this that there is an effective static field along *z* in this frame of $(\omega_0 - \omega_{rf})/\gamma$, rather than zero as in the on-resonance case. The magnetic fields present in the rotating frame are then those shown in Fig. 1.2; there is a field of magnitude $(\omega_0 - \omega_{rf})/\gamma$ along *z* and B_1 along *x* (for a pulse with phase 0°). The nuclear magnetization precesses around the resultant field $\mathbf{B_{eff}}$ shown in Fig. 1.2.

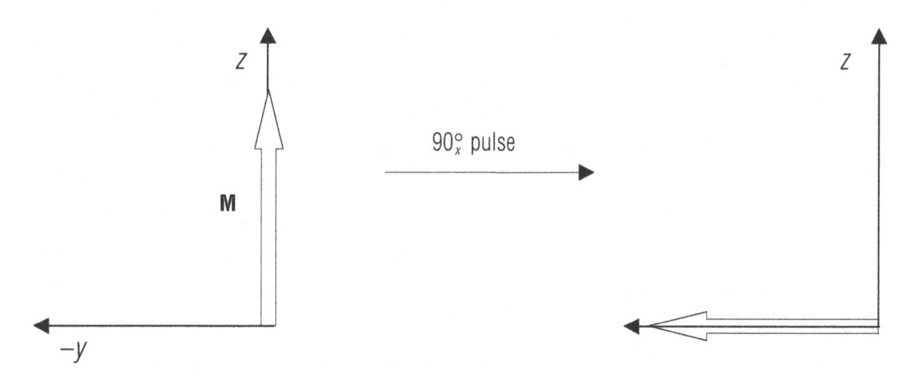

Fig. 1.3 The effect of a 90°_x on-resonance pulse on equilibrium magnetization in the rotating frame. The equilibrium magnetization rotates by 90° about x and ends up along $-y$.

NMR spectroscopists talk generally of an rf pulse 'flipping' the magnetization. The *flip angle* or *nutation angle*, θ_{rf}, of an on-resonance pulse is the angle that the pulse field $\mathbf{B}_1$ turns the magnetization during time τ_{rf}:

$$\theta_{rf} = \omega_1 \tau_{rf} = \gamma B_1 \tau_{rf} \tag{1.8}$$

Thus, a 90° pulse is simply one which has a flip angle of $\theta_{rf} = \pi/2$ radians or 90°. The corresponding pulse length is referred to as the 90° pulse length. Rf pulses along x in the rotating frame are referred to as 'x-pulses', those along y as 'y-pulses', and so on.

By definition, positive rotations are anticlockwise about the given axis. So, after a 90° x-pulse (for shorthand labelled 90°_x), nuclear magnetization $\mathbf{M}$, which started along z, is left lying along $-y$ (Fig. 1.3). From the point at which the rf pulse is turned off, the magnetization acts under the only magnetic field remaining, which is the effective field along z, of magnitude $(\omega_0 - \omega_{rf})/\gamma$, i.e. zero if the rotating frame frequency ω_{rf} is the same as the Larmor frequency, ω_0. If the effective field along z is zero, then the magnetization is stationary in the rotating frame after the pulse is switched off; if non-zero, the magnetization precesses around z from the position it was in at the end of the pulse at frequency $\omega_0 - \omega_{rf}$.

1.2 The quantum mechanical picture: hamiltonians and the Schrödinger equation

In the quantum mechanical picture, we start from a consideration of individual nuclei and, from this, generate a picture for the whole collection of nuclei in a sample. This is called the *ensemble average*. We will often refer to the *spin system* by which we mean a nuclear spin or collection of interacting nuclear spins, in a specified environment, such as a static magnetic field along z, for instance. We will find that the state of a spin system at equilibrium is one of a number of possible states or *eigenstates*, whose specific form depends on the nature of the spin system. In an NMR experiment, we have a sample which is composed of many identical spin systems. For instance, if each spin in the sample is isolated from other nuclei (so that there are no interactions between nuclei) and in the same chemical environment, then the spin system is a single nuclear spin subjected to whatever magnetic fields are present in the NMR experiment. The sample contains many such (identical) spin systems, the state of each being one of the possible eigenstates of the spin system. The proportion of the spin systems in the sample in any one eigenstate is given by a Boltzmann distribution for a sample which is at thermal equilibrium. In an NMR experiment, we measure the behaviour of the whole sample, not of individual spin systems. From the preceding discussion, it is clear that the behaviour of the sample depends not only on the nature of the possible eigenstates of the spin systems in the sample, but also on the population of each eigenstate over the sample as a whole. This will lead us to describe the state of each spin system in the sample as a superposition of the possible eigenstates for the spin system. The superposition state is the same for all identical spin systems and takes into account the probability of occurrence of each eigenstate, so leading to a proper description of the behaviour of the whole sample of spin systems.

Box 1.1: Quantum mechanics and NMR

In this box, we describe the key concepts in quantum mechanics which are used when discussing NMR and define the terms which will be used throughout this book.

Wavefunctions

In quantum mechanics, the state of a system, such as a nuclear spin or collection of spins in some specified environment, is described by a wavefunction, a mathematical function which depends on the spatial and spin coordinates of the nuclei in the system. We shall denote wavefunctions by the symbol ψ or Ψ. The wavefunctions themselves can often be specified by a set of *quantum numbers*, which in turn determine the values of the physical observables for the system.

Operators, physical observables and expectation values

We actually determine the value of any physical observable for the system by using the appropriate *operator* corresponding to the observable. An operator is simply something which acts on a function to produce another function, e.g. multiply by y, d/dx and so on. Throughout this work, we shall denote operators by '^', to distinguish them from functions, etc.

The value of a physical observable in a state described by ψ is equal to the *expectation value* of the corresponding operator. For an operator $\hat{A}$, the expectation value $\langle \hat{A} \rangle$ is

$$\langle \hat{A} \rangle = \frac{\int \psi^* \hat{A} \psi \, d\tau}{\int \psi^* \psi \, d\tau} \tag{i}$$

where the integrals are over all the spatial and spin coordinates of the wavefunction ψ.

Schrödinger's equation, eigenfunctions and eigenvalues

The wavefunction for a nuclear spin system in which all interactions are time invariant is the solution of the time-independent Schrödinger equation:

$$\hat{H}\psi = E\psi \tag{ii}$$

where $\hat{H}$ is the energy operator for the system, called the *hamiltonian*. The form of the hamiltonian varies from system to system and depends on the interactions in the spin system. The quantity E is the energy of the system. Equation (ii) is that of an *eigenvalue* equation; the general form of such an equation for an operator $\hat{A}$ is

$$\hat{A}f = af \tag{iii}$$

where f is an eigenfunction of the operator $\hat{A}$ and a is the eigenvalue of $\hat{A}$ corresponding to the eigenfunction f. Thus the problem of finding the wavefunction for a system is one of finding the eigenfunctions of the relevant hamiltonian operator. In general, there are a set of eigenfunctions which solve (iii) for any given operator. The set of all the functions which solve (iii) is called a *complete set*. That is, any other function can be written as some linear combination of this complete set.

Continued on p. 8

Box 1.1 *Cont.*

If the spin system is subject to a time-dependent interaction (as will often be the case in NMR experiments), then the wavefunction describing the state of the system is necessarily time dependent. We must then solve the time-dependent Schrödinger equation:

$$\hat{H}(t)\psi(t) = i\hbar\frac{\partial\psi(t)}{\partial t} \qquad \text{(iv)}$$

where the time-dependent interaction is represented by a time-dependent hamiltonian.

Spin operators and spin states

The total nuclear wavefunction Ψ can be approximately factorized into a spatial and a spin part:

$$\Psi = \psi_{\text{spin}}\psi_{\text{space}} \qquad \text{(v)}$$

where ψ_{spin} is a function, the *nuclear spin wavefunction*, describing the spin state of the nucleus, which depends only on the nuclear spin coordinates, and ψ_{space} is a function depending only on the spatial coordinates of the nucleus. We shall be dealing almost exclusively with the wavefunctions describing the *spin states* of nuclei. Fortunately, the spin and spatial parts of a nuclear wavefunction are largely uncoupled; that is, the spatial position of a nucleus is largely unaffected by its spin coordinates and vice versa.[N1]

To determine the spin properties of a nucleus, we operate on the nuclear spin wavefunction with *spin operators*, i.e. operators which act only on the spin coordinates of the nuclear wavefunction.

For a single spin, a consistent set of spin operators which allow all possible spin properties of a nucleus to be determined are: $\hat{I}^2$, the operator for the magnitude of the nuclear spin squared, $\hat{I}_x$, $\hat{I}_y$ and $\hat{I}_z$, which are the operators for the x, y and z components of nuclear spin respectively. These are *single spin operators*, i.e. they act only on wavefunctions describing one spin. These are related via

$$\hat{I}^2 = \hat{I}_x^2 + \hat{I}_y^2 + \hat{I}_z^2 \qquad \text{(vi)}$$

If two operators commute, it can be shown that they have identical eigenfunctions. The *commutator* of two operators $\hat{A}$ and $\hat{B}$, $[\hat{A}, \hat{B}]$, is defined as

$$[\hat{A}, \hat{B}] = \hat{A}\hat{B} - \hat{B}\hat{A} \qquad \text{(vii)}$$

Now $\hat{I}^2$ commutes with $\hat{I}_z$. Hence, $\hat{I}^2$ and $\hat{I}_z$ have identical eigenfunctions; they are specified by the quantum numbers I and m and are denoted ψ_{Im}. The eigenfunctions of $\hat{I}_z$ for $I = \frac{1}{2}$ are often denoted α ($m = \frac{1}{2}$) and β ($m = -\frac{1}{2}$). The quantum number I can take the values 0, $\frac{1}{2}$, 1, $\frac{3}{2}$, 2, . . . , etc. and m, the values I, $I - 1$, . . . , $-I$. Hence, we talk of ^{1}H being a 'spin-$\frac{1}{2}$' nucleus, i.e. $I = \frac{1}{2}$, ^{23}Na, a spin-$\frac{3}{2}$ nucleus ($I = \frac{3}{2}$) and so on.

The eigenvalues corresponding to the ψ_{Im} are defined by the eigenvalue equations:

$$\begin{aligned} \hat{I}^2\psi_{Im} &= I(I+1)\hbar\psi_{Im} \\ \hat{I}_z\psi_{Im} &= m\hbar\psi_{Im} \end{aligned} \tag{viii}$$

However, in NMR, the factor of $\hbar$ in these eigenvalues is commonly ignored, and included instead as part of the operator, i.e.

$$\begin{aligned} \hat{I}^2\psi_{Im} &= I(I+1)\psi_{Im} \\ \hat{I}_z\psi_{Im} &= m\psi_{Im} \end{aligned} \tag{ix}$$

or

$$\begin{aligned} \hbar\hat{I}^2\psi_{Im} &= I(I+1)\hbar\psi_{Im} \\ \hbar\hat{I}_z\psi_{Im} &= m\hbar\psi_{Im} \end{aligned} \tag{x}$$

We shall adopt this latter convention also. The expectation values of $\hat{I}^2$ and $\hat{I}_z$ (see Equation (i)), i.e. the magnitude of the nuclear spin angular momentum squared and the z-component of the nuclear spin angular momentum respectively, when the wavefunction describing the system is an eigenfunction of $\hat{I}^2$ and $\hat{I}_z$ are given by:

$$\begin{aligned} \langle\hat{I}^2\rangle &= \frac{\int\psi_{Im}^*\hat{I}^2\psi_{Im}\mathrm{d}\tau}{\int\psi_{Im}^*\psi_{Im}\mathrm{d}\tau} \\ &= I(I+1)\frac{\int\psi_{Im}^*\psi_{Im}\mathrm{d}\tau}{\int\psi_{Im}^*\psi_{Im}\mathrm{d}\tau} \\ &= I(I+1) \end{aligned} \tag{xi}$$

and

$$\langle\hat{I}_z\rangle = \frac{\int\psi_{Im}^*\hat{I}_z\psi_{Im}\mathrm{d}\tau}{\int\psi_{Im}^*\psi_{Im}\mathrm{d}\tau} = m \tag{xii}$$

Continued on p. 10

Box 1.1 *Cont.*

That is, the expectation values of $\hat{I}^2$ and $\hat{I}_z$, when the spin state is described by one of their eigenfunctions, are simply the eigenvalues of the respective operators.

If the hamiltonian for a spin system can be described entirely in terms of $\hat{I}^2$ and $\hat{I}_z$ operators only, then the spin wavefunctions of the system are eigenfunctions of $\hat{I}^2$ and $\hat{I}_z$, i.e. ψ_{Im}. If not, providing the spin system consists of non-interacting nuclei so that the hamiltonian only involves single spin operators, we can describe the spin wavefunction Ψ as a linear combination of the ψ_{Im} functions, i.e.

$$\Psi = \sum_m c_m \psi_{Im} \tag{xiii}$$

where the c_m are the combination coefficients in the eigenfunction Ψ of the system. The ψ_{Im} are said to be the *basis* or *basis set* for the expansion of Ψ. Substituting Equation (xiii) for Ψ into the time-independent Schrödinger's equation, we have

$$\begin{aligned} \hat{H}\Psi &= E\Psi \\ \Rightarrow \hat{H}\left(\sum_m c_m \psi_{Im}\right) &= E\sum_m c_m \psi_{Im} \\ \Rightarrow \sum_m c_m \hat{H}\psi_{Im} &= E\sum_m c_m \psi_{Im} \end{aligned} \tag{xiv}$$

We find the combination coefficients $\{c_m\}$ by multiplying from the left by each possible ψ^*_{Im} in turn and integrating over all spin space, so generating $(2I + 1)$ simultaneous equations. Each equation so generated has the form

$$\begin{aligned} \sum_m c_m \int \psi^*_{Im'} \hat{H}\psi_{Im} \mathrm{d}\tau &= E\sum_m c_m \int \psi^*_{Im'} \psi_{Im} \mathrm{d}\tau \\ \Rightarrow \sum_m c_m \int \psi^*_{Im'} \hat{H}\psi_{Im} \mathrm{d}\tau - Ec_{m'} &= 0 \end{aligned} \tag{xv}$$

where in the final step we have used the fact that the ψ_{Im} are orthogonal and normalized, i.e. $\int \psi^*_{Im'}\psi_{Im}\mathrm{d}\tau = \delta_{m'm}$. Equation (xv) can be rewritten in matrix form:

$$(\mathbf{H} - E\mathbf{1}) \cdot \mathbf{c} = 0 \tag{xvi}$$

where the elements of the matrix $\mathbf{H}$ are

$$H_{m'm} = \int \psi^*_{Im'} \hat{H}\psi_{Im} \mathrm{d}\tau \tag{xvii}$$

and the vector $\mathbf{c}$ contains the combination coefficients, c_m; $\mathbf{1}$ is a (diagonal) unit matrix. The $(2I + 1)$ simultaneous equations represented by Equation (xvi) can be solved by setting

$$\det|\mathbf{H} - E\mathbf{1}| = 0 \qquad \text{(xviii)}$$

and finding the $(2I + 1)$ values of E. Substituting any one of these back into Equation (xv) allows the c_m coefficients to be determined for the wavefunction corresponding to that value of E.

Dirac's bra-ket notation

When dealing with integrals of the form in Equation (xv), it is often easier to use Dirac's bra-ket notation, and we will use this notation throughout this book. In the notation, the integral in Equation (xv) is written

$$\int \psi_{Im'}^{*} \hat{H} \psi_{Im} \mathrm{d}\tau = \langle I, m' | \hat{H} | I, m \rangle \qquad \text{(xix)}$$

where, the *bra* is

$$\langle I, m' | = \psi_{Im'}^{*} \qquad \text{(xx)}$$

and the *ket* is

$$| I, m \rangle = \psi_{Im} \qquad \text{(xxi)}$$

Note that $\langle I, m|$ is the complex conjugate of $|I, m\rangle$ and that the presence of a ket and bra implies integration over all variables in ψ_{Im}.

Matrices

Matrix representations of operators such as hamiltonians (Equation (xvii)) are frequently needed in NMR. The ψ_{Im} functions involved in the integrals in each matrix element in Equation (xvii), for instance, are the *basis functions* for just such a representation. Below, for future use, are the matrices of the $\hat{I}^2$, $\hat{I}_x$, $\hat{I}_y$, $\hat{I}_z$ operators in the spin-$\frac{1}{2}$ $\Psi_{I,m}$ basis, $\psi_{\frac{1}{2},\frac{1}{2}}$ and $\psi_{\frac{1}{2},-\frac{1}{2}}$, otherwise denoted α and β.

$$\begin{aligned} \mathbf{I}^2 &= \frac{3}{4}\begin{pmatrix} 1 & 0 \\ 0 & 1 \end{pmatrix} & \mathbf{I}_x &= \frac{1}{2}\begin{pmatrix} 0 & 1 \\ 1 & 0 \end{pmatrix} \\ \mathbf{I}_y &= -\frac{i}{2}\begin{pmatrix} 0 & 1 \\ -1 & 0 \end{pmatrix} & \mathbf{I}_z &= \frac{1}{2}\begin{pmatrix} 1 & 0 \\ 0 & -1 \end{pmatrix} \end{aligned} \qquad \text{(xxii)}$$

where each matrix is

Continued on p. 12

Box 1.1 Cont.

$$A = \begin{pmatrix} \langle\alpha|\hat{A}|\alpha\rangle & \langle\alpha|\hat{A}|\beta\rangle \\ \langle\beta|\hat{A}|\alpha\rangle & \langle\beta|\hat{A}|\beta\rangle \end{pmatrix} \quad \text{(xxiii)}$$

The matrix elements of $\hat{I}_x$ and $\hat{I}_y$ are evaluated using the *raising* and *lowering* operators, which are defined as

$$\hat{I}_\pm = \hat{I}_x \pm i\hat{I}_y \quad \text{(xxiv)}$$

such that

$$\hat{I}_x = \frac{1}{2}(\hat{I}_+ + \hat{I}_-) \quad \hat{I}_y = -\frac{i}{2}(\hat{I}_+ - \hat{I}_-) \quad \text{(xxv)}$$

The action of these operators on a function ψ_{Im} is

$$\begin{aligned} \hat{I}_+|I,m\rangle &= (I(I+1) - m(m+1))^{\frac{1}{2}}|I,m+1\rangle \\ \hat{I}_-|I,m\rangle &= (I(I+1) - m(m-1))^{\frac{1}{2}}|I,m-1\rangle \end{aligned} \quad \text{(xxvi)}$$

In other words, $\hat{I}_+$ creates a new wavefunction with the quantum number m raised by one, whilst $\hat{I}_-$ creates one with m lowered by one. Note that if $\hat{I}_+$ operates on a wavefunction with the maximum value of m (for a given I), the result is zero, and similarly for $\hat{I}_-$ operating on a wavefunction with minimum m, i.e.

$$\hat{I}_+|I,I\rangle = 0 \quad \hat{I}_-|I,-I\rangle = 0 \quad \text{(xxvii)}$$

1.2.1 Nuclei in a static, uniform field

The simplest spin system is that consisting of an isolated spin in the static, uniform magnetic field of the NMR experiment, with no other interactions present. The hamiltonian $\hat{H}$ for a nuclear spin in a static field is

$$\hat{H} = -\hat{\boldsymbol{\mu}} \cdot \mathbf{B}_0 \quad (1.9)$$

where $\hat{\boldsymbol{\mu}}$ is the nuclear magnetic moment operator and $\mathbf{B}_0$ is the magnetic field applied in the NMR experiment. This hamiltonian is often referred to as the *Zeeman hamiltonian*. In turn, $\hat{\boldsymbol{\mu}}$ can be written in terms of the nuclear spin operator $\hat{\mathbf{I}}$:[N2]

$$\hat{\boldsymbol{\mu}} = \gamma\hbar\hat{\mathbf{I}} \quad (1.10)$$

The applied field is taken to be along z, so combining Equations (1.9) and (1.10), we have

$$\hat{H} = -\gamma\hbar\hat{I}_z B_0 \tag{1.11}$$

The *eigenfunctions* of $\hat{H}$ are the wavefunctions describing the possible states of the spin system in the $\mathbf{B}_0$ field. Since $\hat{H}$ is proportional to the operator $\hat{I}_z$ in this case, the eigenfunctions of $\hat{H}$ are the eigenfunctions of $\hat{I}_z$, which are simply written as $|I, m\rangle$ in *bra-ket* notation, or alternatively as ψ_{Im}, where I is the nuclear spin quantum number (see Box 1.1). The quantum number m can take $2I + 1$ values: $I, I - 1, I - 2, \ldots, -I$. The *eigenvalues* of $\hat{H}$ are the energies associated with the different possible states of the spin. The eigenvalues are obtained by operating with $\hat{H}$ on the spin wavefunctions:

$$\hat{H}|I,m\rangle = E_{I,m}|I,m\rangle \tag{1.12}$$

where $E_{I,m}$ is the energy of the eigenstate $|I, m\rangle$. Substituting Equation (1.11) for $\hat{H}$ in Equation (1.12) yields

$$\hat{H}|I,m\rangle = -(\gamma\hbar B_0)\hat{I}_z|I,m\rangle = -(\gamma\hbar B_0)m|I,m\rangle \tag{1.13}$$

since $|I, m\rangle$ is an eigenfunction of $\hat{I}_z$, with eigenvalue m, i.e.

$$\hat{I}_z|I,m\rangle = m|I,m\rangle \tag{1.14}$$

The energies of the eigenstates are obtained from comparing Equations (1.12) and (1.13):

$$E_{I,m} = -\gamma\hbar B_0 m \tag{1.15}$$

So for a spin with $I = \frac{1}{2}$, $m = \pm\frac{1}{2}$ there are two possible eigenstates with energies $E_{\frac{1}{2},\pm\frac{1}{2}} = \mp\frac{1}{2}\gamma\hbar B_0$ (Fig. 1.4). These states are frequently referred to as the *Zeeman states*. The transition energy ΔE between the spin states is $\gamma\hbar B_0$. In frequency units, this corresponds to ω_0 $(= \gamma B_0)$, the Larmor frequency in the vector model. Note, however, that the Larmor frequency in the vector model corresponds to a rotation of the net nuclear magnetization vector about $\mathbf{B}_0$ and not to a transition frequency.

So in a sample of non-interacting spin-$\frac{1}{2}$ nuclei, each spin system can exist in one of two possible eigenstates. At equilibrium, there is a Boltzmann distribution of nuclear spins over these two states, the population of each eigenstate ψ being p_ψ given by

$$p_\psi = \frac{\exp(-E_\psi/kT)}{\sum_{\psi'}\exp(-E_{\psi'}/kT)} \tag{1.16}$$

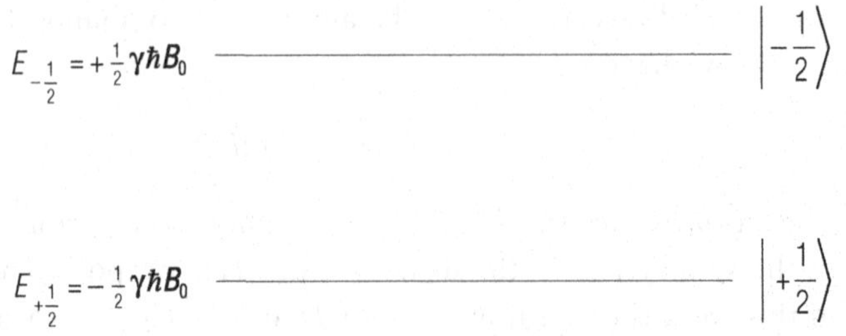

$E_{-\frac{1}{2}} = +\frac{1}{2}\gamma\hbar B_0$ ———————— $\left|-\frac{1}{2}\right\rangle$

$E_{+\frac{1}{2}} = -\frac{1}{2}\gamma\hbar B_0$ ———————— $\left|+\frac{1}{2}\right\rangle$

Fig. 1.4 The energy levels (Zeeman levels) for a spin-$\frac{1}{2}$ nucleus in an applied magnetic field $\mathbf{B}_0$ (positive γ). The levels are labelled according to their magnetic quantum number, *m*.

where E_ψ is the energy of the ψ eigenstate. For spin-$\frac{1}{2}$ nuclei, $\sum_{m=\pm\frac{1}{2}} \exp(-E_m/kT) \approx 2$.

The nature of the eigenstates and their respective populations determines all the properties of the ensemble of spin systems in the sample and hence determines the outcome of any NMR experiment on the sample. For instance, the expectation value of the z-magnetization for the sample is given by a sum of contributions from each possible eigenstate, scaled by the population of each eigenstate. We call this the *ensemble average* of the z-magnetization, and denote it by a bar over the appropriate quantities, i.e. those which are averaged over the sample:

$$\overline{\langle\hat{\mu}_z\rangle} = \gamma\hbar\overline{\langle\hat{I}_z\rangle} = \gamma\hbar\sum_{\psi} p_\psi\langle\psi|\hat{I}_z|\psi\rangle \tag{1.17}$$

where $\gamma\hbar\langle\psi|\hat{I}_z|\psi\rangle$ is the expectation value of the z-magnetization for a spin in eigenstate ψ.

In this picture of the spin ensemble, each spin system is in one of the possible eigenstates of the hamiltonian describing a single spin, the probability of it being in the ψ eigenstate being p_ψ. Alternatively, we can describe each spin system as being in the same superposition state, Ψ, where

$$\Psi = \sum_{\psi} \sqrt{p_\psi}\,|\psi\rangle \tag{1.18}$$

This is a completely equivalent approach as identical spin systems in the sample cannot be distinguished nor their individual spin states observed in the NMR experiment. All we can observe is the ensemble average, and whether we choose to describe the spin systems of the sample as being distributed over a set of eigenstates or in some superposition state, the same ensemble properties are calculated. For instance, if we use the superposition state Ψ to calculate the expectation value of z-magnetization, we obtain

$$\overline{\langle\hat{\mu}_z\rangle} = \gamma\hbar\langle\Psi|\hat{I}_z|\Psi\rangle = \gamma\hbar\sum_{\psi} p_{\psi}\langle\psi|\hat{I}_z|\psi\rangle \tag{1.19}$$

which is the same expression obtained previously (Equation (1.17)) by considering the distribution of spins over the possible eigenstates for each spin.

Expanding Equation (1.17) or equivalently (1.19) for the two level system corresponding to isolated spin-$\frac{1}{2}$ nuclei in the $\mathbf{B}_0$ field, we have

$$\begin{aligned}\overline{\langle\hat{\mu}_z\rangle} &= \gamma\hbar\left(p_{\frac{1}{2}}\left\langle\frac{1}{2},\frac{1}{2}\middle|\hat{I}_z\middle|\frac{1}{2},\frac{1}{2}\right\rangle + p_{-\frac{1}{2}}\left\langle\frac{1}{2},-\frac{1}{2}\middle|\hat{I}_z\middle|\frac{1}{2},-\frac{1}{2}\right\rangle\right)\\ &= \gamma\hbar\left(\frac{1}{2}p_{\frac{1}{2}} - \frac{1}{2}p_{-\frac{1}{2}}\right)\\ &= \frac{1}{2}\gamma\hbar\left(p_{\frac{1}{2}} - p_{-\frac{1}{2}}\right)\end{aligned} \tag{1.20}$$

where $p_{\pm\frac{1}{2}}$ are the populations of the respective spin states. In other words, the z-magnetization corresponds to the population difference between the two spin states.

1.2.2 The effect of rf pulses

An rf pulse introduces an oscillating magnetic field, $\mathbf{B}_1(t)$, into the spin system. The time dependence of the magnetic field in this case means that both the eigenstates of the spin systems and their energies are time-dependent, in contrast to the previous case considered of the nuclei in the static field $\mathbf{B}_0$. We will find that the eigenstates of the hamiltonian describing the spin systems in this case are time-dependent, linear combinations of the Zeeman states found previously, i.e. the eigenstates for spins in a static field, $\mathbf{B}_0$. Thus, we say that the oscillating field $\mathbf{B}_1(t)$ *mixes* the Zeeman states. The hamiltonian, $\hat{H}$, describing a single spin in this situation must now include the interaction of the nuclear spin with both the static $\mathbf{B}_0$ field along z and the oscillating $\mathbf{B}_1(t)$ field of amplitude $2B_1$, which will be taken to oscillate along x. The total field felt by the nucleus is then

$$\mathbf{B}_{\text{total}}(t) = \mathbf{i}2B_1\cos(\omega_{\text{rf}}t) + \mathbf{k}B_0 \tag{1.21}$$

where $\mathbf{i}$ and $\mathbf{k}$ are unit vectors along x and z respectively. The factor of two in the amplitude of the $\mathbf{B}_1$ field is for convenience, as becomes apparent in the discussion in Box 1.2 later. Bearing in mind that the general form for a hamiltonian describing the interaction of a nuclear spin $\mathbf{I}$ with a field $\mathbf{B}$ is (in frequency units)

$$\hat{H} = -\hat{\boldsymbol{\mu}}\cdot\mathbf{B} = -\gamma\hat{\mathbf{I}}\cdot\mathbf{B} \tag{1.22}$$

the hamiltonian for the current case is then

$$\hat{H} = -\gamma(\hat{I}_z B_0 + \hat{I}_x B_1 \cos(\omega_{rf} t)) \tag{1.23}$$

As in the vector model, the oscillating $\mathbf{B}_1$ vector can be written as two counter-rotating components. It can be shown that only one of these components has any significant effect on the spin system, allowing the hamiltonian of Equation (1.23) to be rewritten as (see Box 1.2 below for details)

$$\hat{H} = -\gamma(\hat{I}_z B_0 + B_1 e^{-i\omega_{rf} t \hat{I}_z} \hat{I}_x e^{+i\omega_{rf} t \hat{I}_z}) \tag{1.24}$$

Ultimately, we want to find the (time-dependent) wavefunctions, Ψ, corresponding to $\hat{H}$ in Equation (1.23). We are thus obliged to use the *time-dependent Schrödinger equation* to find the spin system eigenfunctions, rather than the time-independent Schrödinger equation of Equation (1.11). The time-dependent Schrödinger equation is (again in frequency units)

$$i\frac{\partial \Psi(t)}{\partial t} = \hat{H}(t)\Psi(t) \tag{1.25}$$

where $\Psi(t)$ are the (time-dependent) wavefunctions describing the spin system. To proceed, we need to remove the time dependence of the hamiltonian (Equation (1.23)) by transforming into a rotating frame identical to that used in the vector model previously, i.e. one rotating about $\mathbf{B}_0$ at rate ω_{rf}. The hamiltonian in this frame becomes $\hat{H}^*$ (see Box 2.1 for details of transforming hamiltonians into rotating frames) where

$$\hat{H}' = ((\gamma B_0 - \omega_{rf})\hat{I}_z + \gamma B_1 \hat{I}_x) \tag{1.26}$$

and the wavefunction Ψ becomes

$$\tilde{\Psi} = e^{+i\omega_{rf} t \hat{I}_z} \Psi \tag{1.27}$$

in the new rotating frame where the operator $e^{+i\omega_{rf} t \hat{I}_z}$ is the rotation operator required to *rotate the (spin coordinate) axis frame* in which the spin wavefunction is defined about the axis frame z by angle $\omega_{rf} t$ (see Box 1.2 for details). Using these in the time-dependent Schrödinger equation (Equation (1.25)) we obtain

$$i\frac{\partial \tilde{\Psi}}{\partial t} = ((\gamma B_0 - \omega_{rf})\hat{I}_z + \gamma B_1 \hat{I}_x)\tilde{\Psi} \tag{1.28}$$

where the time dependence has been removed by transforming the whole problem to the rotating frame.

We will solve the differential Equation (1.28) for the specific case of a spin system consisting of an isolated spin-$\frac{1}{2}$ nucleus. In the following discussion, we will use the eigenfunctions of the spin operator $\hat{I}_z$, $|\frac{1}{2},\frac{1}{2}\rangle$ and $|\frac{1}{2},-\frac{1}{2}\rangle$, i.e. the Zeeman states, which we shall abbreviate to $|\frac{1}{2}\rangle$ and $|-\frac{1}{2}\rangle$ depicting only the m spin quantum number for clarity. These functions form a complete set for a spin-$\frac{1}{2}$ nucleus, and so any state of a spin-$\frac{1}{2}$ nucleus can be expressed as some linear combination of them, albeit with time-dependent combination coefficients as will be the case here. So we write the eigenfunctions of Equation (1.28) that we seek as (dropping the ~ for convenience)

$$\Psi(t) = c_{\frac{1}{2}}(t)\left|\frac{1}{2}\right\rangle + c_{-\frac{1}{2}}(t)\left|-\frac{1}{2}\right\rangle \tag{1.29}$$

Substituting this into Equation (1.28) we obtain

$$i\left(\left|\frac{1}{2}\right\rangle\frac{\mathrm{d}c_{\frac{1}{2}}}{\mathrm{d}t} + \left|-\frac{1}{2}\right\rangle\frac{\mathrm{d}c_{-\frac{1}{2}}}{\mathrm{d}t}\right) = (\gamma B_1\hat{I}_x)\left(c_{\frac{1}{2}}\left|\frac{1}{2}\right\rangle + c_{-\frac{1}{2}}\left|-\frac{1}{2}\right\rangle\right) \tag{1.30}$$

By multiplying this equation from the left by $\langle\frac{1}{2}|$, integrating over all spin space and using the orthonormality of $|+\frac{1}{2}\rangle$ and $|-\frac{1}{2}\rangle$ we obtain

$$i\frac{\mathrm{d}c_{\frac{1}{2}}}{\mathrm{d}t} = \frac{1}{2}\gamma B_1 c_{-\frac{1}{2}} \tag{1.31}$$

where we have evaluated the matrix elements of $\hat{I}_x$ using the equations in (xxii) in Box 1.1. Now multiplying Equation (1.30) from the left by $\langle-\frac{1}{2}|$ instead, we obtain a second equation:

$$i\frac{\mathrm{d}c_{-\frac{1}{2}}}{\mathrm{d}t} = \frac{1}{2}\gamma B_1 c_{\frac{1}{2}} \tag{1.32}$$

Equations (1.31) and (1.32) are simultaneous equations which we can easily solve to find expressions for $c_{\pm\frac{1}{2}}$, the combination coefficients in Equation (1.29):

$$\begin{aligned} c_{\frac{1}{2}}(t) &= c_{\frac{1}{2}}(0)\cos\left(\frac{1}{2}\omega_1 t\right) + ic_{-\frac{1}{2}}(0)\sin\left(\frac{1}{2}\omega_1 t\right) \\ c_{-\frac{1}{2}}(t) &= c_{-\frac{1}{2}}(0)\cos\left(\frac{1}{2}\omega_1 t\right) + ic_{\frac{1}{2}}(0)\sin\left(\frac{1}{2}\omega_1 t\right) \end{aligned} \tag{1.33}$$

where $\omega_1 = -\gamma B_1$ as usual and $c_{\pm\frac{1}{2}}(0)$ are the combination coefficients at time $t = 0$, i.e. at the start of the pulse. For any given spin system in the sample, the $c_{\pm\frac{1}{2}}(0)$ are simply 1 or 0 depending on which of the two possible initial states, $|\pm\frac{1}{2}\rangle$, the spin system is in at the start of the pulse. Equations (1.33) with the $c_{\pm\frac{1}{2}}(0)$ coefficients set to 1 and 0 or vice versa then describe the two possible time-dependent states of that one spin system. The population of each of these states over the sample as a whole is determined by the populations of the initial starting states, $|\pm\frac{1}{2}\rangle$.

Alternatively, we can find the superposition state (Equation (1.18)) which effectively describes each spin system in the sample by using the superposition state at $t = 0$, i.e. in the absence of the pulse, to determine the coefficients $c_{\pm\frac{1}{2}}(0)$. We have already seen that in the absence of a pulse, each spin system can be described by a single superposition state of the form:

$$\Psi(0) = \sqrt{p_{\frac{1}{2}}}\left|\frac{1}{2}\right\rangle + \sqrt{p_{-\frac{1}{2}}}\left|-\frac{1}{2}\right\rangle \tag{1.34}$$

where $\sqrt{p_{\pm\frac{1}{2}}}$ are the square roots of the populations of the Zeeman states for the spin system, $|\pm\frac{1}{2}\rangle$; these can be substituted for the $c_{\pm\frac{1}{2}}(0)$ coefficients in Equations (1.33) to find the $c_{\pm\frac{1}{2}}(t)$ coefficients which define the time-dependent superposition state describing each of the spin systems during the rf pulse. As in Equation (1.34), these coefficients then correspond to the square roots of the 'populations' of the $|\pm\frac{1}{2}\rangle$ functions. In making this statement, however, we must recognize that the eigenstates of the spin system during an rf pulse are time-dependent mixtures of these functions and not the $|\pm\frac{1}{2}\rangle$ states themselves.

There are a couple of interesting points to note about Equations (1.33). If a 90° pulse is applied, i.e. $\omega_1 t = \pi/2$, then at the end of the pulse, $t = \pi/(2\omega_1) = t_p$, the average spin state coefficients are

$$\begin{aligned} c_{\frac{1}{2}}(t_p) &= \sqrt{p_{\frac{1}{2}}(0)}\cos(\pi/4) + i\sqrt{p_{-\frac{1}{2}}(0)}\sin(\pi/4) \\ c_{-\frac{1}{2}}(t_p) &= \sqrt{p_{-\frac{1}{2}}(0)}\cos(\pi/4) + i\sqrt{p_{\frac{1}{2}}(0)}\sin(\pi/4) \end{aligned} \tag{1.35}$$

so that the populations $p_{\pm\frac{1}{2}}(t_p) = c_{\pm\frac{1}{2}}(t_p)^* c_{\pm\frac{1}{2}}(t_p)$ are simply

$$\begin{aligned} p_{\frac{1}{2}}(t_p) &= \frac{1}{2}\left(p_{\frac{1}{2}}(0) + p_{-\frac{1}{2}}(0)\right) \\ p_{-\frac{1}{2}}(t_p) &= \frac{1}{2}\left(p_{-\frac{1}{2}}(0) + p_{\frac{1}{2}}(0)\right) \end{aligned} \tag{1.36}$$

since $\cos(\pi/4) = \sin(\pi/4) = \frac{1}{\sqrt{2}}$. In other words, the populations of the $|\pm\frac{1}{2}\rangle$ functions are equal after a 90° pulse. A similar analysis for a 180° pulse shows that the populations of the $|\pm\frac{1}{2}\rangle$ functions are inverted at the end of a 180° pulse, i.e.

$$p_{\frac{1}{2}}(t_p) = p_{-\frac{1}{2}}(0) \quad p_{-\frac{1}{2}}(t_p) = p_{\frac{1}{2}}(0) \tag{1.37}$$

Finally, we know from the vector model that an x-pulse creates $-y$-magnetization in the rotating frame. To calculate the expectation value of y-magnetization in the rotating frame in the quantum mechanical model, we use the definition of expectation value (Equation (i) in Box 1.1 above) with Equations (1.29) and (1.33) for the rotating frame eigenfunctions Ψ':

$$\begin{aligned}\langle \hat{\mu}_y(t)\rangle &= \gamma\hbar\langle \hat{I}_y(t)\rangle = \gamma\hbar\langle \Psi'(t)|\hat{I}_y|\Psi'(t)\rangle \\ &= \gamma\hbar\left(c_{-\frac{1}{2}}(t)^* c_{\frac{1}{2}}(t)\left\langle -\frac{1}{2}\middle|\hat{I}_y\middle|\frac{1}{2}\right\rangle + c_{\frac{1}{2}}(t)^* c_{-\frac{1}{2}}(t)\left\langle \frac{1}{2}\middle|\hat{I}_y\middle|-\frac{1}{2}\right\rangle \right) \\ &= -\frac{1}{2}\gamma\hbar\left(c_{\frac{1}{2}}(0)^2 - c_{-\frac{1}{2}}(0)^2 \right)\sin(\omega_1 t)\end{aligned} \tag{1.38}$$

where we have substituted Ψ' (and the coefficients $c_{\pm\frac{1}{2}}(t)$ within Ψ') and evaluated the matrix elements of $\hat{I}_y$, $\langle \pm\frac{1}{2}|\hat{I}_y|\mp\frac{1}{2}\rangle$. The y-magnetization in Equation (1.38) can be rewritten in terms of the populations of the $|\pm\frac{1}{2}\rangle$ functions:

$$\langle \hat{\mu}_y(t)\rangle = -\frac{1}{2}\gamma\hbar\left(p_{\frac{1}{2}}(0) - p_{-\frac{1}{2}}(0) \right)\sin(\omega_1 t) \tag{1.39}$$

Comparing this with Equation (1.20) for the z-magnetization in the $\mathbf{B}_0$ field prior to the rf pulse, we see that the y-magnetization is equal to the initial z-magnetization multiplied by a factor, $-\sin(\omega_1 t)$, which is the same result that the vector model gave us.

Box 1.2: Exponential operators, rotation operators and rotations

In NMR, we frequently use *exponential operators* which have the form $e^{\hat{A}}$ where $\hat{A}$ itself is an operator. An exponential operator is defined through the series expansion for an exponential:

$$e^{\hat{A}} = 1 + \hat{A} + \frac{\hat{A}^2}{2!} + \frac{\hat{A}^3}{3!} + \ldots \tag{i}$$

One of the properties of exponential operators which we will use from time to time is that $e^{(\hat{A}+\hat{B})} = e^{\hat{A}}e^{\hat{B}}$ only if the operators $\hat{A}$ and $\hat{B}$ commute (see Box 1.1 for definition of commutation).

Continued on p. 20

Box 1.2 *Cont.*

Rotation of vectors, wavefunctions and operators (active rotations)

A special class of exponential operators are the *rotation operators* which have the form $e^{-i\phi\hat{L}_\alpha}$ where $\hat{L}_\alpha$ is the operator for the α component of angular momentum. As we will show below, $e^{-i\phi\hat{L}_\alpha}$ is an operator for a rotation about the axis α by angle ϕ. Rotation operators can be used for rotating axis frames, vectors, functions such as wavefunctions and operators. Rotation of an object within an axis frame is called an *active rotation*; in such an operation, the axis frame remains unchanged, but the orientation of the object with respect to the frame changes. Perhaps the simplest such rotation operation is the rotation of a Cartesian vector $\mathbf{v}$ within the axis frame within which the vector is defined. The rotated vector $\mathbf{v}'$ is given by

$$\mathbf{v}' = \hat{R}\mathbf{v} \tag{ii}$$

where $\hat{R}$ is the rotation operator. In this book, we are primarily interested in the rotation of wavefunctions and operators within their defining axis frames. The transformation required to rotate a wavefunction ψ (or indeed any other type of function defined with respect to a Cartesian axis frame) is

$$\psi' = \hat{R}\psi \tag{iii}$$

where ψ' is the rotated wavefunction.

We can easily demonstrate that $e^{-i\phi\hat{L}_\alpha}$ represents a rotation operator for rotation of an object about axis α by angle ϕ by considering a specific example. Consider the rotation of a function $f(x, y, z) = x$ (i.e. the value of the function at all points in space is simply the value of the x-coordinate) by angle ϕ about z. The rotated function, $f'(x, y, z)$ is given by

$$f'(x,y,z) = \hat{R}_z(\phi)f(x,y,z) = e^{-i\phi\hat{L}_z}x \tag{iv}$$

Now, we can use the series expansion of Equation (i) to expand $e^{-i\phi\hat{L}_z}$ in Equation (iv):

$$e^{-i\phi\hat{L}_z}x = x - i\phi\hat{L}_z x - \frac{\phi^2}{2}\hat{L}_z^2 x + \ldots \tag{v}$$

We can then use the definition of the angular momentum operator $\hat{L}_z$ to find how $e^{-i\phi\hat{L}_z}$ operates on x. The definition of $\hat{L}_z$ is[N3]

$$\hat{L}_z = \frac{1}{i}\left(\hat{x}\frac{\partial}{\partial y} - \hat{y}\frac{\partial}{\partial x}\right) \tag{vi}$$

where $\hat{x}$, $\hat{y}$ are the operators for position along x and y respectively and whose operations are multiply by x and y respectively. Using this definition in Equation (v), we obtain

$$e^{-i\phi\hat{L}_z}x = x + y\phi - \frac{\phi^2}{2}x + \ldots \tag{vii}$$

which can be rewritten as

$$e^{-i\phi\hat{L}_z}x = x\cos\phi + y\sin\phi = f'(x,y,z) \tag{viii}$$

where we have used the series expansions for sine and cosine:

$$\cos\phi = 1 - \frac{\phi^2}{2} + \ldots \qquad \sin\phi = \phi - \frac{\phi^3}{6} + \ldots \tag{ix}$$

That Equation (viii) represents the rotated function $f'(x, y, z)$ can be easily seen as follows. Consider a line of points along the x axis, i.e. points with coordinates $(x, 0, 0)$ and their corresponding values of $f(x, y, z)$ (the original function before rotation). The values of f for these points are of course just x, for a function $f(x, y, z) = x$, i.e. the value of the function is the distance from the origin along this line of points. Now consider the rotation of f by angle ϕ about z. The part of the function that lay along x now lies along a line in the x–y plane which is orientated an angle ϕ from the x-axis. The coordinates of points along this line are $(r\cos\phi, r\sin\phi, 0)$ where r is the distance along the line from the origin. If we use these coordinates in the rotated function $f'(x, y, z)$ of Equation (viii), to obtain the values of the rotated function along this line, we get $f'(r\cos\phi, r\sin\phi, 0) = r\cos^2\phi + r\sin^2\phi = r$, i.e. the value of the function is equal to the distance from the origin along this new line. The rotated function at points along this line is thus equal to the original unrotated function at points along x, as it should be.

To rotate an operator within its defining axis frame, we must perform a transformation of the form

$$\hat{B}' = \hat{R}\hat{B}\hat{R}^{-1} \tag{x}$$

where $\hat{R}$ is the rotation operator, $\hat{B}$ the operator being rotated and $\hat{B}'$ the operator after rotation, i.e. the rotated operator.

In NMR, we frequently come across exponential operators of the form $e^{-i\phi\hat{I}_\alpha}$ which have a similar form to a rotation operator with the angular momentum operator replaced with a spin angular momentum operator. Indeed, it can be shown that $e^{-i\phi\hat{I}_\alpha}$ represents a rotation operator which acts on spin coordinates (rather than

Continued on p. 22

Box 1.2 *Cont.*

spatial coordinates as in the example above) because, of course, $\hat{I}_\alpha$ acts only on spin coordinates. We can demonstrate this with an example, using the transformation of Equation (x) to rotate a spin operator, $\hat{I}_x$, with $\hat{R}_z(\phi) = e^{-i\phi\hat{I}_z}$:

$$\begin{aligned}\hat{R}_z(\phi)\hat{I}_x\hat{R}_z(\phi)^{-1} &= e^{-i\phi\hat{I}_z}\hat{I}_x e^{+i\phi\hat{I}_z} \\ &= \left(1 - i\phi\hat{I}_z - \frac{\phi^2}{2}\hat{I}_z^2 + \ldots\right)\hat{I}_x\left(1 + i\phi\hat{I}_z - \frac{\phi^2}{2}\hat{I}_z^2 + \ldots\right) \\ &= \hat{I}_x - i\phi[\hat{I}_z, \hat{I}_x] - \frac{\phi^2}{2}[\hat{I}_z,[\hat{I}_z, \hat{I}_x]] \\ &\quad + \frac{i\phi^3}{6}[\hat{I}_z,[\hat{I}_z,[\hat{I}_z, \hat{I}_x]]] + \ldots \end{aligned} \tag{xi}$$

using the series expansion of the exponential operators (Equation (i)). To proceed further, we need to simplify the commutators in Equation (xi). The Cartesian spin operators do not commute among themselves, but the following commutation relation exists

$$\lfloor \hat{I}_x, \hat{I}_y \rfloor = i\hat{I}_z \tag{xii}$$

and all cyclic permutations of this, i.e.

$$[\hat{I}_z, \hat{I}_x] = i\hat{I}_y \quad [\hat{I}_y, \hat{I}_z] = i\hat{I}_x \tag{xiii}$$

Using the commutation relations of Equations (xii) and (xiii), we can simplify Equation (xi) quite considerably:

$$\begin{aligned}&\hat{I}_x - i\phi[\hat{I}_z, \hat{I}_x] - \frac{\phi^2}{2}[\hat{I}_z,[\hat{I}_z, \hat{I}_x]] + \frac{i\phi^3}{6}[\hat{I}_z,[\hat{I}_z,[\hat{I}_z, \hat{I}_x]]] + \ldots \\ &= \hat{I}_x + \phi\hat{I}_y - \frac{\phi^2}{2}\hat{I}_x - \frac{\phi^3}{6}\hat{I}_y + \ldots \\ &= \hat{I}_x\left(1 - \frac{\phi^2}{2} + \ldots\right) + \hat{I}_y\left(\phi - \frac{\phi^3}{6} + \ldots\right) \\ &= \hat{I}_x \cos\phi + \hat{I}_y \sin\phi \end{aligned} \tag{xiv}$$

where we have used the series expansions for $\cos\phi$ and $\sin\phi$ (Equation (ix)) in the last step.

So the transformation has transformed $\hat{I}_x$ into $\hat{I}_x\cos\phi + \hat{I}_y\sin\phi$. Figure B1.2.1, in which the Cartesian spin operators are represented as vectors along the appropriate axes, shows that this represents a rotation of $\hat{I}_x$ by an angle ϕ about z.

We can generalize this result; if two operators have the commutation relation

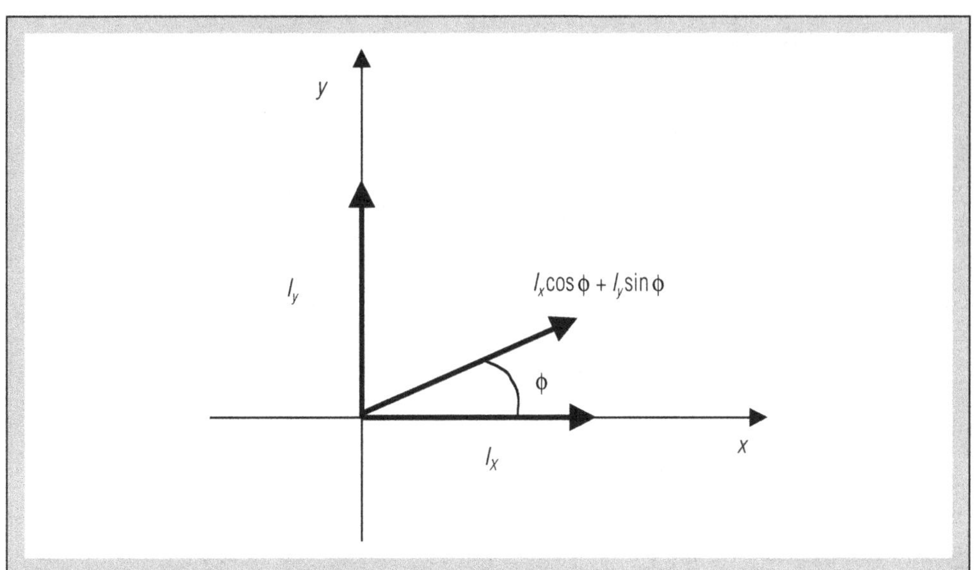

Fig. B1.2.1 The result of rotating I_x by an angle φ about the z-axis.

$$[\hat{A},\hat{B}] = i\hat{C} \tag{xv}$$

then the following transformation exists

$$e^{-i\phi\hat{A}}\hat{B}e^{+i\phi\hat{A}} = \hat{B}\cos\phi + \hat{C}\sin\phi \tag{xvi}$$

Rotation of axis frames

So far we have considered *active* rotations of operators and wavefunctions within their defining axis frames. Often, we will want to rotate an axis frame, and then re-express an operator or wavefunction in the new axis frame, a so-called *passive* rotation; the operator or wavefunction stays still while the axis frame moves (Fig. B1.2.2).

The relationship between a wavefunction ψ, expressed in the '*old*' frame and the same wavefunction expressed with respect to the '*new*' frame is

$$\psi^{new} = \hat{R}^{-1}\psi^{old} \tag{xvii}$$

where $\hat{R}$ is the rotation operator which transforms the *old* axis frame into the *new* axis frame, i.e.

Continued on p. 24

Box 1.2 Cont.

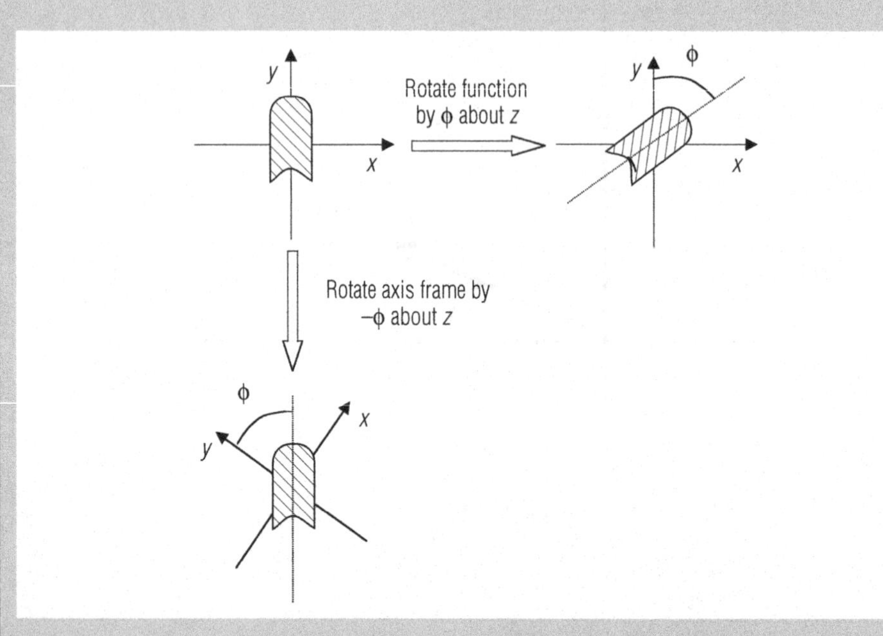

Fig. B1.2.2 Demonstrating that rotating a function by an angle ϕ about an axis is equivalent to rotating the axis frame by $-\phi$ about the same axis. The relationship between the function and the axis frame is the same in both cases.

$$\hat{R}(x^{old}, y^{old}, z^{old}) = (x^{new}, y^{new}, z^{new}) \tag{xviii}$$

The relationship between an operator $\hat{A}$ expressed with respect to its original frame ('*old*') and the same operator expressed in a rotated frame ('*new*') is

$$\hat{A}^{new} = \hat{R}^{-1}\hat{A}^{old}\hat{R} \tag{xix}$$

The rotation operators in all cases have the same form as those used in the active rotation of wavefunctions and operators, i.e. $\hat{R} = e^{-i\phi \hat{I}_\alpha}$, but remember that here $\hat{R}$ describes the rotation of the axis frame from the old frame into the new frame. It is worth comparing the equations for rotation of operators and wavefunctions (active rotations) with those expressing the effect of a rotation of axis frame on operators and wavefunctions (passive rotations), i.e. Equations (iii) and (x) and Equations (xvii) and (xix). Clearly, they have a very similar form, but the operator $\hat{R}$ in the active rotation equations is replaced by $\hat{R}^{-1}$ in the passive rotation equations and $\hat{R}^{-1}$ by $\hat{R}$. Figure B1.2.2 illustrates the reason for this: a rotation of a function (or operator) by an angle ϕ about a given axis (leaving the axis frame unchanged) leads to the same result as leaving the function alone and rotating the axis frame in which the function is defined by $-\phi$ about the same axis, i.e. rotating a function is equivalent to performing the *inverse rotation* on the axis frame in which the function is defined.

Representation of rf fields

In Equation (1.24) we needed to represent an oscillating magnetic field, specifically one of magnitude $2B_1$ which is oscillating at frequency ω_{rf} along the laboratory x-axis. We are now in a position to re-express this field in a more convenient form. The operator describing the interaction of this field with a spin **I** is

$$\hat{H}_{rf} = -2\gamma B_1 \hat{I}_x \cos \omega_{rf} t \qquad \text{(xx)}$$

This can be rewritten as

$$\hat{H}_{rf} = -\gamma B_1 e^{-i\omega_{rf} t \hat{I}_z} \hat{I}_x e^{+i\omega_{rf} t \hat{I}_z} - \gamma B_1 e^{+i\omega_{rf} t \hat{I}_z} \hat{I}_x e^{-i\omega_{rf} t \hat{I}_z} \qquad \text{(xxi)}$$

using Equation (xvi) above. The first term contains rotation operators which rotate $\hat{I}_x$ at frequency ω_{rf} about the laboratory z-axis, while the second term rotates $\hat{I}_x$ at frequency $-\omega_{rf}$ about the same axis, i.e. Equation (xxi) represents two counter-rotating magnetic field components. It can be shown that only the first term of Equation (xxi) has any effect on the nuclear spin system, which is the term used in Equation (1.24). To remove the time dependence of this first term in Equation (xxi), we need to perform a transformation

$$\begin{aligned} \hat{R}_z(-\omega_{rf} t) \hat{H}_{rf} \hat{R}_z(+\omega_{rf} t) &= -\gamma B_1 e^{+i\omega_{rf} t \hat{I}_z} \left(e^{-i\omega_{rf} t \hat{I}_z} \hat{I}_x e^{+i\omega_{rf} t \hat{I}_z} \right) e^{-i\omega_{rf} t \hat{I}_z} \\ &= -\gamma B_1 \hat{I}_x \qquad \text{(xxii)} \end{aligned}$$

which is equivalent to a rotation of the axis frame in which $\hat{H}_{rf}$ is defined about the laboratory z-axis at a rate ω_{rf} as defined in Equation (xix)

Euler angles

In NMR we often define rotation operators in terms of the *Euler angles* between the two frames, (X, Y, Z) and (x, y, z) in this case. Euler angles are generally labelled (α, β, γ) and are defined as follows:

Euler angles The transformation of frame (X, Y, Z) into (x, y, z) is described by a rotation of (X, Y, Z) by angle α about Z. This takes the (X, Y, Z) frame into $(X_2,$

Continued on p. 26

Box 1.2 *Cont.*

Y_2, Z_2). There then follows a rotation of angle β about the Y_2 axis that resulted from the previous rotation, taking the (X_2, Y_2, Z_2) frame into (X_3, Y_3, Z_3). Finally, there is a rotation of angle γ about the Z_3 axis that has resulted from the previous two coordinate rotations. This takes (X_3, Y_3, Z_3) into the (x, y, z) frame.

A completely equivalent definition of Euler angles expresses all rotations with respect to a single axis frame (rather than one which moves with the axis frame being rotated). This feature often makes this definition easier to deal with. It is:

Euler angles The transformation of (X, Y, Z) into (x, y, z) is described by a rotation of a frame coincident with (X, Y, Z) by γ about Z, taking this frame into (X', Y', Z'). There then follows a rotation of (X', Y', Z') by β about Y, i.e. the original axis frame Y-axis, taking the (X', Y', Z') frame into (X'', Y'', Z''). Finally, a rotation of α about Z, i.e. the original axis frame Z-axis, takes (X'', Y'', Z'') into (x, y, z). The frame being rotated in this case (the one initially coincident with (X, Y, Z)) acts like an object being rotated within the axis frame (X, Y, Z) in this definition of Euler angles. This definition thus employs *active* rotations, while the previous definition used passive rotations.

We employ the definition throughout that a positive rotation is that defined by the right-hand thumb rule, i.e. the direction of a positive rotation is the direction of the curl of the fingers when the right-hand thumb is pointed along the positive direction of the required axis.

It is not always easy to identify the Euler angles relating two frames! Polar angles are often easier to visualize; fortunately it is relatively easy to derive Euler angles from some polar angles as follows:

- (α, β) are the polar angles (θ, φ) of the z-axis in the (X, Y, Z) frame.
- (β, 180° − γ) are the polar angles of the Z-axis in the (x, y, z) frame.

Rotations with Euler angles

With the definition of the Euler angles, we can now derive an expression for a rotation operator, $\hat{R}(\alpha, \beta, \gamma)$, which performs the rotation of an axis frame (x, y, z) by the Euler angles (α, β, γ). We imagine this (x, y, z) frame to be attached to an object which is located within an axis frame (X, Y, Z) such that (x, y, z) is initially coincident with (X, Y, Z). $\hat{R}(\alpha, \beta, \gamma)$ is then the operator which performs a rotation of the object and its axis frame by an angle γ about Z, then β about Y

and finally α about Z. The operator $\hat{R}(\alpha, \beta, \gamma)$ can be broken down into its constituent rotations as

$$\hat{R}(\alpha,\beta,\gamma) = \hat{R}_Z(\alpha)\hat{R}_Y(\beta)\hat{R}_Z(\gamma) \qquad \text{(xxiii)}$$

where $\hat{R}_z(\alpha)$, for instance, is a rotation of angle α about the Z-axis. It has already been shown that the operator for a rotation of an object by angle θ about an axis a is

$$\hat{R}_a(\theta) = \exp(-i\theta\hat{L}_a) \qquad \text{(xxiv)}$$

where $\hat{L}_a$ is the operator for angular momentum (or spin) about axis a. So, we can now write down the expression for the operator $\hat{R}(\alpha, \beta, \gamma)$:

$$\hat{R}(\alpha,\beta,\gamma) = \exp(-i\alpha\hat{L}_Z)\exp(-i\beta\hat{L}_Y)\exp(-i\gamma\hat{L}_Z) \qquad \text{(xxv)}$$

Rotation of Cartesian axis frames

It is simple to derive a rotation matrix **R**, the matrix equivalent of the operator $\hat{R}$ above, which describes how to rotate an axis frame (x, y, z) fixed on an object within a frame (X, Y, Z) (and so rotate the object in the process). Consider first the rotation of γ about Z. Figure B1.2.3 below illustrates how this moves the (x, y, z) object-fixed frame.

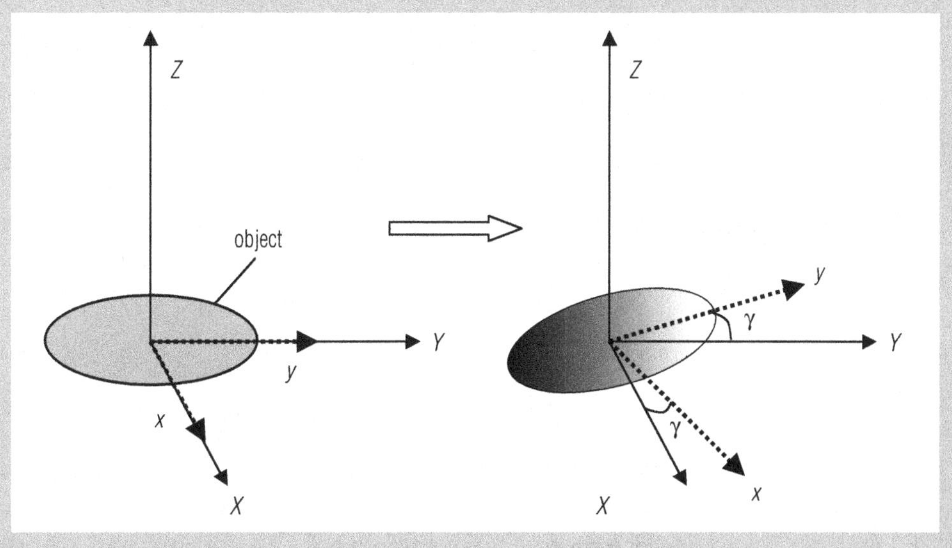

Fig. B1.2.3 The rotation of an object-fixed axis frame (x, y, z) by γ about Z.

Continued on p. 28

Box 1.2 *Cont.*

The rotation matrix which performs this transformation is:

$$\mathbf{R}_Z(\gamma) = \begin{pmatrix} \cos\gamma & -\sin\gamma & 0 \\ \sin\gamma & \cos\gamma & 0 \\ 0 & 0 & 1 \end{pmatrix} \qquad \text{(xxvi)}$$

This can be verified by taking unit vectors along each of the (X, Y, Z) axes, $\mathbf{n}_X$, $\mathbf{n}_Y$, $\mathbf{n}_Z$, which are coparallel with the initial orientation of (x, y, z) and performing the transformation $\mathbf{R}_Z(\gamma)\mathbf{n}_\alpha$ on each of these, where $\alpha = X, Y, Z$. The resultant in each case will be a new unit vector along the appropriate axis of the rotated (x, y, z).

We can then go on to the rotation of β about Y. The rotation matrix describing this rotation is:

$$\mathbf{R}_Y(\beta) = \begin{pmatrix} \cos\beta & 0 & \sin\beta \\ 0 & 1 & 0 \\ -\sin\beta & 0 & \cos\beta \end{pmatrix} \qquad \text{(xxvii)}$$

which can be verified in the same way as the previous rotation matrix. Finally, the last rotation of α about Z is described by:

$$\mathbf{R}_Z(\alpha) = \begin{pmatrix} \cos\alpha & -\sin\alpha & 0 \\ \sin\alpha & \cos\alpha & 0 \\ 0 & 0 & 1 \end{pmatrix} \qquad \text{(xxviii)}$$

Now we can produce the required transformation matrix, $\mathbf{R}(\alpha, \beta, \gamma)$, to go from (X, Y, Z) to the final orientation of (x, y, z) using Equation (xxiii) (in which rotation matrices can be substituted for rotation operators) and Equations (xxvi), (xxvii) and (xxviii):

$$\begin{aligned} \mathbf{R}(\alpha,\beta,\gamma) &= \mathbf{R}_Z(\alpha)\mathbf{R}_Y(\beta)\mathbf{R}_Z(\gamma) \\ &= \begin{pmatrix} \cos\alpha & -\sin\alpha & 0 \\ \sin\alpha & \cos\alpha & 0 \\ 0 & 0 & 1 \end{pmatrix} \begin{pmatrix} \cos\beta & 0 & \sin\beta \\ 0 & 1 & 0 \\ -\sin\beta & 0 & \cos\beta \end{pmatrix} \begin{pmatrix} \cos\gamma & -\sin\gamma & 0 \\ \sin\gamma & \cos\gamma & 0 \\ 0 & 0 & 1 \end{pmatrix} \end{aligned} \qquad \text{(xxix)}$$

which altogether gives:

$$\mathbf{R}(\alpha,\beta,\gamma) = \begin{pmatrix} \cos\alpha\cos\beta\cos\gamma - \sin\alpha\sin\gamma & -\cos\alpha\cos\beta\sin\gamma - \sin\alpha\cos\gamma & \cos\alpha\sin\beta \\ \sin\alpha\cos\beta\cos\gamma + \cos\alpha\sin\gamma & -\sin\alpha\cos\beta\sin\gamma + \cos\alpha\cos\gamma & \sin\alpha\sin\beta \\ -\sin\beta\cos\gamma & \sin\beta\sin\gamma & \cos\beta \end{pmatrix} \qquad \text{(xxx)}$$

Often, we will want to transform a Cartesian tensor **T** describing a physical quantity from being expressed in frame (X, Y, Z) to being expressed in a frame (x, y, z) (a passive rotation). The appropriate transformation under these circumstances is:

$$\mathbf{T}(x,y,z) = \mathbf{R}^{-1}\,\mathbf{T}(X,Y,Z)\mathbf{R} \tag{xxxi}$$

where **R** is the rotation matrix derived above which rotates a frame initially coparallel with (X, Y, Z) into a frame (x, y, z).

1.3 The density matrix representation and coherences

The quantum mechanical description given in the previous section examines spin systems in a sample of many spin systems through a superposition state. This approach is revealing, but time consuming.

A completely equivalent approach is to describe the spin system through a *density operator* or *density matrix*. Our discussion of the density operator which follows uses the ideas behind the superposition state introduced in Section 1.2. However, whereas our previous discussion of the superposition state used the specific example of identical spin–$\frac{1}{2}$ nuclei in various environments, here we keep the discussion completely general.

We start by imagining a collection of identical spin systems (the concept of a spin system was defined previously at the beginning of Section 1.2), each of which can be in any one of N states we label ψ. We do not know which state each individual spin system is in, only the probability p_ψ of it being in a particular state, ψ. This was what led us to describe the state of each spin system with a single superposition state, Ψ, where $\Psi = \sum_\psi \sqrt{p_\psi}\,\psi$. The expectation value of a quantity A with corresponding operator $\hat{A}$ over the sample is given by (see Box 1.1 for definition of expectation value):

$$\overline{\langle \hat{A} \rangle} = \langle \Psi | \hat{A} | \Psi \rangle = \sum_\psi p_\psi \langle \psi | \hat{A} | \psi \rangle \tag{1.40}$$

where the summation is over all the possible states for each spin system, and where we have assumed that the wavefunction ψ is normalized. Now let us write the state of the system in a general form as a sum over functions ϕ_i taken from a complete set of functions; we call this complete set the *basis set* for expressing the states of the spin system. We did this in Section 1.2.2 when finding the wavefunctions for a system consisting of a single spin under the effects of an rf pulse. In practice, the basis set is some convenient complete set, often the eigenfunctions of the Zeeman hamiltonian for the spin system, as in Section 1.2.1. So, we can write the possible states of each spin system in general as

$$\psi = \sum_i c_{\psi i}\phi_i \tag{1.41}$$

Substituting this in Equation (1.40) we have

$$\overline{\langle \hat{A} \rangle} = \sum_\psi p_\psi \sum_{i,j} c^*_{\psi i} c_{\psi j} \langle \phi_i | \hat{A} | \phi_j \rangle \tag{1.42}$$

The advantage of this approach is that the matrix elements of $\hat{A}$ in this basis, i.e. $\langle \phi_i | \hat{A} | \phi_j \rangle$, are the same whichever state ψ we are dealing with. If we now define $\sum_\psi p_\psi\, c_{\psi j}\, c^*_{\psi i}$ to be the jith element of another matrix ρ, then we can see that Equation (1.42) for the expectation value of $\hat{A}$ can be rewritten as

$$\langle \hat{A} \rangle = \mathrm{Tr}(\mathbf{A}\rho) = \sum_i (\mathbf{A}\rho)_{ii} = \sum_i \sum_j A_{ij}\rho_{ji} \tag{1.43}$$

where $\mathbf{A}$ is the matrix of operator $\hat{A}$ in the $\{\phi_i\}$ basis whose ijth element is $\langle \phi_i | \hat{A} | \phi_j \rangle$. The matrix ρ is called the *density matrix*. It has a corresponding operator which can be deduced by inspecting its matrix elements, i.e. $\rho_{ji} = \langle j | \hat{\rho} | i \rangle = \sum_\psi p_\psi c_{\psi j} c^*_{\psi i}$:

$$\hat{\rho} = \sum_\psi p_\psi |\psi\rangle\langle\psi| \tag{1.44}$$

For further details and discussion of the density matrix, the reader is referred to the excellent text by Goldman [2].

1.3.1 Coherences and populations

Let us examine some of the properties of the density operator and its matrix representation. First, the diagonal elements are equal to

$$\rho_{ii} = \sum_\psi p_\psi c^*_{\psi i} c_{\psi i} = \overline{c^*_{\psi i} c_{\psi i}} \tag{1.45}$$

The bar here simply means 'average over all the spin systems in the sample' or *ensemble average*, which is what the weighted sum over all possible states for the spin system ψ in Equation (1.45) represents. We can see from Equation (1.45) that ρ_{ii} is simply the average population of the ϕ_i basis function over the sample, as $c_i^* c_i$ is the population of the ith basis function. Secondly, consider an off-diagonal element of the density matrix:

$$\rho_{ij} = \sum_{\psi} p_{\psi} c^*_{\psi i} c_{\psi j} = \overline{c^*_{\psi i} c_{\psi j}} \tag{1.46}$$

Consider first a sample where all the spin systems in the sample are in the same state, ψ'. Then the diagonal elements of the corresponding density matrix are

$$\rho_{ii} = c^*_{\psi' i} c_{\psi' i} \qquad \rho_{jj} = c^*_{\psi' j} c_{\psi' j} \tag{1.47}$$

and the off-diagonal elements are

$$\rho_{ij} = c^*_{\psi' i} c_{\psi' j} \tag{1.48}$$

The diagonal elements represent the populations of the basis functions in the state ψ'. A non-zero off-diagonal element signifies that both $c_{\psi' i}$ and $c_{\psi' j}$ are non-zero and therefore that the state ψ' contains (possibly among other things) a mixture of ϕ_i and ϕ_j basis functions.

Now consider a sample in which there is a distribution of spin systems among all possible states. Then, the averaging over states which occurs in Equation (1.46) may well cause the off-diagonal elements of the density matrix to vanish. Indeed, the off-diagonal elements will vanish if there is no *correlation over time* between the basis functions from which the spin system states are derived. However, if there is some correlation between the basis functions, the average in Equation (1.46) no longer vanishes, and off-diagonal elements of the density matrix representing the sample will be non-zero. We say in this case that there is a *coherence* between the ϕ_i and ϕ_j functions in the superposition state Ψ, which describes the spin system. An example will help to clarify this; consider the situation dealt with in Section 1.2, where the spin system consisted of a single isolated spin. Let us take as our basis set of functions to describe the possible states of this system in any environment, the Zeeman states for this spin system, i.e. $|+\frac{1}{2}\rangle$ and $|-\frac{1}{2}\rangle$. In the absence of any rf pulses, the two possible states of the system are simply $|+\frac{1}{2}\rangle$ and $|-\frac{1}{2}\rangle$, i.e.

$$\psi_1 = \left|+\frac{1}{2}\right\rangle \quad \psi_2 = \left|-\frac{1}{2}\right\rangle \tag{1.49}$$

We now form the density matrix for this spin system, using Equations (1.45) and (1.46) for the diagonal and off-diagonal matrix elements respectively. The summation in these two equations is over the states ψ_1 and ψ_2. So for instance

$$\rho_{\frac{1}{2},\frac{1}{2}} = p_{\frac{1}{2}}(1 \times 1) + p_{-\frac{1}{2}}(0 \times 0) \tag{1.50}$$

The complete density matrix so formed is calculated to be (where the basis functions forming the elements are arranged horizontally and vertically in the order $|+\frac{1}{2}\rangle$, $|-\frac{1}{2}\rangle$)

$$\rho = \begin{pmatrix} p_{+\frac{1}{2}} & 0 \\ 0 & p_{-\frac{1}{2}} \end{pmatrix} \tag{1.51}$$

where $p_{\pm\frac{1}{2}}$ are the populations of the $|\pm\frac{1}{2}\rangle$ states. So, there are no coherences associated with this spin system, only populations of the basis functions.

Now consider the same spin system but subjected to an rf pulse. The wavefunctions describing the possible states of this spin system are given by Equations (1.29) and (1.33). The two possible wavefunctions are now

$$\begin{aligned} \psi_1 &= \cos\left(\frac{1}{2}\omega_1 t\right)\left|+\frac{1}{2}\right\rangle + i\sin\left(\frac{1}{2}\omega_1 t\right)\left|-\frac{1}{2}\right\rangle \\ \psi_2 &= +i\sin\left(\frac{1}{2}\omega_1 t\right)\left|+\frac{1}{2}\right\rangle + \cos\left(\frac{1}{2}\omega_1 t\right)\left|-\frac{1}{2}\right\rangle \end{aligned} \tag{1.52}$$

which are found by setting $c_{+\frac{1}{2}}(0) = 1$; $c_{-\frac{1}{2}}(0) = 0$ for ψ_1, i.e. initial state of the spin system at the start of the pulse ($t = 0$) is $|+\frac{1}{2}\rangle$ and $c_{+\frac{1}{2}}(0) = 0$; $c_{-\frac{1}{2}}(0) = 1$ for ψ_2, i.e. initial state is $|-\frac{1}{2}\rangle$. The density matrix is then calculated for this situation to be

$$\rho = \begin{pmatrix} P_+ & -i\Delta p\sin(\omega_1 t) \\ i\Delta p\sin(\omega_1 t) & P_- \end{pmatrix} \tag{1.53}$$

where $\Delta p = p_1 - p_2$ and

$$\begin{aligned} P_+ &= p_1\cos^2\left(\frac{1}{2}\omega_1 t\right) + p_2\sin^2\left(\frac{1}{2}\omega_1 t\right) \\ P_- &= p_1\sin^2\left(\frac{1}{2}\omega_1 t\right) + p_2\cos^2\left(\frac{1}{2}\omega_1 t\right) \end{aligned} \tag{1.54}$$

Here, then, there is a coherence between the $|+\frac{1}{2}\rangle$ and $|-\frac{1}{2}\rangle$ basis functions, as well as populations of both functions. In essence, the coherence is an expression of the fact that the wavefunctions describing the spin system in this situation contain mixtures of the $|+\frac{1}{2}\rangle$ and $|-\frac{1}{2}\rangle$ basis functions. Looking at the form of the wavefunctions for this case (Equation (1.51)), we see that over time the ψ_1 wavefunction oscillates between $\psi_1 = |+\frac{1}{2}\rangle$ and $\psi_1 = |-\frac{1}{2}\rangle$ while the ψ_2 wavefunction oscillates in the opposite sense, i.e. between $\psi_2 = |-\frac{1}{2}\rangle$ and $\psi_2 = |+\frac{1}{2}\rangle$. The oscillations of the two wavefunctions ψ_1 and ψ_2 are *coher-*

ent in the sense that as the amount of $|+\frac{1}{2}\rangle$ function (say) increases in one wavefunction over time, so the amount of $|+\frac{1}{2}\rangle$ function in the other wavefunction decreases at the same rate.

To complete this illustration, we use Equation (1.43) to calculate the expectation value of y-magnetization for this situation of isolated spins subjected to an rf pulse. Following Equation (1.42)

$$\langle\hat{\mu}_y\rangle = \mathrm{Tr}(\mu_y\rho) = \gamma\hbar\,\mathrm{Tr}(\mathbf{I}_y\rho) \tag{1.55}$$

The matrix of $\hat{I}_y$ in the basis set of $|\pm\frac{1}{2}\rangle$ functions is given in Equation (xx) in Box 1.1 previously. Using this and the density matrix of Equation (1.53), we find that

$$\langle\hat{\mu}_y\rangle = -\frac{1}{2}\gamma\hbar(p_1 - p_2)\sin(\omega_1 t) \tag{1.56}$$

For short t (i.e. for normal length rf pulses), the populations p_1 and p_2 of the wavefunctions ψ_1 and ψ_2 are the same as the populations of the initial wavefunctions the ψ_1 and ψ_2 wavefunctions started from at $t = 0$, i.e. the $|\pm\frac{1}{2}\rangle$ states, as there has been insufficient time for thermal equilibration to take place. So Equation (1.56) can be rewritten as

$$\langle\hat{\mu}_y\rangle = -\frac{1}{2}\gamma\hbar\left(p_{+\frac{1}{2}}(0) - p_{-\frac{1}{2}}(0)\right)\sin(\omega_1 t) \tag{1.57}$$

which is the same as Equation (1.39) obtained previously using the superposition state approach.

If the set of basis functions $\{\phi\}$ is the Zeeman basis, the *coherence order* of a density matrix element ρ_{ij} is given by $m_j - m_i$, the difference between the z-spin quantum numbers for the i and j basis functions which are mixed in the coherence. The terms coherence and coherence order are used widely and should be understood in the context of the density matrix representation. In the previous example of a spin under the influence of an rf pulse, if we compare the density matrix before the pulse (Equation (1.51)) with that during the pulse (Equation (1.53)), we see that the effect of the pulse is to create ± 1-order coherences.

1.3.2 The density operator at thermal equilibrium

Given the description of the density operator and matrix given above, it is relatively easy to show that the density operator for a spin system at thermal equilibrium (i.e. obeying the Boltzmann distribution) and in which individual spin systems are described by a hamiltonian $\hat{H}$ is

$$\hat{\rho}_{eq} = \frac{1}{Z} e^{-\hat{H}/kT} \tag{1.58}$$

where

$$Z = \mathrm{Tr}\left(e^{-\hat{H}/kT}\right) \tag{1.59}$$

Equation (1.58) involves the exponential of an operator, which is defined in Equation (i) in Box 1.2. If the basis functions chosen to form the density matrix from the density operator in Equation (1.58) are the eigenfunctions of $\hat{H}$, then the equilibrium density matrix is diagonal with elements equal to the populations of the corresponding eigenstates of $\hat{H}$, as predicted by the Boltzmann distribution.

1.3.3 *Time evolution of the density matrix*

Nuclei in a static, uniform field

Taking the simplest case of a spin in a uniform magnetic field $\mathbf{B}_0$, the hamiltonian in question is, as before (in frequency units)

$$\hat{H} = -\gamma \hat{I}_z B_0 = \omega_0 \hat{I}_z \tag{1.60}$$

as previously. Using this hamiltonian, we can approximate ρ_{eq} as

$$\hat{\rho}_{eq} \approx \frac{1}{Z}\left(1 + \frac{\hbar\omega_0}{kT}\hat{I}_z\right) \tag{1.61}$$

neglecting the higher terms in the expansion of $e^{-\hat{H}/kT}$. This approximation is valid for the typical magnetic fields $\mathbf{B}_0$ and temperatures used in NMR experiments. As discussed above, we can think of the density operator as describing a superposition state of a spin system. Any new interaction in the spin system will of course change its state and so the density operator describing it. An interaction in a spin system is described by a hamiltonian and the change it causes on the density operator is given by

$$\frac{d\hat{\rho}}{dt} = -i[\hat{H}, \hat{\rho}] \tag{1.62}$$

This equation is derived from the time-dependent Schrödinger equation (Equation (1.25)). The solution of Equation (1.62) for a time-independent hamiltonian is

$$\hat{\rho}(t) = e^{-i\hat{H}t}\hat{\rho}(0)\, e^{i\hat{H}t} \tag{1.63}$$

where $\hat{\rho}(t)$ is the density operator at time t and $\hat{\rho}(0)$ that at time $t = 0$, i.e. immediately prior to the new interaction described by $\hat{H}$. The term $e^{-i\hat{H}t}$ is often referred to as the *propagator*. In the case of a time-dependent hamiltonian, Equation (1.63) becomes

$$\hat{\rho}(t) = \hat{T}\,e^{-i\int_0^t \hat{H}(t')dt'}\,\hat{\rho}(0)\,e^{i\int_0^t \hat{H}(t')dt'} \tag{1.64}$$

where $\hat{T}$ is the Dyson time-ordering operator. This operator is necessary when $\hat{H}(t)$ does not commute with itself at different times t. The exponential $e^{i\int_0^t \hat{H}(t')dt'}$ can be written as a product

$$e^{i\int_0^t \hat{H}(t')dt'} = \prod_{\substack{n=0 \\ \lim \Delta t \to 0}}^{n=t/\Delta t} e^{i\hat{H}(t_n)\Delta t} \tag{1.65}$$

where $\hat{H}(t_n)$ is the hamiltonian at the nth time interval Δt. The Dyson time-ordering operator ensures that the $e^{i\hat{H}(t_n)\Delta t}$ operators in the series are placed in strict chronological sequence, with the earliest one placed on the right-hand side.

The same approach is used if the hamiltonian describing the spin system is *piecewise constant*; that is, it is $\hat{H}_1$ for a period of time t_1, $\hat{H}_2$ for a period t_2 and so on. This might be the case for instance for a pulse sequence, where there are periods during which the rf pulse is switched on and periods in between where other interactions are present. The hamiltonians describing the spin system during each period depend on the interactions present, and so vary through the pulse sequence. Under these circumstances, the density operator at the end of the pulse sequence is given by

$$\hat{\rho}(t_1 + t_2 + \ldots + t_n) = e^{-i\hat{H}_n t_n} \ldots e^{-i\hat{H}_2 t_2}\, e^{-i\hat{H}_1 t_1}\, \hat{\rho}(0)\, e^{i\hat{H}_1 t_1}\, e^{i\hat{H}_2 t_2} \ldots e^{i\hat{H}_n t_n} \tag{1.66}$$

The effect of rf pulses

We will now consider a specific example, that of an x-pulse acting on a spin system which starts at equilibrium in a uniform magnetic field $\mathbf{B}_0$. We can then compare the results obtained using the density matrix approach with those obtained in Section 1.2.

In the rotating frame, the hamiltonian describing the interaction of the on-resonance rf x-pulse with the spin system is (as shown in Section 1.2 and Box 1.2)

$$\hat{H}_{rf} = \omega_1 \hat{I}_x \tag{1.67}$$

in frequency units, with $\omega_1 = -\gamma B_1$, where B_1 is the magnetic field associated with the rf pulse. Then using Equation (1.63), the density operator $\hat{\rho}(t)$ during the rf pulse is

$$\hat{\rho}(t) = e^{i\omega_1 \hat{I}_x t}\, \hat{\rho}(0)\, e^{-i\omega_1 \hat{I}_x t} \tag{1.68}$$

where $\hat{\rho}(0)$ is the density operator at the start of the pulse. Substituting $\hat{\rho}_{eq}$ (Equation (1.61)), the density operator at equilibrium, for $\hat{\rho}(0)$ we have

$$\begin{aligned}\hat{\rho}(t) &= e^{i\omega_1 \hat{I}_x t}\left(1 - \frac{1}{Z}\frac{\hbar\omega_0}{kT}\hat{I}_z\right)e^{-i\omega_1 \hat{I}_x t} \\ &= \frac{1}{Z} - \frac{1}{Z}\frac{\hbar\omega_0}{kT} e^{i\omega_1 \hat{I}_x t}\hat{I}_z\, e^{-i\omega_1 \hat{I}_x t}\end{aligned} \tag{1.69}$$

The term $e^{i\omega_1 \hat{I}_x t}\hat{I}_z e^{-i\omega_1 \hat{I}_x t}$ represents a rotation of the operator $\hat{I}_z$ about x by angle $-\omega_1 t$ (see Box 1.2); thus we have

$$e^{i\omega_1 \hat{I}_x t}\hat{I}_z e^{-i\omega_1 \hat{I}_x t} = \hat{I}_z \cos\omega_1 t - \hat{I}_y \sin\omega_1 t \tag{1.70}$$

Substituting this in Equation (1.69) for the density operator $\hat{\rho}(t)$

$$\hat{\rho}(t) = \frac{1}{Z} - \frac{1}{Z}\frac{\hbar\omega_0}{kT}(\hat{I}_z \cos\omega_1 t - \hat{I}_y \sin\omega_1 t) \tag{1.71}$$

The y-magnetization is then given by

$$\begin{aligned}\langle\mu_y(t)\rangle &= \mathrm{Tr}(\hat{\mu}_y \hat{\rho}(t)) = \gamma\hbar \mathrm{Tr}(\hat{I}_y \hat{\rho}(t)) \\ &= \frac{\gamma\hbar}{Z}\mathrm{Tr}(\mathbf{I}_y \cdot 1) - \gamma\hbar\frac{1}{Z}\frac{\hbar\omega_0}{kT}\,\mathrm{Tr}(\mathbf{I}_y\mathbf{I}_z \cos\omega_1 t - \mathbf{I}_y^2 \sin\omega_1 t) \\ &= -\frac{1}{2}\gamma\hbar\left(\frac{1}{Z}\frac{\hbar\omega_0}{kT}\right)\sin\omega_1 t\end{aligned} \tag{1.72}$$

In the final step in Equation (1.72), we have used the $\mathbf{I}_y$ and $\mathbf{I}_z$ matrices in the Zeeman basis as given in Box 1.1. Comparing this with Equation (1.39) for the y-magnetization derived from the quantum mechanical picture, where

$$\langle\hat{\mu}_y(t)\rangle = -\frac{1}{2}\gamma\hbar\left(p_{\frac{1}{2}}(0) - p_{-\frac{1}{2}}(0)\right)\sin(\omega_1 t) \tag{1.39}$$

if we write the spin level populations as $p_{\pm\frac{1}{2}}(0) = \frac{1}{Z}\exp\left(\mp\frac{1}{2}(\hbar\omega_0/kT)\right)$ and approximate the exponentials as $\exp\left(\mp\frac{1}{2}(\hbar\omega_0/kT)\right) \approx 1 \mp \frac{1}{2}(\hbar\omega_0/kT)$, as was done in deriving the density operator, we see that Equations (1.39) and (1.72) are identical.

1.4 Nuclear spin interactions

So far the only interactions we have considered on the nuclear spin system are those with externally applied magnetic fields, the static, uniform field $\mathbf{B}_0$ in the NMR experiment and that due to rf pulses, $\mathbf{B}_1$. However, there are sources of magnetic fields internal to the sample which also affect the nuclear spin system and which for solid samples can cause extensive line broadening. Here, we consider briefly the hamiltonian operators which describe the most important of these interactions, chemical shielding, dipole coupling and quadrupole coupling. More detailed discussion of each interaction is to be found in later chapters devoted to each interaction.

In the NMR experiment, the applied static field $\mathbf{B}_0$ is in general orders of magnitude larger than any local fields arising within the sample as a result of other nuclear magnetic dipoles and chemical shielding. As a result, $\mathbf{B}_0$ remains as the quantization axis for the nuclear spins in the sample and many of these local fields have a negligible effect on the spin states (at equilibrium). The only components of local fields which have any significant effect on the spin states are

1. Components parallel or antiparallel to the applied field $\mathbf{B}_0$ and
2. Components precessing in the plane perpendicular to the applied field $\mathbf{B}_0$ at a frequency at or near the Larmor frequency (or other resonance frequencies of the spin system).

Components parallel (or antiparallel) to $\mathbf{B}_0$ add to (or subtract from) $\mathbf{B}_0$ and therefore alter the energies of the spin states which are determined by the total field strength parallel to $\mathbf{B}_0$, the quantization axis.

Components perpendicular to $\mathbf{B}_0$ precessing near the Larmor frequency are akin to the rf x-pulse considered in Section 1.3. As seen there, such fields mix the Zeeman spin states which exist in the $\mathbf{B}_0$ field alone.

The hamiltonian describing the interaction betweeen any local field $\mathbf{B}_{\text{loc}}$ and a nuclear spin $\mathbf{I}$ takes the usual form:

$$\begin{aligned} \hat{H}_{\text{loc}} &= -\gamma \hat{\mathbf{I}} \cdot \mathbf{B}_{\text{loc}} \\ &= -\gamma \left(\hat{I}_x B_x^{\text{loc}} + \hat{I}_y B_y^{\text{loc}} + \hat{I}_z B_z^{\text{loc}} \right) \end{aligned} \tag{1.73}$$

When we consider specific nuclear spin interactions in the following pages, we will find that we can always express the local magnetic field $\mathbf{B}_{\text{loc}}$ in the interaction as

$$\mathbf{B}_{\text{loc}} = \mathbf{A}_{\text{loc}} \cdot \mathbf{J} \tag{1.74}$$

where $\mathbf{A}_{\text{loc}}$ is a second-rank Cartesian tensor, often referred to as the *coupling tensor*, which describes the nuclear spin interaction and its orientation dependence. This is discussed further below. The vector $\mathbf{J}$ is the ultimate source of the $\mathbf{B}_{\text{loc}}$ magnetic field at the nucleus, e.g. another nuclear spin in the case of dipole–dipole coupling, or the $\mathbf{B}_0$ field itself in the case of chemical shielding. Hence, we can write a general contribution to the nuclear spin hamiltonian from an interaction A by reference to Equations (1.73) and (1.74) as (in frequency units)

$$\hat{H}_A = -\gamma \hat{\mathbf{I}} \cdot \mathbf{B}_{\text{loc}} = \hat{\mathbf{I}} \cdot \mathbf{A}_{\text{loc}} \cdot \mathbf{J} \tag{1.75}$$

where the factor of $-\gamma$ has been incorporated into $\mathbf{A}_{\text{loc}} \cdot \mathbf{J}$. The relevance of the vector $\mathbf{J}$ and the tensor $\mathbf{A}_{\text{loc}}$ will become clear as we consider some examples.

For chemical shielding, the source of the chemical shift, the interaction hamiltonian is

$$\hat{H}_{\text{cs}} = \gamma \hat{\mathbf{I}} \cdot \boldsymbol{\sigma} \cdot \mathbf{B}_0 \tag{1.76}$$

in frequency units, where $\boldsymbol{\sigma}$ is the shielding tensor and $\mathbf{B}_0$ is the applied magnetic field in the NMR experiment. It is the perturbation of the electrons around a nuclear spin $\mathbf{I}$ by the applied field $\mathbf{B}_0$ which gives rise to the shielding effect. The tensor $\boldsymbol{\sigma}$ describes the extent of the effect and how it depends on molecular orientation. The orientation dependence arises essentially because in the NMR experiment, the applied field $\mathbf{B}_0$ effectively orientates the spin vectors, $\mathbf{I}$, and also determines the shielding field (strength and direction) from the electrons with which the spin $\mathbf{I}$ interacts. Thus the strength of the interaction between the spin $\mathbf{I}$ and shielding field depends on the relative position of the spin $\mathbf{I}$ and the electrons giving rise to the shielding field with respect to $\mathbf{B}_0$. Figure 1.5 illustrates this schematically. This is all described in much more detail in Chapter 3.

For dipole–dipole coupling, the interaction hamiltonian, again in frequency units, is

$$\hat{H}_{\text{dd}} = -2\hat{\mathbf{I}} \cdot \mathbf{D} \cdot \hat{\mathbf{S}} \tag{1.77}$$

where now the dipolar-coupled spin $\mathbf{S}$ is the source of the local magnetic field acting on spin $\mathbf{I}$. $\mathbf{D}$ is the dipolar-coupling tensor which describes the strength and orientation dependence of the interaction between $\mathbf{I}$ and $\mathbf{S}$. The interaction between the two spins arises by virtue of the magnetic field each creates around itself. The orientation dependence arises because the applied magnetic field in the NMR experiment, $\mathbf{B}_0$, effectively orientates both spins $\mathbf{I}$ and $\mathbf{S}$, and hence the fields that each creates. Hence the interaction between

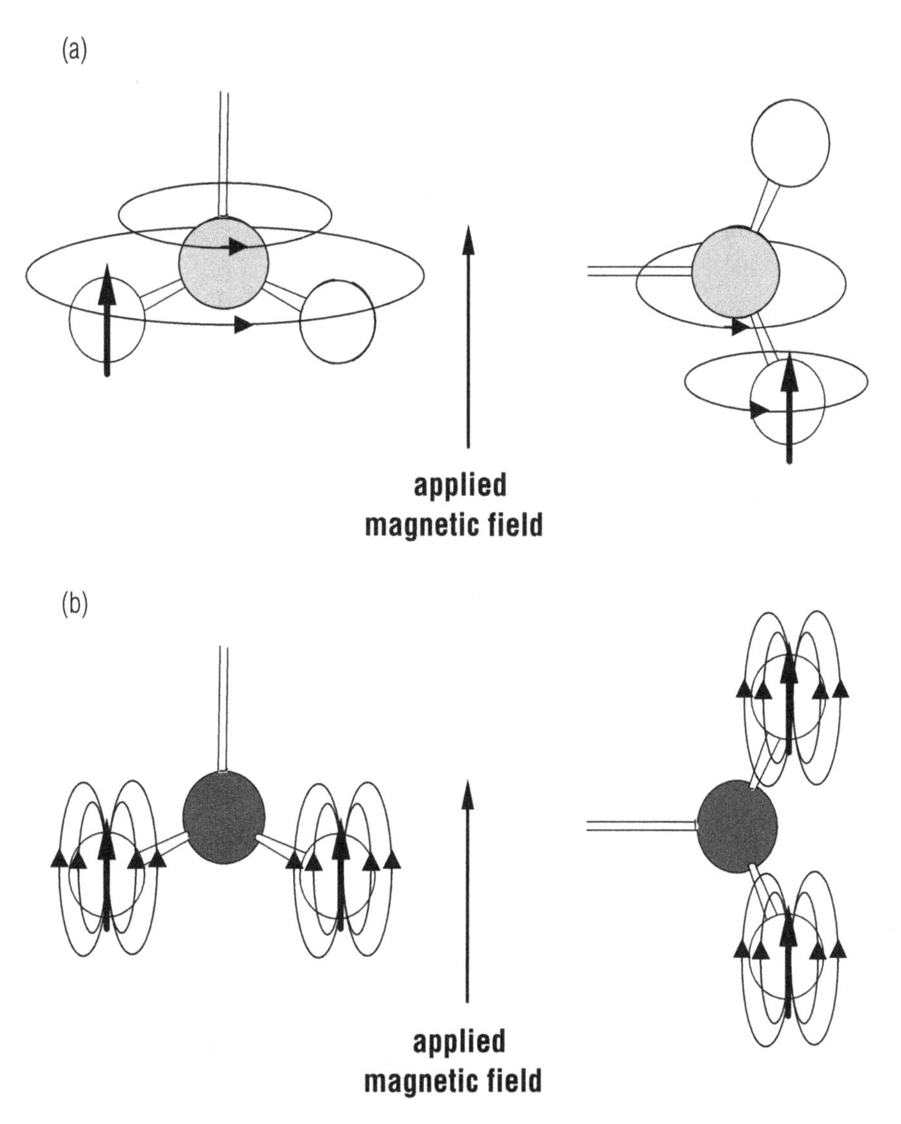

Fig. 1.5 Illustrating the orientation dependence of local nuclear spin interactions. (a) The shielding interaction. The electron circulation about the applied magnetic field which (in part) gives rise to a shielding field is shown schematically only. The size (and direction) of the shielding field felt at the nuclear spin (black arrow) depends on the orientation of the electron density with respect to the applied field. (b) The dipolar coupling between two nuclear spins. The z-component of each spin is orientated by the applied magnetic field of the NMR experiment, which effectively orientates the local magnetic field each gives rise to. Hence, the interaction between these local fields depends on the relative position in space of the two spins with respect to the applied field.

them then depends on the relative position of the two spins in space with respect to $\mathbf{B}_0$ in a similar manner to the shielding effect described above (Fig. 1.5). This is discussed further in Chapter 4.

The quadrupole-coupling interaction is slightly different as it is an interaction between a nuclear *electric* quadrupole moment and an electric field gradient, rather than an interaction between a nuclear magnetic dipole moment and a magnetic field as the other interactions are. Nevertheless, the hamiltonian describing the quadrupole interaction can be expressed in a similar form to those for the chemical shielding and dipole–dipole coupling:

$$\hat{H}_Q = \frac{eQ}{2I(2I-1)\hbar}\mathbf{I}\cdot\mathbf{V}\cdot\mathbf{I} \tag{1.78}$$

where $\mathbf{V}$ is the electric field gradient tensor and Q is the nuclear electric quadrupole moment. This is discussed in detail in Chapter 5.

An alternative way of assessing the effect of a nuclear spin interaction uses *perturbation theory*. The Zeeman interaction for the spin system in question, i.e. interaction of the spin system (whatever that might be) with the applied field, $\mathbf{B}_0$, is considered to be the dominant interaction in the system and is described by the hamiltonian $\hat{H}_0$. The nuclear spin interaction, described by a hamiltonian, $\hat{H}_1$, is then considered as a perturbation on the spin system, so that the total hamiltonian $\hat{H}$ for the spin system is given by

$$\hat{H} = \hat{H}_0 + \hat{H}_1 \tag{1.79}$$

Now perturbation theory shows that if we are interested in the energy of the perturbed system to first order, we need to consider the wavefunctions of the perturbed system to *zeroth order*. The wavefunctions of a perturbed system described by $\hat{H}$ are, to zeroth order, simply the eigenfunctions of the dominant interaction hamiltonian, $\hat{H}_0$. These eigenfunctions are of course just the Zeeman states for the spin system, $\psi_{I,m}$ (where I denotes the total spin for the system and m its component along $\mathbf{B}_0$). The parts of $\hat{H}_1$ which affect the spin system wavefunctions to zeroth order must also have eigenfunctions $\psi_{I,\,m}$, i.e. must have the same eigenfunctions as $\hat{H}_0$. As discussed in Box 1.1, two operators can only have the same eigenfunctions if they commute. Thus, the only parts of $\hat{H}_1$ which affect the spin system to zeroth order (and so its energy levels to first order) are those parts which commute with $\hat{H}_0$.[N4] That is not to say that an interaction hamiltonian, $\hat{H}_1$, which does *not* commute with $\hat{H}_0$, may not cause higher-order energy corrections. However, these higher-order energy corrections are unlikely to be important in determining the energy levels of the spin system and, therefore, in determining the NMR spectrum, unless the interaction has a large amplitude relative to the Zeeman interaction. Hamiltonians describing nuclear spin interactions within a spin system which commute with the Zeeman hamiltonian for that spin system are said to be *secular*.

1.4.1 Interaction tensors

As we have already mentioned, nuclear spin interactions such as chemical shielding depend on molecular orientation and, hence, their size cannot be described by a single number, but, as demonstrated below, must be described by a second-rank tensor. We have been describing nuclear spin systems in terms of Cartesian spin operators, such as $\hat{I}_x$, $\hat{I}_y$, $\hat{I}_z$, and so we need to use *Cartesian* second-rank tensors to describe the orientation dependence of nuclear spin interactions acting on such spin operators. There are other ways of describing nuclear spin systems, such as by using spherical tensor operators, in which case we would need to use spherical tensors to describe the orientation dependence of the nuclear spin interactions. This formalism is described in Box 4.1 in Chapter 4. Here, however, we concentrate on the approach using Cartesian spin operators and tensors and discuss interaction tensors by using the shielding tensor as an example.

In general, a second-rank Cartesian tensor, such as the shielding tensor $\boldsymbol{\sigma}$, is represented by a 3×3 matrix:

$$\boldsymbol{\sigma} = \begin{pmatrix} \sigma_{xx} & \sigma_{xy} & \sigma_{xz} \\ \sigma_{yx} & \sigma_{yy} & \sigma_{yz} \\ \sigma_{zx} & \sigma_{zy} & \sigma_{zz} \end{pmatrix} \tag{1.80}$$

where x, y, z is some, as yet unspecified, axis frame. The meaning of the shielding tensor becomes clearer when we express the shielding tensor in the laboratory frame (which is defined by $\mathbf{B}_0$ being along z). The local magnetic field $\mathbf{B}_{\text{loc}}$ at a nucleus with a shielding tensor σ^{lab}, i.e. a shielding tensor expressed within the laboratory frame, is

$$\begin{aligned} \mathbf{B}_{\text{loc}} &= \boldsymbol{\sigma}^{\text{lab}} \cdot \mathbf{B}_0 \\ &= \begin{pmatrix} \sigma_{xx}^{\text{lab}} & \sigma_{xy}^{\text{lab}} & \sigma_{xz}^{\text{lab}} \\ \sigma_{yx}^{\text{lab}} & \sigma_{yy}^{\text{lab}} & \sigma_{yz}^{\text{lab}} \\ \sigma_{zx}^{\text{lab}} & \sigma_{zy}^{\text{lab}} & \sigma_{zz}^{\text{lab}} \end{pmatrix} \cdot \begin{pmatrix} 0 \\ 0 \\ B_0 \end{pmatrix} \\ &= \begin{pmatrix} \sigma_{xz}^{\text{lab}} B_0 \\ \sigma_{yz}^{\text{lab}} B_0 \\ \sigma_{zz}^{\text{lab}} B_0 \end{pmatrix} \end{aligned} \tag{1.81}$$

for an applied field $\mathbf{B}_0$ in the laboratory z direction. So, for instance, $\sigma_{xz}^{\text{lab}} B_0$ is the local shielding field in the laboratory x-direction, when $\mathbf{B}_0$ is applied along z. The dipolar- and quadrupole-coupling tensors can be understood in a similar manner. For instance, $D_{xz}^{\text{lab}} \hat{S}_z$ (where $\mathbf{D}$ is the dipolar coupling tensor describing the interaction between two spins $\mathbf{I}$ and $\mathbf{S}$) is the local magnetic field at spin $\mathbf{I}$ in the laboratory x-direction arising from the z-

component of the S spin. Now we can see why we need a second-rank tensor to describe the strength of the nuclear spin interaction with locally arising magnetic fields. In general, the feature causing the extra local field is itself a vector quantity, i.e. $\mathbf{B}_0$ in the case of shielding, S in the case of dipolar coupling. It is thus described by a three-component Cartesian vector. The local field it causes is another three-component Cartesian vector. Thus to describe the relationship between the feature causing the local field and the local field itself, we have to describe the relationship between two three-dimensional vectors (which are not in general co-parallel). This requires a 3×3 matrix, or in general a second-rank tensor.

It is possible to choose the axis frame that an interaction tensor such as σ is defined with respect to so that the interaction tensor is diagonal. This axis frame is the *principal axis frame*, designated 'PAF', or $x^{\text{PAF}}, y^{\text{PAF}}, z^{\text{PAF}}$. The numbers along the resulting diagonal of σ^{PAF} are the *principal values* of the shielding tensor, so, for instance, σ_{xx}^{PAF} is the principal value associated with the principal axis frame x-axis. The orientation of the principal axis frame is determined by the local environment of the nucleus to which the interaction pertains. For instance, the shielding tensor principal axis frame is determined by the electronic structure of the molecule that contains the nucleus in question and is fixed with respect to the molecule. This latter point is very important. We can picture an interaction tensor as being represented by an ellipsoid fixed within the molecule and centred on the nucleus it applies to. The principal axes of the ellipse coincide with the principal axis frame of the shielding tensor, and the length of each principal axis of the ellipsoid is proportional to the magnitude of the principal value of the interaction tensor associated with that principal axis. If the molecular orientation in the laboratory frame changes, then so does the orientation of the interaction tensor, as illustrated in Fig. 1.6. If the nucleus is at a crystallographic site of symmetry, then the interaction tensor reflects this symmetry. For instance, the shielding tensor for a nucleus at a site of axial symmetry has a principal axis frame in which the z^{PAF}-axis coincides with the symmetry axis and the principal values are such that $\sigma_{xx}^{\text{PAF}} = \sigma_{yy}^{\text{PAF}} \neq \sigma_{zz}^{\text{PAF}}$.

The three principal values of an interaction tensor $A_{\alpha\alpha}^{\text{PAF}}$ are frequently expressed instead as the isotropic value A_{iso}, the anisotropy Δ_A, and the asymmetry η_A of the interaction. These quantities are defined from the principal values as follows:

$$
\begin{aligned}
A_{\text{iso}} &= \frac{1}{3}\left(A_{xx}^{\text{PAF}} + A_{yy}^{\text{PAF}} + A_{zz}^{\text{PAF}}\right) \\
\Delta_A &= A_{zz}^{\text{PAF}} - A_{\text{iso}} \\
\eta_A &= \frac{A_{xx}^{\text{PAF}} - A_{yy}^{\text{PAF}}}{\Delta_A}
\end{aligned}
\tag{1.82}
$$

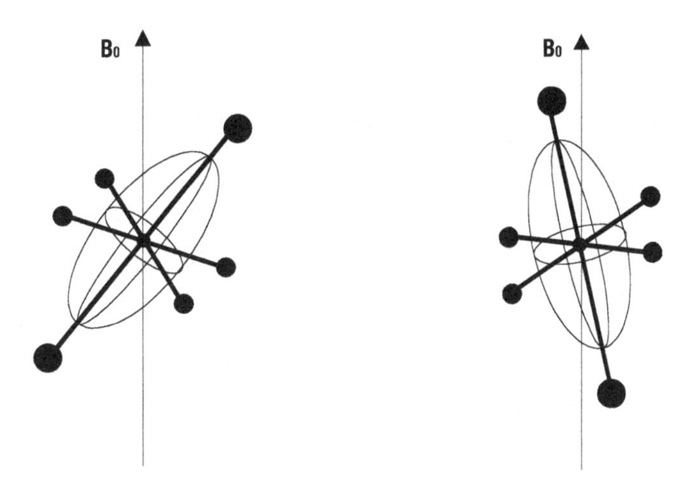

Fig. 1.6 Illustrating the ellipsoid representation of an interaction tensor using the shielding tensor as an example. The principal axes of the ellipsoid coincide with the shielding tensor principal axis frame, which, in turn, is fixed in orientation with respect to the molecule.

In effect, the anisotropy is related to the length of the ellipsoid describing the interaction tensor and the asymmetry is a measure of how far a cross-section through the ellipsoid (parallel to the x^{PAF}–y^{PAF} plane) deviates from circular.

1.5 General features of Fourier transform NMR experiments

One of the most important features of the Fourier transform (FT) NMR experiment is that we do not measure absolute spectral frequencies, but measure all frequencies relative to some carrier or spectral frequency, as outlined in Box 1.3, which is constant for a given nuclear species. We refer to the frequency relative to the spectral frequency as the *frequency offset*. We may then reference all signals in the spectrum relative to a specific signal given by a standard or reference compound, in order to quote the positions of the signals in our spectrum as *chemical shifts* (see Section 3.1.3), which are dimensionless quantities. We shall often refer to frequency offsets throughout this book, and they should be understood in terms of the concept outlined above and detailed in Box 1.3.

1.5.1 Multidimensional NMR

As the preceding sections have hinted, what we observe in an (FT) NMR experiment are not transitions between energy levels as we do in other forms of spectroscopy. Rather, we observe the evolution over time of the ensemble average state or superposition state of the spin system, having first created a non-equilibrium (ensemble average) state in order to force the spin system to change with time; an equilibrium state is of course stationary. The evolution of the spin system in its non-equilibrium state takes place at characteristic frequencies determined by the various interactions in the spin system discussed in Section 1.4, and it is these frequencies which we hope to extract from our observations of the spin system.

So, for instance, in the absence of any rf pulses or other interactions, nuclear spins feel only the applied field $\mathbf{B}_0$, and the ensemble average spin state is described by a linear combination of the Zeeman spin states, with combination coefficients determined by the Boltzmann populations of these levels (providing the spin system is at equilibrium). This is the equilibrium spin state for this situation. If we then apply an rf pulse, the Zeeman functions are mixed as described in Section 1.2. After the pulse, the situation of the spin system returns to that before the pulse, i.e. interaction with the $\mathbf{B}_0$ field only. However, the superposition state describing the spin system at this point is some new linear combination of the Zeeman functions determined by the rf pulse as well as the Boltzmann population distribution before the pulse and is (in general) a non-equilibrium state for the spin system in this environment.

Another way of putting this is to say that, in NMR, we observe the time evolution of coherences between the Zeeman functions of the spin system. As we saw in Section 1.3, the idea of a coherence arises directly from the density matrix description of the state of a spin system; this is the reason for the immense importance of the density matrix in NMR. The time evolution of coherences takes a simple form when the Zeeman functions for the spin system are the eigenfunctions of the hamiltonian describing the spin system. This will be the case if the hamiltonian commutes with the Zeeman hamiltonian for the spin system (see Box 1.1). Under these circumstances, we can show that the ijth element of a density matrix in the Zeeman basis, i.e. the coherence between Zeeman levels i and j, evolves according to $\exp(i\omega_{ij}t)$, where ω_{ij} is the energy gap between the j and i energy levels of the spin system in frequency units, i.e. $\omega_{ij} = (E_i - E_j)/\hbar$, where E_i, and E_j are the energies of the i and j levels respectively. The only coherence we can observe directly is coherence of order −1, i.e. evolution of coherent superpositions of Zeeman functions whose magnetic quantum numbers m differ by −1. The transverse magnetization of the classical vector model of NMR is equivalent to such coherences. The Fourier transform of $\exp(i\omega_{ij}t)$ is a δ function at frequency ω_{ij}. Thus if we can measure the $\exp(\omega_{ij}t)$ evolution over time of the desired coherence, Fourier transformation of the resulting time domain series will produce a frequency spectrum of that coherence.

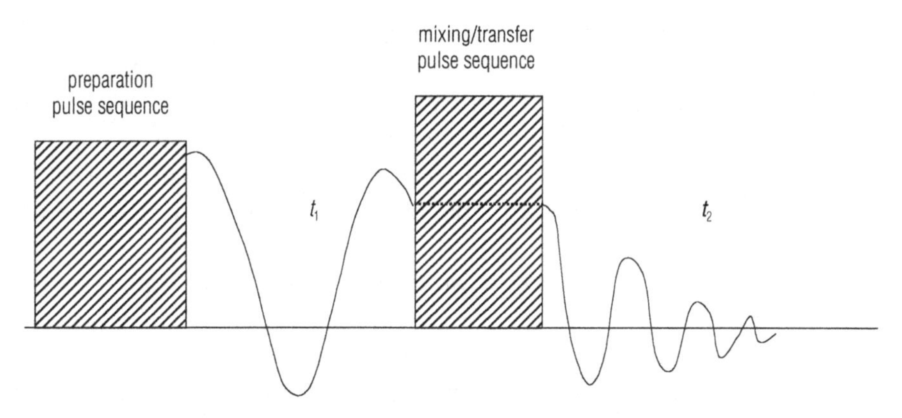

Fig. 1.7 The general form of a two-dimensional NMR pulse sequence. Selected coherences evolve at frequency ω_1 during t_1 and at ω_2 during t_2.

This, in turn, contains signals at frequencies which correspond to the energy gaps between energy levels in the system.

Although we can only observe −1-order coherence, we can create many other orders and observe their evolution too, providing we do it indirectly. This is the idea behind two-dimensional (or in general multidimensional) NMR experiments. Consider the pulse sequence in Fig. 1.7. An initial sequence of rf pulses (the *preparation sequence*) creates a non-equilibrium spin state. In the subsequent period t_1, the desired coherence of order n is allowed to evolve. In practice, selecting the desired coherence order is performed using phase cycling (see below for details). The mixing or transfer pulse sequence at the end of t_1 transforms the selected n-order coherence into observable −1 coherence for observation during t_2. The signal in t_2 has either its amplitude or phase, or both, modulated according to the evolution in t_1. Thus by repeating the experiment for successive different values of t_1, we end up with a two-dimensional time-domain dataset which contains in the t_2 dimension, information on the evolution of the −1-order coherence and in the t_1 dimension, information on the evolution of the n-order coherence. The experiment can also be arranged so that −1-order coherence is monitored in both dimensions. This is the case in exchange experiments, where, during the mixing period, the frequency of evolution of the t_1 −1-order coherence changes as a result of chemical or spin exchange, the new frequency then being recorded in t_2.

Whatever the coherences involved, the result and interpretation of the final spectrum is the same. The two-dimensional time-domain dataset can be transformed to the frequency domain (usually by Fourier transformation) to produce a two-dimensional frequency spectrum, in which the f_1

dimension contains the characteristic evolution frequencies of the n-order coherences in t_1 and the f_2 dimension contains those of the -1-order coherences present in t_2. The two-dimensional spectrum is then a correlation map in which a peak in the two-dimensional spectrum at (ω_1, ω_2) indicates that, in t_1, there was n-order coherence evolving at frequency ω_1 which was then transformed by the mixing period into -1-order coherence with frequency ω_2.

In order to obtain this two-dimensional frequency spectrum, we need to understand how to use phase cycling to select our desired coherences and how to transform from a time-domain dataset to a frequency spectrum containing pure absorption line shapes. This latter part comes down to being able to record the two time dimensions in *quadrature*. Both of these features are discussed in the following sections.

1.5.2 Phase cycling

A typical pulse sequence is shown in Fig. 1.8. It consists of three pulses, with phases ϕ_1, ϕ_2 and ϕ_3 respectively. Also shown in Fig. 1.8 is the desired *coherence order pathway*, the sequence of coherence orders required at each stage during the pulse sequence. The desired coherence order pathway in Fig. 1.8 is $0 \rightarrow +2 \rightarrow +1 \rightarrow -1$. We use rf pulses to change the order of coherence present at the different stages in a pulse sequence, with each of the three pulses endeavouring to execute the three changes in

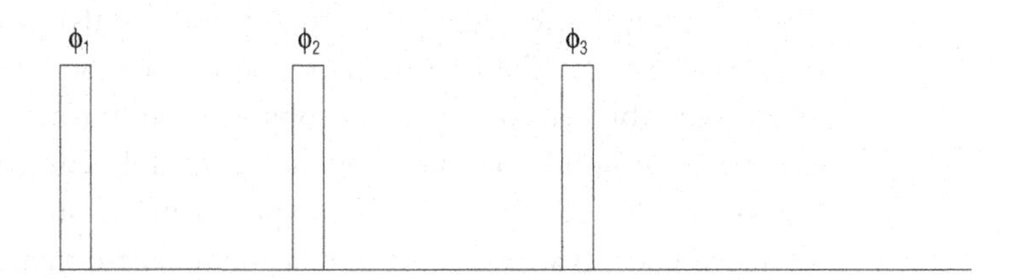

Desired coherence pathway:

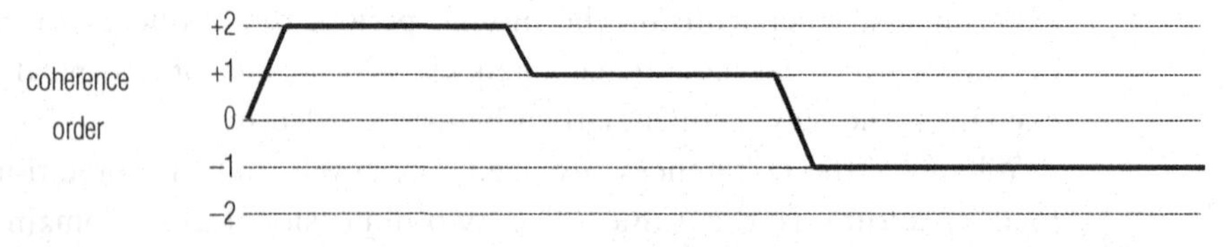

Fig. 1.8 A typical pulse sequence with pulse phases ϕ_1, ϕ_2 and ϕ_3. Each pulse causes a change in coherence order; the coherences evolve in the periods between the pulses. The desired coherence orders at each stage of the pulse sequence are shown in the diagram at the bottom. Selection of this pathway is achieved with phase cycling as discussed in the text.

coherence order. We will label the change in coherence order caused by a pulse as Δp.

The key rule we use in phase cycling is the following [3]

> *If the phase of a pulse is changed by* ϕ*, a coherence undergoing a change in coherence level of* Δp *acquires a phase shift of* $-\Delta p \cdot \phi$.

Thus the overall phase acquired by the desired coherence pathway in Fig. 1.8 is $-2\phi_1 + \phi_2 + 2\phi_3$, since the change in coherence order at the first pulse (phase ϕ_1) is +2, at the second (phase ϕ_2), −1 and at the third (phase ϕ_3), −2. What we now have to do is to arrange a phase cycle of the pulses in the sequence, while always setting the receiver phase to follow the overall phase acquired by the desired coherence order pathway, so that the signal due to this pathway adds up over the phase cycle. The phase cycle of *N* steps needs to be arranged so that the signals due to unwanted pathways arrive at the receiver with phases such that the signals exactly cancel over the cycle. In order to achieve this, the following rules can be employed:

> *If a phase cycle uses steps of 360/*N *degrees, then in addition to the pathway with change in coherence order* Δp*, pathways with* $\Delta p \pm nN$*, where* $n = 1, 2, 3, \ldots$ *are also selected.*
>
> *If a particular value of* Δp *is to be selected from* m *consectutive values, then* N *must be at least* m.

In order to derive an effective phase cycle, we must know what other pathways and coherence orders are likely to be present in addition to the one we want, so that we know what we must aim to exclude. The phase cycle can be simplified by noting that as we can only observe −1-order coherence in the final observation, there is no need to deliberately select −1-order coherence in the final step of the pulse sequence; the experiment is already self-selecting in this respect.

An example will suffice to show how a suitable phase cycle may be constructed. Suppose in the example in Fig. 1.8, another unwanted coherence pathway $0 \rightarrow -1$ (rather than +2) $\rightarrow +1 \rightarrow -1$ can occur in addition to the desired pathway. We can exclude this pathway by phase cycling the first pulse so as to select only $\Delta p = +2$ at this step. We need to select +2 from possible −1, 0 and +1 coherence order changes at this step. Actually, we do not know if the 0 and +1 coherence order changes can occur, but we do know −1 can, so we include all the other coherence order changes between that and the desired one. So, we need to select the Δp from four consecutive values of possible Δp. Thus the number of steps in the phase cycle, *N*, needs to be at least four. Table 1.1 follows what happens to the overall phase of coherence along the desired pathway and the unwanted pathway. We

Table 1.1 The effects of phase cycling on the pulse sequence shown in Fig. 1.7. The first pulse only is cycled in four steps. The phases of the other two pulses are assumed to be zero. The phases shown in brackets are the equivalent phase angles between 0° and 360°. The receiver is stepped to follow the phase of the desired coherence throughout the cycle, so that signal from this coherence adds throughout the cycle.

Desired pathway $0 \rightarrow +2 \rightarrow +1 \rightarrow -1$

Cycle step	Pulse phase (pulse 1)	Phase of coherence after pulse $\Delta p = +2$	Receiver phase (following desired coherence)	Phase of final coherence relative to receiver
1	0°	0°	0°	0°
2	90°	−180° (180°)	180°	0°
3	180°	−360° (0°)	0°	0°
4	270°	−540° (180°)	180°	0°

Unwanted pathway $0 \rightarrow -1 \rightarrow +1 \rightarrow -1$

Cycle step	Pulse phase (pulse 1)	Phase of coherence after pulse $\Delta p = -1$	Receiver phase (following desired coherence)	Phase of final coherence relative to receiver
1	0°	0°	0°	0°
2	90°	90°	180°	90°
3	180°	180°	0°	180°
4	270°	270°	180°	270°

assume that the phases of pulses 2 and 3 remain constant through the cycles; we have set them to zero so that they do not contribute to the overall phase of the coherence at the end of the sequence, but this does not affect the outcome of the phase cycle. The unwanted signal has a different relative phase at the receiver on each step of the phase cycle; adding these four signals, which have a 90° phase shift between each of them, gives a net signal of zero. Thus, as required, the unwanted signal cancels over the phase cycle, while the desired one adds. In general, a signal cancels exactly if in an N-step phase cycle, its phase relative to the receiver acquires each of the possible values $360/N$ over the phase cycle.

This method of phase cycling is what is currently used routinely in solid-state NMR. However, it is not the only method of phase cycling which will produce the required result. Much shorter schemes are possible using a protocol termed *cogwheel phase cycling* [4]. This method may well become popular in the future.

1.5.3 Quadrature detection

In NMR, we measure all time-evolving signals relative to a rotating frame of reference rotating at the *carrier frequency*, which is normally the same as the frequency of the rf pulse. Any coherence evolves in general accord-

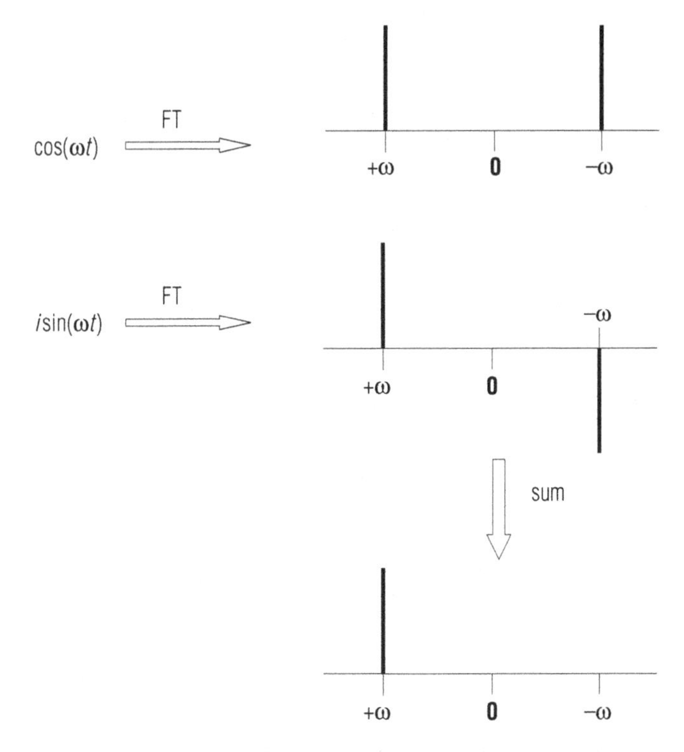

Fig. 1.9 The Fourier transformation of $\exp(i\omega t) = \cos(\omega t) + i\sin(\omega t)$.

ing to $C\exp(i\omega t)$, where ω is the evolution frequency offset in the rotating frame and C is the overall amplitude of the coherence, which may be time-dependent, for example, due to the loss of the coherence through relaxation. $C\exp(i\omega t)$ can be rewritten in terms of its real and imaginary components:

$$C\exp(i\omega t) = C\cos(\omega t) + iC\sin(i\omega t) \qquad (1.83)$$

Both the real and imaginary components of the evolution must be measured if Fourier transformation of the time-domain signal is to produce an unambiguous frequency spectrum, as illustrated in Fig. 1.9. The measurement of these two components is known as *quadrature detection.*

We should also consider the decay of the coherence. The effect of decay is shown in Fig. 1.10. If the amplitude of the coherence is multiplied by a decaying function, $\exp(-t/T_2)$, there are two principal effects we should note. (1) The lines in the frequency spectrum acquire a linewidth rather than being δ functions and (2) the imaginary part of the frequency spectrum is no longer zero (although its net integral is zero).

Fourier transformation of $C\exp(i\omega t)$, where C includes a decaying term $\exp(-t/T_2)$, results in a complex frequency spectrum $A + iD$, as described in Fig. 1.10. The real part of this frequency spectrum, A, is a pure absorption

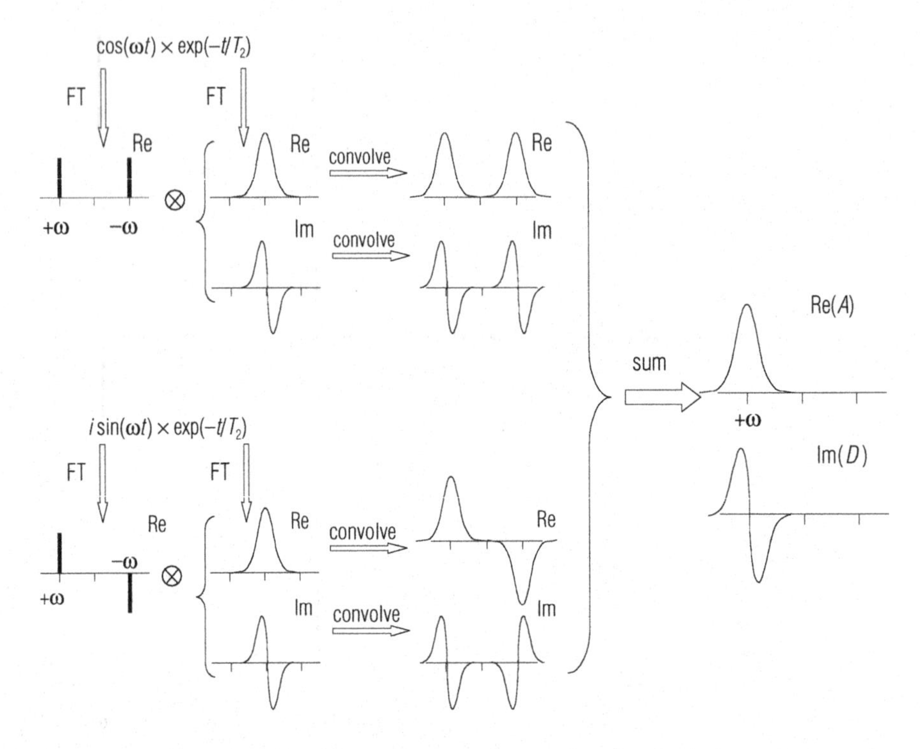

Fig. 1.10 The effects of multiplying a time domain signal exp($i\omega t$) by a decaying function exp($-t/T_2$) on the frequency spectrum. The frequency spectrum is found by Fourier transforming the product function, exp($i\omega t$) exp($-t/T_2$), which is the process shown above. The Fourier transforms of the sine and cosine components of exp($i\omega t$) (= cos(ωt) + $i$sin(ωt)) are shown in Fig. 1.9. The Fourier transform of exp($-t/T_2$) is shown above; the real part is an absorption mode lorenzian of linewidth $2/T_2$ at half height and the imaginary part is a dispersive mode lorenzian. The frequency spectrum of exp($i\omega t$) exp($-t/T_2$) is found by convolving the frequency spectra of exp($i\omega t$) and exp($-t/T_2$). The final frequency spectrum has real and imaginary parts. The real part is a pure absorption spectrum, denoted as A, and the imaginary part is a pure dispersive spectrum, denoted D. It is worth noting that in the absence of the decaying function, the Fourier transform of exp($i\omega t$) alone has no imaginary part. The non-zero imaginary (and dispersive) part of the frequency spectrum arises from the decaying function, exp($-t/T_2$).

lineshape and is what spectroscopists usually refer to as 'the spectrum'. The imaginary part is a purely dispersive lineshape (Fig. 1.10).

Quadrature detection is performed in practice using the scheme outlined in Box 1.3. The two time-domain signals arising from quadrature detection, S_1 and S_2, should be directly proportional to the real ($F_{Re} = C\cos(\omega t)$) and imaginary ($F_{Im} = C\sin(\omega t)$) parts of the coherence evolution in Equation (1.83). However, there may by a phase offset (determined by the *receiver phase*) between S_1/S_2 and F_{Re}/F_{Im}, such that they are in fact related by (ignoring the proportionality constant between S_1/S_2 and F_{Re}/F_{Im} for convenience):

$$\begin{aligned} Re &= S_1 \cos\phi + S_2 \sin\phi \\ Im &= -S_1 \sin\phi + S_2 \cos\phi \end{aligned} \tag{1.84}$$

Fourier transformation of $S_1 + iS_2$ (i.e. treating S_1/S_2 as F_{Re}/F_{Im}) with no phase correction results in a *phase-distorted* lineshape:

$$S_1 + iS_2 \xrightarrow{FT} (A\cos\phi + D\sin\phi) + i(-A\sin\phi + D\cos\phi) \qquad (1.85)$$

where the real part of the frequency spectrum is now a mixture of absorptive (A) and dispersive (D) lineshapes, $A\cos\phi + D\sin\phi$. This can be corrected according to Equation (1.84) in either the time or frequency domains; this is the so-called *phasing* of the spectrum. The phase factor ϕ is adjusted until the real part of the frequency spectrum is purely absorptive, and the imaginary part purely dispersive.

In a two-dimensional NMR experiment, as well as recording separately the $\sin(\omega_2 t_2)$ and $\cos(\omega_2 t_2)$ signals arising in t_2, we must also arrange that the sin/cos components arising in t_1 are recorded separately as well. If we do not do this, but acquire a signal of the form $\exp(i\omega_1 t_1)\exp(i\omega_2 t_2)$, subsequent Fourier transformation in t_1 and t_2 has the following effect:

$$\begin{aligned}\exp(i\omega_1 t_1)\exp(i\omega_2 t_2) &\xrightarrow{FT} (A_1 + iD_1)(A_2 + iD_2)\\ &= (A_1A_2 - D_1D_2) + i(A_1D_2 - D_1A_2)\end{aligned} \qquad (1.86)$$

The real part of the resulting frequency spectrum contains a mixture of absorptive and dispersive signal being of the form $(A_1A_2 - D_1D_2)$, which is a *phase-twisted lineshape*.

A method of producing purely absorptive lineshapes in two-dimensional frequency spectra is to perform two experiments, with their respective preparation pulse sequences phase shifted so that the desired coherences in t_1 have a 90° phase shift between the two experiments. This is done by shifting the phases of all the preparation pulses by $90°/(n_0 - n)$, where n_0 is the initial coherence order present at the start of the preparation pulse sequence (usually 0) and n is the desired coherence order in t_1. Both experiments must be arranged so that both $+n$ and $-n$ coherences are selected in t_1, with equal amplitude. The evolution frequency of the $-n$-order coherence is minus that of the $+n$-order coherence, so the $+n$-order coherence evolves according to $\exp(i\omega_n t_1)$ and the $-n$-order coherence according to $\exp(-i\omega_n t_1)$. The signals arising in the two experiments are then

1. $[\exp(i\omega_n t_1) + \exp(-i\omega_n t_1)]\exp(i\omega_2 t_2) = 2\cos\omega_n t_1 \cdot \exp(i\omega_2 t_2)$
2. $[\exp(i\pi/2)\exp(i\omega_n t_1) + \exp(-i\pi/2)\exp(-i\omega_n t_1)]\exp(i\omega_2 t_2)$
   $= 2i\sin\omega_n t_1 \cdot \exp(i\omega_2 t_2)$

where in signal 2 a phase shift of $\pi/2$ has been applied to the t_1 signal. These two datasets are then processed as shown in Fig. 1.11, and result in a purely absorptive two-dimensional frequency spectrum.

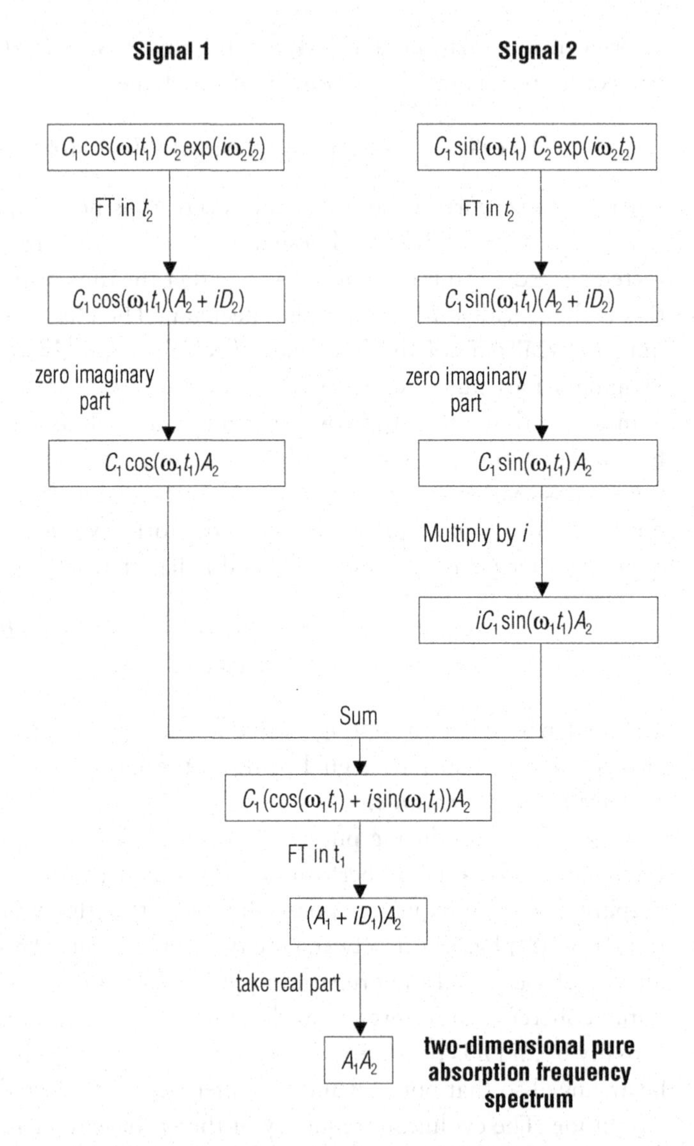

Fig. 1.11 Processing of the two signals from a two-dimensional NMR experiment so as to produce pure absorption line shapes in the two-dimensional frequency spectrum. The two signals are produced in two separate experiments and have the forms indicated at the top of the diagram (see text for details). The C_1 and C_2 coefficients in the signals are the amplitudes of the signals during the t_1 and t_2 periods of the experiment respectively, and normally contain decay terms which are functions of time. A_1 and A_2 are the amplitudes of the real parts of the frequency spectra arising from Fourier transformation (FT) of the t_1 and t_2 time-domain signals and D_1 and D_2 are their imaginary counterparts. The A terms correspond to absorption mode lineshapes, while the D terms correspond to dispersive mode lineshapes.

This technique relies on the $+n$- and $-n$-order coherence pathways being selected with the same amplitude; sometimes this is not achieved and so phase-twisted lineshapes are still observed in the spectrum. What prevents their amplitude being equal is usually that the final mixing steps $\pm n \to -1$, which cause the transformation to observable coherence, do not occur with equal efficiency. One way around this is to amend the pulse sequence so that the coherence order pathway is $\pm n \to 0 \to -1$ instead. The symmetrical routes $\pm n \to 0$ should occur with equal efficiencies. This procedure is often called a *z-filter* as coherence order zero also corresponds to a state with spin polarization along z.

Box 1.3: The NMR spectrometer

The detailed working of a solid-state NMR spectrometer varies considerably between spectrometers, so only a general outline which is (usually) common to all spectrometers is given here.

Generating rf pulses

The spectral frequency rf wave, or *carrier wave*, (frequency ω_0) required to irradiate the sample is usually produced by 'mixing' two (or more) rf waves with frequencies ω_{IF} and ω_{mix} such that $\omega_{IF} + \omega_{mix} = \omega_0$ (although some spectrometers will use the difference signal which arises from the mixing process, so that $\omega_{IF} - \omega_{mix} = \omega_0$). The ω_{IF} is the *intermediate frequency* and is a constant whatever the nucleus being studied. ω_{mix} is variable depending on the required spectral frequency ω_0. The reason for this manner of producing the irradiating rf wave becomes clearer later. Essentially, the same waves with frequencies ω_{IF} and ω_{mix} are used in the detection of the response of the sample to the rf pulse and this ensures phase coherence between the rf pulses and the measured response of the sample.

The rf pulse of the required length for the experiment is produced in several ways, usually by chopping the ω_{IF} wave into the required length, prior to mixing with ω_{mix} to produce the spectral frequency wave. The rf wave within the pulse is also phase shifted relative to the original carrier wave by the required 'phase' of the rf pulse. 'Phase' here is the phase of the rf wave relative to the carrier wave. In turn the phase of a wave can be understood by considering an rf wave with a

Continued on p. 54

Box 1.3 *Cont.*

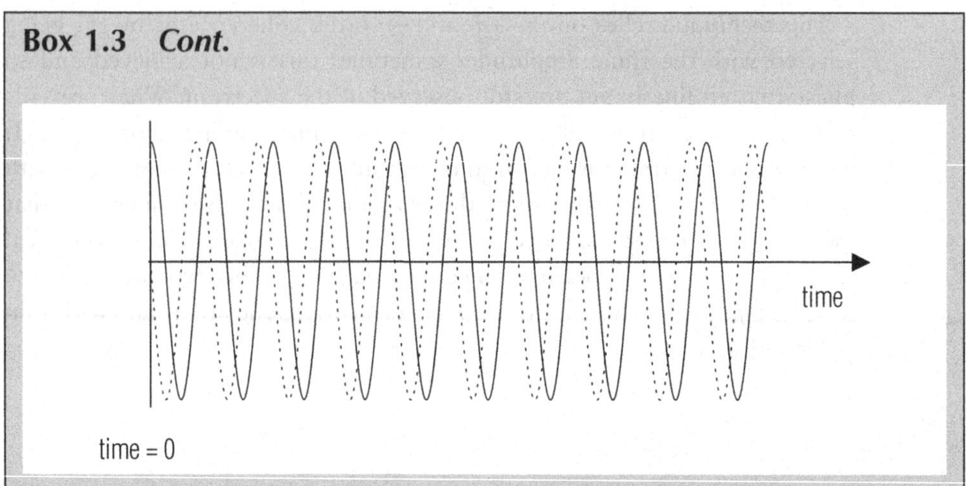

Fig. B1.3.1 Illustrating the concept of the phase of a wave. Both waves here have the general form $\cos(\omega_{rf}t + \phi_{rf})$ where ω_{rf} is the frequency of the oscillation of the wave and ϕ_{rf} is its phase. The wave with the solid line has a phase $\phi_{rf} = 0°$ and the dashed one $\phi_{rf} = 90°$. This can be ascertained by looking at the position of the wave at time $t = 0$ – see text for details.

time dependence of $\cos(\omega_{rf}t + \phi_{rf})$, where ϕ_{rf} is the phase of the wave and ω_{rf} its oscillation frequency. If $\phi_{rf} = 0$, then at t (time) = 0, the rf wave is at a maximum in magnitude, i.e. $\cos(\omega_{rf}t + \phi_{rf}) = 1$. On the other hand, if $\phi_{rf} = 90°$, then at $t = 0$, the magnitude of the wave is zero. Figure B1.3.1 shows two waves which differ in phase by 90°. Figure B1.3.2 shows the effect of producing an rf pulse from the carrier wave.

This manner of producing the rf pulse has several consequences, the most important being on the shape of the pulse. A rectangular pulse consists of, not just the spectral frequency, but a range of frequencies with amplitudes given by a sinc function, $\sin(\frac{1}{2}(\omega - \omega_0)\tau_p)/\frac{1}{2}(\omega - \omega_0)\tau_p$ (see Fig. B1.3.2). The range of frequencies arises from the sharp edges of the pulse. An rf pulse is not a continuous wave but a wave with zero amplitude outside the pulse time and then the wave of the required frequency during the pulse; only by adding together waves of many different frequencies can a wave with such a shape be produced. The sinc function distribution of frequencies is simply the Fourier transform of the rectangular pulse shape; the width of the central part of the distribution is proportional to τ_p^{-1}, where τ_p is the pulse length. Now if the final pulse spectral frequency wave is produced by mixing a pulse at frequency ω_{IF} and another wave of frequency ω_{mix}, some filtering is required after mixing to ensure that we only have the wave with frequency $\omega_{IF} + \omega_{mix} = \omega_0$ and none of the harmonics which will also be produced by mixing two waves. Filtering inevitably truncates the distribution of frequencies within the rf pulse, as illustrated in Fig. B1.3.3. The final pulse shape is then not strictly rectangular, but is derived from the Fourier transform of the truncated frequency distribution as shown in Fig. B1.3.3. This means that the pulse does not finish at the

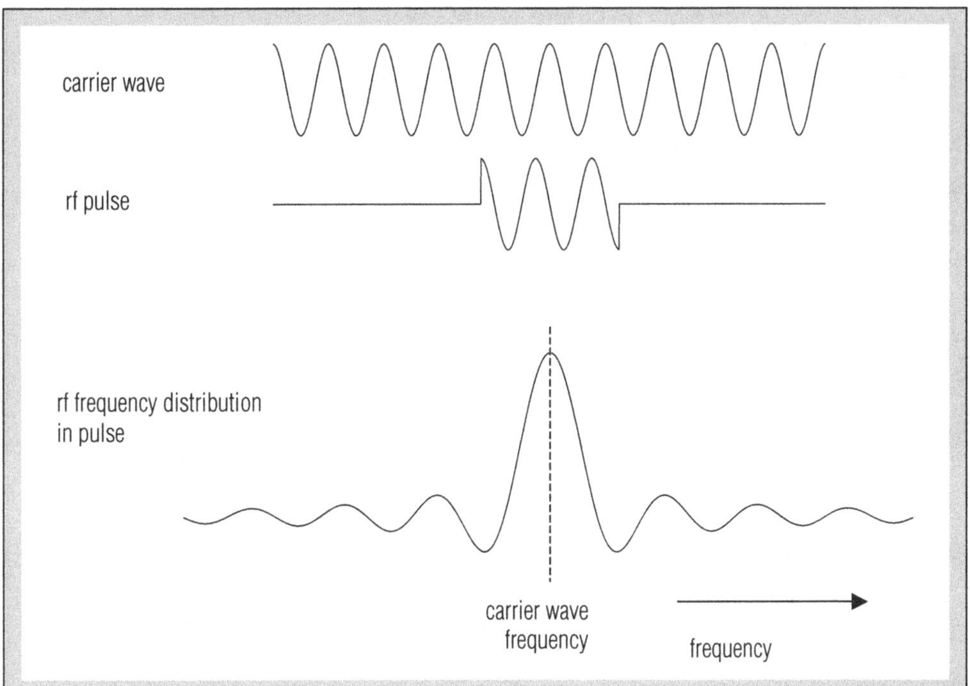

Fig. B1.3.2 Illustrating the concept of the phase of an rf pulse. The phase in this context is the phase of the rf wave during the pulse (bottom trace) relative to the phase of the carrier wave (top) from which the rf pulse is generated. In the example shown, the rf pulse has a 90° phase shift from the carrier wave. At the bottom, the distribution of frequencies arising from the pulse is shown. The width of the distribution depends on τ_p^{-1}, where τ_p is the length of the pulse.

point at which the transmitter is switched off, but necessarily continues for some time after (Fig. B1.3.3). It is important to remember this point if a pulse sequence is required which places rf pulses very close together; simply programming the spectrometer to produce pulses at the required timings may not produce the required result. In practice, such a pulse sequence should be optimized on the sample.

A further consequence of the mode of production of rf pulses is that the production takes time. Most spectrometers take their pulse timings, etc. from a clock which delivers instructions to the transmitters, pulse programmers, etc. at the clock rate, i.e. a 10 MHz clock means that instructions can be delivered every 100 ns, *but no quicker*. Thus, an instruction to the spectrometer to change the phase of a pulse takes 100 ns, then to change its frequency, another 100 ns, then to tell it to form a pulse, another 100 ns. It then takes time for the mixing and filtering required to form the pulse which has been instructed. All this makes for potential time lags in the pulse sequence. This should be checked for when implementing a new pulse sequence by first observing the output pulse sequence from the transmitters on an oscilloscope.

Continued on p. 56

Box 1.3 *Cont.*

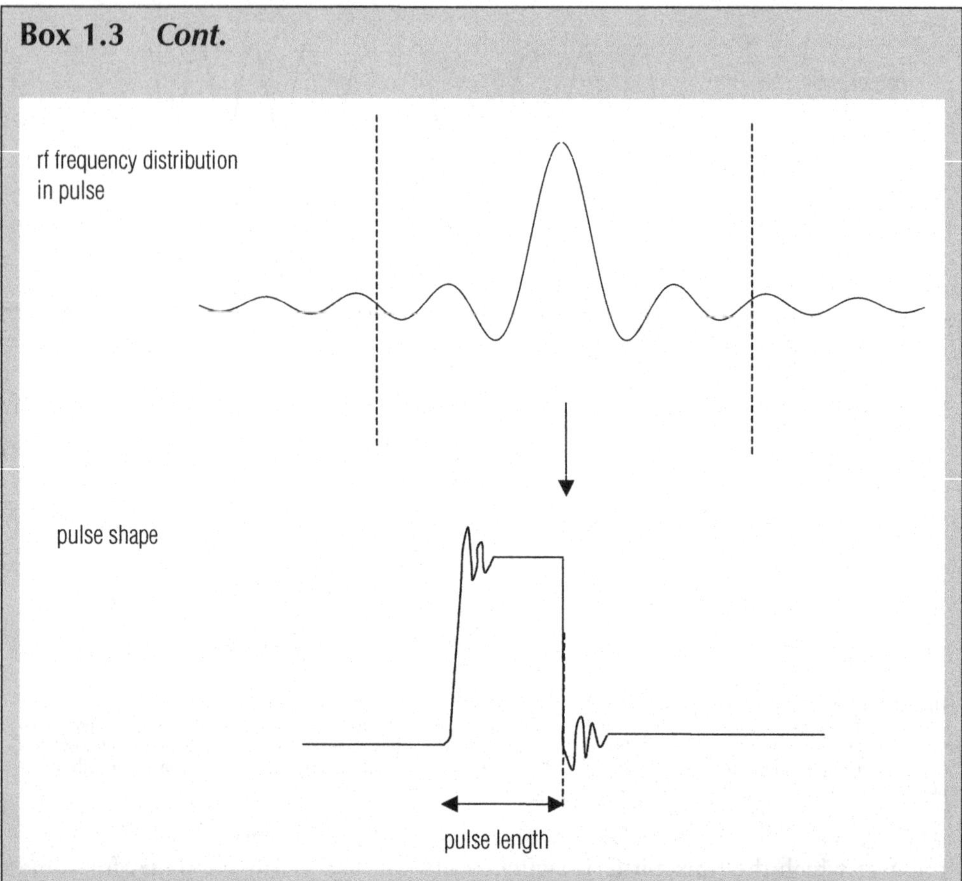

Fig. B1.3.3 Illustrating the effect of filtering on the final pulse shape. The frequency distribution in the pulse is truncated by the filtering. Thus the final pulse shape (derived from the Fourier transform of the frequency distribution) is no longer rectangular. In particular, it has a tail after the end of the pulse. The pulse length is defined as the length of time for which the transmitter is switched on.

Detecting the NMR signal

After irradiation of the sample by the required pulse sequence, we need to measure the response of the sample. This part of the NMR spectrometer is illustrated schematically in Fig. B1.3.4. The signal coming from the NMR probehead (detected by the coil in the probe) is of the form $2c \exp(i(\omega_0 + \Delta\omega)t)$, where $2c$ is the amplitude of the signal, ω_0 is the *carrier* or *spectral frequency* and $\Delta\omega$ is the offset, which may include chemical shift, as well as the effects of other nuclear spin interactions. The first step in the recording of this signal is to amplify the whole signal (which also has the effect of amplifying any noise present as well); the *pre-amplifier* which performs this step may be tuned to the approximate frequency expected for the signal beforehand. In the next step, the signal is mixed down to the *intermediate frequency*, or *IF*, so that the signal now has frequency $\omega_{IF} + \Delta\omega$.

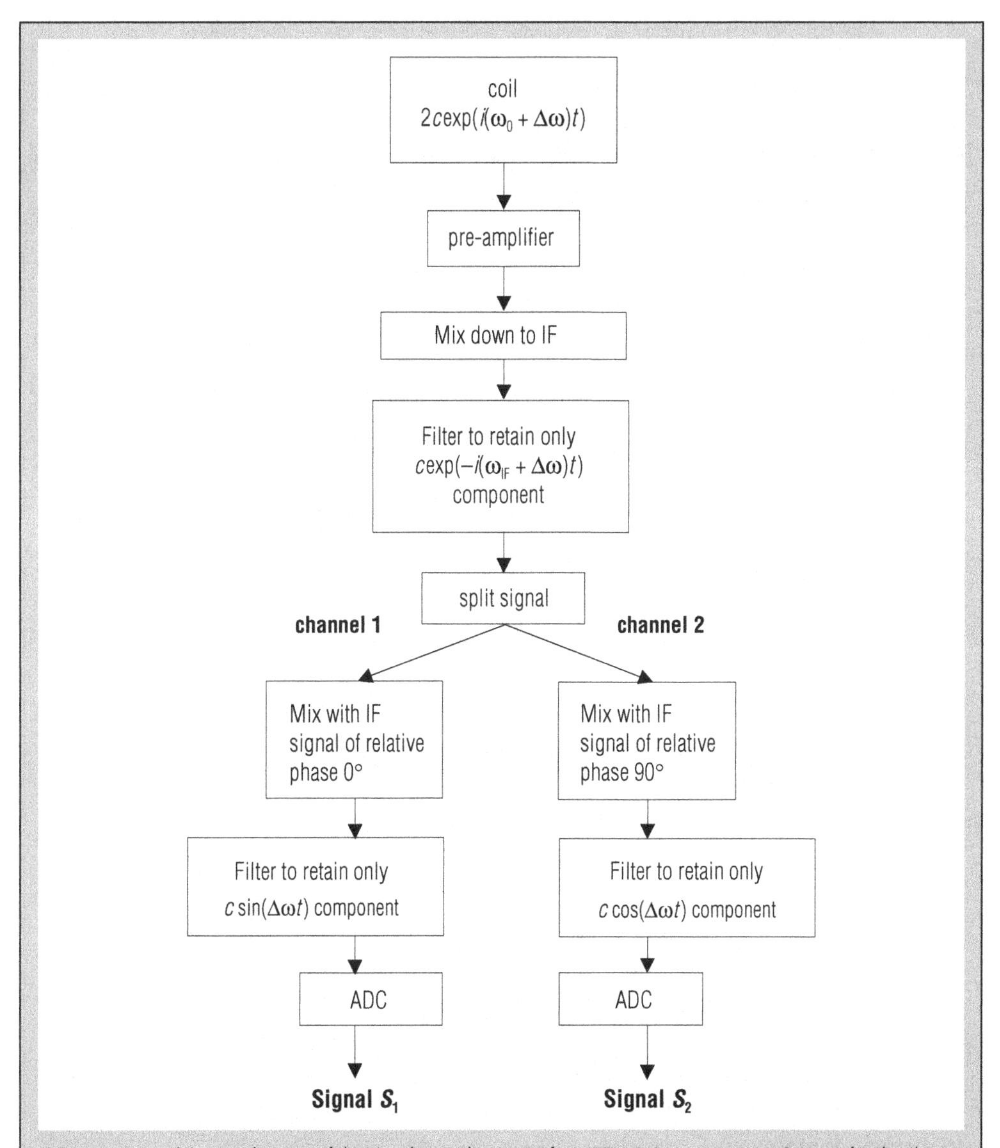

Fig. B1.3.4 A schematic diagram of the signal-recording part of an (FT) NMR spectrometer. See text for details.

The same IF is used whatever nucleus is being studied and, therefore, whatever the initial spectral frequency might have been. The reason for performing this step is that the spectrometer now only has to deal with frequencies around ω_{IF}, a relatively small range, no matter what nucleus is being studied. Thus, constant frequency filters, etc. can be used in the subsequent steps, which has tremendous advantages over variable frequency filtration, which would have to be used otherwise. The frequency of the IF is chosen to minimize noise and optimize

Continued on p. 58

Box 1.3 *Cont.*

frequency accuracy and phase stability. The mixing-down step is actually performed by multiplying the incoming signal with a spectrometer-generated signal of frequency $\omega_0 - \omega_{IF} = \omega_{mix}$. This spectrometer-generated rf wave is produced from the same wave which generated the rf pulses used in the irradiation of the sample to ensure phase coherence between the rf pulse sequence and the measured signal. The mixing of this spectrometer-generated signal and the incoming signal produces a new signal whose time variation is of the form

$$\begin{aligned} 2c\exp(i(\omega_0+\Delta\omega)t)\sin(\omega_{mix}t) &= 2c[\cos((\omega_0+\Delta\omega)t)+i\sin((\omega_0+\Delta\omega)t)]\sin(\omega_{mix}t) \\ &= 2c[\cos((\omega_0+\Delta\omega)t)\sin(\omega_{mix}t)+i\sin((\omega_0+\Delta\omega)t) \\ &\quad \sin(\omega_{mix}t)] \\ &= ci[\exp(i(\omega_0+\omega_{mix}+\Delta\omega)t)-\exp(-i(\omega_{IF}+\Delta\omega)t)] \end{aligned}$$

Filtering the resulting signal with a narrow band filter then selects the component oscillating around ω_{IF}. The form of the signal at this point is then $\exp(-i(\omega_{IF} + \Delta\omega)t)$. The next step is to separate the $\sin(\omega_{IF} + \Delta\omega)$ (real) and $\cos(\omega_{IF} + \Delta\omega)$ (imaginary) components of this complex signal. This is done as follows. The signal is first divided into two. The two halves are routed through different channels (1 and 2) and are each mixed down with a spectrometer-generated signal at the IF frequency, one of which differs in phase by 90° relative to the other. Filters are applied to both channels so that only the difference frequency between the input signal and the IF, i.e. $\Delta\omega$, remains, in a similar manner to the mixing down from the spectral frequency to the IF. The signal is now said to be at *baseband*. The signal in channel 1 mixed with the reference IF signal generates a $\sin(\Delta\omega t)$ or *in-phase* component, while the signal in channel 2 mixed with the signal 90° phase-shifted from the reference IF generates a $\cos(\Delta\omega t)$ or *quadrature* component. The final step is to put the baseband signals from both channels through analogue-to-digital converters (ADCs) to produce digitized signals from the analogue waveforms. These digitized signals can then be stored in a computer for subsequent processing to the frequency domain.

Notes

1. The spatial part of the nuclear wavefunction deals with the molecular vibrations and rotations which affect the spatial position of the nucleus. As implied, there is a small coupling between the spatial and spin coordinates of the nuclear wavefunction and this becomes important when considering nuclear spin relaxation.
2. Note that here we are using the definition that the operator for the nuclear spin angular momentum is $\hbar\hat{\mathbf{I}}$ – see Box 1.1 for more details.
3. The definition of $\hat{L}_z$ comes from the classical expression for angular momentum about z and replacing all the terms in that expression with the corresponding quantum mechanical operators.

4. It is worth noting for later use that the first-order energy correction to the energy of a Zeeman state due to a perturbation described by a hamiltonian, $\hat{H}_1$, is from perturbation theory $E_i^{(1)} = \langle \psi_{I,m} | \hat{H}_1 | \psi_{I,m} \rangle$.

References

1. Levitt, M.H. (2001) *Spin Dynamics: Basics of Nuclear Magnetic Resonance*. John Wiley and Sons, Chichester.
2. Goldman, M. (1988) *Quantum Description of High-Resolution NMR in Liquids*. Clarendon Press, Oxford.
3. Keeler, J.H. (1988) *Multinuclear Magnetic Resonance in Liquids and Solids – Chemical Applications* (eds P. Granger & R.K. Harris). NATO ASI Series C, **322**, 103.
4. Levitt, M.H., Madhu, P.K. & Hughes, C.E. (2002) *J. Magn. Reson*, **155**, 300.

Essential Techniques for Solid-State NMR 2

2.1 Introduction

This chapter describes those techniques which are the bread and butter of solid-state NMR. There are detailed descriptions of how to implement the techniques practically as well as their theoretical basis.

In solid-state NMR, we generally deal with *powder samples*; that is, samples consisting of many crystallites with random orientations. The nuclear spin interactions which affect solid-state NMR spectra, chemical shielding, dipole–dipole coupling and quadrupole coupling, are all dependent on the crystallite orientation; they are said to be *anisotropic*. This is described in much more detail in the chapters dealing specifically with the various nuclear spin interactions. As a result of this, the NMR spectrum of a powder sample contains broad lines, or *powder patterns*, as all the different molecular orientations present in the sample give rise to different spectral frequencies. As outlined in Chapter 1, studying nuclear interactions through their NMR spectra can give useful information on a chemical system. However, when there are several inequivalent nuclear sites in a sample, the powder patterns from each may overlap. The consequent lack of resolution in the NMR spectrum obscures any information that the spectrum might contain. In addition to this, dipolar coupling to an abundant nucleus, for instance to ^{1}H in organic compounds, leads to considerable line broadening, as seen in Chapter 1.

Hence, it is necessary in solid-state NMR to apply techniques to achieve high resolution in spectra. So in this chapter, we will first discuss how to achieve high-resolution NMR spectra and then go on to examine two further techniques, cross-polarization and echo pulse sequences, which are both commonly used in solid-state NMR, either on their own or as part of more complicated pulse sequences.

2.2 Magic-angle spinning (MAS)

Magic-angle spinning is used routinely in the vast majority of solid-state NMR experiments, where its primary task is to remove the effects of chemical shift anisotropy and to assist in the removal of heteronuclear dipolar coupling effects. It is also used to narrow lines from quadrupolar nuclei and is increasingly the method of choice for removing the effects of homonuclear dipolar coupling from NMR spectra. This latter application requires very high spinning rates however, and so is not yet routine in all laboratories.

In solution NMR spectra, effects of chemical shift anisotropy, dipolar coupling, etc., are rarely observed. This is because the rapid isotropic tumbling of the molecules in a solution averages the molecular orientation dependence of the transition frequencies (see Section 1.4 for details) to zero on the NMR timescale, i.e. the rate of change of molecular orientation is fast relative to the magnitude of the chemical shift anisotropy, dipole–dipole coupling, etc., in frequency units.

Magic-angle spinning achieves the same result for solids. Consider the following experimental setup in Fig. 2.1. The molecular orientation dependence of the nuclear spin interactions is discussed in detail in the respective chapters which deal with each interaction. Here, suffice to say that the molecular orientation dependence is of the form $3\cos^2\theta - 1$, where θ is an angle which describes the orientation of the spin interaction tensor, i.e. shielding

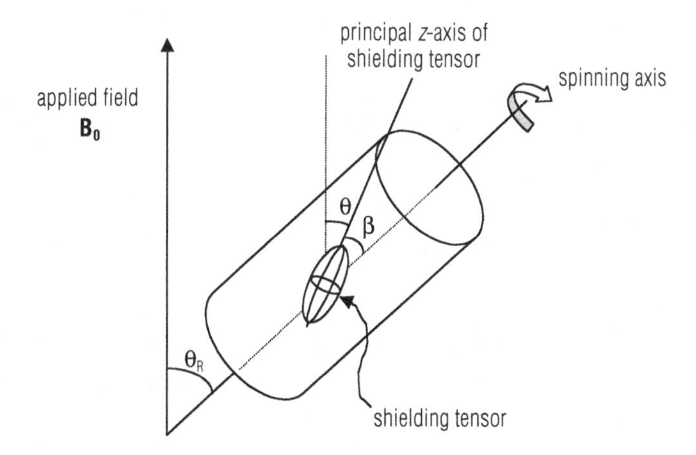

Fig. 2.1 The magic-angle spinning experiment. The sample is spun rapidly in a cylindrical *rotor* about a spinning axis orientated at the magic angle ($\theta_R = 54.74°$) with respect to the applied magnetic field $\mathbf{B_0}$. Magic-angle spinning removes the effects of chemical shielding anisotropy and heteronuclear dipolar coupling. The chemical shielding tensor is represented here by an ellipsoid; it is fixed in the molecule to which it applies and so rotates with the sample. The angle θ is the angle between $\mathbf{B_0}$ and the principal z-axis of the shielding tensor; β is the angle between the z-axis of the shielding tensor principal axis frame and the spinning axis.

tensor in the case of chemical shielding, dipolar coupling tensor in the case of dipole–dipole coupling and so on. In a powder sample, the angle θ effectively takes on all possible values, as the interaction tensor orientation is fixed within the molecule to which it pertains (see Section 1.4) and in powder samples, all molecular orientations are represented.

If we spin the sample about an axis inclined at an angle θ_R to the applied field, then θ varies with time as the molecule rotates with the sample. The average of the orientation dependence of the nuclear spin interation, $(3\cos^2\theta - 1)$, under these circumstances can be shown to be

$$\langle 3\cos^2\theta - 1\rangle = \frac{1}{2}(3\cos^2\theta_R - 1)(3\cos^2\beta - 1) \tag{2.1}$$

where the angles β and θ_R are defined in Fig. 2.1. The angle β is between the principal z-axis of the shielding tensor and the spinning axis, θ_R is the angle between the applied field and the spinning axis and θ is the angle between the principal z-axis of the interaction tensor and the applied field $\mathbf{B}_0$. The angle β is obviously fixed for a given nucleus in a rigid solid, but like θ it takes on all possible values in a powder sample. The angle θ_R is under the control of the experimenter, however. If θ_R is set to 54.74°, then $(3\cos^2\theta_R - 1) = 0$, and so the average, $\langle 3\cos^2\theta - 1\rangle$, is zero also. Therefore, provided that the spinning rate is fast so that θ is averaged rapidly compared with the anisotropy of the interaction, the interaction anisotropy averages to zero.

This technique averages the anisotropy associated with *any* interaction which causes a shift in the energies of the Zeeman spin functions, but no mixing between Zeeman functions (to first order), such as chemical shift anisotropy, heteronuclear dipolar coupling and quadrupole coupling. However, it also has an effect on secular interactions which mix Zeeman functions, i.e. homonuclear dipolar coupling. This topic is discussed in Section 2.2.5.

2.2.1 Spinning sidebands

In order for magic-angle spinning to reduce a powder pattern to a single line at the isotropic chemical shift, the rate of the sample spinning must be fast in comparison to the anisotropy of the interaction being spun out. 'Fast' in this context means around a factor of 3 or 4 greater than the anisotropy.

Slower spinning produces a set of *spinning sidebands* in addition to the line at the isotropic chemical shift (Fig. 2.2). The spinning sidebands are sharp lines, set at the spinning rate apart, and radiate out from the line at isotropic chemical shift. Note that the line at the isotropic chemical shift is *not* necessarily the most intense line. The only characteristic feature of the

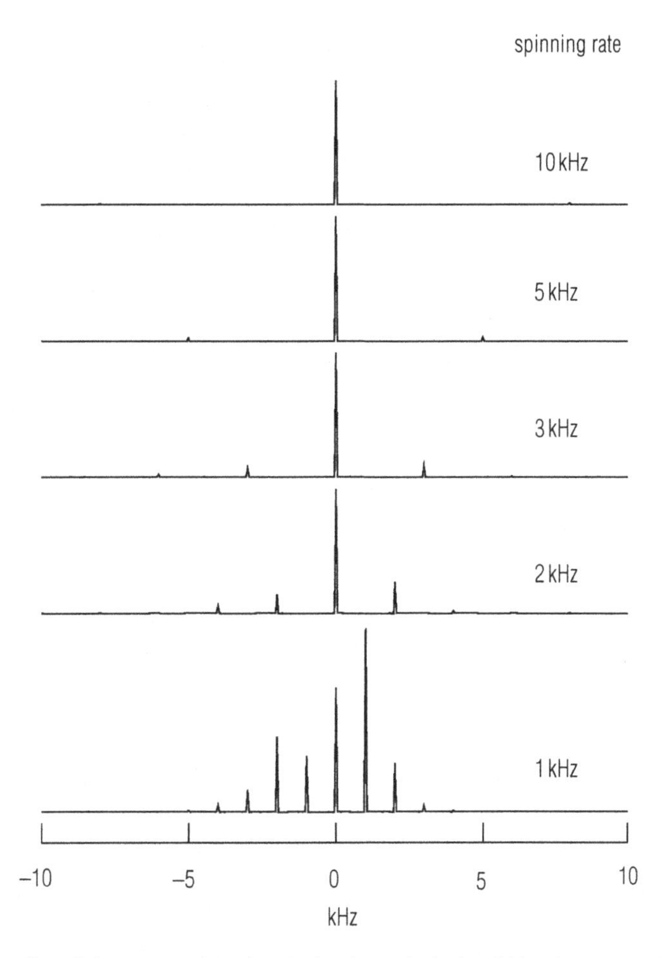

Fig. 2.2 The effect of slow rate magic-angle spinning. A set of spinning sidebands appears, with a centreband at the isotropic chemical shift and further lines spaced at the spinning frequency. The intensities of the sidebands change with spinning rate, with higher order sidebands (i.e. those further away from the centreband) becoming less intense as the spinning rate increases. The chemical shift parameters used in the calculation of these sideband patterns are: isotropic chemical shift offset, 0 Hz, chemical shift anisotropy, 5 kHz, asymmetry, 0.

isotropic chemical shift line is that it is the only line that does not change position with spinning rate; this feature is the only reliable way of identifying it. At the time of writing, spinning rates of up to 50 kHz are achievable, with 30 kHz being routine on a modern spectrometer. At this spinning rate, there will rarely be more than one or two small spinning sidebands in a spectrum, if that, for ^{13}C, ^{15}N and ^{31}P, for instance. However, nuclei which have large numbers of electrons associated with them, such as ^{195}Pt, ^{207}Pb and ^{117,119}Sn, can have very large chemical shift anisotropies, so that even at very high spinning rates many spinning sidebands are still observed. It is worth noting at this point that the chemical shift anisotropy (in frequency

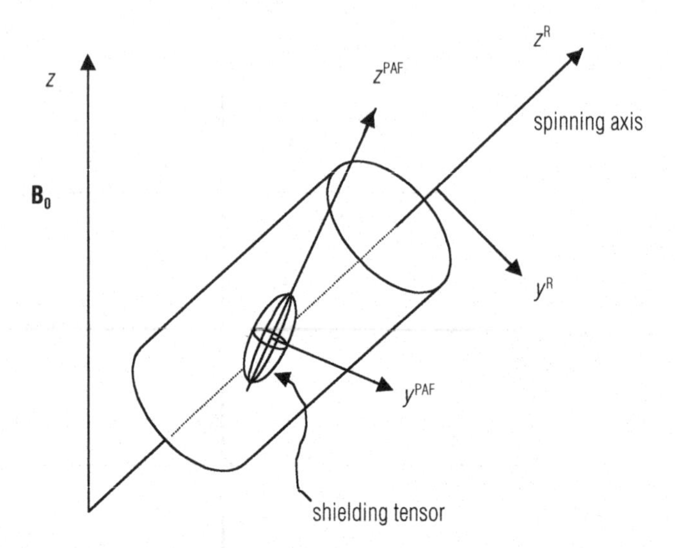

Fig. 2.3 Illustrating the axis frames used in the discussion of magic-angle spinning. Shown here is the so-called *laboratory frame* whose z-axis defines the direction of $\mathbf{B}_0$, the *rotor frame* (superscript^R) whose z^R-axis is parallel to the spinning axis and the *principal axis frame* (superscript^{PAF}), the frame in which the shielding tensor is diagonal. The shielding tensor is represented here by an ellipsoid.

units) is proportional to B_0, and so as the trend towards ever larger magnetic fields increases, these nuclei are likely to become more difficult to deal with, rather than less.

However, as we shall see, spinning sidebands can be very useful. They can be used to determine details of the anisotropic interactions which are being averaged by the magic-angle spinning. In order to do this we must first describe a mathematical analysis of these spinning sidebands.

Fortunately, a mathematical description of magic-angle spinning is easily given. Consider the case of chemical shielding, where the shielding tensor is expressed with respect to an axis frame fixed on the rotor (the 'rotor axis frame') and is labelled σ^R, for a given crystallite orientation with respect to the rotor axis frame (Fig. 2.3).

As shown in Chapter 3, the contribution of the chemical shielding to the observed frequency, from a molecular orientation defined by Ω at a time t, is

$$\omega(\Omega;t) = -\omega_0 \mathbf{b}_0^R \sigma^R \mathbf{b}_0^R \tag{2.2}$$

where $\mathbf{b}_0^R$ is the unit vector in the direction of $\mathbf{B}_0$ in the rotor axis frame R. In turn, $\mathbf{b}_0^R$ is given by

$$\mathbf{b}_0^R = (\sin\theta_R \cos\omega_R t, \sin\theta_R \sin\omega_R t, \cos\theta_R) \tag{2.3}$$

where θ_R is the angle of the rotor axis frame z-axis with respect to $\mathbf{B}_0$ and ω_R is the spinning rate. (θ_R, $\omega_R t$) are the polar angles which describe the orientation of $\mathbf{B}_0$ in the rotor axis frame R. Substituting for $\mathbf{b}_0^{\mathrm{R}}$ in Equation (2.2) for the spectral frequency and after some (considerable) rearrangement using trigonometric identities, we obtain [1, 2]

$$\begin{aligned}\omega(\Omega;t) = -\omega_0 \Big\{ \sigma_{\mathrm{iso}} &+ \frac{1}{2}(3\cos^2\theta_{\mathrm{R}} - 1)(\sigma_{zz}^{\mathrm{R}} - \sigma_{\mathrm{iso}}) \\ &+ \sin^2\theta_{\mathrm{R}}\left[\frac{1}{2}(\sigma_{xx}^{\mathrm{R}} - \sigma_{yy}^{\mathrm{R}})\cos(2\omega_{\mathrm{R}}t) + \sigma_{xy}^{\mathrm{R}}\sin(2\omega_{\mathrm{R}}t)\right] \\ &+ 2\sin\theta_{\mathrm{R}}\cos\theta_{\mathrm{R}}[\sigma_{xz}^{\mathrm{R}}\cos(\omega_{\mathrm{R}}t) + \sigma_{yz}^{\mathrm{R}}\sin(\omega_{\mathrm{R}}t)]\Big\}\end{aligned} \tag{2.4}$$

where σ_{iso} is the isotropic shielding as defined by:

$$\sigma_{\mathrm{iso}} = \frac{1}{3}(\sigma_{xx} + \sigma_{yy} + \sigma_{zz}) \tag{2.5}$$

In this definition of isotropic shielding, σ can be measured in any frame, as the trace of a tensor is invariant to rotation of the defining axis frame.

Now, σ^{R} can be written in terms of σ^{PAF}, the shielding tensor expressed in the principal axis frame (defined in Chapter 1, Section 1.4) as

$$\boldsymbol{\sigma}^{\mathrm{R}} = \mathbf{R}^{-1}(\alpha,\beta,\gamma)\begin{pmatrix} \sigma_{xx}^{\mathrm{PAF}} & 0 & 0 \\ 0 & \sigma_{yy}^{\mathrm{PAF}} & 0 \\ 0 & 0 & \sigma_{zz}^{\mathrm{PAF}} \end{pmatrix}\mathbf{R}(\alpha,\beta,\gamma) \tag{2.6}$$

where $\mathbf{R}(\alpha, \beta, \gamma)$ is the rotation matrix corresponding to the rotation operator which rotates the PAF into the rotor axis frame via Euler angles α, β and γ. Euler angles are defined in Box 1.2 in Chapter 1, as is the effect of frame rotation on tensor quantities. Using this definition of σ^{R} and the fact that $\theta_{\mathrm{R}} = 54.74°$ (the magic angle), after much rearrangement, we find

$$\begin{aligned}\omega(\alpha,\beta,\gamma;t) = -\omega_0\{\sigma_{\mathrm{iso}} &+ [A_1\cos(\omega_{\mathrm{R}}t+\gamma) + B_1\sin(\omega_{\mathrm{R}}t+\gamma)] \\ &+ [A_2\cos(2\omega_{\mathrm{R}}t+2\gamma) + B_2\sin(2\omega_{\mathrm{R}}t+2\gamma)]\}\end{aligned} \tag{2.7}$$

with

$$\begin{aligned} A_1 &= \frac{2}{3}\sqrt{2}\sin\beta\cos\beta[\cos^2\alpha(\sigma_{xx}^{\mathrm{PAF}} - \sigma_{zz}^{\mathrm{PAF}}) + \sin^2\alpha(\sigma_{yy}^{\mathrm{PAF}} - \sigma_{zz}^{\mathrm{PAF}})] \\ B_1 &= \frac{2}{3}\sqrt{2}\sin\alpha\cos\alpha\sin\beta(\sigma_{xx}^{\mathrm{PAF}} - \sigma_{yy}^{\mathrm{PAF}}) \\ A_2 &= \frac{1}{3}((\cos^2\beta\cos^2\alpha - \sin^2\alpha)(\sigma_{xx}^{\mathrm{PAF}} - \sigma_{zz}^{\mathrm{PAF}}) \\ &\quad + (\cos^2\beta\sin^2\alpha - \cos^2\alpha)(\sigma_{yy}^{\mathrm{PAF}} - \sigma_{zz}^{\mathrm{PAF}})) \\ B_2 &= -\frac{2}{3}\sin\alpha\cos\alpha\cos\beta(\sigma_{xx}^{\mathrm{PAF}} - \sigma_{yy}^{\mathrm{PAF}}) \end{aligned} \tag{2.8}$$

The first term in Equation (2.7) is obviously just the isotropic term. The remaining terms, however, oscillate at frequencies ω_R and $2\omega_R$ respectively. When the spinning rate ω_R is much greater than the chemical shift anisotropy Δ_{cs} (see Section 3.1.2, Equation (3.4) for definition) these terms have a negligible effect on the NMR spectrum. However, when $\omega_R \leq \Delta_{cs}$, these terms create the spinning sidebands.

It is a simple matter to calculate the spinning sideband pattern using Equation (2.7). The free induction decay (FID) $g(t)$ in an NMR experiment where the evolution frequency of the observed coherence/magnetization is time-dependent is

$$g(t) = \frac{1}{8\pi^2}\int_0^{2\pi}\int_0^{\pi}\int_0^{2\pi}\exp\left(i\int_0^t \omega(\alpha,\beta,\gamma;t)\,\mathrm{d}t\right)\sin\beta\,\mathrm{d}\alpha\,\mathrm{d}\beta\,\mathrm{d}\gamma \qquad (2.9)$$

for a powder sample, where the integrals over α, β and γ sum the FIDs from all possible molecular/crystallite orientations in the powder sample. The Euler angles (α, β, γ) define the orientation of the molecule-fixed principal frame (PAF) relative to the rotor frame. $\omega(\alpha, \beta, \gamma; t)$ is the spectral or evolution frequency of the observed coherence/magnetization for the molecular orientation defined by (α, β, γ). Substituting Equation (2.7) for $\omega(\alpha, \beta, \gamma; t)$ in Equation (2.9) and performing the integrations over molecular orientation numerically allows the full FID for the sample to be calculated. From that, it is a simple matter to obtain the frequency spectrum by Fourier transformation.

The chemical shift anisotropy and asymmetry (see Section 3.1.2, Equation (3.4) for definitions) can be obtained from an experimentally observed spinning sideband pattern by calculating the expected spectrum for different anisotropy and asymmetry values until reasonable agreement with the experimental pattern is obtained [1, 2]. Spinning sideband patterns of the type described here can be produced from heteronuclear dipolar coupling and quadrupolar coupling as well as chemical shielding. By substituting the relevant interaction tensor in Equations (2.7) and (2.8), the spinning sideband patterns arising from them can be calculated. There is a small modification in the case of dipolar coupling. If we consider the dipolar coupling between two spins I and S for instance, the FID arising from spin I, $g^I(t)$, is as in Equation (2.9) but in addition must be summed over the $(2S + 1)$ possible I-spin transitions which can occur, each corresponding to a different possible S-spin z-magnetization. This simply affects the evolution frequency $\omega(\alpha, \beta, \gamma; t)$ via a factor of m_s (the z-component of the S spin) which occurs in the equation for transition frequency in this case (see Equations 4.20 and 4.21 in Chapter 4).

2.2.2 Rotor or rotational echoes

Solid-state NMR literature often refers to *rotational echoes* in association with magic-angle spinning. These are simply explained as follows. Consider a component of magnetization in the x–y plane of the rotating frame (resulting from a 90° pulse, for instance); the evolution of this magnetization is what is recorded in the FID in the NMR experiment. This magnetization has an evolution frequency determined by the applied field $\mathbf{B}_0$ and chemical shielding. Suppose this component of magnetization arises from a principal axis frame of orientation (α, β, γ) with respect to the rotor axis frame and that the whole sample is spun at the magic angle. As the sample spins, the evolution frequency (as recorded in the FID) varies, because the crystallite orientation changes with respect to the $\mathbf{B}_0$ field as the sample rotates, and so the chemical shielding changes also. However, when the sample returns to its starting position, the evolution frequency returns to its starting value and then goes through the same cycle of values all over again. Thus, the FID corresponding to the magnetization component consists of a sequence of repeated 'sub-FIDs', or *rotational echoes* (Fig. 2.4).

Summing over all the magnetization components arising from the whole sample gives a similar result. Fourier transformation of any one of the rotor echoes results in a powder pattern identical to that from a non-spinning sample. This is because rotating the sample alters the γ angle for each crystallite, but as one crystallite moves from $\gamma \rightarrow \gamma + \Delta\gamma$ due to the sample rotation, so another moves from $\gamma - \Delta\gamma \rightarrow \gamma$, so that the distribution of crystallites over the γ at any point during the rotor cycle is as it would be in a static sample (assuming the sample has cylindrical symmetry).

If a new time series is made up of the rotational echo maxima (the dots in Fig. 2.4) and Fourier transformed, a purely isotropic spectrum free of spinning sidebands is obtained. This is explained more fully in Section 2.2.3.

2.2.3 Removing spinning sidebands

Although spinning sidebands are very useful for determining anisotropies and asymmetries of nuclear spin interactions, they can obscure other signals in the spectrum. Several sets of overlapping spinning sidebands can also result in a very confusing spectrum. Therefore, we need some way of removing spinning sidebands.

One of the best ways of removing spinning sidebands is simply to spin the sample faster! At the time of writing, rates of 50 kHz are achievable with rotors of 1.2 mm external diameter. Where this is still not fast enough, or not possible, there are two other options. The first is to record the FID synchronously with the sample spinning, i.e. set the FID dwell time equal to

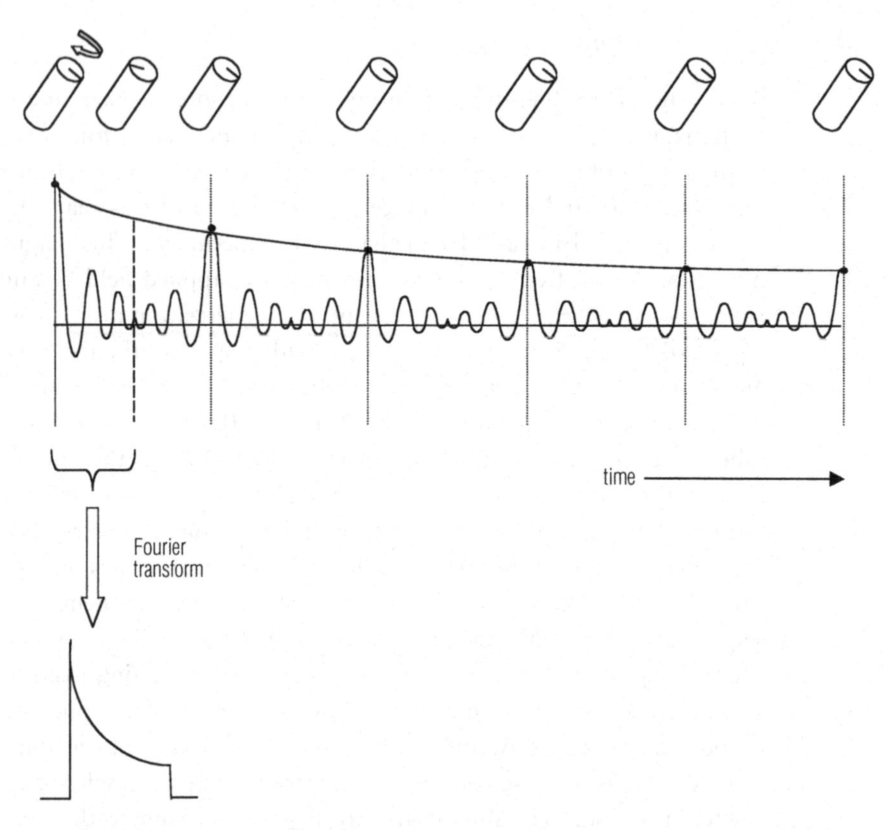

Fig. 2.4 The formation of rotor echoes. The FID shown is that formed under magic-angle spinning (on resonance); essentially the FID repeats every rotor period (marked with dotted lines) and, within each rotor period, is symmetric about the half period point. Fourier transformation of one half of a single rotor period results in the powder pattern that would be formed in the absence of magic-angle spinning. Fourier transformation of the echo maxima (denoted by the black dots) gives a single line at the isotropic chemical shift. Fourier transformation of the entire FID gives a line at the isotropic chemical shift flanked by spinning sidebands as described in the text. The rotor echoes arise as a result of the orientation dependence of the chemical shift (or other time-independent interaction); the chemical shift of a given crystallite changes as the rotor changes position during spinning. As the rotor returns to its original position at the end of each rotor period, so the chemical shift returns to its original value and an echo forms in the FID.

the rotor period, or equivalently, spectral width equal to rotor spinning frequency. This works because all the anisotropic components of $\omega(\alpha, \beta, \gamma; t)$ in Equation (2.9) for the FID, $g(t)$, integrate over complete rotor periods to zero, leaving only the isotropic component at the sampling points $t = n\,\tau_R$, where n is an integer and τ_R is the rotor period. This is equivalent to sampling at the points indicated by dots in Fig. 2.4. This method, however, is not often used as it requires the use of what is often a restrictively small spectral width (which can lead to problems with folding if it results in part of the spectrum falling outside the spectral width).

The preferred method is to use a special pulse sequence known as TOtal Suppression of Spinning sidebands (TOSS) [3]. This method applies a series of 180° pulses at precisely placed points in a period *prior* to acquisition.

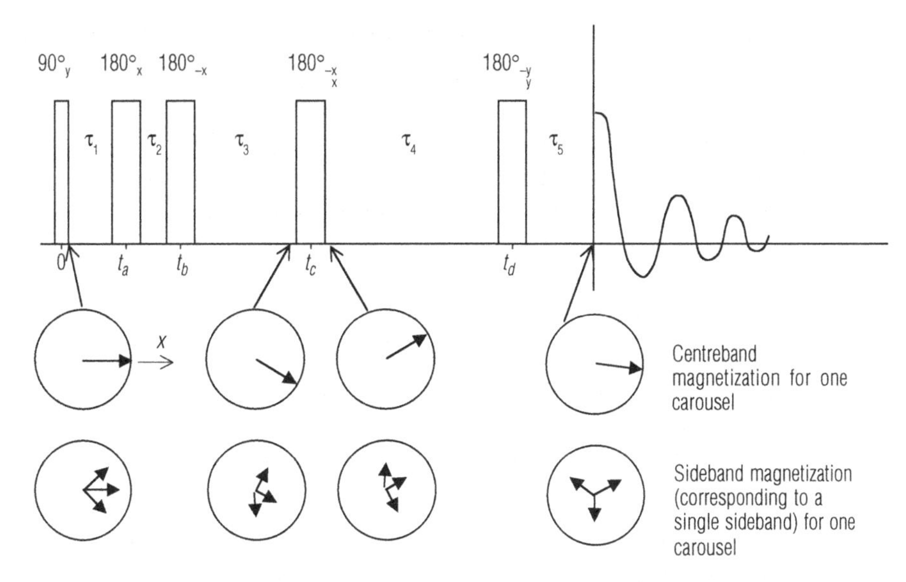

Fig. 2.5 The TOSS (TOtal Suppression of Spinning sidebands) pulse sequence consists of four 180° pulses in a period prior to acquisition. The phase cycling of the 180° pulses shown compensates for pulse imperfections. The initial 90° pulse simply generates transverse magnetization. An example of the pulse spacings is $\tau_1 = 0.1885\tau_R$; $\tau_2 = 0.0412\tau_R$; $\tau_3 = 0.5818\tau_R$; $\tau_4 = 0.9588\tau_R$; $\tau_5 = 0.2297\tau_R$, where τ_R is the rotor period. The schematics below the pulse sequence indicate the precessional motion of the centreband magnetization and a single sideband magnetization for one carousel of crystallite orientations (see text for details). At the end of the TOSS sequence, the phases of the sideband magnetization components from the different crystallite orientations within the carousel are random, so that there is no net sideband magnetization. The diagram shows the minimum phase cycling required in the pulse sequence.

Usually four 180° pulses are used with phase cycling to compensate for pulse imperfections. The pulse sequence is shown in Fig. 2.5. Finite 180° pulse lengths reduce the efficiency of this technique, so sometimes composite 180° pulses are used to compensate. The technique assumes the sample has cylindrical symmetry, which is normally the case, providing the crystallites in the sample are small and truly randomly distributed. If there are many spinning sidebands, TOSS does not work well and small residual spinning sidebands (which may show phase distortions) are observed. It is worth noting that the isotropic signal intensity under TOSS is different from the isotropic signal intensity in the sideband pattern and that, in general, TOSS signals are not quantitative. The source of this is revealed in the discussion below. The original TOSS sequence [3] involves two 180° pulses particularly close together. At high spinning rates these two pulses can become so close that they merge, in which case the pulse sequence is ineffective. For this reason, there has been some effort towards producing improved TOSS sequences [4], although all work on the same principles.

The way a TOSS pulse sequence works can be understood qualitatively as follows. An initial $90°_y$ pulse creates transverse magnetization along x. The transverse magnetization associated with each different crystallite orientation in the sample begins to precess about the applied field $\mathbf{B}_0$. Before we look at the TOSS sequence in detail, we need first to describe further the transverse magnetization components from each crystallite orientation. The precession frequency of each such component will change with time as the rotor rotates and the crystallites change orientation with respect to the applied field. The *phase* of a magnetization component is the angle between the magnetization and the x-axis of the transverse plane (Fig. 2.5). The spectrum associated with the transverse magnetization of each crystallite consists of a centreband at the isotropic chemical shift and a series of spinning sidebands radiating out from the centreband, as we have already seen. Thus we can decompose the time-varying transverse magnetization for each crystallite into a component precessing at the isotropic chemical shift frequency, plus a series of other components precessing at the sideband frequencies, i.e. $\omega_{iso} + m\omega_R$, where m is an integer (positive or negative). We will refer to these components as centreband magnetization and sideband magnetization respectively, with mth order sideband magnetization being that associated with the mth spinning sideband. The initial phase (i.e. at time $t = 0$) and the amplitude of each sideband and centreband magnetization component depend on crystallite orientation. The sum of centreband and sideband magnetization components for a single crystallite orientation will be orientated along x however, representing the initial transverse magnetization after a $90°_y$ pulse. This is a useful picture from which to understand the mechanism of the TOSS experiment.

We then need to define the concept of a *carousel* of crystallite orientations [4]. A crystallite orientation with respect to a rotor-fixed axis frame is defined by the set of Euler angles (α, β, γ), as in Section 2.2.1 previously. A carousel of crystallite orientations is the set of orientations with the same α and β angles, but different γ, with all γ values from 0 to 2π being present in the set. The significance of this is that a crystallite in one orientation within the carousel will visit all other orientations in the carousel during the course of one rotor cycle. This in turn means that the transverse magnetization associated with all crystallite orientations within the carousel goes through the same precessional frequencies during one rotor cycle, albeit at different times. The initial phase of the mth-order sideband magnetization within a carousel depends on γ, while the initial phase of the centreband magnetization is independent of γ and thus constant within a carousel of orientations.

The effect of the TOSS pulse sequence is to make the phases of the mth order sideband magnetizations for the crystallite orientations within one carousel random at the end of the pulse sequence. Thus, the net mth order sideband magnetization when summed over all the orientations of the carousel is zero at the end of the TOSS pulse sequence, i.e. at the start of the FID whose acquisition follows the TOSS sequence (Fig. 2.5). At the same time, the phases of the centreband magnetizations within the carousel are arranged to be all the same, and so the centreband magnetizations add constructively and are present at the end of the TOSS sequence with the intensity that would be expected in the absence of the TOSS preparation sequence.

The reason that for a powder sample the intensity of the net centreband tends to be reduced after a TOSS preparation sequence (in comparison to the centreband intensity in the absence of TOSS) is that most TOSS sequences result in the centreband magnetizations having a phase which is carousel-dependent. In a powder sample, the centreband signal is summed over all possible carousels, and so there is inevitably some destructive interference between signals from different carousels.

The phases acquired by the sideband and centreband magnetizations are determined by the 180° pulses in the TOSS sequence and their positions in time. The effect of a $180°_\alpha$ pulse on transverse magnetization is to rotate all magnetization components by 180° about the pulse axis α, as illustrated in Fig. 2.5. In between the 180° pulses of the TOSS sequence, the transverse magnetization components continue to precess at their sideband and centreband frequencies respectively. Thus, the net phase acquired by a magnetization component by the end of the pulse sequence depends on their precessional frequency and the number and timing of the 180° pulses and is given by $\phi(\alpha, \beta, \gamma)$:

$$\phi(\alpha,\beta,\gamma) = \int_0^{t_a} \omega(\alpha,\beta,\gamma;t)\mathrm{d}t - \int_{t_a}^{t_b} \omega(\alpha,\beta,\gamma;t)\mathrm{d}t + \int_{t_b}^{t_c} \omega(\alpha,\beta,\gamma;t)\mathrm{d}t - \ldots \quad (2.10)$$

where t_a, t_b, etc., are the timings of the 180° pulses. The 180° pulses are assumed to be perfect and effectively just change the sign of the precession frequency of the transverse magnetization, given by $\omega(\alpha, \beta, \gamma; t)$ for a magnetization component arising from a crystallite orientation (α, β, γ). This is what gives rise to the alternation in sign of the integral components of Equation (2.10). The series in Equation (2.10) continues over all the periods of free precession in the TOSS sequence. The net phase acquired by a carousel of γ-orientations, $\Phi(\alpha, \beta)$, is then simply

$$\Phi(\alpha,\beta) = \int_0^{2\pi} \phi(\alpha,\beta,\gamma)\mathrm{d}\gamma \quad (2.11)$$

As shown in Equation (2.7), the precession frequency $\omega(\alpha, \beta, \gamma; t)$ is a sum of a constant isotropic term, ω_{iso} $(= -\sigma_{iso}\,\omega_0)$ and time-varying terms (due to magic-angle spinning) which depend on the chemical shift anisotropy. A suitable TOSS sequence has pulse timings t_a, t_b, . . . such that the phase acquired at the end of the sequence for each carousel of orientations, $\Phi(\alpha, \beta)$, depends only on ω_{iso}. Further details of how to generate other solutions can be found in reference [4].

2.2.4 Setting the magic angle and spinning rate

It is very important that the magic angle is set accurately; a mis-setting of even a few thousandths of a degree can lead to noticeable linebroadening. Spinning the sample about an angle θ_R with respect to the applied field in effect scales the anisotropy of any interaction by a factor $(3\cos^2\theta_R - 1)$ (see Equation (2.1)). This factor is zero of course at the magic angle, but is non-zero if the angle is mis-set. Under these circumstances, the spectrum will consist not of sharp lines but of a series of powder patterns (albeit very narrow ones, if the spinning angle is close to the magic angle) with effective anisotropies of $\Delta(3\cos^2\theta_R - 1)$, where Δ is the true anisotropy of the interaction. This is the source of the linebroadening when the magic angle is mis-set.

To set the magic angle, a sample is required which gives a good signal in a single scan and which gives lots of sharp spinning sidebands at moderate spinning rates, i.e. the observed nucleus should be affected by an inhomogeneous interaction with an anisotropy of the order of 100 kHz. Ideally, the sample should not require any form of decoupling to achieve a sharp spectrum, i.e. no ^{1}H or other abundant spin dipolar-coupled to the observed nucleus. It will also be much easier to set the magic angle if the spectrum can be repeatedly scanned with only a short interval between scans, i.e. the observed nucleus should have a short spin-lattice relaxation time. The most commonly used standard for setting the magic angle is the ^{79}Br ($I = \frac{3}{2}$, $\omega_0(^{79}\mathrm{Br}) = 0.2513\ \omega_0(^{1}\mathrm{H})$) resonance in solid KBr. This quadrupolar nucleus relaxes quickly and successive FIDs can be collected at one second intervals. The quadrupole coupling constant for ^{79}Br in this compound is such that the first-order quadrupole coupling on the satellite transitions ($\pm\frac{3}{2} \leftrightarrow \pm\frac{1}{2}$) gives rise to a good set of spinning sidebands, which can be contained in a spectral width of 100 kHz at spinning rates of 3–8 kHz.

To set the magic angle in practice, start by setting the sample spinning at a rate of around 3 kHz. Once the spinning rate is stable, record a ^{79}Br spectrum with a few scans using a conventional single pulse/acquire pulse sequence; you should only need the number of scans it takes to complete one phase cycle of the pulse sequence for this. Keep the excitation pulse short, a few microseconds at most, to avoid the spectral distortions which will otherwise occur as the quadrupole coupling competes with the rf pulse during the excitation period. Fourier transform and phase the spectrum,

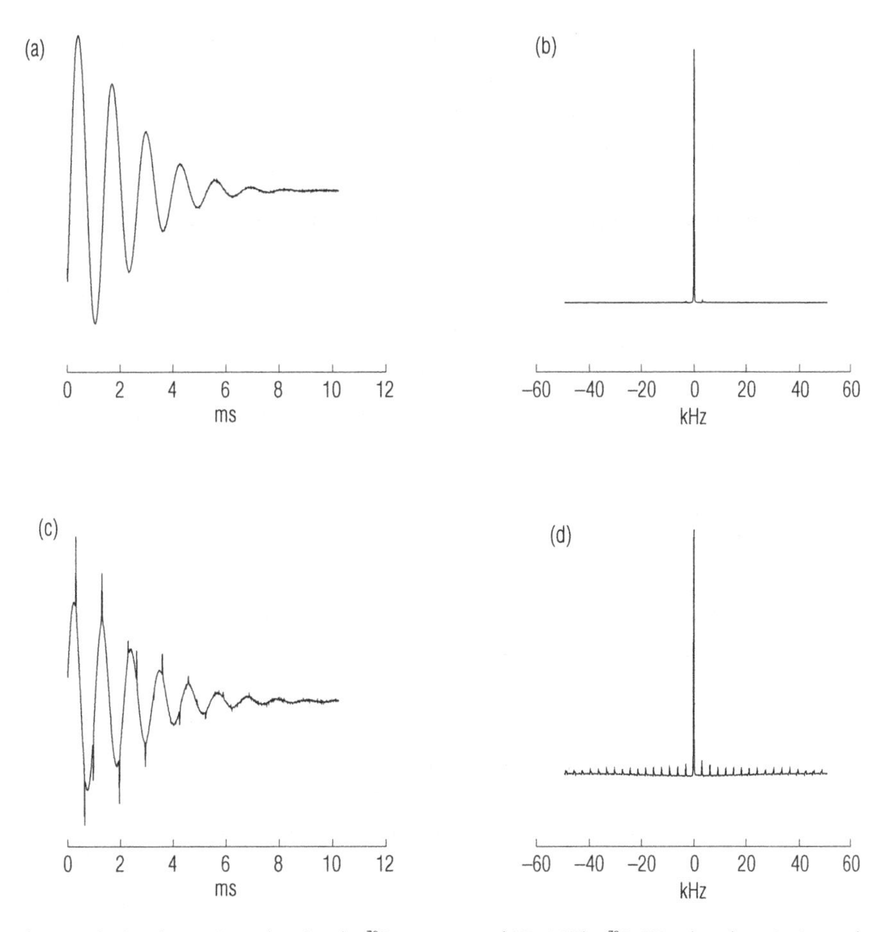

Fig. 2.6 Setting the magic angle using the ^{79}Br resonance of KBr. (a) The ^{79}Br FID when the spinning angle is away from the magic angle. (b) The ^{79}Br spectrum from (a). (c) The ^{79}Br FID when the magic angle is well set. (d) The ^{79}Br spectrum from (c). The spinning rate is 1.5 kHz in all cases. The sharp features in the FID of (c) are rotational echoes, which give rise to the spinning sidebands in spectrum (d).

then set spectral frequency so that the centreband of the spectrum (the largest peak in this case) is close to resonance, and re-record the spectrum. When the magic angle is accurately set, you should be able to identify around 10 or more spinning sidebands either side of the centreband, when spinning at 3 kHz (see Fig. 2.6).

To adjust the magic angle to its optimum value, set the spectrometer to collect single scan spectra or FIDs at one second intervals. If observing the FID, you should see sharp rotational echoes, as in Fig. 2.6. Now adjust the spinning angle a little, and check the effect on the FID/spectrum. If more rotational echoes/spinning sidebands appear towards the end of the FID/edges of the spectrum, then you are moving closer to the magic angle;

if rotational echoes towards the end of the FID or spinning sidebands at the outer edges of the spectrum disappear, then you are moving away from the magic angle. Adjust the spinning angle so as to maximize the number and height of rotational echoes/spinning sidebands observed. Whether you work from the FID or the spectrum is a matter of personal preference, although most spectroscopists use the FID from habit. This method should lead to a very accurate setting of the magic angle, sufficient for most experiments. A few experiments, such as the Satellite Transition Magic-Angle Spinning experiment (STMAS) for quadrupolar nuclei, are extremely sensitive to the setting of the spinning angle and may require further optimization using the desired experiment itself. This will be covered in more detail for the STMAS experiment in Chapter 5.

How fast to spin?

If you are aiming to produce a purely isotropic spectrum, with no spinning sidebands, then clearly, the spinning rate needs to be fast enough to average effectively the largest anisotropic (inhomogeneous) interaction in the sample, i.e. the spinning rate needs to be around a factor of 3 greater than the largest anisotropy. Where this is not possible, there are pulse sequences which can assist in removing spinning sidebands (see Section 2.2.4). One further cautionary note should be mentioned. If there is any molecular motion in the sample which causes the resonance frequency of a spin to change in time, even in the absence of magic-angle spinning, then magic-angle spinning will not effectively average the interaction anisotropy if the time variation of the resonance frequency is at a similar rate to the magic-angle spinning. For instance, rotational motion of a molecule or part of a molecule containing an observed nucleus leads to a time-dependent resonance frequency because of the changing orientation (with respect to the applied field) of the nuclear spin interaction tensor associated with the nucleus as the molecule rotates. If the correlation time for the motion is of the order 10^{-2}–10^{-4}s, then the resonance frequency is changing significantly during one rotor period and so the magic-angle spinning cannot be effective in averaging the interaction to zero. This phenomenon is discussed further in Chapter 6. Suffice to say here that if, under magic-angle spinning, the spectrum contains unexpectedly broad lines (or perhaps even missing lines if the linebroadening is extreme), it could be due to molecular motion in the sample. To test if this is the case, change the spinning rate to a much lower and/or much higher rate. If the lineshape changes significantly, then there is probably molecular motion in the sample affecting the spins giving rise to the broadened lines. Alternatively, one can change the temperature of the sample and see if the lineshape changes.

2.2.5 *Magic-angle spinning for homonuclear dipolar couplings*

As mentioned at the beginning of this section, magic-angle spinning can be used for removing the effects of homonuclear dipolar coupling providing the spinning rate is high enough. The dipolar-coupling hamiltonian for a homonuclear-coupled spin pair, *I* and *S*, is shown in Chapter 4 to be (in angular frequency units):

$$\hat{H}_{\rm dd}^{\rm homo} = -d \cdot \frac{1}{2}(3\cos^2\theta - 1)[3\hat{I}_z\hat{S}_z - \hat{\mathbf{I}}\cdot\hat{\mathbf{S}}] \qquad (2.12)$$

with $d = \left(\frac{\mu_0}{4\pi}\right)\frac{\gamma_I\gamma_S\hbar}{r^3}$, the dipolar-coupling constant, and θ being the angle between the *I–S* internuclear axis and the applied field $\mathbf{B}_0$. For a general multispin system, the corresponding hamiltonian is simply

$$\hat{H}_{\rm dd}^{\rm homo} = -\sum_{i>j} d_{ij} \cdot \frac{1}{2}(3\cos^2\theta_{ij} - 1)[3\hat{I}_z^i\hat{S}_z^j - \hat{\mathbf{I}}^i\cdot\hat{\mathbf{I}}^j] \qquad (2.13)$$

where i and j label the spins, so that $d_{ij} = \left(\frac{\mu_0}{4\pi}\right)\frac{\gamma_i\gamma_j\hbar}{r_{ij}^3}$ where r_{ij} is the internuclear distance between spins i and j and θ_{ij} is the angle between the i–j internuclear vector and the applied magnetic field $\mathbf{B}_0$.

From Equations (2.12) and (2.13), the homonuclear dipolar coupling quite clearly depends on the geometric factor $(3\cos^2\theta - 1)$ and so is averaged to zero by magic-angle spinning (see Section 2.2 for details of magic-angle spinning), providing the rate of spinning is fast compared to the homonuclear dipolar-coupling linewidth. Nowadays, spinning rates of 30–50 kHz can be achieved on commercially available probes. This is fast enough to produce high-resolution spectra for ^{1}H in many organic solids for instance, where otherwise homonuclear dipolar linebroadening (linewidth at half maximum height) would be of the order of 20–50 kHz.

At spinning rates much less than the dipolar linewidth, magic-angle spinning has very little effect on the NMR spectrum and the broad dipolar line that would be observed in the absence of spinning is little altered.

At intermediate spinning rates (rates around a quarter to a half of the dipolar linewidth), spinning sidebands appear, but these spinning sidebands are very different in character to those arising from incompletely spun-out chemical shift anisotropy or heteronuclear dipolar coupling (Fig. 2.7). The spinning sidebands associated with chemical shift anisotropy and heteronuclear dipolar coupling are all sharp lines. Those associated with homonuclear dipolar coupling are usually broad. It is difficult to describe this regime in a rigorously quantitative fashion. The behaviour can, however, be understood qualitatively at least, as follows. The term $\hat{\mathbf{I}}\cdot\hat{\mathbf{S}}$ in

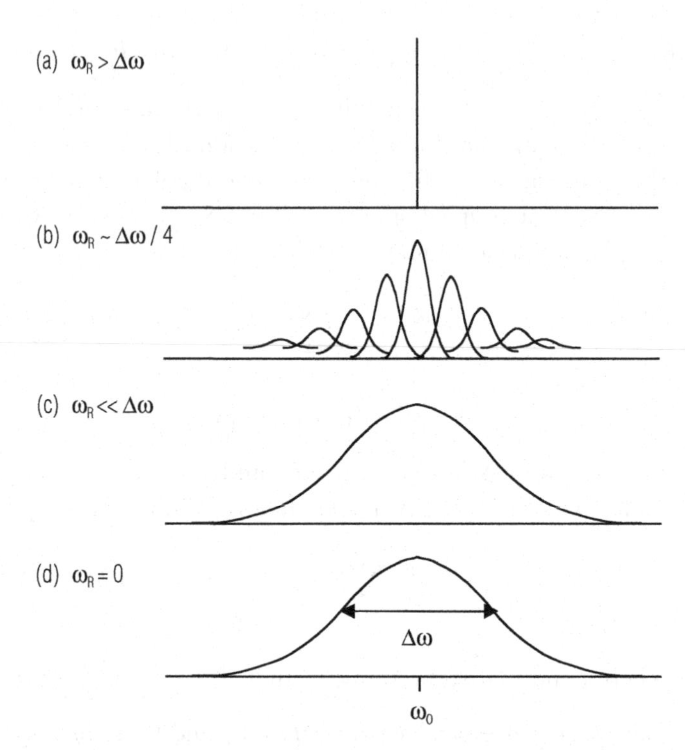

Fig. 2.7 The effects of magic-angle spinning on homonuclear dipole–dipole coupling. (a) The broad line which arises in the NMR spectrum (shown in (d)) from this interaction is only narrowed by magic-angle spinning for spinning rates of the order of the linewidth or greater. (b) For slower rates, broad spinning sidebands appear and (c) at still slower rates, there is little effect on the spectrum at all.

the homonuclear dipolar coupling hamiltonian mixes the degenerate Zeeman functions associated with the collection of spins in the spin system (see Section 4.1.2 for further details). As shown in Chapter 4 (Section 4.1.2), this mixing results in the eigenfunctions for the spin system being linear combinations of the degenerate Zeeman functions. However, when we observe the spin system in a state described by such an eigenfunction, the state of the system collapses to a single Zeeman function. Thus, if we make repeated observations of the spin system over time, each spin in the dipolar-coupled network varies its spin state between α and β (for a spin-$\frac{1}{2}$ system). In effect, magic-angle spinning continually 'observes' each spin and 'sees' its z-component of spin vary with time. The time evolution between the different degenerate Zeeman, and hence between the different possible spin states for each spin, is governed by the time-dependent Schrödinger equation (see Box 1.1, Equation (iv)). Thus, the z-component of each spin appears to oscillate at a frequency governed by the various dipolar couplings affecting each spin. To average an interaction to zero through magic-angle spinning, the state of each spin in the system needs to be constant over the

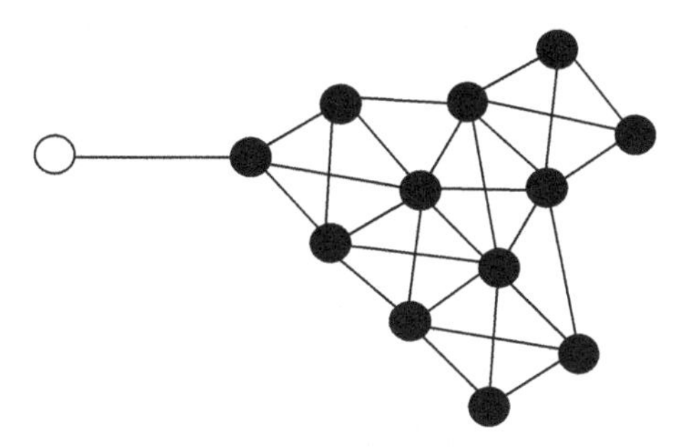

Fig. 2.8 Heteronuclear dipolar coupling between an isolated spin *I* (observed) and a spin *S* which is part of a homonuclear dipolar-coupled network of *S* spins. Such a situation leads to time-dependent dipolar coupling between *I* and *S*, which renders magic-angle spinning largely ineffective.

time for one period of the sample rotation. However, in the case of homonuclear dipolar coupling, the state of the spin system is changing on the timescale of the sample rotation, for intermediate sample rotation rates. This prevents the complete averaging of the dipolar interaction in the spin system.

At lower spinning rates, however, the rate at which the state of the system changes is rapid compared with the sample spinning. In this case then, the spinning does not have a chance to alter the time-dependent state of the spin system; hence the NMR spectrum of the system is unaffected by the spinning. This does not happen in simple heteronuclear spin systems, as there are no degenerate spin levels in such a system. However, complications do arise when an observed spin, *I*, is dipolar coupled to a heteronuclear spin *S*, which is itself part of a network of homonuclear *S* spins, the situation shown in Fig. 2.8. If the *S* spins have significant (homonuclear) dipolar couplings between them, then the state of each spin in the *S* spin system is effectively time dependent, which means that the dipolar coupling between the *I* spin and the *S* spin it is coupled to is also time dependent. This is discussed further in Section 4.1.5. This situation means that magic-angle spinning will not effectively average to zero the heteronuclear *I–S* dipolar coupling, and extensive linebroadening may be seen in the *I* spin spectrum. This occurs frequently, for instance when recording ^{13}C spectra of organic solids in which the ^{13}C spins are coupled to an extensive network of ^{1}H spins. In such situations, it is necessary to use other decoupling methods to achieve resolution, as described in the next section.

2.3 Heteronuclear decoupling

When observing a dilute spin, e.g. ^{13}C with 1.1% abundance, with ^{1}H or other abundant spins nearby, broadening due to heteronuclear dipolar cou-

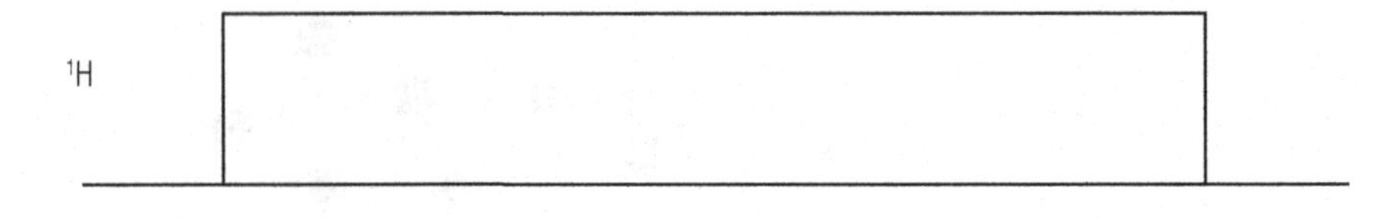

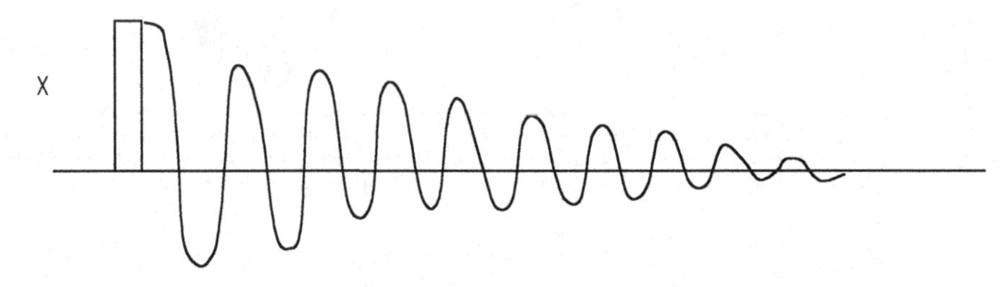

Fig. 2.9 High-power decoupling. High-power decoupling removes the effects of ^{1}H dipolar coupling from the NMR spectrum of X in this case; it can of course be applied to any abundant spin in place of ^{1}H in the same manner. High-power irradiation is simply applied to the ^{1}H spins during the acquisition of the X spin spectrum. Here a single pulse is used to generate the X transverse magnetization; this can of course be replaced with a more complicated preparation sequence. ^{1}H decoupling can be used in the preparation sequence if necessary as well.

pling often causes a serious problem to an already weak spectrum, even under magic-angle spinning, as described in the previous section. Fortunately, the effects of this coupling are easily removed by rf irradiation in one form or another.

2.3.1 High-power decoupling

High-power decoupling is the simplest technique which removes the effects of heteronuclear coupling. Consider the case of dipolar-coupled ^{1}H and ^{13}C spins, where the ^{13}C spins are to be observed. The method (Fig. 2.9) consists of applying continuous irradiation of very high power (100–1000 watts) at the frequency of the proton resonance. The required pulse sequence for the ^{13}C nuclei is then applied and the ^{13}C FID measured while continuing the ^{1}H irradiation.

It is worth noting that the pulses used for decoupling ^{1}H in samples with abundant ^{1}H spins do not need to be particularly broad banded, i.e. they do not need to cover a wide frequency range. Because of the homonuclear dipolar coupling between the ^{1}H nuclei in such samples, the effects of an rf pulse applied to any part of the ^{1}H spectrum are transmitted among all coupled ^{1}H. This is because the rf irradiation applied to the ^{1}H spins affects the z-component of spin of each ^{1}H nucleus which is close to resonance with the irradiation. The z-component of spin of any one ^{1}H spin affects that of all nearby ^{1}H spins through the $\hat{I}_+\hat{S}_- + \hat{I}_-\hat{S}_+$ terms in the homonuclear dipolar coupling hamiltonian (see Section 4.1.2 for details). The close-to-resonance spins which are affected by the rf irradiation then affect the z-component of spin of all their neighbours, and so on, so that the effect of rf irradiation on just one ^{1}H spin is transmitted throughout the ^{1}H dipolar-coupled network.

Any proper explanation of how high-power decoupling in solids works requires the use of *average hamiltonian theory*, which is described in Box 2.1 below. However, for our purposes here, the effect of high-power decoupling on the spin system can be described rather more simply. The effect of the close-to-resonance rf irradiation is to cause the ^{1}H spins to undergo repeated transitions ($\alpha \leftrightarrow \beta$) at a rate determined by the amplitude of the rf irradiation, ω_1 (in angular frequency units). The strength of the ^{1}H–^{13}C dipolar coupling is determined in part by the factor m_H, the z-component of spin of the ^{1}H spin in the dipolar-coupled spin pair (see Equations (4.20) and (4.21) in Chapter 4). The ^{13}C spectrum will be affected by the time-averaged dipolar coupling only, providing the rate of transition $\alpha \leftrightarrow \beta$ on the ^{1}H spin is fast relative to the strength of the ^{1}H – ^{13}C dipolar coupling. In turn, the time-averaged dipolar coupling is clearly zero, as m_H oscillates rapidly between $\pm\frac{1}{2}$.

Implementing high-power decoupling

Setting up high-power decoupling is very straightforward. First of all, record a simple pulse-acquire spectrum of the nucleus to be decoupled (for instance ^{1}H) and set the decoupling channel frequency approximately on resonance. For nuclei in a strongly homonuclear dipolar-coupled network, the spectrum is generally very broad, even under magic-angle spinning, and placing the decoupling frequency anywhere in this broad signal is generally sufficient, although the decoupling is usually more effective if the decoupling frequency is close to the centre of this signal. In cases where the homonuclear dipolar coupling is weaker, more care needs to be taken to set the decoupling channel close to resonance, although it should also be noted that in the case of weak homonuclear dipolar coupling among the spins to be decoupled, fast magic-angle spinning is usually relatively effective in decoupling the heteronuclear couplings between the observed spin and the homonuclear network.

All that remains is to set the amplitude of the rf irradiation in the decoupling. For effective decoupling, the rf amplitude (which determines the rate of spin transitions for the nucleus to be decoupled, i.e. ^{1}H) needs to be around three times greater than the largest heteronuclear, i.e. ^{1}H–^{13}C, dipolar coupling to be eradicated. A directly-bonded ^{1}H–^{13}C spin pair has a dipolar coupling constant of around 22 kHz, so a minimum ^{1}H decoupling amplitude of 66 kHz is required for effective decoupling.[N1] The ^{1}H nutation frequency is equal to the rf amplitude applied to the ^1H spins in frequency units when the irradiation is on resonance, so the ^1H rf amplitude is easily calibrated by measuring the ^1H 180° pulse length, for instance. The ^1H nutation frequency is just $(2\tau_{180})^{-1}$ where τ_{180} is the 180° pulse length. Note that this is the minimum rf amplitude which is required, and if the observed spins are strongly dipolar-coupled to several ^1H spins, then a larger decoupling rf amplitude may be needed. Modern probeheads are capable of delivering 100–200 kHz of rf amplitude to the sample, with higher amplitudes possible for smaller rf coil diameters.

Care should be taken when setting the decoupling rf amplitude that the amplitude chosen does not exceed the specification of the probehead being used in the spectrometer. Even if it does not, it is very important that the probehead is very well tuned for both the decoupling frequency and the observed frequency before high-power rf is put into it. This is especially important for decoupling, as the rf power is turned on for relatively long periods at a time. In solution-state NMR spectroscopy, it is relatively rare that such high rf amplitudes are used and so probe tuning is rather less critical to the health of the probehead. It is a common mistake when moving from solution- to solid-state NMR to assume that the same principle applies. If high-power decoupling is attempted in a poorly tuned probehead, electrical arcing will occur which may irreparably damage the probehead (and lead to a spectrum containing noise which has a greater amplitude than the required signal).

The high-power decoupling irradiation needs to run for the length of the FID collection in a simple pulse-acquire experiment (and longer in more complicated experiments where decoupling is required in other periods of the experiment). The longer the decoupling period, the greater the potential to damage the probehead. Thus, it is important to set the FID length to the minimum required to obtain the necessary signal and no longer. If arcing does occur, it will be apparent from the acquiring FID which will contain very large, random spikes, often towards the end of the FID. If this is observed, first of all, check the tuning of the probehead. Secondly, check that the sample rotor/holder and the coil in which it sits in the probehead are spotlessly clean; finger grease is excellent for promoting arcing. If arcing still occurs, you can try shortening the FID and so shortening the period for which the decoupler is turned on and/or increasing the length of time between scans. If this still does not solve the problem, you have no option but to turn down the decoupler power and accept less efficient decoupling. It is worth noting that the damage caused to a probehead by arcing itself promotes arcing, so if arcing has occurred to a significant extent or for an extended period of time, it may be necessary to replace or repair arc-damaged components in the probe. When cleaning arc damage from a component in the probehead, remember that a smooth surface must be maintained on the component to minimize arcing.

Other considerations

High-power decoupling can be very effective and is simple to set up. However, it does involve relatively large amounts of rf power going into the sample, which causes heating of the sample, particularly if it contains water. For many samples this does not matter, but for some, for instance proteins, it can cause irreversible structural changes at a molecular level, or phase changes.

Molecular motion in the sample can also interfere with high-power decoupling (or, indeed, any decoupling method), for much the same reason that it can interfere with magic-angle spinning. High-power decoupling works by averaging the heteronuclear dipolar coupling to zero at a rate determined by the decoupling rf amplitude. If molecular motion occurs on a similar timescale (i.e. the inverse of the correlation time for the motion is of the order of the decoupling rf amplitude), then it will interfere with the averaging by the decoupling irradiation. The result is usually significantly broadened signals in the spectrum, where the decoupling has been ineffective. If such a thing is suspected, it may be confirmed by changing the decoupling rf amplitude.

If high-power decoupling is used in conjunction with magic-angle spinning, some consideration needs to be given to the relative decoupling amplitude and magic-angle spinning rate. If the decoupling amplitude (in frequency units) and spinning rate are similar, the two processes interfere.[N2] Obviously for moderate spinning rates (5–20 kHz) and high decoupling amplitudes (>60 kHz), there should be no problems. However, for the very high spinning rates which are sometimes used (>40 kHz), it can be more effective to use low-power decoupling, so as to avoid interference between the two processes [5].

2.3.2 Other heteronuclear decoupling sequences

As already mentioned in the previous section, high-power decoupling can require excessively high rf powers for effective decoupling. In this case, an alternative decoupling pulse sequence must be used to achieve the desired resolution within an acceptable rf power range. Two particularly effective decoupling sequences for this circumstance are shown in Fig. 2.10. Both are designed to be used with fast magic-angle spinning.

Both the Two Pulse Phase Modulation (TPPM) [6] and XiX [7] schemes involve continuous, high-power irradiation on the spins to be decoupled during collection of the FID or other period of a pulse sequence where

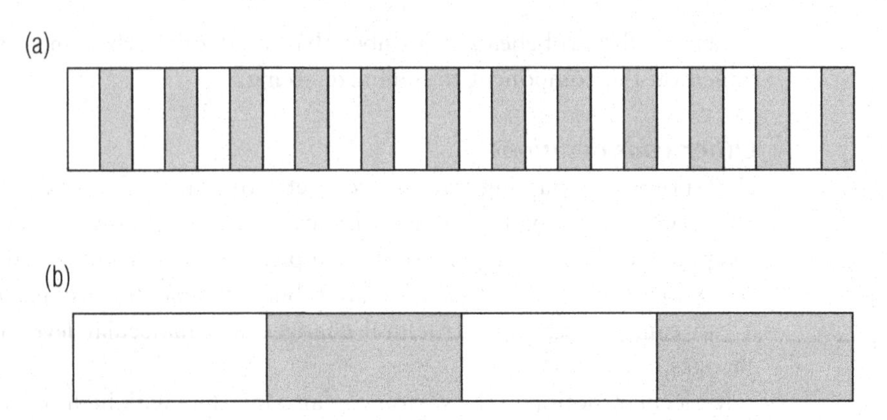

Fig. 2.10 Decoupling pulse sequences. (a) The two pulse phase modulation (TPPM) scheme. The scheme consists of alternating pulses of phase ϕ_p and $\phi_p + \Delta\phi_p$, each of flip angle θ_p. The phase difference between the pulses, $\Delta\phi_p$, and the pulse flip angle are determined experimentally. (b) The XiX scheme. The scheme consists of alternating pulses with phase differing by 180°. The ratio t_p/τ_R, where τ_R is the rotor period and t_p is the pulse length, is the important parameter in this case and is determined experimentally.

decoupling is required. The advantage over the simple high-power decoupling described previously is that these schemes often result in better decoupling for the same rf power. The TPPM sequence consists of continuous irradiation with two pulses of flip angle θ_p and phases which differ by $\Delta\phi_p$. The optimal values for θ_p and $\Delta\phi_p$ are found experimentally, although θ_p is often close to 180° and $\Delta\phi_p$ is usually in the range 10–70° . The scheme uses high rf powers, typically between 100 and 150 kHz. The optimal values of θ_p and $\Delta\phi_p$ depend on the sample and the spinning rate under magic-angle spinning, so the decoupling sequence should be optimized on the sample of interest at the desired spinning rate. The sensitivity of the scheme to the pulse flip angle θ_p means that the sequence is relatively sensitive to rf inhomogeneities and rf instability. Nevertheless, the TPPM scheme usually gives much better performance than simple high-power decoupling for the same rf power and is used extensively.

The XiX scheme also uses continuous, high-power rf irradiation with two pulses, but this time the pulses differ in phase by 180°. The performance of this scheme depends only on the ratio of the pulse length (t_p) to the rotor period (τ_R) of the sample spinning. In general, the scheme is effective for t_p/τ_R greater than one, but values of t_p/τ_R which are integer multiples of $\tau_R/4$ should be avoided, as these points correspond to 'resonances' at which the decoupling efficiency is severely reduced. As with the TPPM sequence, the optimal value of t_p/τ_R is determined experimentally, although a suitable start point can be predicted from calculations performed by the inventors of this sequence [7]. This scheme has the advantage of being relatively insensitive to rf inhomogeneities and instability. Its performance is particularly good at high spinning rates (>30 kHz), where it often outperforms TPPM.

Other high-power decoupling schemes for use with magic-angle spinning which have appeared in the literature are FMPM [8], SPARC [9], $C12_2^{-1}$[10], SPINAL [11] and amplitude-modulated TPPM [12]. These have similar efficiency to TPPM.

If high-power decoupling schemes need to be avoided to prevent sample heating, then fast magic-angle spinning (>40 kHz) and low-power, continuous wave decoupling may also be used effectively [5].

2.4 Homonuclear decoupling

As described in Section 2.2.4, magic-angle spinning can be used to remove the effects of homonuclear dipolar coupling from NMR spectra, providing the rate of sample spinning is fast relative to the homonuclear dipolar linewidth. Where this is not achievable, the effects of homonuclear dipolar coupling may be removed instead by special pulse sequences. Many such are known and are often collectively referred to as *multiple pulse sequences.*

Most multiple pulse sequences are arranged in such a way that at certain windows within the pulses sequence, the effect of the dipolar hamiltonian on the nuclear magnetization is zero. If the nuclear magnetization is detected only at these points, the effects of dipolar coupling are removed from the spectrum. There are many useful sequences in the literature. The first, and one of the simplest, is the WAHUHA sequence, shown in Fig. 2.11 [13]. The MREV-8 sequence [14] is also widely used (also shown in Fig. 2.11). There are a whole host of other sequences which have been devised since the pioneering WAHUHA sequence, all of which aim to achieve the same type of result (see for instance reference [15]). In all, the pulsing continues in cycles throughout the period of the free induction decay, with one detection point per cycle at the appropriate point until the magnetization has decayed completely from the x–y plane. Rather than average the geometrical parts of $\hat{H}_{dd}^{homo}$ to zero in the way that magic-angle spinning does, these pulse sequences average the spin factors of $\hat{H}_{dd}^{homo}$ to zero. Any explanation of these sequences needs the application of average hamiltonian theory; this and its application to the analysis of the WAHUHA pulse sequence are outlined in Box 2.1.

2.4.1 Implementing homonuclear decoupling sequences

Unfortunately, implementing homonuclear decoupling pulse sequences is far from trivial. First, FID points have to be collected at particular points between cycles of pulses; the implementation of this is spectrometer specific. Second, very high power pulses are needed, with short rise times if the pulse sequence is to be effective, as the pulse length must be short relative to the

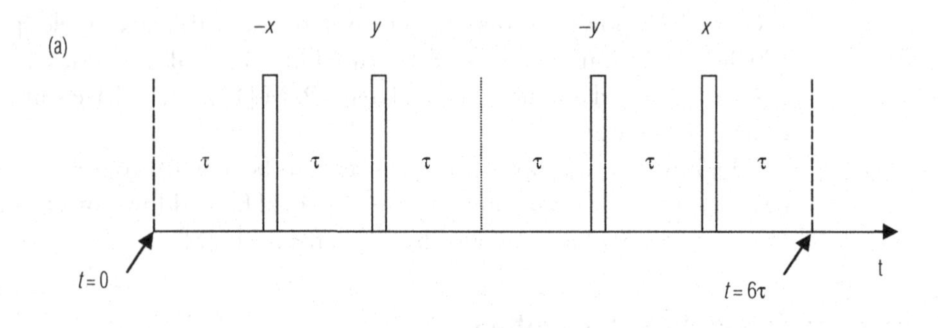

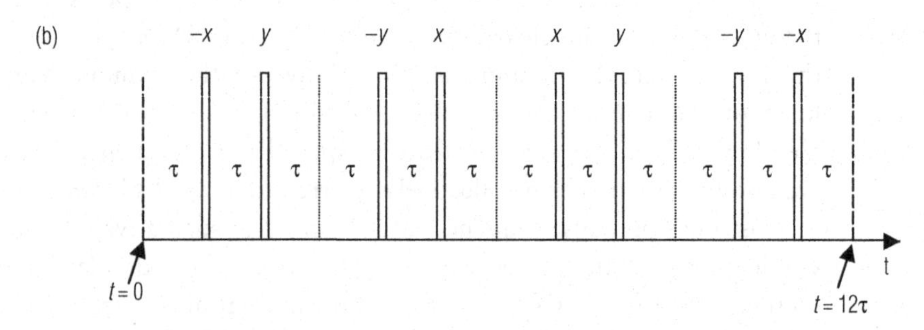

Fig. 2.11 Homonuclear decoupling pulse sequences for removing the effects of homonuclear dipolar coupling from the NMR spectrum. All pulses are 90° pulses with the phase indicated. (a) WAHUHA sequence and (b) MREV-8 sequence. The sequences are repeated throughout acquisition, with FID points being collected at the end of each cycle, indicated by an arrow in each case. The MREV-8 sequence is in fact two phase-permuted WAHUHA sequences and has the effect of zeroing higher-order terms in the average hamiltonian.

cycle time. In turn, short cycle times are needed, as the spectral width of the resulting spectrum is the inverse of the cycle time, and the spectral width must be large enough to contain all the signals. This can put considerable strain on the probehead electronics. Finally, magic-angle spinning is often used with homonuclear decoupling sequences to remove the effects of other anisotropic interactions from the spectrum, but the averaging effects of magic-angle spinning can potentially interfere with the decoupling sequence, so it is important to synchronize the sample spinning and the decoupling sequence accurately (sometimes known as CRAMPS – Combined Rotation And Multiple Pulse Sequence). The first and most serious problem, that of synchronizing the pulse sequence and the collection of the FID, is avoided if the homonuclear decoupling pulse sequence is used in the t_1 dimension (the indirectly detected dimension) of a two-dimensional pulse sequence. This protocol has been used [16] to generate high-resolution ^{1}H spectra from two-dimensional experiments in which a homonuclear decoupling sequence is run during t_1 and a normal ^{1}H spectrum is collected under magic-angle spinning in t_2. In the resulting two-dimensional spectrum, the ν_2 spectral

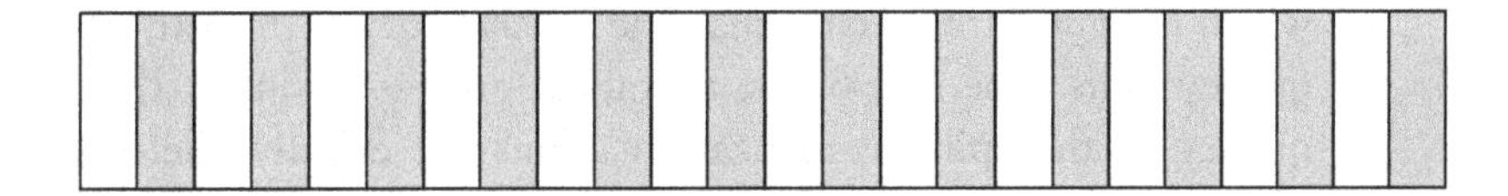

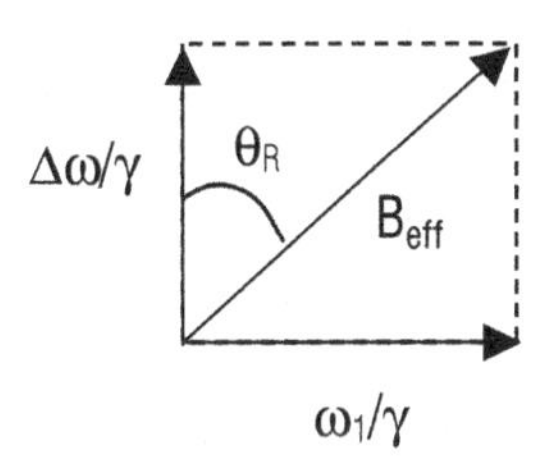

Fig. 2.12 The frequency-switched Lee–Goldburg homonuclear decoupling sequence. The pulses are 360° pulses with rf amplitude ω_1 (in frequency units) alternating in offset from resonance by $\pm\Delta\omega$. The off-resonance condition is set so that the effective field ($\mathbf{B}_{eff}$) that the spins feel in their rotating frame is at the magic angle ($\theta_R = 54.74°$) to the applied field.

dimension (which contains broad signals typical of ^{1}H magic-angle spinning spectra) can be discarded; it is only the ν_1 spectrum which is useful. It may seem long-winded to acquire a two-dimensional dataset just to obtain a one-dimensional spectrum, but in practice it is much quicker to do this than to spend time setting up a one-dimensional homonuclear decoupling experiment and, in addition, the results are usually better too. The author would strongly advise this approach where high resolution ^{1}H spectra are needed and very fast magic-angle spinning is not an option or not sufficient.

If the homonuclear decoupling is required only in an indirectly-detected dimension of a multidimensional experiment, then there are further pulse sequences which can be considered. One popular one is the frequency-switched Lee–Goldburg scheme [17, 18], shown in Fig. 2.12. Here, off-resonance 360° pulses are applied continuously to the spins throughout the period in which decoupling is required. The pulses are set off resonance by an amount $\Delta\omega$ such that the effective field that the spins experience in the rotating frame (see Chapter 1, Section 1.1.2 for discussion of effective field and off-resonance effects) is orientated at the magic angle with respect to the applied field, $\mathbf{B}_0$. A 360° pulse at this offset then causes the spin magnetization to precess completely by one turn around this effective field. The average homonuclear dipolar coupling between the spins over this period is then zero if the effective field is at the magic angle, as can be shown by application of average hamiltonian theory, as in Box 2.1. The effects of homonuclear dipolar coupling are removed from the spectrum under this scheme, providing the averaging of the dipolar coupling takes place at a rate which is significantly faster than the strength of the dipolar coupling in frequency units. The rate of the averaging process is simply the nutation frequency of the spin magnetization about the effective field, which is the effective field strength in frequency units, i.e. $\sqrt{\omega_1^2 + \Delta\omega^2}$ where ω_1 is the rf amplitude of the 360° pulses and $\Delta\omega$ is the offset of the pulse frequency from resonance, so providing reasonably high-power pulses are used, this condition is easily met. The 360° pulses are alternated in phase to reduce errors which occur from the fact that it is not possible to set the off-resonance condition required exactly for all parts of the sample.

Box 2.1: Average hamiltonian theory and the toggling frame

Average hamiltonian theory

As shown in Chapter 1, we can calculate the density operator describing a spin system at time t, $\hat{\rho}(t)$ from that at time 0 via the equation

$$\hat{\rho}(t) = \hat{U}(t)\hat{\rho}(0)\hat{U}(t)^{-1} \quad \text{(i)}$$

where the so-called propagator $\hat{U}(t)$ is given by

$$\hat{U}(t) = \exp(-i\hat{H}t) \quad \text{(ii)}$$

in which $\hat{H}$ is the hamiltonian operator which describes the spin system between 0 and t. This formulation of the propagator assumes that the hamiltonian is constant over the time period. Frequently, however, this is not the case. For instance, in the WAHUHA pulse sequence (Fig. 2.11) described above, the hamiltonian changes when rf pulses are applied from that which operates during the time gaps τ between pulses. Under these circumstances, Equation (ii) for the propagator becomes

$$\hat{U}(t) = \exp(-i\hat{H}_n t_n)\ldots\exp(-i\hat{H}_1 t_1) \quad \text{(iii)}$$

where the hamiltonian which operates in the first time period t_1 is $\hat{H}_1$ and so on. Note that the hamiltonians in Equation (iii) appear in strict chrononlogical order. In cases like this, it would be much more convenient to replace the series of exponential functions in Equation (iii) for the propagator with a single exponential relying on some *average hamiltonian* $\overline{H}$ which has the same effect as the series of hamiltonians $\hat{H}_1 \ldots \hat{H}_n$ (Fig. B2.1.1), i.e.

$$\hat{U}(t) = \exp(-i\hat{H}_n t_n)\ldots\exp(-i\hat{H}_1 t_1) = \exp(-i\overline{H}t)$$

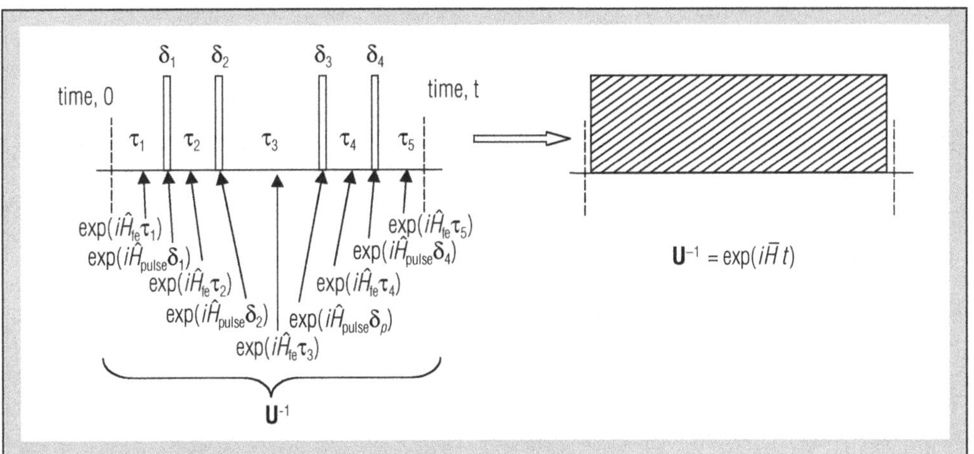

Fig. B2.1.1 The idea behind average hamiltonian theory. The propagator **U** which describes a pulse sequence consists of a series of exponentials, $\exp(-i\hat{H}_i\tau_i)$ and $\exp(-i\hat{H}_i\delta_i)$, where the hamiltonian $\hat{H}_i$ describes the spin system during the period τ_i and δ_i respectively. In this diagram, the hamiltonian $\hat{H}_{fe}$ operates during periods of free precession between the pulses and $\hat{H}_{pulse}$ operates during the pulses. The lengths of the pulses are given by δ_i for the ith pulse and the lengths of the periods of free precession by τ_i. In average hamiltonian theory, the series of exponentials is replaced by a single exponential $\exp(-i\overline{H}t)$ employing an *average hamiltonian* $\overline{H}$, which describes the net effect of the pulse sequence at its end. $\overline{H}$ can be envisaged as a 'black box', which accounts for the effects of the whole pulse sequence but does not allow us to see what happens at points during the pulse sequence.

Of course, this is always possible, but not, unfortunately, always useful, as in general the appropriate average hamiltonian $\overline{H}$ will depend on t, the time at which we wish to know the propagator.

However, in the case where the hamiltonian (or series of hamiltonians) describing the spin system is *periodic* in time, and we only wish to know about the state of the spin system at specific time points spaced by the period of the hamiltonian, we can always calculate a single average hamiltonian which correctly describes the behaviour of the spin system at these points.

One way of calculating this average hamiltonian, a rather 'brute force' approach, is to calculate the propagator $\hat{U}(t)$ by simply evaluating the series of exponentials in Equation (iii) appropriate to the particular periodic hamiltonian of interest. This then equates to $\exp(-i\overline{H}t)$ (= $\hat{U}(t)$), so that $\overline{H}$ may be found by diagonalizing the matrix $\mathbf{U}(t)$ formed from $\hat{U}(t)$ in some appropriate basis. The eigenvalues resulting from this process are then $\exp(-i\overline{H}_{jj}t)$ for the jth eigenvalue, where $\overline{H}_{jj}$ is the matrix element of $\overline{H}$ in the eigenvector basis arising from the diagonalization. However, this process rarely leads to much physical insight.

An alternative approach evaluates the series of exponential operators in Equation (iii) using the Magnus expansion:

$$e^{\hat{A}}e^{\hat{B}} = \exp\left\{\hat{A}+\hat{B}+\frac{1}{2!}[\hat{A},\hat{B}]+\frac{1}{3!}([\hat{A},[\hat{A},\hat{B}]]+[[\hat{A},\hat{B}],\hat{B}])+\ldots\right\} \quad \text{(iv)}$$

Continued on p. 88

Box 2.1 *Cont.*

If this is applied to the series of exponential operators in Equation (iii) for a periodic hamiltonian of period t_p, i.e. $t_1 + t_2 + \ldots + t_n = t_p$ to evaluate $\hat{U}(t_p)$ and we then equate

$$\hat{U}(t_p) \equiv \exp(-i\overline{H}t_p) \tag{v}$$

we find that

$$\overline{H}(t_p) = \overline{H}^{(0)} + \overline{H}^{(1)} + \overline{H}^{(2)} + \ldots \tag{vi}$$

where

$$\begin{aligned}
\overline{H}^{(0)} &= \frac{1}{t_p}\{\hat{H}_1 t_1 + \hat{H}_2 t_2 + \ldots + \hat{H}_n t_n\} \\
\overline{H}^{(1)} &= -\frac{i}{2t_p}\{[\hat{H}_2 t_2, \hat{H}_1 t_1] + [\hat{H}_3 t_3, \hat{H}_1 t_1] + [\hat{H}_2 t_2, \hat{H}_3 t_3] + \ldots\} \\
\overline{H}^{(2)} &= -\frac{1}{6t_p}\{[\hat{H}_3 t_3, [\hat{H}_2 t_2, \hat{H}_1 t_1]] + [[\hat{H}_3 t_3, \hat{H}_2 t_2], \hat{H}_1 t_1] \\
&\quad + \frac{1}{2}[\hat{H}_2 t_2, [\hat{H}_2 t_2, \hat{H}_1 t_1]] + \frac{1}{2}[[\hat{H}_2 t_2, \hat{H}_1 t_1], \hat{H}_1 t_1] + \ldots\}
\end{aligned} \tag{vii}$$

Equations (vi) and (vii) are not as daunting as they at first appear; the first-order term $\overline{H}^{(0)}$ (first order in the hamiltonian) is simply the average of the piecewise hamiltonians $\hat{H}_1, \hat{H}_2, \ldots, \hat{H}_n$ which operate during one period. In the case where these hamiltonians all commute with each other, or nearly so, all higher-order terms can be neglected and the first-order term is a good description of the average hamiltonian. Clearly, if this is the case, the analytic form of the approximate average hamiltonian is easily determined and can often give useful physical insight into how a pulse sequence works.

In cases where the hamiltonians which operate over the time period do not commute with each other, it is often possible to transform them into a new frame, the so-called *toggling frame*, in which non-commuting terms disappear. It is then possible to approximate the average hamiltonian in this frame by the first-order term of Equation (vi).

As we shall see shortly, the WAHUHA pulse sequence creates an average hamiltonian which to first order, i.e. for $\overline{H} = \overline{H}^{(0)}$, has no contribution from homonuclear dipolar coupling. The other pulse sequences for the removal of homonuclear dipolar coupling effects (see reference [7], for instance) create average hamiltonians for which this is the case to higher orders of $\overline{H}^{(i)}$ in $\overline{H}$.

The toggling frame and the WAHUHA pulse sequence

The WAHUHA pulse sequence in Fig. 2.11 is a typical series of rf pulses with gaps between them (of length τ) where whatever coherences are present are simply allowed to evolve under the effects of whatever internal spin interactions are present, in this case, homonuclear dipolar coupling.

We will assume all pulses are of negligible length compared with τ, the gaps between them, and are strong compared with the dipolar coupling, so that we can ignore the coupling during the pulse. We also assume that the pulses are all on resonance and are defined with respect to the rotating frame, so that throughout the pulse sequence, the effect of the $\mathbf{B}_0$ field apparently vanishes, i.e. the Zeeman term in the hamiltonian in this frame is zero. Then the hamiltonian for the system (in angular frequency units) during a pulse is simply

$$\hat{H}^{\phi}_{\text{pulse}} = \omega_1(\hat{I}_x \cos\phi + \hat{I}_y \sin\phi) \tag{viii}$$

where ϕ is the phase of the pulse and

$$\hat{I}_x = \sum_i \hat{I}^i_x \quad \text{and} \quad \hat{I}_y = \sum_i \hat{I}^i_y$$

with $\hat{I}^i_x$, $\hat{I}^i_y$ being the single spin operators, i.e. $\hat{I}_x$, $\hat{I}_y$ are the sums over the single spin operators for all the homonuclear spins of interest. The hamiltonian during the periods of free evolution (labelled *fe*) is simply that due to the homonuclear dipolar coupling, in the absence of other interactions (we will consider the effects of chemical shift and other inhomogeneous interactions at the end), namely (in angular frequency units)

$$\hat{H}_{fe} = -\sum_{i>j} C_{ij}(3\hat{I}^i_z\hat{I}^j_z - \hat{\mathbf{I}}^i \cdot \hat{\mathbf{I}}^j) \tag{ix}$$

where

$$C_{ij} = \left(\frac{\mu_0}{4\pi}\right)\frac{\gamma_i\gamma_j}{r_{ij}^3}\hbar\frac{1}{2}(3\cos^2\theta - 1) \tag{x}$$

In all these hamiltonians, x, y and z refer to the rotating frame axes. Now, the hamiltonians describing the pulses and those describing the periods of free precession do not commute with each other, so the first-order term in the average hamiltonian is *not* a good approximation to the full average hamiltonian. In particular, the pulse hamiltonians are a problem, as they do not commute among each

Continued on p. 90

Box 2.1 *Cont.*

other either, i.e. $\hat{I}_x$ in an x-pulse operator does not commute with $\hat{I}_y$ in a y-pulse operator for instance, and the WAHUHA sequence employs both x- and y-pulses. We need to transform to a new frame where the terms in the hamiltonian due to the pulses disappear and the remaining dipolar terms still commute with each other. The average hamiltonian in this new frame is then simply the first-order term of Equation (vi), which is easy to calculate.

In order to find this new frame, we start by asking, what is the effect of a pulse on the density operator describing this spin system? Consider a situation, in the rotating frame, of an on-resonance x-pulse, with no other interactions present, so that the hamiltonian is simply

$$\hat{H} = \hat{H}^{x}_{\text{pulse}} = \omega_1 \hat{I}_x \tag{xi}$$

The density operator after time t is given by the usual expression:

$$\hat{\rho}(t) = \exp(-i\hat{H}t)\hat{\rho}(0)\exp(i\hat{H}t) = \exp(i\omega_1\hat{I}_x t)\hat{\rho}(0)\exp(-i\omega_1\hat{I}_x t) \tag{xii}$$

Previous discussion of exponential operators (Chapter 1, Box 1.2) explained that an expression of the form $\exp(i\theta\hat{I}_x)\,\hat{O}\,\exp(-i\theta\hat{I}_x)$, where $\hat{O}$ is an operator, is simply

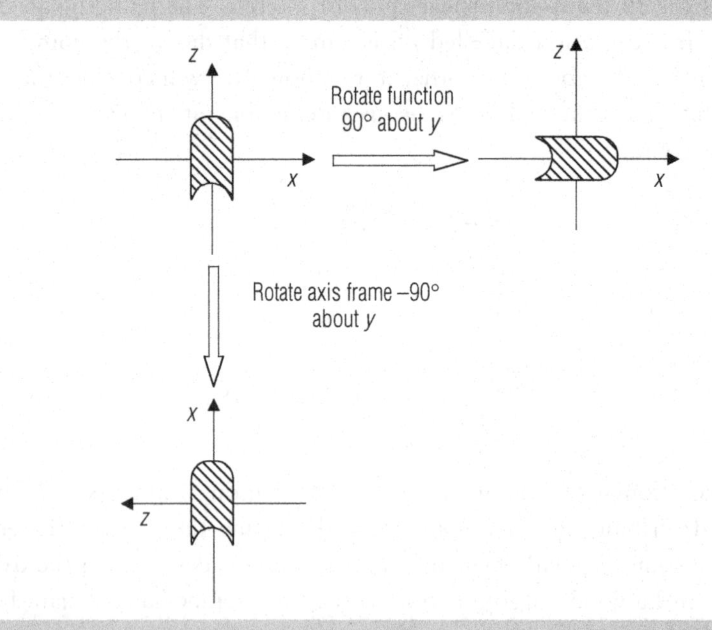

Fig. B2.1.2 Rotating a function (represented here by the hashed shape) is equivalent to rotating the defining axis frame through the same angle, but in the opposite direction.

a rotation of the operator $\hat{O}$ by θ about x. As shown in Fig. B2.1.2, rotating an operator (or function) by θ about a given axis is equivalent to leaving the operator (or function) where it is and, instead, rotating the axis frame in which the operator is defined by $-\theta$ about the same axis.

So the effect of the on-resonance x-pulse is to rotate the density operator by $+\omega_1 t$ about x or, equivalently, to rotate the axis frame that the density operator $\hat{\rho}$ is defined in by $-\omega_1 t$ about x, where x in every case refers to the normal, rotating frame axis. Thus, when considering the effect of a pulse sequence, every time we get to a pulse, rather than rotate the density operator according to the pulse, we can instead rotate the axis frame in which the density operator is defined. A rotation of the axis frame by $-\omega_1 t$ about the pulse axis creates the same effect as rotating the density operator by $+\omega_1 t$ about the pulse axis.

We call the new, transformed frame, the *toggling frame*. Expressing the density operator in this way, in a frame which moves with the effect of a pulse, is often called the *interaction representation*.

The hamiltonian $\hat{H}^*(t)$ after a time t in such a toggling frame is in general

$$\hat{H}^*(t) = \hat{R}_x^{-1}\hat{H}\hat{R}_x - \omega_1\hat{I}_x \tag{xiii}$$

where $\hat{R}_x = \exp(-i\omega_1\hat{I}_x t)$ represents the rotation operator for rotation of the original axis frame by $\omega_1 t$ about the rotating frame x-axis and $\hat{H}$ is the hamiltonian in the rotating frame. See Box 1.2 in Chapter 1 for further discussion of frame transformations.

The form of Equation (xiii) is found by insisting that the time-dependent Schrödinger equation (see Box 1.1 in Chapter 1) is invariant to the frame transformation, as we shall now show.

The time-dependent Schrödinger equation is $i(\mathrm{d}\psi/\mathrm{d}t) = \hat{H}\psi$, expressed in angular frequency units, where ψ is the wavefunction describing the spin system. The time-dependent Schrödinger equation in the toggling frame must be invariant to the frame transformation and so is simply

$$i\frac{\mathrm{d}\psi^{\mathrm{tog}}}{\mathrm{d}t} = \hat{H}^{\mathrm{tog}}\psi^{\mathrm{tog}} \tag{xiv}$$

where all quantities are referred to the toggling frame. Under the frame transformation described by the rotation operator $\hat{R}_x$, the wavefunction ψ becomes ψ^{tog} in the toggling frame, where

$$\psi^{\mathrm{tog}} = \hat{R}_x^{-1}\psi = \exp\left(+i\omega_1\hat{I}_x t\right)\psi \tag{xv}$$

Thus $\mathrm{d}\psi^{\mathrm{tog}}/\mathrm{d}t$ is given by

Continued on p. 92

Box 2.1 Cont.

$$\begin{aligned}\frac{d\psi^{tog}}{dt} &= \exp\left(+i\omega_1\hat{I}_x t\right)\frac{d\psi}{dt} + i\omega_1\hat{I}_x \exp\left(+i\omega_1\hat{I}_x t\right)\psi \\ &= \exp\left(+i\omega_1\hat{I}_x t\right)\frac{d\psi}{dt} + i\omega_1\hat{I}_x\psi^{tog} \\ &= \hat{R}_x^{-1}\frac{d\psi}{dt} + i\omega_1\hat{I}_x\psi^{tog} \qquad \text{(xvi)}\end{aligned}$$

where all quantities are expressed with respect to the toggling frame. Substituting for $d\psi^{tog}/dt$ in Equation (xiv) gives

$$i\left(\hat{R}_x^{-1}\frac{d\psi}{dt} + \left(i\omega_1\hat{I}_x\right)\psi^{tog}\right) = H^{tog}\psi^{tog} \qquad \text{(xvii)}$$

We then use the fact that $i(d\psi/dt) = \hat{H}\psi$ and $\psi = \hat{R}_x\psi^{tog}$ to rewrite the left-hand side of this equation as

$$i\hat{R}_x^{-1}\hat{H}\hat{R}_x\psi^{tog} + \left(i\omega_1\hat{I}_x\right)\psi^{tog} = \hat{H}^{tog}\psi^{tog} \qquad \text{(xviii)}$$

By comparing the left- and right-hand sides of this equation, we see that $\hat{H}^{tog}$, the hamiltonian in the rotating frame, is

$$\hat{H}^{tog} = \hat{R}_x^{-1}\hat{H}\hat{R}_x - \omega_1\hat{I}_x \qquad \text{(xix)}$$

So, in the case where $\hat{H} = \hat{H}_{pulse}^{x} = \omega_1\hat{I}_x$, the toggling frame hamiltonian is simply

$$\hat{H}^{*}(t) = \omega_1\hat{R}_x^{-1}\hat{I}_x\hat{R}_x - \omega_1\hat{I}_x; \qquad R_x = \exp(-i\omega_1\hat{I}_x t) \qquad \text{(xx)}$$

The first term represents a rotation of the operator $\hat{I}_x$ about x, which of course leaves the x-direction, and so $\hat{I}_x$, unchanged. Thus $\hat{H}^{*}(t)$ becomes

$$\hat{H}^{*}(t) = \omega_1\hat{I}_x - \omega_1\hat{I}_x = 0 \qquad \text{(xxi)}$$

As expected, in this toggling frame, the effect of the rf pulse is nulled. The same frame transformation on the density operator takes account of the pulse effects.

Now suppose we have a hamiltonian in the rotating frame

$$\hat{H} = \hat{H}_{int} + \hat{H}_{pulse}^{x} \qquad \text{(xxii)}$$

where $\hat{H}_{int}$ describes some spin interaction and $\hat{H}_{pulse}^{x}$ describes an x-pulse. Transforming this hamiltonian to the toggling frame again using Equation (xiii):

$$\begin{aligned}\hat{H}^{*}(t) &= \hat{R}_x^{-1}\hat{H}\hat{R}_x - \omega_1\hat{I}_x \\ &= \hat{R}_x^{-1}(\hat{H}_{int} + \hat{H}_{pulse}^{x})\hat{R}_x - \omega_1\hat{I}_x \\ &= \hat{R}_x^{-1}\hat{H}_{int}\hat{R}_x + \hat{R}_x^{-1}\hat{H}_{pulse}^{x}\hat{R}_x - \omega_1\hat{I}_x \\ &= \hat{R}_x^{-1}\hat{H}_{int}\hat{R}_x \qquad \text{(xxiii)}\end{aligned}$$

since

$$\hat{R}_x^{-1}\hat{H}_{\text{pulse}}^x\hat{R}_x = \omega_1 \exp(i\omega_1\hat{I}_x t)\hat{I}_x \exp(-i\omega_1\hat{I}_x t) = \omega_1\hat{I}_x$$

as before. In other words, the toggling frame hamiltonian depends only on the spin interaction hamiltonian $\hat{H}_{\text{int}}$ and not on the pulse part.

The resulting toggling frame hamiltonian, $\hat{H}^*$, can then be used to calculate $\hat{\rho}^*$, the toggling frame density operator using an equivalent expression to that in Equation (xii):

$$\hat{\rho}^*(t) = \exp(-i\hat{H}^*t)\hat{\rho}^*(0)\exp(i\hat{H}^*t) \qquad \text{(xxiv)}$$

where $\hat{\rho}^*(0)$ is the initial density operator at time $t = 0$ in the toggling frame. Calculating the average hamiltonian within the toggling frame truncates the expression (vi) for the average hamiltonian to the first-order term $\overline{H}_0$ only, providing that the toggling frame hamiltonians which occur at different times in the time period considered, commute with each other in this frame. The toggling frame density operator calculated in this way is completely equivalent to the more usual rotating frame density operator, but simply expressed with respect to a different frame.

So, we now apply the principle of the toggling frame to the particular sequence at hand for the calculation of $\hat{H}^{(0)}$. At the start of the pulse sequence, there is a period of free evolution where the system evolves under the dipolar coupling with the usual hamiltonian:

$$\hat{H}(0\rightarrow\tau)_{fe} = -\sum_{i>j} C_{ij}(3\hat{I}_z^i\hat{I}_z^j - \hat{\mathbf{I}}^i\cdot\hat{\mathbf{I}}^j) = \hat{H}_{zz} \qquad \text{(xxv)}$$

Next, we have a 90° $-x$ pulse (see Fig. B2.1.3).

So, we first transform to a new toggling frame which is rotated about the rotating frame $-x$ axis by 90°. Now we can ignore the pulse and continue on. Next is another period of free evolution under the dipolar coupling. We must transform the hamiltonian for this period to the new toggling frame we are in, or equivalently rotate the hamiltonian within the original rotating frame. To follow this latter route, we must rotate the operator $\hat{H}_{fe} = -\sum_{i>j} C_{ij}(3\hat{I}_z^i\hat{I}_z^j - \hat{\mathbf{I}}^i\cdot\hat{\mathbf{I}}^j)$ by rotating it $-90°$ about rotating frame $-x$ axis.[N3] The coefficients C_{ij}s are just numbers, so do not need to be transformed. $\hat{\mathbf{I}}^i\cdot\hat{\mathbf{I}}^j$ are scalar products, so rotation has no effect on them. We only have to consider the $\hat{I}_z^i\hat{I}_z^j$ terms. Rotating $\hat{I}_z^i$ by $-90°$ about $-x$ gives $\hat{I}_y^i$, where y refers to the rotating frame y-axis. Thus, the dipolar hamiltonian as viewed from the toggling frame appropriate for the second τ period is

Continued on p. 94

Box 2.1 *Cont.*

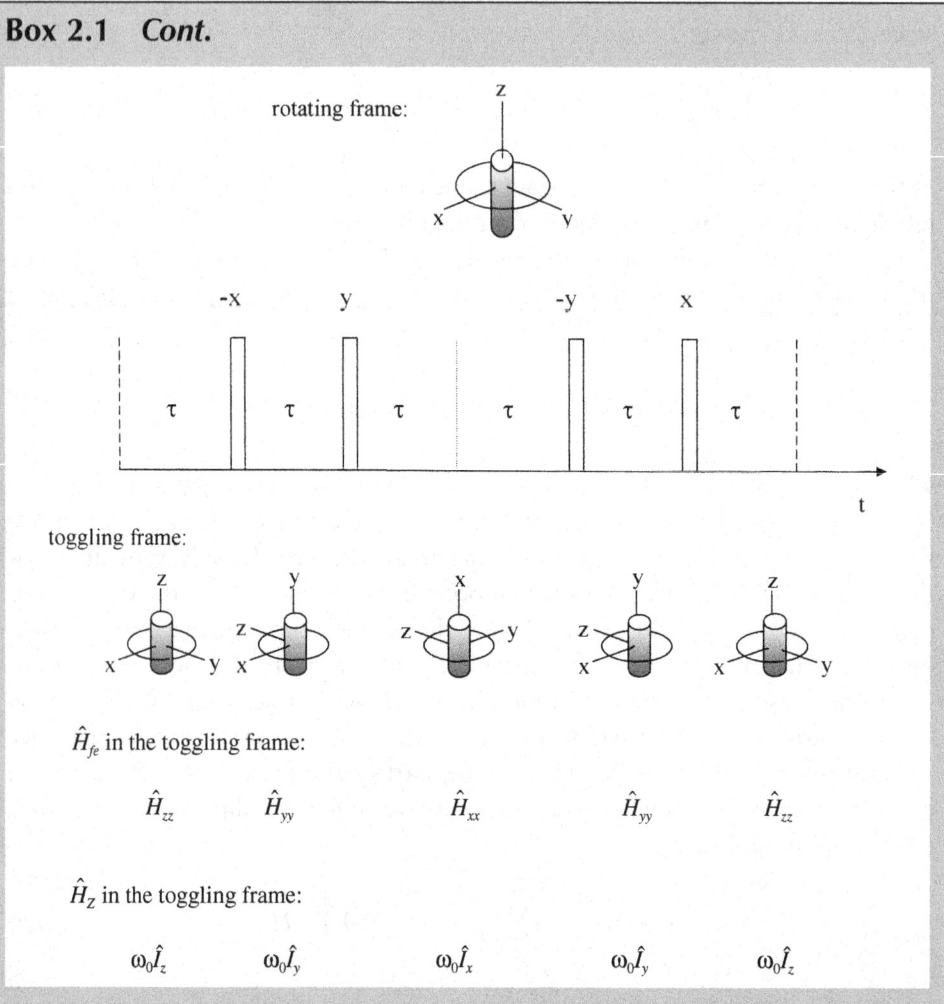

Fig. B2.1.3 The toggling frame for the WAHUHA pulse sequence. The pulse phases in the sequence are referred to the rotating frame indicated. The frame in which the homonuclear dipolar coupling is considered is the toggling frame. The toggling frame is determined by the phase of the rf pulse before each τ period. The dipolar hamiltonian which acts in the periods of free precession, $\hat{H}_{fe}$, and the Zeeman hamiltonian, $\hat{H}_Z$, in the toggling frame are given in the diagram. The dipolar hamiltonian is represented in the figures by a cylinder (which has the same symmetry as the hamiltonian).

$$\hat{H}(\tau\rightarrow 2\tau)_{\text{fe}} = -\sum_{i>j} C_{ij}(3\hat{I}_y^i\hat{I}_y^j - \hat{\mathbf{I}}^i\cdot\hat{\mathbf{I}}^j) \equiv \hat{H}_{yy} \qquad \text{(xxvi)}$$

The next pulse requires a toggling frame which is rotated from the previous toggling frame by 90° about y of the original rotating frame. We have to accumulate the effects of the frame changes on the dipolar hamiltonian in the τ period which follows subsequently, so to take account of this new frame change we must rotate

$\hat{H}_{yy}$, the dipolar hamiltonian within the previous toggling frame, by rotating it −90° about the rotating frame *y* axis. The rotating frame *y*-axis corresponds to the −*z*-axis of the toggling frame in which we perform this rotation of the dipolar hamiltonian. The result gives $\hat{H}_{xx}$ (where *x* again refers to the rotating frame *x*-axis) which has an analogous definition to $\hat{H}_{yy}$ and $\hat{H}_{zz}$).

We continue in this fashion through the entire pulse sequence and end up with the first-order average hamiltonian (first order in the hamiltonian):

$$\overline{H}^{(0)} = \frac{\hat{H}_{zz}\tau + \hat{H}_{yy}\tau + 2\hat{H}_{xx}\tau + \hat{H}_{yy}\tau + \hat{H}_{zz}\tau}{6\tau} \qquad \text{(xxvii)}$$

Now: $\hat{H}_{xx} + \hat{H}_{yy} + \hat{H}_{zz}$

$$= -\sum_{i>j} C_{ij}\left\lfloor(3\hat{I}^i_x\hat{I}^j_x - \hat{\mathbf{I}}^i\cdot\hat{\mathbf{I}}^j) + (3\hat{I}^i_y\hat{I}^j_y - \hat{\mathbf{I}}^i\cdot\hat{\mathbf{I}}^j) + (3\hat{I}^i_z\hat{I}^j_z - \hat{\mathbf{I}}^i\cdot\hat{\mathbf{I}}^j)\right\rfloor \equiv 0 \qquad \text{(xxviii)}$$

using:

$$\hat{\mathbf{I}}^i\cdot\hat{\mathbf{I}}^j = \hat{I}^i_x\hat{I}^j_x + \hat{I}^i_y\hat{I}^j_y + \hat{I}^i_z\hat{I}^j_z \qquad \text{(xxix)}$$

In other words, there is no net interaction acting on the spin system at the end of the pulse sequence to first order in the dipolar-coupling constant (d_{ij} which is contained in C_{ij}); the effects of the dipolar coupling have been averaged to zero to first order. Note that, throughout this analysis, the toggling frame hamiltonian has been expressed in terms of operators defined with respect to the usual rotating frame. This allows the summation of hamiltonians from different toggling frames (as in Equation (xxviii)). Furthermore, the toggling frame density operator calculated from the final average hamiltonian analysed in this way is then expressed in terms of rotating frame operators and so is identical to the rotating frame density operator.

It is interesting to determine what happens to any chemical shift terms under the WAHUHA pulse sequence. The hamiltonian describing the chemical shift is the usual

$$\hat{H}_{cs} = \omega_{cs}\hat{I}_z \qquad \text{(xxx)}$$

where ω_{cs} is the chemical shift and $\hat{I}_z = \sum_i \hat{I}^i_z$. Going through the same toggling frame sequence described above for the WAHUHA pulse sequence (and shown in Fig. B2.1.3), we obtain the first-order average hamiltonian for the chemical shift interaction as

Continued on p. 96

Box 2.1 Cont.

$$\overline{H}_{cs}^{(0)} = \omega_{cs}\left(\frac{\hat{I}_z\tau + \hat{I}_y\tau + 2\hat{I}_x\tau + \hat{I}_y\tau + \hat{I}_z\tau}{6\tau}\right) = \frac{1}{3}\omega_{cs}(\hat{I}_x + \hat{I}_y + \hat{I}_z) \qquad \text{(xxxi)}$$

The chemical shift hamiltonian has the general form $-\gamma\mathbf{B}_{eff}\cdot\hat{\mathbf{I}}$ with a characteristic frequency $\omega = -\gamma B_{eff}$. Comparing this with Equation (xxxi), we see that $\mathbf{B}_{eff}$ must lie in the direction (1, 1, 1) of the (x, y, z) rotating frame (in which $\hat{I}_x$, etc., are defined). The characteristic frequency ω is $(1/\sqrt{3})\omega_{cs}$ (bearing in mind that a unit vector in the direction (1, 1, 1) is in fact $(1/\sqrt{3})$ (1, 1, 1). Thus, all I-spin chemical shifts in a spectrum recorded while using WAHUHA are scaled by a factor of $1/\sqrt{3}$. Indeed, the size of any interaction linear in $\hat{I}_z$ will be scaled by this same factor, and this includes any heteronuclear dipolar couplings acting on the I-spins. All the pulse sequences are designed to average away the effects of homonuclear dipolar coupling scale chemical shifts, although the particular scaling factor depends on the particular pulse sequence. When finite length pulses are used, these change the first-order-average hamiltonian from that derived using infinitely sharp pulses, and so the scaling factor is changed. It is not always easy to predict the true scaling factor in a real multiple pulse experiment where finite length pulses are used.

2.5 Cross-polarization

Cross-polarization [19] is usually used to assist in observing dilute spins, such as ^{13}C, although it can also be used to perform some spectral editing, and to obtain information on which spins are close in space. After magic-angle spinning, cross-polarization is one of the most widely used techniques in solid-state NMR.

Observing dilute spins such as ^{13}C presents a number of problems including:

1. The low abundance of the nuclei means that the signal-to-noise ratio is inevitably poor.
2. The relaxation times of low abundance nuclei tend to be very long. This is because strong homonuclear dipolar interactions which can stimulate relaxation transitions are largely absent. Only weaker heteronuclear dipolar interactions are present. The long relaxation times mean that long gaps must be left between scans, often of the order of minutes. When several thousand scans are required to lower the noise to a suitable level, the spectra can take a very long time to collect!

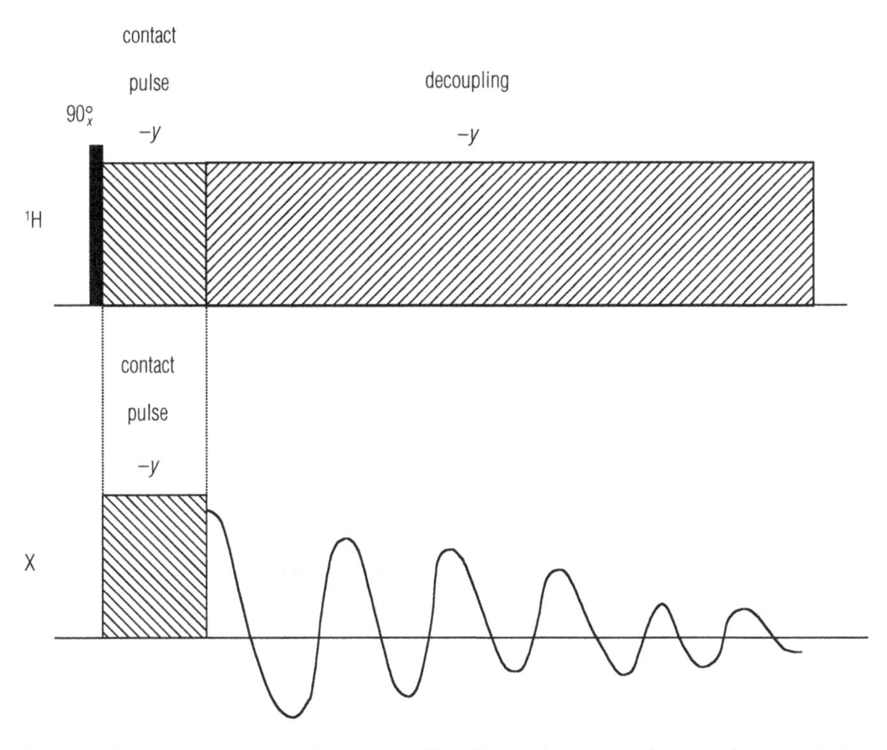

Fig. 2.13 The cross-polarization pulse sequence. The effect of the sequence is to transfer magnetization from the abundant ^{1}H spins in the sample to the X spin via the agency of the dipolar coupling between ^{1}H and X spins.

Both problems can be solved with the pulse sequence in Fig. 2.13, in which the dilute nucleus X derives its magnetization from a nearby network of abundant spins, in this case assumed to be ^{1}H, although it can be any abundant spin-$\frac{1}{2}$ nucleus. Abundant quadrupolar nuclei can also be used although the experiment is slightly different in these cases and is discussed in Chapter 5.

2.5.1 Theory

A proper explanation of cross-polarization involves the use of average hamiltonian theory and is described in Box 2.2 below. Fortunately, however, it is rarely necessary to know the details of the explanation for the everyday practice of the technique. There are, however, several important points to note, most particularly that the cross-polarization transfer is mediated by the dipolar interaction between ^{1}H and X spins. The following is a simplistic explanation of the polarization transfer, but it does serve to highlight the main important features of the experiment. It relies on the transformation of the whole problem to a *doubly rotating frame*, that is one in which the ^1H spins are considered in a frame in which all the magnetic fields due to ^1H pulses appear static, and the X spins are considered in a frame in which all the fields due to X pulses likewise appear static. Thus, the ^1H spins are considered in a frame which rotates about $\mathbf{B}_0$ (the laboratory z-axis) at the frequency of the ^1H rf irradiation and the X spins are considered in a frame which rotates about $\mathbf{B}_0$ at the frequency of the X rf irradiation. We shall also simplify the procedure slightly by assuming that all the pulses are exactly on resonance for the spins to which they are applied.[N4]

An initial ^1H 90°_x pulse creates ^1H magnetization along $-y$ in the ^1H rotating frame (Fig. 2.14). An on-resonance, $-y$ ^1H *contact pulse* is then applied. The field due to this pulse (along $-y$) is known as the *spin-lock field* and is labelled $\mathbf{B}_1(^1H)$. It acts on the rotating frame ^1H magnetization in the same way as $\mathbf{B}_0$ does in the laboratory frame on the equilibrium ^1H magnetization in the absence of rf pulses. Thus it acts to maintain, to some extent, the ^1H magnetization along $-y$. $\mathbf{B}_1(^1H)$ is the only field acting on the ^1H spins in the ^1H rotating frame, since we consider the contact pulse it arises from to be on resonance, so the effects of the $\mathbf{B}_0$ field vanish. $\mathbf{B}_1(^1H)$ thus acts as a quantization axis for the ^1H spins in the rotating frame during the contact pulse. We can then describe the (uncoupled) ^1H spin states in the rotating frame during the pulse as α^*_H and β^*_H, where α^*_H corresponds to a state with a quantized spin component parallel to $\mathbf{B}_1(^1H)$ and β^*_H corresponds to a state with a quantized spin component antiparallel to $\mathbf{B}_1(^1H)$. We include the asterisks on the spin states simply to remind us that they are rotating frame spin states.

We then consider the X spins during the simultaneous X contact pulse (with concomitant spin lock field $\mathbf{B}_1(X)$) in a rotating frame rotating at the X rf irradiation frequency. In a similar manner, the $\mathbf{B}_1(X)$ field provides the quantization axis for the X spins in the rotating frame during the X contact pulse; the X spin states in the rotating frame are α^*_X and β^*_X in analogy with the ^1H spin states, but defined now with respect to the $\mathbf{B}_1(X)$ field as the quantization axis.

Now, the amplitudes of the two contact pulses in the cross-polarization experiment have to be carefully set so as to achieve the Hartmann–Hahn matching condition [20]:

$$\gamma_H B_1(^1H) = \gamma_X B_1(X) \tag{2.14}$$

In this simplistic approach, this sets the energy gaps between the respective rotating frame spin states of ^1H and X spins to be equal. Now we need to consider the effect of dipolar coupling between the ^1H and X spins. The dipolar coupling operator which describes this interaction has the usual

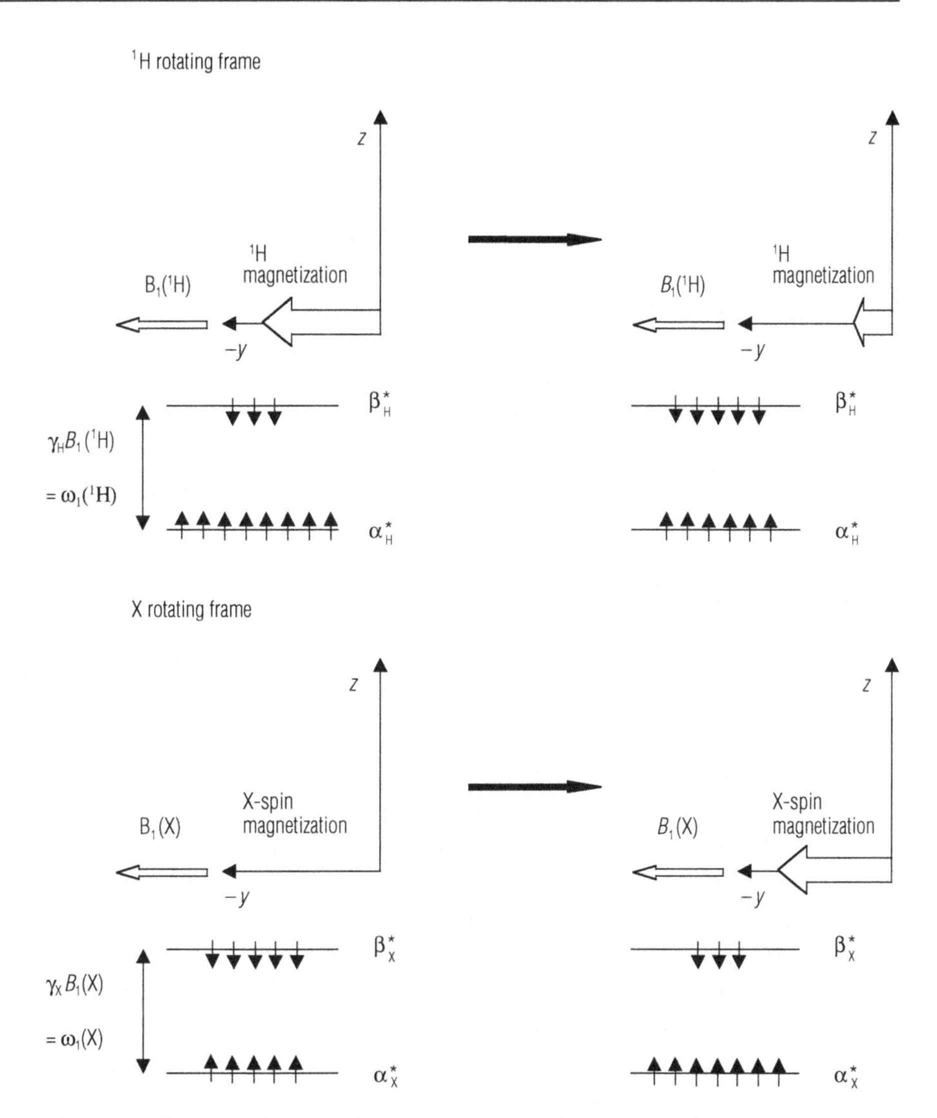

Fig. 2.14 Explanation of the cross-polarization experiment. In this experiment, ^{1}H magnetization is transferred to the X spins via the dipolar coupling between ^{1}H and X during a contact pulse applied simultaneously to both spin types after an initial 90°_x pulse to the ^{1}H spins. The figure here deals with what happens to the spins during the contact pulse. We consider a doubly rotating frame of reference such that the pulses applied to both spins appear stationary along $-y$. The pulses are on resonance for the spins they are applied to, so the only magnetic fields present in the rotating frames are those due to the fields arising from the respective pulses. These fields constitute the quantization axes for the spins in the rotating frame. Also shown in the figure are the spin polarizations, again in their respective rotating frames. Note that at the start of the contact pulse, there is no X-spin polarization in this frame. The α^* and β^* states are the spin states of the spins in their respective rotating frames in the absence of any other interactions. The dipolar coupling between ^{1}H and X spins depends on operators like $\hat{I}^H_{iz}\hat{S}^X_z$, which act in a direction perpendicular to the quantization axes of the spins. Thus, the effects of this operator cannot change the *net* spin polarization along the quantization axes, nor can it change the net energy of the combined ^{1}H, X spin system. Thus, for every ^{1}H spin which flips, an X spin must flip in the opposite sense, but this should be a zero-energy process. When the Hartmann–Hahn condition is met, i.e. $\gamma_H B_1(^1H) = \gamma_X B_1(X)$, the energy gaps between the rotating frames ^{1}H and X spin states are the same, so the flipping process results in no net energy change as required.

form for a heteronuclear dipolar interaction (in angular frequency units) (Section 4.1.3):

$$\hat{H}_{\mathrm{HX}} = -\sum_i d_i \cdot (3\cos^2\theta_i - 1) \cdot \hat{I}_{iz}^{\mathrm{H}} \hat{S}_z^{\mathrm{X}} \tag{2.15}$$

where d_i is the dipolar-coupling constant for the coupling between the *i*th ^{1}H and the X spin. This operator is unaffected by transformation to the doubly rotating frame, as it only contains $\hat{I}_z$ and $\hat{S}_z$ operators, which are obviously unaltered by rotations about *z* (of the laboratory frame). Now, in their respective rotating frames, both the ^{1}H and X spins are quantized in directions which are perpendicular to *z*; thus the dipole–dipole coupling operator cannot affect the net energy of the spin system in the rotating frame, as the energy of the spin system is determined by fields (and thus operators) which are parallel to the spins' quantization axes, which in turn are in the *x–y* plane of the rotating frame. For a similar reason, the dipole–dipole coupling operator cannot alter the net spin polarization (sum of ^{1}H and X spin polarizations) parallel to the quantization axis. So the dipole–dipole coupling operator, which, from its form, clearly couples the ^{1}H and X spins, has to act in such a way as to conserve both the energy and the angular momentum of the total ^{1}H network–X spin system. When the Hartmann–Hahn condition is met, the energy gaps between ^{1}H and X rotating frame spin states are equal, so a transition absorbing energy on a ^{1}H spin, for instance, can be exactly compensated by a transition releasing energy on an X spin. Under these circumstances, the dipolar coupling between ^{1}H and X can allow a redistribution of energy between ^{1}H and X spins while maintaining the total energy of the system as a constant, as required. Because every transition on a ^{1}H spin is compensated for by a transition in the opposite direction on an X spin, the net spin polarization of the system is also necessarily preserved, as required. The nature of the redistribution of energy in the spin system is determined by the initial distribution of spins over the rotating frame states. The initial magnitude of ^{1}H magnetization in the direction of the spin-lock field $\mathbf{B}_1(^1\mathrm{H})$ is the same as that in the laboratory frame parallel to $\mathbf{B}_0$, since it was produced by a 90° rotation of the $\mathbf{B}_0$-generated ^{1}H magnetization. It is thus too big to be sustained by the much smaller $\mathbf{B}_1(^1\mathrm{H})$ field. The ^{1}H magnetization thus reduces by $\alpha^* \rightarrow \beta^*$ rotating frame spin state transitions, while at the same time X spin $\beta^* \rightarrow \alpha^*$ transitions occur to conserve energy and lead ultimately to a large X magnetization in the direction of the $\mathbf{B}_1(\mathrm{X})$ field in the rotating frame.

The only thing this analysis does not consider explicitly is the effect of dipolar coupling between ^{1}H spins (we assume that between the rare X nuclei to be negligible, because of the unlikelihood of finding two X spins in close spatial proximity). Clearly ^1H–^1H dipolar coupling cannot be ignored and in fact acts to redistribute energy within the ^1H network. This has important implications in the cross-polarization experiment, as it means that as the ^1H magnetization is transferred to X from one ^1H spin, the ^1H magnetization within the coupled ^1H network adjusts to the new circumstances.

2.5.2 Setting up the cross-polarization experiment

It is always prudent to set up the cross-polarization experiment on a standard or reference compound before applying it to the sample of interest. A suitable standard obviously has to have the two nuclei to be used in the cross-polarization experiment present in reasonable concentration and in spatially close proximity to each other, to ensure good cross-polarization transfer. Ideally, the abundant nucleus (usually ^1H) from which the cross-polarization occurs should have a reasonably short spin-lattice relaxation time so that trial spectra can be collected quickly. The X nucleus to which cross-polarization occurs should give signals over a similar range to those expected in the samples to be run. Table 2.1 lists suitable standards for common cross-polarization combinations. In the following, we shall assume that cross-polarization takes place from ^1H to X, but ^1H may be replaced by any other (usually abundant) nucleus.

If magic-angle spinning is to be used in the experiment, this should be set up first and the spinning rate which will be required in the cross-polarization experiments established. As discussed later, it is important that the spinning rate is not too fast with respect to the dipolar couplings which mediate the polarization transfer, or the transfer efficiency is severely reduced. As with high-power decoupling, a well-tuned probe is essential from the outset, as the cross-polarization experiment potentially involves high rf powers in relatively long bursts on two channels simultaneously. Most standard cross-polarization pulse programs on modern spectrometers also include high-power decoupling of some sort during the FID collection.

The next step is then to set the ^1H channel on (or at least close to) resonance and to measure an approximate 90° pulse length for the ^1H spins. It only needs to be approximate, as it can be better refined later in a cross-polarization experiment, so a value from a previous experiment is usually fine.

The next stage is to set the Hartmann–Hahn match. Setting this accurately is critical to the success of the experiment. The following procedure assumes that the spectroscopist has no prior knowledge of the Hartmann–Hahn match conditions; if the experiment has been previously

Table 2.1 Commonly-used compounds for setting cross-polarization experiments and their NMR parameters.

	Compound	Spectrum	Suggested contact time	Suggested recycle delay	Comments
$^{1}H \rightarrow {}^{13}C$	Hexamethylbenzene	17.35 ppm, 132.2 ppm relative to TMS	8 ms	10 s	Useful as spectrum contains two peaks which cover a large part of the total ^{13}C chemical shift range. Long contact time required as the only ^{1}H are contained in rapidly spinning Me groups.
	Adamantane	38.4 ppm	5 ms	10 s	Relatively long contact time required due to mobility of molecules.
$^{1}H \rightarrow {}^{15}N$	Glycine	Single peak at 109 ppm wrt liquid ammonia	10 ms	2 s	
$^{1}H \rightarrow {}^{29}Si$	Kaolinite	Two strongly overlapping peaks centred at −91.2 ppm relative to TMS	3 ms	5 s	Relatively cheap material
$^{1}H \rightarrow {}^{31}P$	$NH_4H_2PO_4$	Single peak near 0 ppm wrt 85% H_3PO_4	5 ms	5 s	Can be used as a static sample if required.
$^{1}H \rightarrow {}^{113}Cd$	$3\ CdSO_4 \cdot 8H_2O$	−61, −73 ppm from 0.1 M $Cd(NO_3)_2$	4 ms	10 s	
$^{15}N \rightarrow {}^{13}C$	Glycine	^{15}N: 109 ppm ^{13}C: 43.6, 176.9 ppm	14 ms	5 s	Use tangential contact pulses (see Fig. 2.16).
$^{31}P \rightarrow {}^{13}C$	Phosphorylated serine	^{13}C: 170, 64, 54 ppm ^{31}P: 2 ppm	3 ms	2 s	Longer contact times are required to observe the carbonyl ^{13}C (170 ppm) with any significant intensity.

run, then the Hartmann–Hahn match conditions are obviously most easily found from refinement of previously used conditions. Set the rf power level for one of the nuclei, say X, at a suitable level for the probehead being used, and then measure the nutation frequency of the X magnetization under this rf amplitude, i.e. by measuring the 180° pulse length and the rise time of the rf irradiation. This enables you to calculate the effective rf amplitude $\omega_1(X)$ being felt by the X spins, i.e. $\omega_1(X) = (2\tau_{180})^{-1}$, where τ_{180} is the 180° pulse length. The Hartmann–Hahn match occurs at $\omega_1(X) = \omega_1(^1H)$ if X is a spin-$\frac{1}{2}$ nucleus (see Chapter 5 for discussion of the Hartmann–Hahn matching condition for quadrupolar nuclei), so the next step is to set the ^1H rf amplitude to this value. This can most easily be done by noting that you are looking for the ^1H rf amplitude which gives a ^1H 180° pulse length of $(2\omega_1(X))^{-1}$. So simply adjust the ^1H rf amplitude to achieve the desired ^1H 180° pulse length.

Having found this rf amplitude, you should be close to the optimal Hartmann–Hahn match. Refinements will need to be made because signals are not exactly on resonance and a whole host of other factors. However, refinements are easily made by recording spectra through a single phase cycle for different ^1H or X rf amplitudes around the previously found starting point and finding the rf amplitudes which give the maximum signal intensity. Where there are several signals in the spectrum, it may be that one rf setting gives the maximum intensity for one signal, while a slightly different one gives the maximum intensity for another. Obviously in these circumstances, a sensible judgement has to be made about the optimal Hartmann–Hahn setting.

The phase cycling employed in the cross-polarization experiment (alternation of the ^1H 90° pulse phase and receiver phase between successive scans) is normally such that no signal will arise from direct excitation of the X spins, so any signal observed arises via cross-polarization from ^1H. This is important, as otherwise significant intensity can arise from direct polarization from the contact pulse if using a labelled sample, for instance, where the X spin is relatively abundant. The phase of the signal arising from such magnetization will in general be different from that from the cross-polarization process and causes spectral distortion.

Having optimized the Hartmann–Hahn match, the ^1H 90° pulse length can be optimized in the usual way, i.e. by finding the ^1H 180° and 360° pulse lengths as explained in Section 1.5, but using the cross-polarization pulse sequence to find these by utilizing the fact that the X cross-polarization signal intensity is proportional to the amount of ^1H transverse magnetization after the initial ^1H pulse in the cross-polarization pulse sequence, i.e. the X cross-polarization signal intensity is zero if the initial ^1H pulse is a 180° or a 360° pulse.

Having optimized these parameters, the cross-polarization experiment as set up can be performed on the samples of interest. For each new sample, the ^{1}H and X channels need to be set close to resonance for the signals in that sample and the optimal contact time needs to be determined experimentally.

The optimal contact time depends on the cross-polarization dynamics for each particular sample. The detailed dynamics of cross-polarization are complicated, especially under magic-angle spinning [21]. The initial cross-polarization transfer involves only the closest ^{1}H and X spins. If there are isolated ^{1}H–X spin pairs, the magnetization simply oscillates back and forth between the ^{1}H and X spin at a rate determined by the dipolar coupling between them. If, however, the ^{1}H spin is part of a network of coupled ^{1}H spins, subsequently the ^{1}H magnetization redistributes among the network of coupled ^{1}H spins and 'tops up' the ^{1}H spins close to X nuclei so that the transfer ^{1}H $\rightarrow$ X continues, rather than oscillates. An extra complication is the fact that the spin-locked ^{1}H magnetization relaxes during the contact pulses, with a characteristic time $T_{1\rho}$. This arises because the spin-locking field is too small to support the initially large ^{1}H magnetization. It has the effect of reducing the degree of cross-polarization transfer.

In experiments where cross-polarization is simply being used to increase signal intensity, the contact times should be varied until maximum signal intensity in the X spin spectrum is achieved. For ^{1}H–^{13}C cross-polarization in organic solids, contact times of a few milliseconds are usually optimal. Generally speaking, there will be rather little signal if the contact time in these cases is less than about 0.5 ms, due to lack of polarization transfer, or more than around 20 ms, because of $T_{1\rho}$ relaxation of the ^{1}H magnetization.[N5] The rate of magnetization transfer in a cross-polarization experiment depends on the strength of the dipolar coupling between ^{1}H and X – the stronger the coupling, the faster the rate of transfer. The coupling, in turn, gets stronger with shorter internuclear distances and larger γ for the nuclei concerned. The rate of transfer also depends on how quickly the ^{1}H magnetization redistributes among the ^{1}H network of coupled spins; this can usually be assumed to be fast because of the strong ^{1}H–^{1}H dipolar interactions.

Anything which disrupts the dipolar coupling also disrupts cross-polarization transfer. So, for instance, molecular motion which averages the dipolar coupling to a smaller value reduces the rate of the cross-polarization transfer and means that longer contact times are generally required for reasonable signal intensity to be obtained. Likewise, magic-angle spinning, which averages the dipolar coupling to zero, can severely disrupt cross-polarization transfer if the spinning rate approaches the dipolar coupling constants governing the cross-polarization transfer. Magic-angle spinning also complicates the Hartmann–Hahn match condition [21],

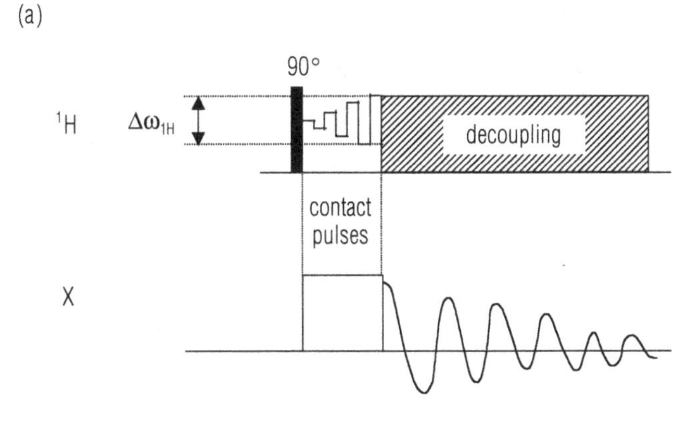

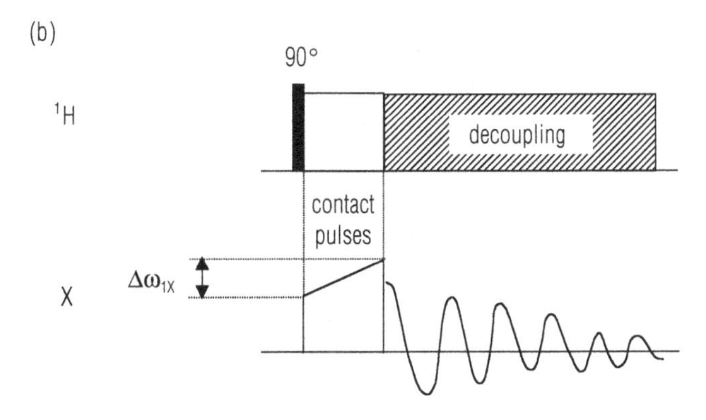

Fig. 2.15 Pulse sequences for cross-polarization under rapid magic-angle spinning. (a) Variable Amplitude Cross-Polarization (VACP) [22]. The 'contact pulse' on the ^1H spins consists of a series of pulses of different amplitude (but same phase). There is a fixed step size in amplitude between the pulses. The overall amplitude variation, $\Delta\omega_{1H}$, is generally of the order of $2\omega_R$. (b) Ramped-amplitude cross-polarization [23]. The contact pulse on one of the spins (it can be either) is steadily increased in amplitude over the contact period. The size of the amplitude increase, $\Delta\omega_{1X}$, is generally of the order of ~2–3 ω_R. For both experiments, the total length of the contact time is of the order of a few milliseconds; the pulse sequence parameters depend on the nature of the sample.

as described in Box 2.2 below. Variable amplitude contact pulses on one of the spins can be used to increase the cross-polarization efficiency under magic-angle spinning [22, 23]. Pulse sequences using these are shown in Fig. 2.15. In many laboratories the ramped cross-polarization pulse sequence of Fig. 2.15 is used routinely when magic-angle spinning is used. Not only does this give significantly greater intensity, but it is less sensitive to small mis-settings of the Hartman–Hahn match.

In cases where the spins involved in cross-polarization are strongly dipolar-coupled to each other, but only weakly coupled to other spins, i.e. the nucleus from which cross-polarization occurs is not part of a strongly-coupled network, it has been found [24] that contact pulses (on both spins)

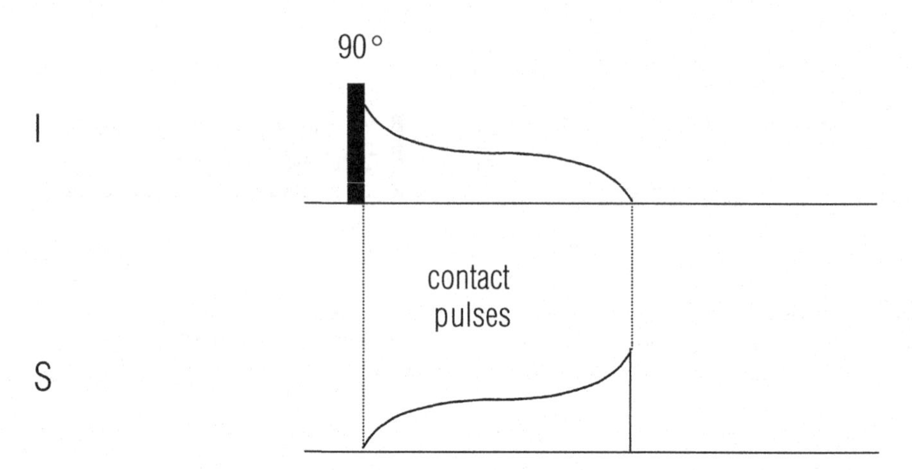

Fig. 2.16 Cross-polarization employing contact pulses which have rf amplitudes which vary tangentially with time [24]. This is useful when the spin I from which cross-polarization occurs is strongly dipolar-coupled to the recipient spin, S, but only weakly coupled to other I spins. The amplitude of the contact pulses vary in time (t) such that $\omega_{1I} - \omega_{1S} = d \tan(\alpha(\frac{1}{2}\tau_p - t))$ where d is the dipolar coupling constant between the I and S spins, τ_p is the total length of the contact time and $\alpha = 2/\tau_p$ (arctan (ω_{1I}^0/d)), ω_{1I}^0 being the rf amplitude at the Hartman–Hahn match.

which have an rf amplitude with a tangential time dependence (Fig. 2.16) perform much better than rectangular or ramped contact pulses. The aim of such a scheme is to effect an adiabatic transfer of magnetization between the spins. This is of considerable use, for instance, in experiments on labelled proteins where cross-polarization between $^{15}N \rightarrow {}^{13}C$ is required.

Box 2.2: Cross-polarization and magic-angle spinning

During the contact pulses of the cross-polarization experiment, the hamiltonian acting on the ^1H–X spin system is (in angular frequency units) [25]:

$$\hat{H} = \hat{H}_Z + \hat{H}_{HH} + \hat{H}_{HX} + \hat{H}^x_{pulse} \quad \text{(i)}$$

with

$$\hat{H}_Z = \omega_0^H \sum_i \hat{I}_{iz}^H + \omega_0^X \hat{S}_z^X \quad \text{(ii)}$$

$$\hat{H}_{HH} = -\sum_{i>j} C_{ij}^{HH} (3\hat{I}_{iz}^H \hat{I}_{jz}^H - \hat{\mathbf{I}}_i^H \cdot \hat{\mathbf{I}}_j^H) \quad \text{(iii)}$$

$$\hat{H}_{HX} = -2\sum_i C_i^{HX} (\hat{I}_{iz}^H \hat{S}_z^X) \quad \text{(iv)}$$

$$\hat{H}^{x}_{\text{pulse}} = \omega_1^{\text{H}} \sum_i \hat{I}^{\text{H}}_{ix} \cos \omega_0^{\text{H}} t + \omega_1^{\text{X}} \hat{S}^{\text{X}}_x \cos \omega_0^{\text{X}} t \qquad \text{(v)}$$

and

$$C^{\text{HH}}_{ij} = \left(\frac{\mu_0}{4\pi}\right) \frac{\gamma_{\text{H}}^2}{r_{ij}^3} \frac{1}{2} (3\cos^2 \theta_{ij} - 1) \qquad \text{(vi)}$$

$$C^{\text{HX}}_{i} = \left(\frac{\mu_0}{4\pi}\right) \frac{\gamma_{1\text{H}} \gamma_{\text{X}}}{r_{i}^3} \frac{1}{2} (3\cos^2 \theta_{i} - 1) \qquad \text{(vii)}$$

where r_i is the distance between the ith ^{1}H spin and the X spin and θ_i is the angle between the ith ^{1}H spin–X spin internuclear axis and the applied field $\mathbf{B}_0$. $\hat{H}_Z$ represents the Zeeman terms of both ^{1}H and X spins; $\hat{H}_{\text{HH}}$ represents the dipolar coupling between the abundant ^{1}H spins and $\hat{H}_{\text{HX}}$ represents the dipolar coupling between the ^{1}H and X spins. We assume that the X spins are low in abundance so that we can ignore dipolar coupling between the X spins. $\hat{H}^{x}_{\text{pulse}}$ describes the effects of the contact pulse applied to each spin; these pulses are assumed to be on resonance for the spins they are applied to. ω_1^{H} and ω_1^{X} are the amplitudes of the ^{1}H and X contact pulses respectively, and ω_0^{H} and ω_0^{X} are their frequencies.

Now we perform a series of transformations to remove the effects of the contact pulses, so that we are left with a hamiltonian in some new frame which only contains terms to do with the various dipolar couplings. It is then much easier to assess the whole cross-polarization process. A similar procedure was adopted in the discussion of the WAHUHA pulse sequence in Box 2.1. First we transform the hamiltonian to a doubly rotating frame, so that the frame precesses about z ($\mathbf{B}_0$) at ω_0^{H} for the ^{1}H spin operators and at ω_0^{X} for the X spin operators, the rf pulse frequencies for each spin. Mathematically, this is accomplished by the transformation

$$\hat{H}^{\text{rot}} = \hat{R}_0^{-1} \hat{H} \hat{R}_0 - \omega_0^{\text{H}} \sum_i \hat{I}^{\text{H}}_{iz} - \omega_0^{\text{X}} \hat{S}^{\text{X}}_z \qquad \text{(viii)}$$

where $\hat{H}^{\text{rot}}$ is the hamiltonian in the doubly rotating frame and the rotation operator $\hat{R}_0$ is given by $\hat{R}_0 = \exp\left(-i\omega_0^{\text{H}} \sum_i \hat{I}^{\text{H}}_{iz} t\right) \exp(-i\omega_0^{\text{X}} \hat{S}^{\text{X}}_z t)$. In this rotation operator, the first term necessarily acts only on the ^{1}H spin operators and the second on the X spin operators. The form of Equation (viii) is found by insisting that the time-dependent Schrödinger equation (see Box 1.1 in Chapter 1) is invariant to the frame transformation, as we shall now show (a similar procedure was used in Box 2.1 earlier in this chapter).

Continued on p. 108

Box 2.2 *Cont.*

The time-dependent Schrödinger equation is $-i(\mathrm{d}\psi/\mathrm{d}t) = \hat{H}\psi$ expressed in frequency units, where ψ is the wavefunction describing the spin system. The time-dependent Schrödinger equation in the rotating frame is invariant to the frame transformation and so is simply

$$i\frac{\mathrm{d}\psi^{\mathrm{rot}}}{\mathrm{d}t} = \hat{H}^{\mathrm{rot}}\psi^{\mathrm{rot}} \tag{ix}$$

where all quantities are referred to the rotating frame. Under the frame transformation described by the rotation operator $\hat{R}_0$, the wavefunction ψ becomes ψ^{rot} in the rotating frame, where

$$\psi^{\mathrm{rot}} = \hat{R}_0^{-1}\psi = \exp\left(+i\omega_0^{\mathrm{H}}\sum_i \hat{I}_{iz}^{\mathrm{H}}t\right)\exp\left(+i\omega_0^{\mathrm{X}}\hat{S}_z^{\mathrm{X}}t\right)\psi \tag{x}$$

Thus $\mathrm{d}\psi^{\mathrm{rot}}/\mathrm{d}t$ is given by

$$\begin{aligned}
\frac{\mathrm{d}\psi^{\mathrm{rot}}}{\mathrm{d}t} &= \exp\left(+i\omega_0^{\mathrm{H}}\sum_i \hat{I}_{iz}^{\mathrm{H}}t\right)\exp\left(+i\omega_0^{\mathrm{X}}\hat{S}_z^{\mathrm{X}}t\right)\frac{\mathrm{d}\psi}{\mathrm{d}t} \\
&\quad + i\omega_0^{\mathrm{X}}\hat{S}_z^{\mathrm{X}}\exp\left(+i\omega_0^{\mathrm{H}}\sum_i \hat{I}_{iz}^{\mathrm{H}}t\right)\exp\left(+i\omega_0^{\mathrm{X}}\hat{S}_z^{\mathrm{X}}t\right)\psi \\
&\quad + i\omega_0^{\mathrm{H}}\sum_i \hat{I}_{iz}^{\mathrm{H}}\exp\left(+i\omega_0^{\mathrm{H}}\sum_i \hat{I}_{iz}^{\mathrm{H}}t\right)\exp\left(+i\omega_0^{\mathrm{X}}\hat{S}_z^{\mathrm{X}}t\right)\psi \\
&= \exp\left(+i\omega_0^{\mathrm{H}}\sum_i \hat{I}_{iz}^{\mathrm{H}}t\right)\exp\left(+i\omega_0^{\mathrm{X}}\hat{S}_z^{\mathrm{X}}t\right)\frac{\mathrm{d}\psi}{\mathrm{d}t} + \left(i\omega_0^{\mathrm{X}}\hat{S}_z^{\mathrm{X}} + i\omega_0^{\mathrm{H}}\sum_i \hat{I}_{iz}^{\mathrm{H}}\right)\psi^{\mathrm{rot}} \\
&= \hat{R}_0^{-1}\frac{\mathrm{d}\psi}{\mathrm{d}t} + \left(i\omega_0^{\mathrm{X}}\hat{S}_z^{\mathrm{X}} + i\omega_0^{\mathrm{H}}\sum_i \hat{I}_{iz}^{\mathrm{H}}\right)\psi^{\mathrm{rot}}
\end{aligned} \tag{xi}$$

where all quantities are expressed with respect to the rotating frame. Substituting for $\mathrm{d}\psi^{\mathrm{rot}}/\mathrm{d}t$ gives

$$i\left(\hat{R}_0^{-1}\frac{\mathrm{d}\psi}{\mathrm{d}t} + \left(i\omega_0^{\mathrm{X}}\hat{S}_z^{\mathrm{X}} + i\omega_0^{\mathrm{H}}\sum_i \hat{I}_{iz}^{\mathrm{H}}\right)\psi^{\mathrm{rot}}\right) = H^{\mathrm{rot}}\psi^{\mathrm{rot}} \tag{xii}$$

We then use the fact that $i(\mathrm{d}\psi/\mathrm{d}t) = \hat{H}\psi$ and $\psi = \hat{R}_0\psi^{\mathrm{rot}}$ to rewrite the left-hand side of this equation as

$$\hat{R}_0^{-1}\hat{H}\hat{R}_0\psi^{\mathrm{rot}} + i\left(i\omega_0^{\mathrm{X}}\hat{S}_z^{\mathrm{X}} + i\omega_0^{\mathrm{H}}\sum_i \hat{I}_{iz}^{\mathrm{H}}\right)\psi^{\mathrm{rot}} = H^{\mathrm{rot}}\psi^{\mathrm{rot}} \tag{xiii}$$

By comparing the left- and right-hand sides of this equation, we see that $\hat{H}^{\mathrm{rot}}$, the hamiltonian in the rotating frame, is

$$\hat{H}^{\mathrm{rot}} = \hat{R}_0^{-1}\hat{H}\hat{R}_0 - \omega_0^{\mathrm{X}}\hat{S}_z^{\mathrm{X}} - \omega_0^{\mathrm{H}}\sum_i \hat{I}_{iz}^{\mathrm{H}} \tag{xiv}$$

Substituting Equation (i) for $\hat{H}$ into Equation (xiv) for the rotating frame hamiltonian, we obtain

$$\hat{H}_{\mathrm{rot}} = \hat{H}_{\mathrm{HH}} + \hat{H}_{\mathrm{HX}} + \hat{H}^{x}_{\mathrm{pulse,rot}} \tag{xv}$$

with

$$\hat{H}^{x}_{\mathrm{pulse,rot}} = +\omega_1^{\mathrm{H}}\sum_i \hat{I}_{iz}^{\mathrm{H}} + \omega_1^{\mathrm{X}}\hat{S}_x^{\mathrm{X}} \tag{xvi}$$

Now we transform $\hat{H}^{\text{rot}}$ further into a toggling frame as described in Box 2.1. The toggling frame, in effect, follows the effects of the contact pulses so as to null the hamiltonians describing them. As shown in Box 2.1, the toggling frame which nulls the terms in the rotating frame hamiltonian due to ^{1}H and X pulses is one which rotates the ^{1}H spins' frame of reference about the x-axis (the direction of the ^{1}H contact pulse) at rate ω_1^{H} and the X spin frame of reference about x at rate ω_1^{X}. This transforms $\hat{H}^{\text{rot}}$ into $\hat{H}^*$ in a similar manner to the rotating frame transformation described previously:

$$\hat{H}^*(t) = \hat{R}_1^{-1}\hat{H}^{\text{rot}}\hat{R}_1 - \omega_1^{\text{H}}\sum_i \hat{I}_{ix}^{\text{H}} - \omega_1^{\text{X}}\hat{S}_x^{\text{X}} \tag{xvii}$$

with $\hat{R}_1 = \exp\left(-i\omega_1^{\text{H}}\sum_i \hat{I}_{ix}^{\text{H}}t\right)\exp(-i\omega_1^{\text{X}}\hat{S}_x^{\text{X}}t)$, the rotation operator required to transform to the toggling frame (cf $\hat{R}_0$ above). The resulting toggling frame hamiltonian $\hat{H}^*(t)$ is

$$\hat{H}^*(t) = \hat{H}_{\text{HH}}^* + \hat{H}_{\text{HX}}^* \tag{xviii}$$

where

$$\hat{H}_{\text{HH}}^* = -\frac{1}{2}\sum_{i>j} C_{ij}^{\text{HH}}(\hat{\mathbf{I}}_i^{\text{H}}\cdot\hat{\mathbf{I}}_j^{\text{H}} - 3\hat{I}_{ix}^{\text{H}}\hat{I}_{jx}^{\text{H}}) \tag{xix}$$

$$\hat{H}_{\text{HX}}^* = -\sum_i C_i^{\text{HX}}\left[\left(\sum_i \hat{I}_{iz}^{\text{H}}\hat{S}_z^{\text{X}} + \sum_i \hat{I}_{iy}^{\text{H}}\hat{S}_y^{\text{X}}\right)\cos(\omega_1^{\text{H}} - \omega_1^{\text{X}})t + \left(\sum_i \hat{I}_{iz}^{\text{H}}\hat{S}_y^{\text{X}} + \sum_i \hat{I}_{iy}^{\text{H}}\hat{S}_z^{\text{X}}\right)\sin(\omega_1^{\text{H}} - \omega_1^{\text{X}})t\right] \tag{xx}$$

The expression for $\hat{H}^*(t)$ in Equation (xviii) ignores any terms which are non-secular or which oscillate at ω_1^{H} or ω_1^{X}, as these are assumed to average to zero over the time of the contact pulses. $\hat{H}_{\text{HH}}^*$ only acts on ^{1}H spins and is a constant. When $\omega_1^{\text{H}} \neq \omega_1^{\text{X}}$, the term $\hat{H}_{\text{HX}}^*$ has little effect on the spin system as it contains oscillating terms which roughly average to zero over the contact pulse. However, when $\omega_1^{\text{H}} \cong \omega_1^{\text{X}}$ (the Hartmann–Hahn match), the oscillating terms disappear[N6] and $\hat{H}_{\text{HX}}^*$ becomes important in causing a double resonance effect between the ^{1}H and X spins. In particular, the $\hat{I}_{iy}^{\text{H}}\hat{S}_Y^{\text{X}}$ term contains terms of the form $\hat{I}_{i+}^{\text{H}}\hat{S}_-^{\text{X}}$ and $\hat{I}_{i-}^{\text{H}}\hat{S}_+^{\text{X}}$, which cause magnetization transfer between the ^{1}H and X spins. It is important to appreciate that these terms arise ultimately from the $\hat{I}_{iz}^{\text{H}}\hat{S}_Z^{\text{X}}$ term in the ^{1}H–X dipolar-coupling hamiltonian, the A-type terms, and not the B-type, so-called 'flip-flop' terms, as is often mistakenly thought. These latter terms have little effect on a heteronuclear spin system, as will be described in Section 4.1.3.

Continued on p. 110

Box 2.2 Cont.

Now we must consider the effect of magic-angle spinning on the experiment. Magic-angle spinning has the effect of introducing time dependence into the dipolar coupling. This appears in the toggling frame (or other frame) hamiltonian via the C_{ij}^{HH} and C_i^{HX} terms which contain the geometric terms describing the dipolar coupling between ^1H spins and between ^1H and X spins respectively. Section 2.2 describes magic-angle spinning and the time dependence it induces in nuclear spin interactions. In essence, the strength of an interaction under magic-angle spinning is described by terms which oscillate at $\pm\omega_R$ and $\pm 2\omega_R$, where ω_R is the spinning rate. For dipolar coupling, there is no isotropic (constant) term which remains under magic-angle spinning (although there is for the chemical shift). It is the $\hat{H}^*_{HX}$ term which induces cross-polarization between the ^1H and X spins. However, under magic-angle spinning the C_i^{HX} terms in $\hat{H}^*_{HX}$ are no longer constant, but oscillate at $\pm\omega_R$ and $\pm 2\omega_R$. This means that at the normal Hartmann–Hahn match, $\omega_1^H \cong \omega_1^X$, the net cross-polarization term in the hamiltonian averages to zero, or at least a small value, over the time of the contact pulse [25] if the rotor period is small relative to the contact time. However, $\hat{H}^*_{HX}$ contains other oscillatory terms, $\cos(\omega_1^H - \omega_1^X)t$ and $\sin(\omega_1^H - \omega_1^X)t$. The net oscillatory behaviour of the $\hat{H}^*_{HX}$ term under magic-angle spinning can be cancelled by matching $(\omega_1^H - \omega_1^X)$ to $\pm\omega_R$ or $\pm 2\omega_R$.

Thus the match condition under magic-angle spinning is $(\omega_1^H - \omega_1^X) = \pm\omega_R$ or $\pm 2\omega_R$, the so-called *sideband match conditions*. Cross-polarization intensity is found at the normal Hartmann–Hahn match, $(\omega_1^H - \omega_1^X) = 0$, but as explained above, this decreases as the spinning rate is increased and the dipole coupling is more effectively averaged.

2.6 Echo pulse sequences

Broad lines, such as those arising from chemical shift anisotropy, dipolar coupling, etc., have rapidly decaying FIDs. 'Ringing' in the coil which is used to measure the FID prevents measurement of the signal until a short time (the *dead time*) after a pulse. This delay means that a significant part of the rapidly decaying FID is not recorded (Fig. 2.17). This not only leads to the obvious loss of intensity but to spectral distortion, as is now described. During the dead time the transverse magnetization which ultimately creates the FID is not stationary but evolves under the various nuclear spin interactions. Components of transverse magnetizations arising

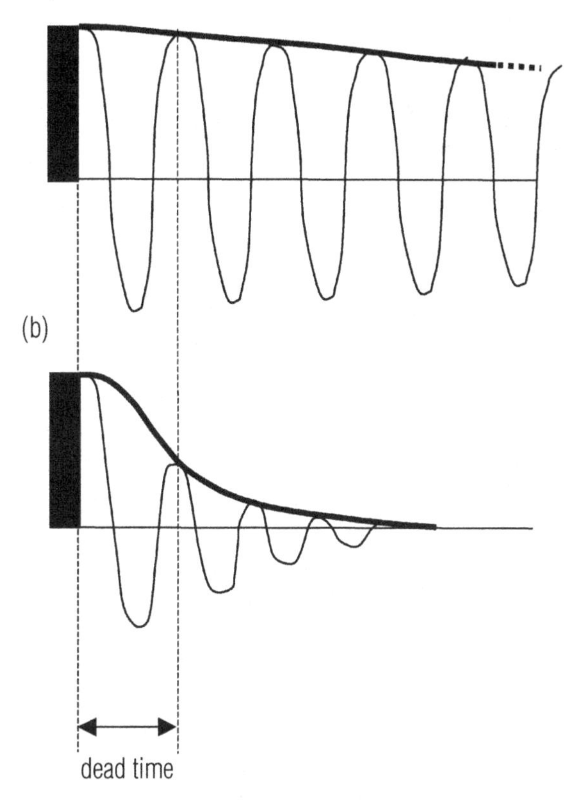

Fig. 2.17 Illustrating the effect of the dead time for cases where the amplitude of the FID decays (a) slowly and (b) quickly and in the manner typical of a homonuclear dipolar-coupled spin system. The thick line indicates the amplitude of the FID in each case. In (a), the recorded FID (that part of the FID after the dead time) closely approximates the full FID, but with a shift of time origin. Such a shift of time origin can be correctly accounted for by a first-order phase correction of the FID or spectrum [24]. In (b), the amplitude of the FID decays approximately as a half-gaussian function. It is clear that the shape of the amplitude function following the dead time is a very poor approximation to the shape of the amplitude function immediately following the pulse. In this case, the loss of signal in the dead time cannot be accounted for by a first-order phase correction, as the recorded FID differs from the full FID by more than a simple time shift of its origin.

from crystallites with different orientations in a powder sample will in general evolve at different rates, reflecting the fact that the nuclear spin interactions which cause the evolution have strengths which depend on molecular orientation. This has the result that at the point where the FID acquisition starts, the different components of transverse magnetization will have acquired a phase which is different (in general) for each component.

This leads to severely distorted lineshapes after Fourier transformation in the case where the FID decays rapidly, i.e. the NMR resonances are broad,

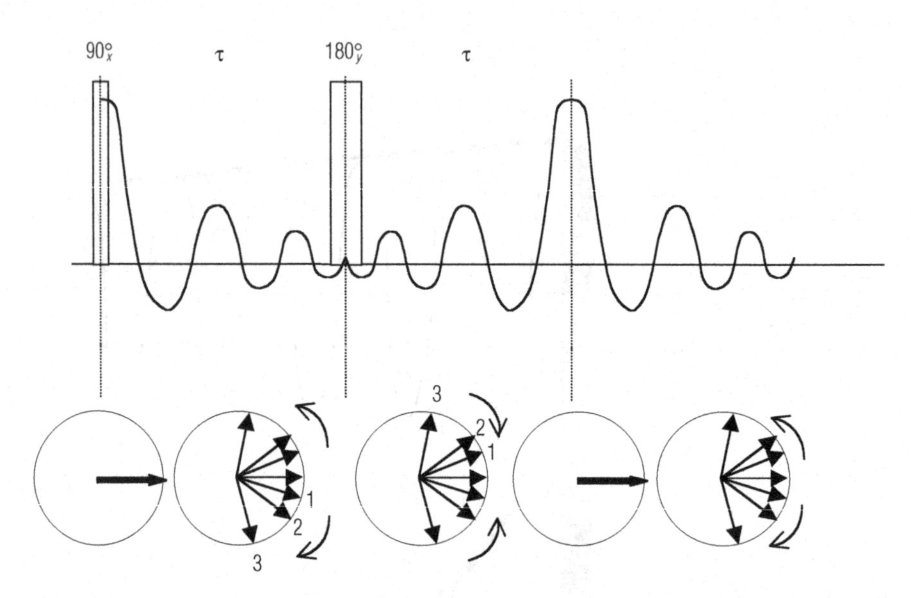

Fig. 2.18 The spin-echo pulse sequence. The behaviour of the transverse magnetization components is shown. Magnetization components dephase under the effects of chemical shift anisotropy or heteronuclear dipolar coupling during the first τ period. The subsequent 180° pulse rotates the magnetization components 180° about the pulse axis (y in this case) so that the components refocus after a further τ period. The τ delay is chosen to be long enough to include the dead time for the probe. Then the FID can be completely recorded from the true echo maximum. If the FID were simply recorded after a single 90° pulse, a significant part of the FID would be lost due to the dead time required before recording can begin.

as there is a wide range of phases of transverse magnetization by the time acquisition starts. Where the loss of signal is relatively small, i.e. the dead time is small and the signal does not decay rapidly, the effects of the truncation of the FID can be largely overcome by a first-order phase correction, which, in effect, takes account of a shift of the time origin of the transverse magnetization evolution [26]. The effect of the dead time is usually small in solution NMR, for instance, since there the FID signals decay much more slowly, so the proportion of the total FID lost in the dead time is only very small. In solid-state NMR, when recording broad powder lineshapes the problem is often much more serious, and cannot be correctly accounted for by a simple first-order phase correction. Essentially, a first-order phase correction is sufficient if the recorded FID closely approximates what would be the full FID if it could be recorded, but simply time-shifted (Fig. 2.17). Where this is not the case (Fig. 2.17), an alternative approach is required.

The problem can be overcome by use of the *spin-echo pulse sequence* (Fig. 2.18), 90°_x–τ–180°_y–τ–*acquire*, for lines broadened by chemical shift anisotropy or heteronuclear dipole–dipole coupling, i.e. for any interaction linear in the observed spin operators, and by the *solid-* or *quadrupole-echo pulse sequence*, 90°_x–τ–90°_y–τ–*acquire*, for lines broadened by quadrupole coupling or homonuclear dipole–dipole coupling, i.e. any interaction bilinear in the observed spin's spin operators.

This latter pulse sequence operates in a similar fashion to the spin-echo pulse sequence in that it refocuses the dispersing transverse magnetization components a time τ after the refocusing pulse. Note that the refocusing pulse in the solid-echo pulse sequence is a 90° pulse rather than the 180° pulse in the spin-echo sequence. Using the above sequence, acquisition of the FID signal necessarily begins a time τ after the last pulse in the sequence and, hence, no signal is lost due to a dead time delay. The only signal loss is due to relaxation of the transverse magnetization during the two τ periods.

In more complicated pulse sequences which generate broad lines, an 'echo' sequence, τ–180°–τ– or τ–90°–τ–, as appropriate, is often added to the end of the sequence to circumvent dead time problems.

Setting up an echo pulse sequence

The setting up of an echo pulse sequence is generally very straightforward, requiring only the accurate calibration of the necessary rf pulse lengths. The recorded FID should begin at the exact echo maximum, otherwise the resulting spectrum will be distorted. In order to ensure that the echo maximum is not missed, the second echo delay in the pulse sequence is generally set to be several microseconds, or more, shorter than the first. The resulting FID dataset is then left-shifted to the echo maximum before Fourier transformation. Details of this can be found in Section 6.5.1.

It should be noted that if the experiment is being recorded with magic-angle spinning, then the echo delay τ should be an integral number of rotor periods, as the sample spinning induces rotor echoes (see Section 2.2.2) which refocus every rotor period. In setting the echo delay here, however, allowance should be made for the finite lengths of the pulses in the echo sequence, i.e. the interval between the centre of the pulses in the echo sequence should be an integral number of rotor periods Otherwise, the echo delay should be set to be longer than the ringdown time required by the probehead being used, but not unduly longer than this as transverse relaxation during the echo delays will diminish the final signal intensity. Section 6.5.1 gives further details to optimize the setting of the echo delay.

Notes

1. Note that 66 kHz is also the magic-angle spinning rate that would be required to remove the effects of this dipolar coupling from the ^{13}C spectrum; such rates are not yet achievable in practice, while an rf amplitude of 66 kHz is easily achievable.
2. There are two main rotary resonances which can occur, the so-called HORROR conditions. The $n\,\omega_R = \omega_1$, $n = 1, 2$, condition recouples heteronuclear dipolar couplings; the $\omega_R/2 = \omega_1$

condition recouples homonuclear dipolar couplings. It is important that both these conditions are avoided when setting up high-power decoupling. Recoupling of dipolar coupling and rotational resonance is discussed in Chapter 4.

3. The advantage of rotating the hamiltonian operator rather than transforming its frame of reference is that we end up with an operator that is expressed with respect to the original rotating frame. When we come to sum operators expressed with respect to different toggling frames (for instance, when forming the first-order component of the average hamiltonian), this is a distinct advantage. In essence, we are expressing the dipolar hamiltonian in a toggling frame in terms of operators defined with respect to the rotating frame.
4. Transformation to a singly rotating frame in the quantum mechanical picture was discussed in Box 2.1. The transformation to the doubly rotating frame required here is achieved in a similar manner to that used in Box 2.1. Now, however, the wavefunction describing the spin system is transformed by $\Psi' = \exp(i\omega_{1_H}\hat{I}_{Hz}t)\exp(i\omega_X\hat{I}_{Xz}t)\Psi$ where ω_{1_H} and ω_X are the frequencies of the ^{1}H and X rf irradiations respectively; $\hat{I}_{Hz}$ and $\hat{I}_{Xz}$ are sums of $\hat{I}_z$ operators over the ^{1}H and X spins respectively, and only operate on these spins respectively. Hence, it appears as though the ^{1}H spin components of the total wavefunction are in a frame which rotates at the ^{1}H Larmor frequency, while the X spin components are in one which rotates at their Larmor frequency (assuming that the total spin wavefunction Ψ can be written as linear combinations of products of functions containing ^{1}H spin variables and X spin variables respectively).
5. The X spin transverse magnetization generated by cross-polarization will also be subject to $T_{1\rho}$ relaxation. However, this process is usually relatively slow compared with the ^{1}H $T_{1\rho}$ relaxation, providing the X spin is not abundant, as it is generally strong homonuclear dipolar interactions which promote rapid $T_{1\rho}$ relaxation.
6. Small oscillating terms remain if the ^{1}H and/or X spin contact pulses are not exactly on resonance.

References

1. Maricq, M.M. & Waugh, J.S. (1979) *J. Chem. Phys.* **70** 3300.
2. Herzfeld, J. & Berger, A.E. (1980) *J. Chem. Phys.* **73** 6021.
3. Dixon, W.T. (1982) *J. Chem. Phys.* **77** 1800.
4. Antzutkin, O.N., Song, Z., Feng, X. & Levitt, M.H. (1994) *J. Chem. Phys.* **100** 130.
5. Ernst, M. & Meier, B.H. (2001) *Chem. Phys. Lett.* **348** 293.
6. Bennett, A.E., Rienstra, C.M., Auger, M., Lakshmi, K.V. & Griffin, R.G. (1995) *J. Chem. Phys.* **103** 6951.
7. Detken, A., Hardy, E.H., Ernst, M. & Meier, B.H. (2002) *Chem. Phys. Lett.* **356** 298.
8. Gan, Z.H. & Ernst, R.R. (1997) *Solid-State NMR* **8** 153.
9. Yu, Y. & Fung, B.M. (1998) *J. Magn. Reson.* **130** 317.
10. Eden, M. & Levitt, M.H. (1999) *J. Chem. Phys.* **111** 1511.
11. Fung, B.M., Khitrin, A.K. & Ermolaev, K. (2000) *J. Magn. Reson.* **142** 97.
12. Takegoshi, K., Mizokami, J. & Terao, T. (2001) *Chem. Phys. Lett.* **341** 540.
13. Haeberlen, U. & Waugh, J.S. (1968) *Phys. Rev.* **175** 453.
14. Rhim, W.-K., Elleman, D.D. & Vaughan, R.W. (1973) *J. Chem. Phys.* **59** 3740.
15. Burum, D.P. & Rhim, W.-K. (1979) *J. Chem. Phys.* **71** 944.
16. Vinogradov, E., Madhu, P.K. & Vega, S. (1999) *Chem. Phys. Lett.* **314** 443.
17. Bielecki, A., Kolbert, A.C. & Levitt, M.H. (1989) *Chem. Phys. Lett.* **155** 341.
18. Bielecki, A., Kolbert, A.C., de Groot, H.J.M., Griffin, R.G. & Levitt, M.H. (1990) *Adv. Magn. Reson.* **14** 111.
19. Pines, A., Gibby, M.G. & Waugh, J.S. (1973) *J. Chem. Phys.* **59** 569.
20. Hartmann, S.R. & Hahn, E.L. (1962) *Phys. Rev.* **128** 2042.

21. Xiaoling, W., Shanmin, Z. & Xuewen, W. (1988) *Phys. Rev. B.* 37 9827; Wu, X. & Zilm, K.W. (1993) *J. Magn. Reson.* A **102** 205.
22. Peersen, O.B., Wu, X., Kustanovich, I. & Smith, S.O. (1993) *J. Magn. Reson. A* **104** 334.
23. Metz, G., Wu, X. & Smith, S.O. (1994) *J. Magn. Reson. A* **110** 219.
24. Hediger, S., Meier, B.H., Kurur, N.D., Bodenhausen, G. & Ernst, R.R. (1994) *Chem. Phys. Lett.* **223** 283.
25. Stejskal, E.O., Schaefer, J. & Waugh, J.S. (1977) *J. Magn. Reson.* **28** 105.
26. Levitt, M.H. (2001) *Spin Dynamics: Basics of Nuclear Magnetic Resonance*. J. Wiley and Sons, Chichester. (Chapter 5).

Shielding and Chemical Shift: Theory and Uses 3

3.1 Theory

3.1.1 Introduction

The electrons that surround a nucleus are not impassive in the magnetic field used in the NMR experiment, but react to produce a secondary field. This secondary field contributes to the total field felt at the nucleus, and therefore has the potential to change the resonance frequency of the nucleus. This interaction of the secondary field produced by the electrons with the nucleus is the *shielding interaction*. The frequency shift that this interaction causes in an NMR spectrum is the *chemical shift*. The shielding interaction may be considered as being composed of two components:

1. The external magnetic field causes all electrons to circulate around it. This produces a secondary field which opposes the applied field at the centre of motion. Hence, this field tends to shield the nucleus. This is known as the *diamagnetic* contribution to the shielding. This contribution varies as $1/r_i^3$, where r_i is the distance of the ith electron from the nucleus. Hence, it arises principally from the core electrons. In consequence of this, the diamagnetic shielding arising from an atom in a molecule is considered fairly constant for a given atom type, whatever its environment. However, it should be noted that all atoms surrounding the atom in question also generate diamagnetic electron currents, and so also contribute to the total diamagnetic field felt at the nucleus.
2. The external magnetic field also distorts the electron distribution. This distortion of the ground electronic state can be described by mixing excited electronic states into the original ground state. Some of these excited electronic states may possess paramagnetic properties, and so create a small amount of paramagnetism in the electronic ground state of the molecule while it is in the magnetic field. This creates a field

which supports the applied field at the nucleus, and so tends to deshield the nucleus. This is known as the *paramagnetic* contribution and varies considerably with the nuclear environment. The degree of mixing of the excited electronic states with the ground state depends on the energy difference between the mixed states, and is known as Temperature Independent Paramagnetism (TIP).

3.1.2 The chemical shielding hamiltonian

The chemical shielding hamiltonian acting on a spin **I** is (in frequency units)

$$\hat{H}_{cs} = \gamma \hat{\mathbf{I}} \cdot \boldsymbol{\sigma} \cdot \mathbf{B}_0 \tag{3.1}$$

which has the general form proposed in Equation (1.75) in Section 1.4. $\mathbf{B}_0$ is the applied field, the equivalent of **A** in Equation (1.75); $\mathbf{B}_0$ is the ultimate source of the shielding magnetic field as it is $\mathbf{B}_0$ that generates the electron current, which in turn generates the shielding magnetic field. The term $\boldsymbol{\sigma}$ is a second-rank Cartesian tensor, called the chemical shielding tensor. The concept of a second-rank Cartesian tensor was discussed in Section 1.4.1 in Chapter 1.

It is useful at this point to decompose the shielding tensor into a *symmetric* ($\boldsymbol{\sigma}^s$) and *antisymmetric* ($\boldsymbol{\sigma}^{as}$) component:

$$\boldsymbol{\sigma} = \boldsymbol{\sigma}^s + \boldsymbol{\sigma}^{as} \tag{3.2}$$

where

$$\begin{aligned}
\boldsymbol{\sigma}^s &= \begin{pmatrix} \sigma_{xx} & \frac{1}{2}(\sigma_{xy}+\sigma_{yx}) & \frac{1}{2}(\sigma_{xz}+\sigma_{zx}) \\ \frac{1}{2}(\sigma_{xy}+\sigma_{yx}) & \sigma_{yy} & \frac{1}{2}(\sigma_{yz}+\sigma_{zy}) \\ \frac{1}{2}(\sigma_{xz}+\sigma_{zx}) & \frac{1}{2}(\sigma_{yz}+\sigma_{zy}) & \sigma_{zz} \end{pmatrix} \\
\boldsymbol{\sigma}^{as} &= \begin{pmatrix} 0 & \frac{1}{2}(\sigma_{xy}-\sigma_{yx}) & \frac{1}{2}(\sigma_{xz}-\sigma_{zx}) \\ \frac{1}{2}(\sigma_{yx}-\sigma_{xy}) & 0 & \frac{1}{2}(\sigma_{yz}-\sigma_{zy}) \\ \frac{1}{2}(\sigma_{zx}-\sigma_{xz}) & \frac{1}{2}(\sigma_{zy}-\sigma_{yz}) & 0 \end{pmatrix}
\end{aligned} \tag{3.3}$$

The reason for this decomposition is that only the symmetric part of the shielding tensor, $\boldsymbol{\sigma}^s$, turns out to affect the NMR spectrum to any great extent;[N1] this will become clear as we examine how the shielding interaction affects the NMR spectrum in the next section. Henceforth, we will only concern ourselves with the symmetric part of the shielding tensor, and 'σ' should be taken to mean the symmetric part of the shielding tensor in what follows.

As discussed in Section 1.4.1, it is possible to choose the axis frame that σ is defined with respect to so that the shielding tensor is diagonal. This axis frame is the *principal axis frame*, designated 'PAF', or x^{PAF}, y^{PAF}, z^{PAF}. The numbers along the resulting diagonal of σ^{PAF} are the *principal values* of the shielding tensor. The three principal values of the shielding tensor $\sigma_{\alpha\alpha}^{PAF}$ are frequently expressed instead as the isotropic value σ_{iso}, the anisotropy Δ_{cs}, and the asymmetry η_{cs}. These quantities are defined from the principal values as follows:

$$\begin{aligned} \sigma_{iso} &= \frac{1}{3}(\sigma_{xx}^{PAF} + \sigma_{yy}^{PAF} + \sigma_{zz}^{PAF}) \\ \Delta &= \sigma_{zz}^{PAF} - \sigma_{iso} \\ \eta &= \frac{\sigma_{xx}^{PAF} - \sigma_{yy}^{PAF}}{\Delta} \end{aligned} \tag{3.4}$$

To deal with the chemical shielding hamiltonian of Equation (3.1), we must first transform the spin coordinates to a rotating frame of reference, one which rotates about the laboratory z-axis at frequency ω_0, the Larmor frequency of the spin. The reason for this is that we measure all our NMR spectra in FT NMR relative to a carrier rf wave (see Box 1.3 in Chapter 1) and so we effectively record our NMR spectra as if the spins were within such a frame of reference. Thus if we wish to calculate the outcome of an NMR experiment, we must first transform the hamiltonian governing the spin system to the appropriate frame of reference.

This frame transformation is easily done using the procedures in Box 2.1. We see the spin in a rotating frame x', y', z', which rotates about the laboratory z-axis at frequency ω_0. Thus, the spins we observe have components (or spin coordinates) $I_{x'}$, $I_{y'}$, $I_{z'}$, rather than the laboratory I_x, I_y, I_z, and it is these we see interacting with the shielding field ($\boldsymbol{\sigma}\mathbf{B}_0$) arising from the electron density. The rotating frame hamiltonian is the operator describing this interaction between the rotating frame spin components and the shielding field. Our mode of observation in the NMR experiment does not affect the shielding tensor $\boldsymbol{\sigma}$ or the applied field $\mathbf{B}_0$; these quantities are naturally described in the laboratory frame. Thus, a potential problem appears: the shielding field $\boldsymbol{\sigma}\mathbf{B}_0$ is described in the laboratory frame, while the observed spin components $I_{x'}$, $I_{y'}$, $I_{z'}$ that it acts on are described with respect to the rotating frame. The obvious way to deal with this is to write the rotating frame spin components in terms of the laboratory frame ones. This is, in effect, what the following details.

The total hamiltonian governing the spin system in this case is the sum of the Zeeman and chemical shielding terms:

$$\hat{H} = \hat{H}_0 + \hat{H}_{cs} \tag{3.5}$$

where $\hat{H}_0 = \omega_0 \hat{I}_z$, the Zeeman hamiltonian. The resulting rotating frame hamiltonian $\hat{H}^*$ is given by (see Box 2.1 for details)

$$\hat{H}^*(t) = \hat{R}_z^{-1}(\omega_0 t)\hat{H}_0 \hat{R}_z(\omega_0 t) + \hat{R}_z^{-1}(\omega_0 t)\hat{H}_{cs}\hat{R}_z(\omega_0 t) - \omega_0 \hat{I}_z \tag{3.6}$$

where the rotation operator $\hat{R}_z(\omega_0 t)$ describes the rotation of an object in spin space by an angle $\omega_0 t$ about the laboratory z-axis and acts only on the spin coordinates, $\hat{I}$. This z-rotation has no effect on the spin operators of $\hat{H}_0$, so the first and third terms of Equation (3.6) cancel, leaving just the second term, which, substituting from Equation (3.1) for the shielding hamiltonian, yields

$$\hat{H}^*(t) = \gamma(\hat{I}_x \cos\omega_0 t + \hat{I}_y \sin\omega_0 t, \hat{I}_y \cos\omega_0 t - \hat{I}_x \sin\omega_0 t, \hat{I}_z) \cdot \sigma \cdot B_0 \tag{3.7}$$

The frame transformation has led to a time dependence in the hamiltonian describing the spin system which can be difficult to deal with. The best way to deal with it is to use average hamiltonian theory, as in Box 2.1. Average hamiltonian theory approximates a periodically time-dependent hamiltonian such as that in Equation (3.7) as a sum of successively higher-order terms. The lowest-order term, confusingly denoted $\overline{H}^{(0)}$, but correctly called the first-order average hamiltonian (as it is first order in the interaction hamiltonian) is simply given by (see Box 2.1):

$$\overline{H}^{(0)} = \frac{1}{t_p}\int_0^{t_p} \hat{H}(t)\mathrm{d}t \tag{3.8}$$

and is a sufficiently good approximation to the full hamiltonian for the purposes of understanding the effect of the shielding hamiltonian in NMR spectra, providing the magnitude of the chemical shielding interaction in frequency units is small relative to the frequency of the frame rotation, the Larmor frequency ω_0.[N2] Using Equation (3.8), we find the chemical shielding contribution to the first-order average hamiltonian in the rotating frame to be:

$$\overline{H}_{cs}^{(0)} = \gamma(0, 0, \hat{I}_z) \cdot \boldsymbol{\sigma} \cdot \mathbf{B}_0 \tag{3.9}$$

where the time-dependent parts of the rotating frame hamiltonian of Equation (3.7) have integrated to zero. If the applied field $\mathbf{B}_0$ is along z, then the shielding contribution to the first-order average rotating frame hamiltonian in Equation (3.9) becomes (in frequency units)

$$\overline{H}_{cs}^{(0)} = \gamma \hat{I}_z \sigma_{zz}^{\text{lab}} B_0 \tag{3.10}$$

where the z is the laboratory frame z-axis. Note that only the zz-component of the shielding tensor is required, i.e. that which governs the shielding field in the direction of the applied field $\mathbf{B}_0$, and the quantization axis for the spins. We will henceforth drop the bar and (0) superscript notation on this average hamiltonian for ease and simply refer to Equation (3.10) as 'the chemical shielding hamiltonian'. However, it is important to remember that it is the chemical shielding contribution to the first-order average rotating frame hamiltonian.

The shielding hamiltonian of Equation (3.10) commutes with the Zeeman hamiltonian. The shielding hamiltonian is therefore described as an *inhomogeneous* interaction (see Section 1.4). The question of how to generate the laboratory frame shielding tensor, σ^{lab}, from that expressed in its principal axis frame, σ^{PAF}, is discussed in the next section.

3.1.3 *Experimental manifestations of the shielding tensor*

The shielding anisotropy has its most important consequences for powder samples, although it does of course influence the frequencies of the resonances observed for single crystal samples too.

The effect of the shielding hamiltonian of Equation (3.10) on the (rotating frame) Zeeman spin levels is found by applying $\hat{H}_{cs}$ to the spin levels described by $|I, m\rangle$, in order to find the first-order energy shift on these levels. The first-order contribution to the energy of the spin levels from chemical shielding, E_{cs}, is given by

$$\begin{aligned} E_{cs} &= \gamma\hbar\sigma_{zz}^{lab}B_0\langle I, m|\hat{I}_z|I, m\rangle \\ &= \gamma\hbar\sigma_{zz}^{lab}B_0 m \end{aligned} \tag{3.11}$$

The contribution to the frequency of the observed NMR signal from the chemical shielding, ω_{cs}, is easily found from Equation (3.11) to be

$$\begin{aligned} \omega_{cs} &= \gamma\sigma_{zz}^{lab}B_0 \\ &= -\omega_0\sigma_{zz}^{lab} \end{aligned} \tag{3.12}$$

To examine ω_{cs} further, we need to rewrite the σ_{zz}^{lab} term in terms of the principal values of the shielding tensor and the orientation of its principal axis frame in the laboratory frame. We find σ_{zz}^{lab} from

$$\sigma_{zz}^{lab} = (0 \quad 0 \quad 1)\boldsymbol{\sigma}^{lab}\begin{pmatrix} 0 \\ 0 \\ 1 \end{pmatrix} \tag{3.13}$$

where the unit vector along z (laboratory frame) (0 0 1) ensures that it is the zz component of $\boldsymbol{\sigma}^{lab}$ which is projected out in Equation (3.13).[N3] If the shielding tensor is expressed in some other axis frame f, the equivalent expression is

$$\sigma_{zz}^{\mathrm{lab}} = \mathbf{b}_0^f \sigma^f \mathbf{b}_0^f \tag{3.14}$$

where $\mathbf{b}_0^f$ is the unit vector in the direction of $\mathbf{B}_0$ in the frame f and σ^f is the shielding tensor in the same frame. Thus, combining Equations (3.12) and (3.14), we have for the chemical contribution to the observed signal frequency

$$\omega_{\mathrm{cs}} = -\omega_0 \mathbf{b}_0^f \sigma^f \mathbf{b}_0^f \tag{3.15}$$

If the direction of $\mathbf{B}_0$ in the shielding tensor principal axis frame, $\mathbf{b}_0^{\mathrm{PAF}}$, is described by the polar angles (θ, ϕ) then (see Fig. 3.1 for definitions of these angles)

$$\mathbf{b}_0^{\mathrm{PAF}} = (\sin\theta\cos\phi, \sin\theta\sin\phi, \cos\theta) \tag{3.16}$$

Using this equation in (3.15) for ω_{cs} we find

$$\omega_{\mathrm{cs}}(\theta, \phi) = \omega_0 - (\sigma_{xx}^{\mathrm{PAF}} \sin^2\theta \cos^2\phi + \sigma_{yy}^{\mathrm{PAF}} \sin^2\theta \sin^2\phi + \sigma_{zz}^{\mathrm{PAF}} \cos^2\theta) \tag{3.17}$$

For a shielding tensor with axial symmetry, $\sigma_{xx}^{\mathrm{PAF}} = \sigma_{yy}^{\mathrm{PAF}}$, and so equation (3.17) simplifies to

$$\omega_{\mathrm{cs}}(\theta) = -\omega_0 \sigma_{zz}^{\mathrm{PAF}} \frac{1}{2}(3\cos^2\theta - 1) \tag{3.18}$$

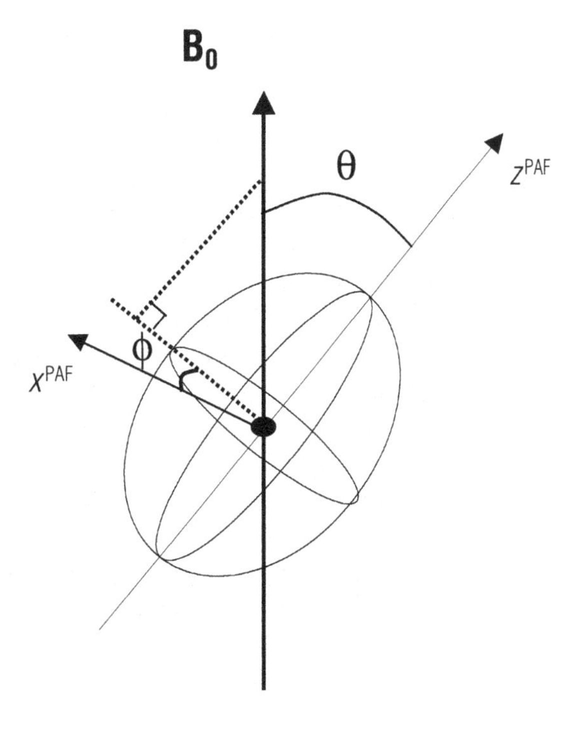

Fig. 3.1 Definitions of the angles θ and ϕ, which are the polar angles defining the orientation of the $\mathbf{B}_0$ field in the $(x^{\mathrm{PAF}}, y^{\mathrm{PAF}}, z^{\mathrm{PAF}})$ axis frame, the principal axis frame of the shielding tensor.

Often we define the shielding tensor relative to the isotropic shielding. The isotropic shielding, σ_{iso}, is simply the average of the diagonal elements of the shielding tensor (in any frame since the trace of a tensor is invariant to rotation of the axis frame defining the tensor). Thus

$$\sigma_{iso} = \frac{1}{3}(\sigma^f_{xx} + \sigma^f_{yy} + \sigma^f_{zz}) \tag{3.19}$$

The *anisotropic* part of the shielding tensor when the shielding tensor is expressed in frame f is then

$$\sigma^f_{aniso} = \sigma^f - \mathbf{1} \cdot \sigma_{iso} \tag{3.20}$$

and the anisotropic part of the chemical shift frequency for an axially symmetric shielding tensor is

$$\begin{aligned} \omega^{aniso}_{cs}(\theta) &= -\omega_0(\sigma^{PAF}_{zz} - \sigma_{iso})\frac{1}{2}(3\cos^2\theta - 1) \\ &= -\omega_0\Delta\frac{1}{2}(3\cos^2\theta - 1) \end{aligned} \tag{3.21}$$

in terms of the chemical shift anisotropy Δ. For the non-axially symmetric case, the chemical shift frequency can be expressed in terms of the isotropic component (ω_{iso}), shielding anisotropy (Δ) and asymmetry (η) (see Equation (3.4) for their definition):

$$\omega_{cs}(\theta, \phi) = -\omega_0\sigma_{iso} - \frac{1}{2}\omega_0\Delta\{3\cos^2\theta - 1 + \eta\sin^2\theta\cos 2\phi\} \tag{3.22}$$

where the quantity $-\omega_0\sigma_{iso} = \omega_{iso}$ is the isotropic chemical shift frequency, relative to the Larmor frequency ω_0.

So what are the implications of Equations (3.21) and (3.22) for the chemical shift frequency? In a *powder* sample, all molecular orientations are present. Remembering that the shielding principal axis frame is fixed in the molecule, this means that in a powder sample, all values of the angle θ (and ϕ in non-axial symmetry) are possible. Each different molecular orientation implies a different orientation of principal axis frame with respect to the applied field, $\mathbf{B}_0$, and so has a different chemical shift associated with it, from Equation (3.21). The spectrum will therefore take the form of a *powder pattern* with lines from the different molecular orientations covering a range of frequencies. The lines from the different orientations overlap and form a continuous lineshape. As can be seen from Equation (3.21), some molecular orientations yield the same chemical shift, and so there may be a number of lines at the same chemical shift. The resultant intensity at any given frequency in the powder pattern is therefore proportional to the number of molecular orientations which have that particular chemical shift.

This means that the powder pattern lineshape has a very distinctive shape, which depends on the symmetry of the shielding tensor, which, in turn, depends on the site symmetry at the nucleus, as we have already seen (Fig. 3.2).

The discontinuities in the lineshapes give the principal values of the shielding tensor, which may be thus read directly off the spectrum. The directions of the principal axes have no effect on a powder spectrum, providing there is true random distribution of molecular orientations in the sample. The isotropic frequency ω_{iso} is at the centre of 'mass' of the powder pattern, i.e. one third of the way between $\omega_{\perp}$ and $\omega_{\parallel}$ in this case.

In the axial symmetry case in Fig. 3.2, the frequencies of the discontinuities are given by (again, relative to the Larmor frequency)

$$\begin{aligned} \nu_{\perp} &= -\omega_0(\sigma_{xx}^{PAF} - \sigma_{iso})/2\pi = -\omega_0(\sigma_{yy}^{PAF} - \sigma_{iso})/2\pi \\ \nu_{\parallel} &= -\omega_0(\sigma_{zz}^{PAF} - \sigma_{iso})/2\pi \end{aligned} \tag{3.23}$$

The lineshape in this case has a much larger intensity at the frequency $\omega_{\perp}$ than at $\omega_{\parallel}$, reflecting the larger number of molecular orientations which contribute to the lineshape at this frequency; all molecules orientated so that the x^{PAF}–y^{PAF} plane is parallel to $\mathbf{B}_0$ contribute to the lineshape at $\omega_{\perp}$, while those orientated so that z^{PAF} is parallel to $\mathbf{B}_0$ contribute to the lineshape at $\omega_{\parallel}$. There are an infinite number of molecular orientations with the x^{PAF}–y^{PAF} plane parallel to $\mathbf{B}_0$, but only one with z^{PAF} parallel to $\mathbf{B}_0$.

For the less than axial symmetry case, ω_{11}, ω_{22} and ω_{33} correspond to principal values of the shielding tensor in a similar manner to $\omega_{\perp}$ and $\omega_{\parallel}$ in the axial symmetry case, i.e.

$$\omega_{ii} = -\omega_0(\sigma_{\alpha\alpha}^{PAF} - \sigma_{iso}) \tag{3.24}$$

but we cannot tell from the powder pattern which axis of the principal axis frame each value corresponds to.

3.1.4 Definition of the chemical shift

The total spectral frequency in absolute units is the Larmor frequency plus the chemical shift contribution. When performing NMR experiments, absolute frequencies are not measured; a reference substance is used and frequencies of lines are measured relative to a specific line in the spectrum of the reference substance and quoted as *chemical shifts with respect to that substance* or *offset frequencies*. The chemical shift δ_{iso} is defined in Equation (3.25).

$$\delta_{iso} = \frac{\nu - \nu_{ref}}{\nu_{ref}} = \frac{(\sigma_{zz}^{lab}(\text{ref}) - \sigma_{zz}^{lab})}{1 - \sigma_{zz}^{lab}(\text{ref})} \tag{3.25}$$

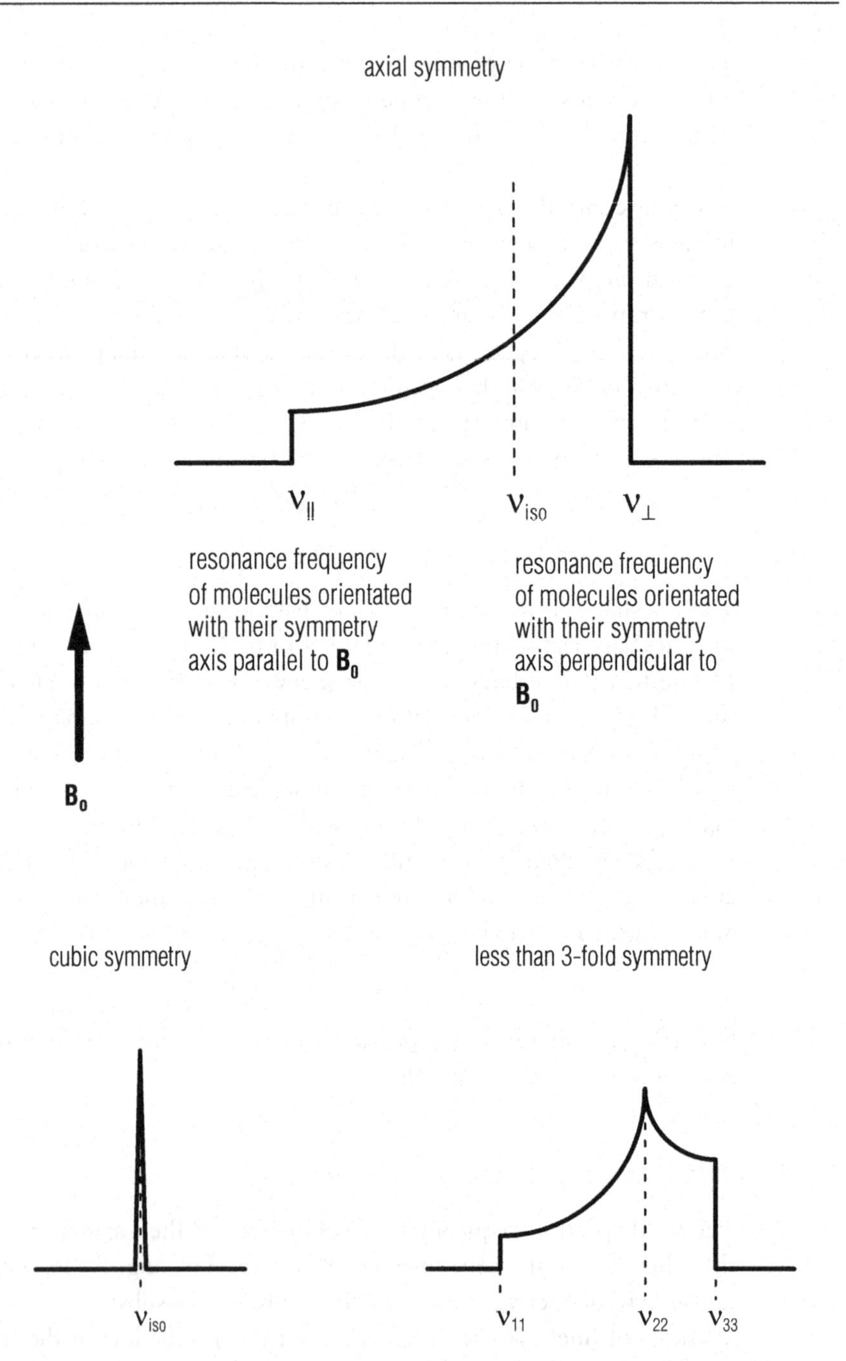

Fig. 3.2 Chemical shift anisotropy powder patterns. Powder patterns arise from samples where there are many crystallites randomly orientated, so that all possible molecular orientations are present with random distribution. Powder patterns arise because each different molecular orientation with respect to $\mathbf{B}_0$ has its own spectral frequency. Each different orientation thus gives rise to its own (sharp) spectral line; the lines from different orientations overlap continuously, giving rise to the observed powder pattern. These powder lineshapes are thus *inhomogeneous*; that is, they may be considered as made up of many independent contributions from different parts of the sample. For the axial case, the nucleus is considered to be at a site of axial symmetry; the unique principal axis of the shielding tensor for that nucleus then lies along the symmetry axis.

where ν is the spectral frequency (in Hz) of the signal for the spin of interest (in its given chemical site) and ν_{ref} is the resonance frequency of the same spin in some reference compound. In the case where $1 \gg \sigma_{zz}^{lab}(\text{ref})$, this reduces to

$$\delta \approx \sigma_{zz}^{lab}(\text{ref}) - \sigma_{zz}^{lab} \tag{3.26}$$

The corresponding *chemical shift tensor* has elements

$$\delta_{\alpha\beta} = \frac{(\sigma_{\alpha\beta}(\text{ref}) - \sigma_{\alpha\beta})}{1 - \sigma_{\alpha\beta}} \tag{3.27}$$

where $\sigma_{\alpha\beta}$ is the $\alpha\beta$ element of the shielding tensor for the spin of interest in some axis frame and $\delta_{\alpha\beta}$ is the corresponding element of the chemical shift tensor for that spin in the same frame. The chemical shift anisotropy and asymmetry are defined in an analogous manner to the shielding anisotropy and asymmetry of Equation (3.4). Often, however, the principal values of the chemical shift tensor are known, but not its principal axis frame. The principal values are then labelled by convention as δ_{11}^{PAF}, δ_{22}^{PAF}, δ_{33}^{PAF}, where $\delta_{11}^{PAF} \geq \delta_{22}^{PAF} \geq \delta_{33}^{PAF}$. The *chemical shift* anisotropy (Δ_{cs}) and asymmetry (η_{cs}) are defined from these as

$$\begin{aligned} \Delta_{cs} &= \delta_{11}^{PAF} - \delta_{iso} \\ \eta_{cs} &= \frac{\delta_{33}^{PAF} - \delta_{22}^{PAF}}{\Delta_{cs}} \end{aligned} \tag{3.28}$$

Observed chemical shifts are related to the chemical shift tensor through

$$\delta = \delta_{iso} + \frac{1}{2}\Delta_{cs}\{3\cos^2\theta - 1 + \eta_{cs}\sin^2\theta\cos 2\phi\}$$

where the isotropic chemical shift, δ_{iso}, is

$$\begin{aligned} \delta_{iso} &= \frac{\sigma_{iso}(\text{ref}) - \sigma_{iso}}{1 - \sigma_{iso}(\text{ref})} \\ &\approx \sigma_{iso}(\text{ref}) - \sigma_{iso} \\ &= \frac{1}{3}(\delta_{11}^{PAF} + \delta_{22}^{PAF} + \delta_{33}^{PAF}) \end{aligned} \tag{3.29}$$

3.2 The relationship between the shielding tensor and electronic structure

As we have already seen, the net shielding tensor, $\boldsymbol{\sigma}$, is given by the sum of the diamagnetic and paramagnetic terms [1]:

$$\boldsymbol{\sigma} = \boldsymbol{\sigma}^{d} + \boldsymbol{\sigma}^{p} \tag{3.30}$$

where components of the constituent tensors are given by:

$$\sigma^{\mathrm{d}}_{\gamma\gamma} = \frac{\mu_0}{4\pi}\frac{e^2}{2m_e}\left\langle \Psi_0 \left| \sum_i \frac{\alpha_i^2 + \beta_i^2}{r_i^3} \right| \Psi_0 \right\rangle \tag{3.31}$$

and

$$\sigma^{\mathrm{p}}_{\alpha\beta} = -\frac{\mu_0}{4\pi}\frac{e^2}{2m_e}\sum_{k\neq 0}\left[\frac{\left\langle \Psi_0 \left| \sum_i \hat{l}_{i\alpha} \right| \Psi_k \right\rangle \left\langle \Psi_k \left| \sum_i \frac{\hat{l}_{i\beta}}{r_i^3} \right| \Psi_0 \right\rangle + \text{c.c}}{E_k - E_0}\right] \tag{3.32}$$

when the gauge origin (i.e. the origin from which the electron distances r_i are measured) is at the nucleus in question. α, β, γ denote Cartesian axes. The Ψ are electronic wavefunctions; the suffix i denotes the ith electron. The ground electronic state is denoted 0 and an excited electronic state, by k. r_i is the distance of the ith electron from the nucleus and α_i, β_i its coordinates along the α and β directions respectively. $\hat{l}_{i\alpha}$, $\hat{l}_{i\beta}$ are components of the electronic orbital angular momentum operator, $\hat{\mathbf{l}}_i$ for the ith electron; 'c.c' denotes the complex conjugate of the first term in the summation of Equation (3.32).

The diamagnetic term (Equation (3.31)) depends only on the ground electronic state. Furthermore, its dependence on r_i^{-3} means that it is very much a *local* contribution, arising largely from the core electrons surrounding the nucleus.

The paramagnetic term (Equation (3.32)) provides more of a commentary on bonding [1]. It depends on integrals involving components of the electronic orbital angular momentum operator, $\hat{l}_\alpha$, the ground electronic state and various excited electronic states. This equation can be written in a more tractable form with the following considerations. The electronic wavefunction Ψ_0 is a product wavefunction involving all the occupied molecular orbitals in the ground electronic state and the excited states Ψ_k are similarly product wavefunctions, but involving electrons in molecular orbitals not occupied in the ground state. The integrals in Equation (3.32) over the excited states Ψ_k are only non-zero for Ψ_k with *one* of the electrons from a ground state occupied molecular orbital excited to an unoccupied orbital. Thus integrals between Ψ_0 and Ψ_k have the effect of mixing a ground state occupied molecular orbital with one that is unoccupied in the ground state and, indeed, we may rewrite Equation (3.32) so that only the two molecular orbitals[N4] which differ between the ground and excited state configurations appear in the integrals. The equation for $\sigma^{\mathrm{p}}_{\alpha\beta}$ then contains terms of the form

$$\frac{\left\langle \psi_{\text{occc}} \left| \sum_i \hat{l}_{i\alpha} \right| \psi_{\text{unocc}} \right\rangle \left\langle \psi_{\text{unocc}} \left| \sum_i \frac{\hat{l}_{i\beta}}{r_i^3} \right| \psi_{\text{occ}} \right\rangle + \text{c.c}}{\varepsilon_{\text{unocc}} - \varepsilon_{\text{occ}}} \tag{3.33}$$

where ψ_{occ} is the molecular orbital which is occupied in the ground electronic state, Ψ_0, but empty in the excited electronic state, Ψ_k; ψ_{unocc} is the molecular orbital which is unoccupied in the ground electronic state, but occupied in the excited electronic state Ψ_k.

The $(E_k - E_0)$ energy factor has been replaced by $(\varepsilon_{unocc} - \varepsilon_{occ})$, the energy difference between the occupied and unoccupied molecular orbitals of (3.33) molecular orbitals, which is very approximately equal to $(E_k - E_0)$. This energy denominator means that the only important integrals in the sum of Equation (3.32) are those which involve high-lying occupied molecular orbitals and low-lying unoccupied molecular orbitals. The problem can be further broken down by approximating the molecular orbitals as linear combinations of atomic orbitals. The r_i^{-3} factor in Equation (3.32) means that only those electrons near the nucleus are important in determining σ^p. Therefore the only parts of the molecular orbitals which make a significant contribution to the integrals involving that factor are those which have significant contributions from atomic orbitals which are centred on the nucleus in question. Symmetry arguments allow further insight; the integrals in the expression of (3.33) are only non-zero if their integrands transform as the totally symmetric representation of the molecular point group. The orbital angular momentum operators $\hat{l}_\alpha$ transform as the rotation operators for rotation about axis α, so the requirement for non-zero integrals can be written as

$$\Gamma_{\text{occ}} \otimes \Gamma_{\hat{R}_\alpha} \otimes \Gamma_{\text{unocc}} \subset \Gamma_0 \tag{3.34}$$

where Γ_{occ} is the point group representation of the occupied molecular orbital of (3.33), Γ_{unocc} is that of the unoccupied molecular orbital and $\Gamma_{\hat{R}_\alpha}$ is the symmetry of the $\hat{R}_\alpha$ rotation operator for rotation about axis α. An equivalent expression to (3.34) is

$$\Gamma_{\text{occ}} \otimes \Gamma_{\hat{R}_\alpha \psi_{\text{unocc}}} \subset \Gamma_0 \tag{3.35}$$

where $\Gamma_{\hat{R}_\alpha \psi_{unocc}}$ is the point group representation of $\hat{R}_\alpha \psi_{unocc}$, i.e. the ψ_{unocc} molecular orbital rotated about axis α. In other words, for the integrals in (3.33) to be non-zero, the occupied molecular orbital must have the same symmetry as (and significant overlap with) the unoccupied molecular orbital, *after it is rotated about axis α.*

The formulation in Equation (3.33), coupled with simple qualitative molecular orbital pictures describing the local bonding around a nucleus, allows

Table 3.1 The principal values of the ^{13}C chemical shielding tensor and isotropic chemical shift (δ_i) relative to TMS for the C_2H_4 molecule and the C_2H_4 ligand in Zeise's salt. Note that in Zeise's salt, the two ethene carbons are inequivalent.

	σ_{11} (ppm)	σ_{22} (ppm)	σ_{33} (ppm)	δ_i (ppm)	Ref.
Free C_2H_4	-234	-120	-24	123	[2]
C_2H_4 in Zeise's salt	-135	-67	10	63	[3]
	−133	−64	12	61	[3]

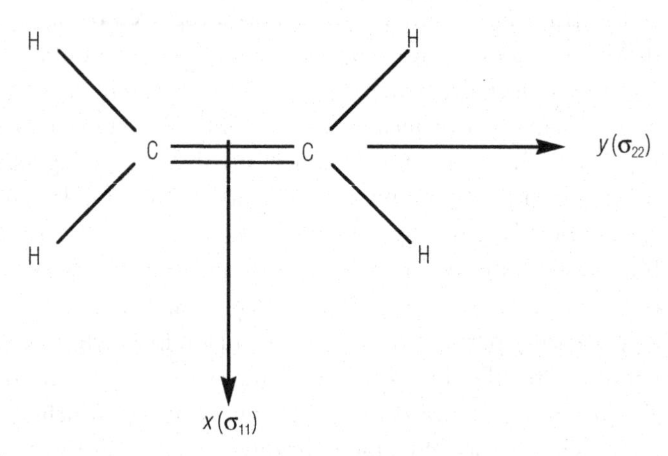

Fig. 3.3 The principal directions of the ^{13}C shielding tensor in ethene [2]. The σ_{33} (z) component is perpendicular to the plane of the molecule.

a qualitative understanding of the shielding tensor in many compounds. From this basis, correlations between the bonding in different compounds can be made on the basis of simple chemical shift measurements. As an example consider the ^{13}C shielding tensor for ethene. Table 3.1 shows the principal shielding values determined by experiment [2]. The principal axes associated with these values are parallel to the molecular two-fold axes, as shown in Fig. 3.3.

The shielding component, σ_{11}, is in the molecular plane; σ_{22} is parallel to the C=C bond, and σ_{33} is perpendicular to the molecular plane.

The contribution to the paramagnetic component of the shielding tensor, $\sigma^{p}_{\alpha\alpha}$ depends upon matrix elements of the form in expression (3.33). As Fig. 3.4(a) illustrates, rotating the C–C σ^* antibonding molecular orbital of the C_2H_4 molecule through 90° about the axis associated with the $\sigma_{11}(x)$ component transforms it into a new orbital which has a non-zero overlap with the C–C π bonding molecular orbital. Similarly, rotating C–C π^* about x gives an orbital that has a non-zero overlap with the C–C σ molecular orbital. These two interactions are the dominant contributions to the paramagnetic part of σ_{11}.

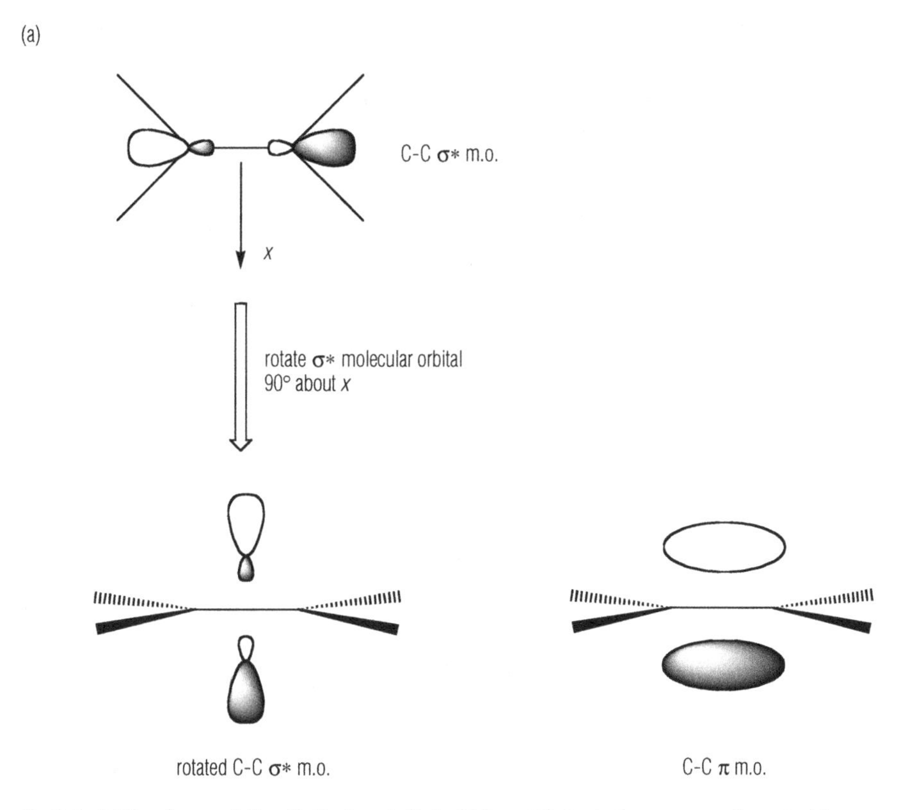

Fig. 3.4 (a) The charge rotation (C–C σ^* → C–C π) which contributes to the paramagnetic part of the σ_{11} component of the ^{13}C shielding tensor in ethene. The C–C π molecular orbital is also shown; there is clearly a non-zero overlap between this orbital and the rotated orbital.

Rotating the C–H σ^* antibonding molecular orbitals about the axis associated with the σ_{22} component (y) transforms it into an orbital which has non-zero overlap with the C–C π bonding molecular orbital (Fig. 3.4(b)). Rotating C–C π^* gives an orbital with a non-zero overlap with the C–H σ molecular orbital. These are the dominant contributions to the paramagnetic part of σ_{22}.

A similar analysis shows that only C–C σ^* → C–H σ and C–H σ^* → C–C σ molecular orbital interactions are important for the σ_{33} component of the shielding tensor.

To estimate the relative sizes of the shielding tensor principal values, we need to consider the relative sizes of the energy factors $(\varepsilon_{\text{unocc}} - \varepsilon_{\text{occ}})^{-1}$ involved for each shielding tensor component, and the relative sizes of the matrix elements in expression (3.33). The energy factors for the dominant terms in the σ_{11} component are $(\varepsilon_{\text{C-C}\sigma^*} - \varepsilon_{\pi})^{-1}$ and $(\varepsilon_{\pi^*} - \varepsilon_{\text{C-C}\sigma})^{-1}$; for the σ_{22} component they are $(\varepsilon_{\text{C-H}\sigma^*} - \varepsilon_{\pi})^{-1}$ and $(\varepsilon_{\pi^*} - \varepsilon_{\text{C-H}\sigma})^{-1}$ and for σ_{33} they are $(\varepsilon_{\text{C-C}\sigma^*} - \varepsilon_{\text{C-H}\sigma})^{-1}$ and $(\varepsilon_{\text{C-H}\sigma^*} - \varepsilon_{\text{C-C}\sigma})^{-1}$. The energy differences between the σ

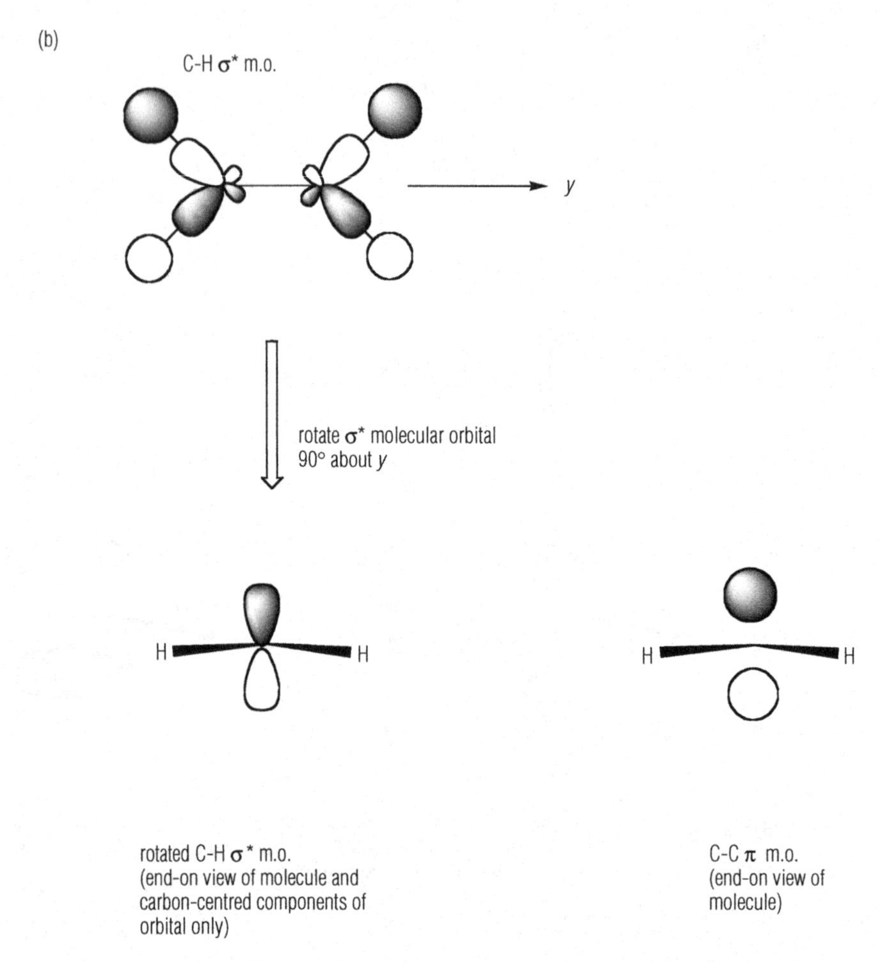

Fig. 3.4 *Continued* (b) The charge rotation (C–H $\sigma^* \rightarrow$ C–C π), which contributes to the paramagnetic part of the σ_{22} component of the ^{13}C shielding tensor in ethene. In the rotated orbital, the ethene molecule is shown end-on (viewed along the C–C bond) and only the parts of the rotated orbital close to the ^{13}C nuclei are shown. Again, the C–C π molecular orbital is shown. Once again, the rotated orbital has a non-zero overlap with this latter orbital.

and σ^* molecular orbitals involved in σ_{33} are likely to be significantly larger than the π/σ^* or σ/π^* energy differences involved in σ_{11} and σ_{22}. This would render the paramagnetic part of the σ_{33} component much smaller than the paramagnetic parts of the other shielding tensor components. The net shielding is of course a sum of paramagnetic and diamagnetic parts, which contribute with opposite signs; the larger the paramagnetic part, the more negative the overall shielding. Thus a small paramagnetic contribution to σ_{33} for the ^{13}C in ethene results in the small negative value for the σ_{33} value observed experimentally.

In comparing the values of σ_{11} and σ_{22}, we note that the paramagnetic part of σ_{11} depends on the mixing of the C–C σ^* molecular orbital after rotation about x and the C–C π molecular orbital and the mixing of the rotated C–C π^* and C–C σ molecular orbitals. The paramagnetic part of σ_{22} depends on the mixing of the C–H σ^* molecular orbital after rotation about y and the C–C π molecular orbital and of the rotated C–C π^* and C–H σ molecular orbitals. The mixing of the C–C σ^* and C–C π orbitals involved in σ_{11} is between molecular orbitals spanning the same two atoms and so is likely to be larger than the mixing of the C–H σ^* and C–C π molecular orbitals involved in σ_{22}, which involves orbitals in relatively different regions of space. Hence, the largest paramagnetic contribution is likely to be in σ_{11}, which explains the very large negative value observed for this component experimentally.

Also shown in Table 3.1 are the ^{13}C shielding tensor principal values for ethene in Zeise's salt, $\{K[PtCl_3(\eta^2\text{-}C_2H_4)]\}$ [3]; the principal axes of the shielding tensor are assumed to be similar to those for the isolated ethene molecule. A comparison of the principal values for the free ethene molecule and the ethene ligand shows that it is the σ_{11} and σ_{22} components which are most changed by the complexation to the Pt metal centre. From the above discussion, it is clear that these are the components which have significant contributions from the C–C π and π^* molecular orbitals of the ethene ligand. In turn, it is these orbitals which are chiefly involved in bonding to the Pt centre, so it is not surprising that the shielding components they contribute to are the most altered by the complexation.

The dependence of the shielding tensor on electronic structure inevitably means that it depends on the geometry of bonds around the nucleus in question. This fact can be used to determine molecular geometry, either through experimental correlations between chemical shift parameters and known bond angles or through *ab-initio* calculations of the shielding tensor for different molecular geometries and comparing these with experimentally-derived shielding tensor principal values. This is already proving to be useful in structure determination in small organic molecules and proteins [4–14].

3.3 Measuring chemical shift anisotropies

Chemical shift anisotropies can be measured directly from simple one-dimensional NMR spectra of powder samples by observing the powder patterns in the spectra, providing that (1) only shielding effects contribute to the powder pattern, and not dipole–dipole or quadrupole coupling, and (2) powder patterns from inequivalent nuclear sites are resolved. When either of these two conditions is not met, it is still possible to extract the anisotropic shielding parameters from the one-dimensional spectrum by simulating the spectrum for all nuclear sites involved and allowing contributions from all nuclear spin interactions. However, in such cases, it is highly likely that the experimental spectrum can be reproduced by more than one set of interaction parameters leading to unacceptable ambiguities. It is then necessary to resort to two-dimensional techniques to separate the signals from different sites. The following techniques all perform such separations. They are often combined with decoupling techniques to remove the effects of heteronuclear dipole coupling. For low abundance nuclei, it is simple to use cross-polarization from a more abundant spin in the sample to generate initial transverse magnetization in the experiments. These techniques have to date only been performed on spin-$\frac{1}{2}$ nuclei; for spins with I greater than $\frac{1}{2}$, quadrupole-coupling effects tend to dominate any spectrum and render the observation of shielding effects difficult.

We deal mainly in these following sections with the principles of the techniques; when recording two-dimensional NMR spectra, it is important to arrange the experiment in such a way as to end up with pure absorption lineshapes in the final two-dimensional frequency spectrum. This was dealt with in Section 1.5.3, and is not explored further here.

3.3.1 Magic-angle spinning with recoupling pulse sequences

There are a whole family of techniques which use relatively fast magic-angle spinning to remove the effects of chemical shift anisotropy and which apply an rf pulse sequence during the t_1 interval of the two-dimensional experiment (Fig. 3.5(a)) to prevent the averaging of the chemical shift anisotropy by the magic-angle spinning during this period. To use current parlance, we refer to techniques which have the effect of *recoupling* the chemical shift anisotropy which would otherwise be averaged to zero by the magic-angle spinning.

Perhaps one of the best of these techniques is that due to Tycko *et al.* [15], which uses $(2n + 2)$ 180° pulses per rotor period during t_1, where n is $1, 2, 3, \ldots$. These pulses are applied at specific times during the rotor period, the precise positions being determined by the analysis outlined below; t_1 is restricted to being an integral number of rotor periods and is generally incremented in steps of the rotor period. The initial transverse magnetization in the experiment is generated by an initial 90° pulse or cross-polarization from an abundant nucleus (Fig. 3.5(b)). In the rotating frame, this magnetization then precesses under the influence of the anisotropic chemical shift. The effect of the 180° pulses applied during the subsequent t_1 interval is to reverse the sense of the precession of the magnetization components. Thus we can write the precession frequency of the magnetization during t_1 as

$$\omega(\alpha, \beta, \gamma; t) = \omega_{cs}(\alpha, \beta, \gamma; t)f(t) \qquad (3.36)$$

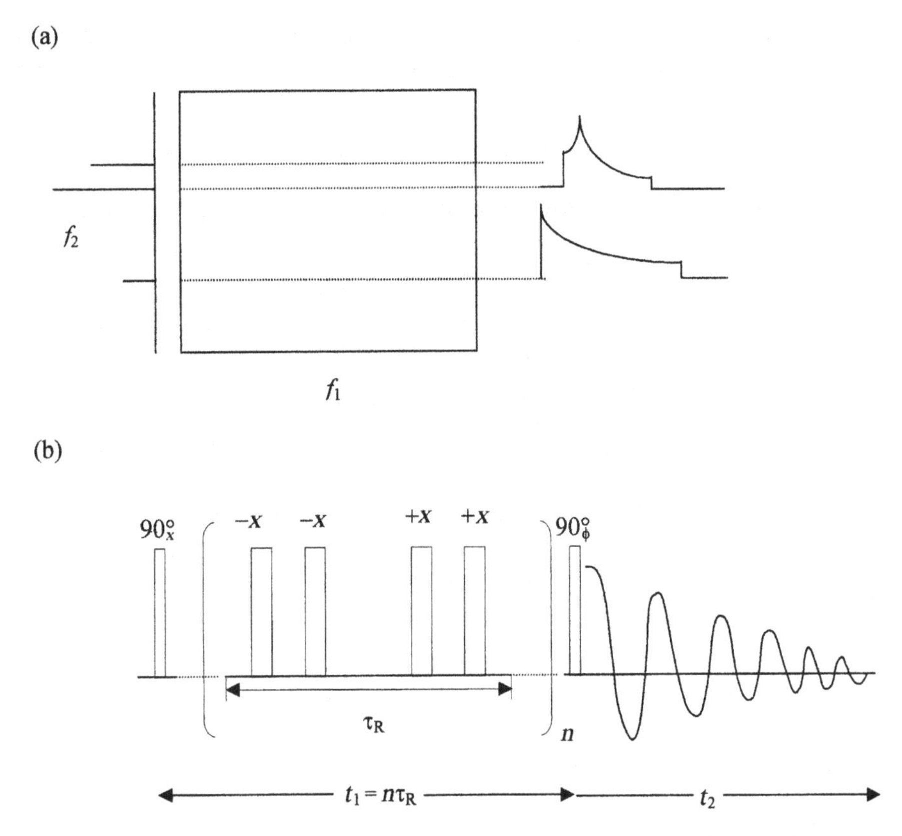

Fig. 3.5 Isotropic/anisotropic chemical shift separation experiment. (a) The form of the two-dimensional spectrum resulting from the pulse sequence due to Tycko *et al.* [15] shown in (b). The experiment is conducted under magic-angle spinning to produce an isotropic spectrum in f_2. An initial 90° pulse produces transverse magnetization in the usual manner. The series of four 180° pulses per rotor period during t_1 has the effect of preventing the averaging of the chemical shift anisotropy by magic-angle spinning during this period, so that powder patterns appear in f_1 of the two-dimensional experiment. To obtain the two-dimensional spectrum, two datasets are recorded; one in which the last 90° pulse in the sequence has the same phase ϕ as the first 90° pulse and a second in which it has its phase shifted by 90°. In this way, the x and y components of the magnetization in t_1 are measured separately so that (with appropriate processing) a pure absorption two-dimensional spectrum can be obtained (see Section 1.5.3).

where $\omega_{cs}(\alpha, \beta, \gamma; t)$ is the precession frequency of the magnetization component arising from a molecular/crystallite orientation defined by the Euler angles (α, β, γ) relative to a rotor-fixed frame of reference; the time dependence of this frequency is induced by the magic-angle spinning. The function $f(t)$ is simply a step function, stepping between the values of +1 and −1 as the 180° pulses are applied. The function $\omega_{cs}(\alpha, \beta, \gamma; t)$ was derived in Section 2.2.1 (Equation 2.7):

$$\omega_{cs}(\alpha, \beta, \gamma; t) = -\omega_0\{\sigma_{iso} + [A_1\cos(\omega_R t + \gamma) + B_1\sin(\omega_R t + \gamma)] + [A_2\cos(2\omega_R t + \gamma) + B_2\sin(2\omega_R t + \gamma)]\} \quad (3.37)$$

where A_1, etc., are functions of (α, β, γ) as defined in Equation (2.8). The precession frequency in a static experiment (no magic-angle spinning, so $\omega_R = 0$ in Equation (3.37)) by way of comparison is

$$\omega_{cs}(\alpha, \beta) = -\omega_0\{\sigma_{iso} + A_1 + A_2\} \tag{3.38}$$

Our aim is to produce a form for the average resonance frequency over one rotor period in the spinning experiment, $\tilde{\omega}(\alpha, \beta, \gamma; t)$, which is proportional to that for the static experiment; if we can do this, the signal in t_1 of the experiment can be Fourier transformed to give a powder pattern which is simply a frequency-scaled version of the static powder pattern. The average resonance frequency over one rotor period is simply

$$\tilde{\omega}(\alpha, \beta, \gamma; t) = \int_0^{\tau_R} \omega(\alpha, \beta, \gamma; t)\mathrm{d}t = \int_0^{\tau_R} \omega_{cs}(\alpha, \beta, \gamma; t) f(t)\mathrm{d}t \tag{3.39}$$

Comparing Equations (3.37) (the precession frequency under magic-angle) and (3.38) (the precession frequency in a static experiment), we see that in Equation (3.39), we need the anisotropic terms A_1 and A_2 to be retained in equal amounts at the end of one rotor period, and the B_1 and B_2 terms to be averaged to zero over each rotor period (as they would normally be in a magic-angle spinning experiment with no rf pulses applied). This is the case if the 180° pulses which define $f(t)$ are arranged such that

$$\int_0^{\tau_R} f(t)\cos(\omega_R t)\mathrm{d}t = \int_0^{\tau_R} f(t)\cos(2\omega_R t)\mathrm{d}t \tag{3.40}$$

If this condition is met, Equation (3.39) for the average precession frequency over the rotor period is just

$$\begin{aligned}
\tilde{\omega}(\alpha, \beta, \gamma; t) &= \int_0^{\tau_R} \omega_{cs}(\alpha, \beta, \gamma; t) f(t)\mathrm{d}t \\
&= -\omega_0\Big\{\sigma_{iso}\int_0^{\tau_R} f(t)\mathrm{d}t + A_1\int_0^{\tau_R} f(t)\cos(\omega_R t + \gamma)\mathrm{d}t \\
&\quad + A_2\int_0^{\tau_R} f(t)\cos(2\omega_R t + \gamma)\mathrm{d}t + \int_0^{\tau_R} f(t) B_1 \sin(\omega_R t + \gamma)\mathrm{d}t \\
&\quad + \int_0^{\tau_R} f(t) B_2 \sin(2\omega_R t + \gamma)\mathrm{d}t\Big\}
\end{aligned} \tag{3.41}$$

The last two integrals in Equation (3.41) are zero providing the step function $f(t)$ is even, so the average precession frequency reduces to

$$\tilde{\omega}(\alpha, \beta, \gamma; t) = -\omega_0\{\sigma_{iso}\delta + \chi(A_1 + A_2)\} \tag{3.42}$$

where the factors δ and χ are given by

$$\begin{aligned}
\delta &= \frac{1}{\tau_R}\int_0^{\tau_R} f(t)\mathrm{d}t \\
\chi &= \frac{1}{\tau_R}\int_0^{\tau_R} f(t)\cos(\omega_R t)\mathrm{d}t
\end{aligned} \tag{3.43}$$

Equation (3.42) for the average precession frequency over a rotor period now has the form of the static precession frequency, but with the isotropic chemical shift scaled by δ and the chemical shift anisotropy by χ. All that remains is to choose step functions $f(t)$ to satisfy Equation (3.40). This is not detailed here, but may be found in reference [15]. Clearly, the scaling factors δ and χ depend on the step function $f(t)$. The spectral width in the f_1 dimension of the two-dimensional spectrum arising from this experiment is necessarily equal to the spinning speed, as the dwell time in the corresponding t_1 dimension is the rotor period. However, this does not provide too great a restriction on the experiment, as one of the strengths of this method is that the chemical shift anisotropy powder patterns (and isotropic chemical shifts) can be scaled to suit the chosen spectral width by selecting the appropriate 180° pulse timings in t_1.

Figure 3.6 shows the results from performing this experiment on ^{13}C in methyl-α-D-glucopyranoside [15].

One potential problem with this experiment is that it recouples any interaction which is described by a hamiltonian which is proportional to $\hat{I}_z$, and thus it also recouples heteronuclear dipolar couplings. Thus for ^{13}C in an organic solid, for instance, this experiment will recouple ^{13}C–^{1}H dipolar coupling. This can of course be removed by high-power decoupling methods (see Section 2.3), but in order to overcome the recoupling effect of this experiment, the rf amplitude applied to the ^{1}H nuclei needs to be around three times greater than that used for the ^{13}C 180° pulses, and this can be excessive. A recent experiment called SUPER [16] is a variation on Tycko's experiment which has much smaller power requirements in the decoupling.

3.3.2 *Variable-angle spinning experiments*

These are experiments in which the sample is spun, but at several different angles, rather than just at the magic angle. To understand these experiments, we first need to consider how transition frequency varies as we spin at different angles. As we know, under magic-angle spinning, spectral frequencies become time-dependent according to Equation (2.4) derived in Section 2.2.1. So for a spin suffering anisotropic chemical shielding described by a shielding tensor σ^R in a rotor-fixed frame, the chemical shift frequency under magic-angle spinning is

$$\begin{aligned}\omega_{cs} = -\omega_0\Big\{&\sigma_{iso} + \frac{1}{2}(3\cos^2\theta_R - 1)(\sigma^R_{zz} - \sigma_{iso}) \\ &+ \sin^2\theta_R\left[\frac{1}{2}(\sigma^R_{xx} - \sigma^R_{yy})\cos(2\omega_R t) + \sigma^R_{xy}\sin(2\omega_R t)\right] \\ &+ 2\sin\theta_R\cos\theta_R\left[\sigma^R_{zz}\cos(\omega_R t) + \sigma^R_{yz}\sin(\omega_R t)\right]\Big\} \qquad (3.44)\end{aligned}$$

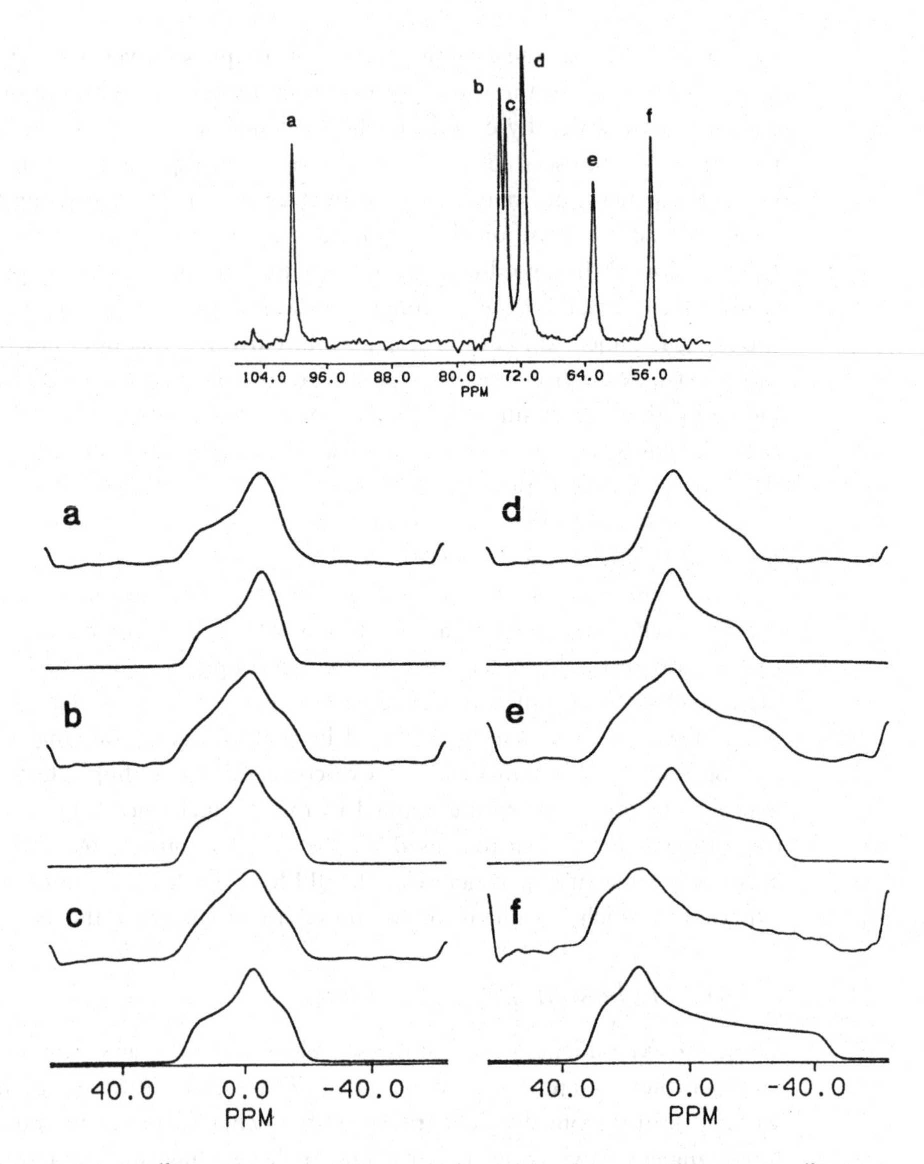

Fig. 3.6 Top: ^{13}C magic-angle spinning spectrum of methyl-α-D-glucopyranoside [15]. Below: ^{13}C chemical shift anisotropy powder patterns taken from the two-dimensional isotropic/anisotropic chemical shift separation experiment of the same compound [15] for the six ^{13}C sites. The top powder pattern in each case is the experimental one. Simulated powder patterns that best fit the experimental ones are shown below each powder pattern.

where ω_R is the spinning rate and θ_R is the spinning angle relative to $\mathbf{B}_0$. If the rate of spinning is much greater than the chemical shift anisotropy, then the time-dependent parts of Equation (3.44) vanish, leaving just

$$\omega_{cs} = -\omega_0\left\{\sigma_{iso} + \frac{1}{2}(3\cos^2\theta_R - 1)\,(\sigma^R_{zz} - \sigma_{iso})\right\} \tag{3.45}$$

At the magic angle, the factor $(3\cos^2\theta_R - 1)$ is zero, leaving just the isotropic term as expected. However, consider what happens if the spinning angle is not the magic angle. Under these circumstances, ω_{cs} consists of two terms, the usual isotropic term and a second anisotropic term scaled by $(3\cos^2\theta_R - 1)$. The anisotropic term can be rewritten in terms of the shielding tensor in its principal axis frame σ^{PAF} so that the chemical shift frequency ω_{cs} becomes

$$\omega_{cs} = -\omega_0\left\{\sigma_{iso} + \frac{1}{2}(3\cos^2\theta_R - 1)(\sigma_{xx}^{PAF}\sin^2\beta\cos^2\alpha + \sigma_{yy}^{PAF}\sin^2\beta\sin^2\alpha + \sigma_{zz}^{PAF}\cos^2\beta - \sigma_{iso})\right\} \quad (3.46)$$

where α, β are Euler angles describing the rotation of the principal axis frame (PAF) into the rotor fixed frame (R). Equation (3.46) is exactly the same form as for a static sample, but with the anisotropic part of the shielding tensor scaled by a factor $(3\cos^2\theta_R - 1)$. In other words, the spectrum arising from sample spinning off the magic angle will be a static powder pattern, but with its chemical shift anisotropy scaled by $(3\cos^2\theta_R - 1)$. Thus, one way of separating chemical shift anisotropy powder patterns is to perform a two-dimensional experiment in which during t_1 (say) the sample is spun at some angle other than the magic angle and then the spinning angle is changed to the magic angle for the subsequent t_2 period. In practice, the angle flipping takes a finite length of time and the magnetization must be stored along $\mathbf{B}_0$ (effected by means of appropriately phased 90° pulses) during the change in spinning angle and then returned to the transverse plane at the start of the t_2 period after the angle adjustment. One problem with this experiment is the obvious one that considerable amounts of signal may be lost through relaxation during the angle-flipping process.

However, there is a much more ingenious way of using variable-angle spinning to achieve isotropic/anisotropic separation, which has become known as Variable-Angle Correlation SpectroscopY or VACSY [17]. Equation (3.46) gives the chemical shift frequency under sample spinning at angle θ_R. The contribution to the FID signal arising from the chemical shift frequency is proportional to $\exp(i\omega_{cs}t)$, which can be rewritten by substituting for ω_{cs} from Equation (3.45):

$$\begin{aligned}\exp(i\omega_{cs}t) &= \exp\left(-i\omega_0\left[\sigma_{iso} + \frac{1}{2}(3\cos^2\theta_R - 1)\sigma_{aniso}\right]t\right) \\ &= \exp(-i\omega_0\sigma_{iso}t)\exp\left(-i\frac{1}{2}\omega_0(3\cos^2\theta_R - 1)\sigma_{aniso}t\right) \\ &= \exp(i\omega_{iso}t)\exp\left(i\frac{1}{2}(3\cos^2\theta_R - 1)\omega_{aniso}t\right)\end{aligned} \quad (3.47)$$

where

$$\sigma_{\text{aniso}} = \sigma_{xx}^{\text{PAF}} \sin^2\beta \cos^2\alpha + \sigma_{yy}^{\text{PAF}} \sin^2\beta \sin^2\alpha + \sigma_{zz}^{\text{PAF}} \cos^2\beta - \sigma_{\text{iso}} \quad (3.48)$$

and $\omega_{\text{iso}} = -\omega_0\sigma_{\text{iso}}$ and $\omega_{\text{aniso}} = -\omega_0\sigma_{\text{aniso}}$. In order to separate isotropic/anisotropic parts in a two-dimensional experiment, we need to obtain a signal of the form

$$\exp(i\omega_{\text{aniso}}t_1)\exp(i\omega_{\text{iso}}t_2) \quad (3.49)$$

from the experiment, i.e. precession at the isotropic frequency in t_2 and at the anisotropic part of the precession frequency in t_1. If we compare the last line of Equation (3.47) with (3.49) we see that, in fact, Equation (3.47) has the desired form if we set $t_2 = t$ and $t_1 = \frac{1}{2}(3\cos^2\theta_R - 1)t$. So we perform a two-dimensional experiment in which we first excite transverse magnetization and then record the FID during a period t_2 while spinning the sample at an angle θ_R. This is the first slice of the two-dimensional dataset. The next slice is obtained by incrementing t_1, which is achieved by changing the spinning angle θ_R and recording the FID again. Continuing in this way gives a two-dimensional dataset, which, after appropriate processing with respect to the t_1 and t_2 domains, gives a two-dimensional frequency spectrum in which isotropic chemical shifts in f_2 are correlated with static-like powder patterns in f_1.

The increments in t_2 $(=\frac{1}{2}(3\cos^2\theta_R - 1)t)$ are not necessarily linear (as they are generally arranged to be in a more conventional two-dimensional NMR experiment). Thus Fourier transformation cannot be used directly to transform from the t_1 domain to the corresponding frequency domain. However, there are other transformation procedures which can be used instead. Alternatively, the t_1 signal can be interpolated and a new t_1 dataset produced in which the sample points are at equal increments of t_1; this time series can then be Fourier transformed to produce the required frequency spectrum.

The great advantage of this experiment is that it gives lots of signal; each t_1 FID acquired is that resulting from a simple pulse-acquire experiment, albeit at a different spinning angle. There is no phase cycling (beyond the usual CYCLOPS), no rotor synchronization and no recoupling pulses. The changes in spinning angle take place between successive experiments, so there is no loss of signal through relaxation as in the previously described variable-angle spinning experiment.

Figure 3.7 shows a ^{13}C VACSY spectrum of *p*-anisic acid [17].

3.3.3 *Magic-angle turning*

Magic-angle turning [18] is a two-dimensional isotropic/anisotropic separation experiment conducted under very slow magic-angle spinning. Under

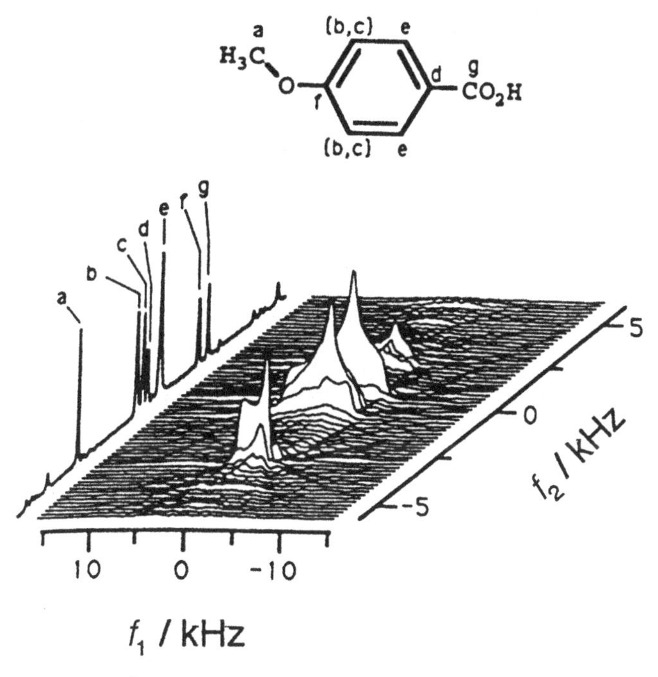

Fig. 3.7 Two-dimensional ^{13}C variable-angle correlation (VACSY) spectrum of *p*-anisic acid [17]. The isotropic ^{13}C spectrum is shown in f_2 and the powder lineshapes dependent on the chemical shift anisotropy and asymmetry for each site in f_1. (Taken from reference [17].)

very slow magic-angle spinning, the spectrum approximates to that of a static (non-spinning) experiment, and it is this feature which produces static-like powder patterns in the magic-angle turning experiment.

The isotropic spectrum in the other dimension of the experiment is produced as follows. Magic-angle spinning averages the anisotropic parts of the chemical shift anisotropy to zero by continuous rotation about a vector orientated at the magic angle with respect to $\mathbf{B}_0$. In fact, however, we do not need to use continuous rotation; we could use discrete hopping instead. Providing we hop the sample between a minimum of three equally-spaced angles about the magic angle, we achieve the same averaging of the anisotropic parts of the interaction. This is shown schematically in Fig. 3.8(a).[N5] In the magic-angle turning experiment, we assume that the magic-angle spinning rate is sufficiently slow compared with the precession of transverse magnetization under the chemical shift anisotropy that the sample is effectively stationary during any reasonably short time delay. The rf pulse sequence used in the magic-angle turning experiment is shown in Fig. 3.8(b). In the experiment, after initial excitation of transverse magnetization, the magnetization is allowed to evolve for a period $t_1/3$. During this period, the rotor is assumed static at (say) position **1** in Fig. 3.8(a). The magnetization is then moved (by a 90° rf pulse) to z to be stored parallel to

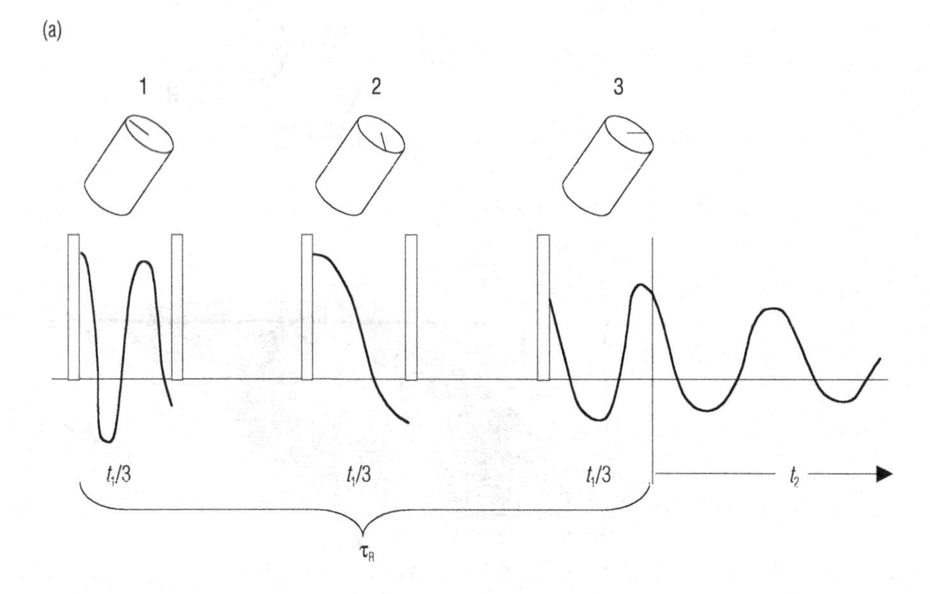

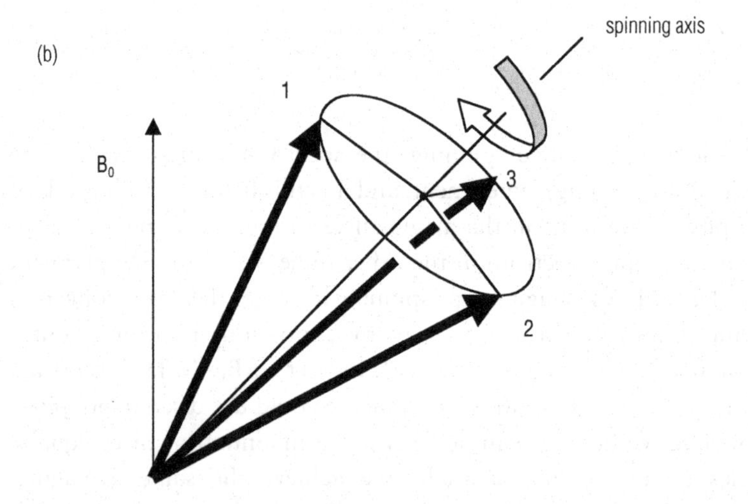

Fig. 3.8 The magic-angle turning experiment. The experiment is conducted under very slow magic-angle spinning, so that during the periods in which the t_1 magnetization is sampled, the rotor appears stationary at the positions **1**, **2** and **3** labelled in (a). The pulse sequence used is also shown in (a). All pulses are 90° pulses. After each $t_1/3$ period, the magnetization is rotated to z ($\mathbf{B_0}$) by a 90° pulse of suitable phase and is stored while the sample rotates to the next position. The net evolution during the whole t_1 period is as if the spins suffered their isotropic chemical shift only; all anisotropic parts cancel identically. During t_2, the spinning is so slow that the spectrum recorded is as if the sample is static. In (b) the position of the shielding tensor principal axis frame z-axis for one molecular/crystallite orientation is shown at the three rotor positions in (a). The average orientation for this axis is the sample spinning axis, which, in turn, is orientated at the magic angle. For this average orientation, the anisotropic chemical shift contribution to the chemical shift frequency is zero. Thus the net evolution of the transverse magnetization in t_1 is governed by the isotropic chemical shift frequency only.

$\mathbf{B}_0$ until the rotor reaches position **2**. At this point, another 90° pulse moves the stored magnetization back to the transverse plane where it is allowed to evolve for a further $t_1/3$. It is then stored along $z/\mathbf{B}_0$ again until the rotor reaches position **3**, where the same process is repeated. Finally, after the last $t_1/3$ period at position **3**, an FID is recorded (as a function of t_2). This whole experiment is repeated for many t_1 increments in the usual manner for a two-dimensional experiment; note, however, that the condition $t_1 \ll \tau_R$, where τ_R is the rotor period, must exist for all t_1 values. Fourier transformation in t_2 gives a static-like powder pattern (due to the ultra-slow spinning). Fourier transformation with respect to t_1 produces an isotropic spectrum, as the net evolution during t_1 depends only on the isotropic chemical shift, the anisotropic parts averaging to zero between the three rotor orientations used in t_1.

The main problem with this experiment is that many modern magic-angle spinning probes do not spin well (or at all!) at the very low rates required in this experiment (generally less than 50 Hz is required to produce quasi-static conditions). An example of a ^{13}C magic-angle turning experiment is shown in Fig. 3.9 for 1,4-dimethoxybenzene [18].

3.3.4 *Two-dimensional separation of spinning sideband patterns*

As explained in Section 2.2.1, magic-angle spinning at a rate which is less than the chemical shift anisotropy gives rise to spinning sidebands in the frequency spectrum. Analysis of the spinning sideband pattern can yield the chemical shift parameters in just the same way as analysis of the static chemical shift anisotropy powder pattern does. Indeed, analysis of spinning sidebands rather than static powder patterns can be preferable as the signal-to-noise ratio is generally better in the magic-angle spinning spectra, since all the spectral intensity is concentrated at a few discrete points in the frequency spectrum rather than being spread out over a powder pattern. Against this, however, is the fact that the simulations required in the analysis of a magic-angle spinning spectrum are more lengthy than those for a static powder pattern.

Often the resolution of different sideband patterns in a simple one-dimensional experiment is sufficient for a reliable analysis to be achieved. However, it is also possible to separate the spinning sideband patterns according to isotropic chemical shift. This can be done in two different ways. The first method separates spinning sideband patterns in f_2 according to the isotropic chemical shift of the pattern in f_1. The pulse sequence used is that in Fig. 3.10 [19]. After initial excitation of the transverse magnetization, the TOSS pulse sequence (see Section 2.2.1) is applied to remove the sideband components during the t_1 period. A 'de-TOSS' pulse sequence

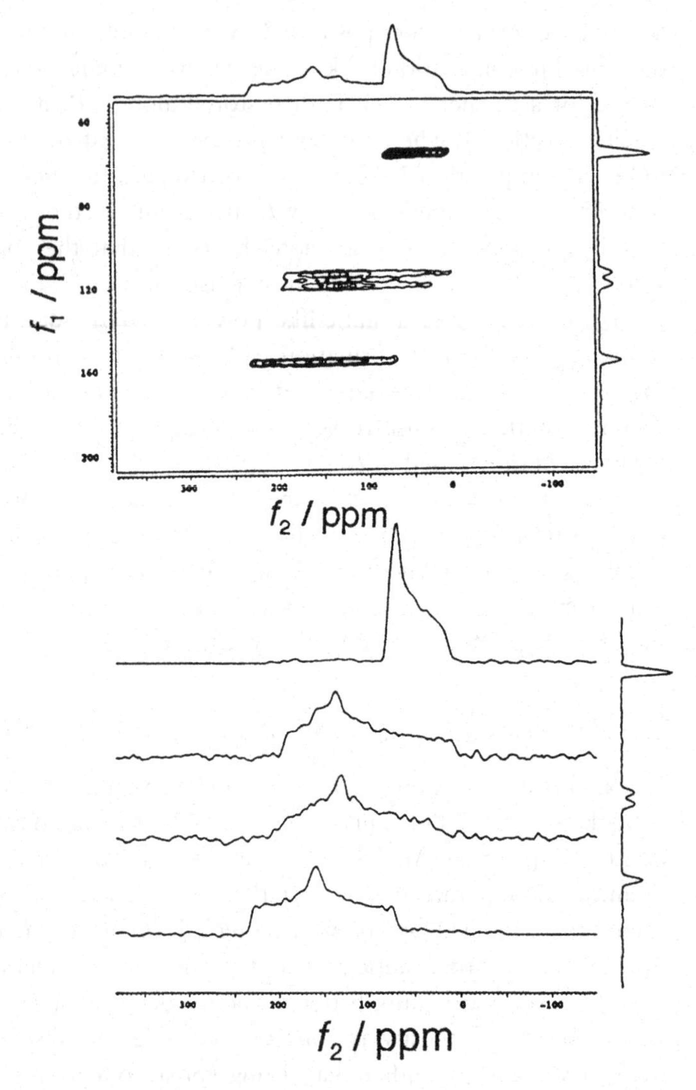

Fig. 3.9 Top: two-dimensional ^{13}C magic-angle turning spectrum (contour plot) of 1,4-dimethoxybenzene [18]. Below: the chemical shift anisotropy powder patterns in f_2 from the two-dimensional spectrum for each ^{13}C site [18].

(reverse TOSS) is then applied to reintroduce the sideband components for the subsequent t_2 period in which the FID is recorded. Subsequent Fourier transformation of the two-dimensional dataset in both dimensions yields the sideband patterns in f_2 separated according to isotropic chemical shift in f_1.

One problem with this method is that many increments are required in t_1 to give the resolution in the f_1 dimension necessary to resolve all the isotropic signals properly and this can lead to very long experiment times.[N6]

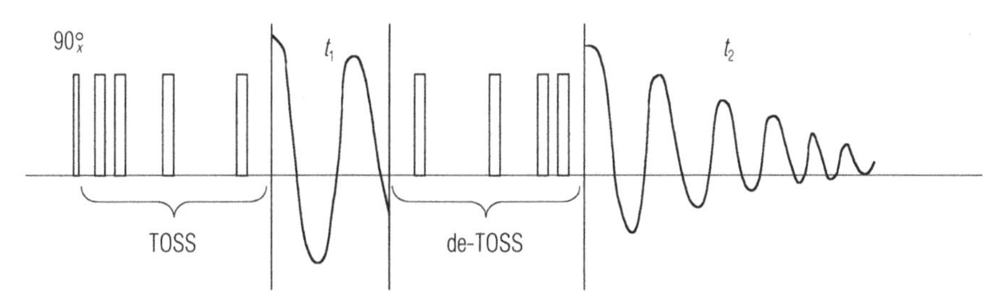

Fig. 3.10 The two-dimensional isotropic/spinning sideband separation experiment based on the TOSS pulse sequence. An initial 90° pulse creates transverse magnetization. The subsequent TOSS sequence, consisting of four 180° pulses, has the effect of removing the spinning sidebands in the variable t_1 period of the two-dimensional experiment. After t_1, a de-TOSS sequence (time-reversed TOSS), in effect, replaces the spinning sideband components for recording in t_2.

Furthermore, each peak in the two-dimensional spectrum has a mixed absorption/dispersion lineshape, which can further complicate matters. Nevertheless, excellent results have been achieved using this method.

The second method of sideband separation results in a different sort of separation; each f_1 slice contains all the spinning sidebands of a given *order*, m, spread along the f_2 axis according to their offset frequency (see Fig. 3.11). The order of a sideband m is its position in the sideband pattern radiating out from the isotropic chemical shift, where the offset frequencies of the sidebands are given by $\Delta\omega_{\text{iso}} + m\omega_{\text{R}}$, where $\Delta\omega_{\text{iso}}$ is the isotropic chemical shift offset and ω_R is the spinning rate. This method is known as the 2D-PASS (Phase Adjusted Spinning Sidebands) method [20]. The basic pulse sequence used is shown in Fig. 3.12. After initial excitation of transverse magnetization via a 90° pulse or cross-polarization, a five 180° pulse sequence (the 'PASS' sequence) is implemented, followed by recording of an FID in t_2. In successive experiments, rather than incrementing a t_1 time period as in a conventional two-dimensional experiment, the timings of the 180° pulses in the PASS sequence are altered. The purpose of the PASS sequence is to impart a specific phase shift to each sideband in the spectrum, that phase shift being dependent only on the sideband order. The way it does this is similar in principle to the working of the TOSS sequence described in Section 2.2.3. The result is that in successive t_1 'increments', the phase of each order of sideband is shifted by an amount dependent only on the sideband order. Thus, the transverse magnetization associated with each sideband (in the terminology used in the discussion of the TOSS sequence in Section 2.2.3) appears to 'evolve' in t_1 at a 'frequency' dependent only on the sideband order. Subsequent Fourier transformation with respect to t_1 thus gives a spectrum of the intensities of the various sideband orders as required.

The advantage of this experiment over the TOSS-based separation of sidebands is that very few t_1 'increments' are required. The corresponding f_1

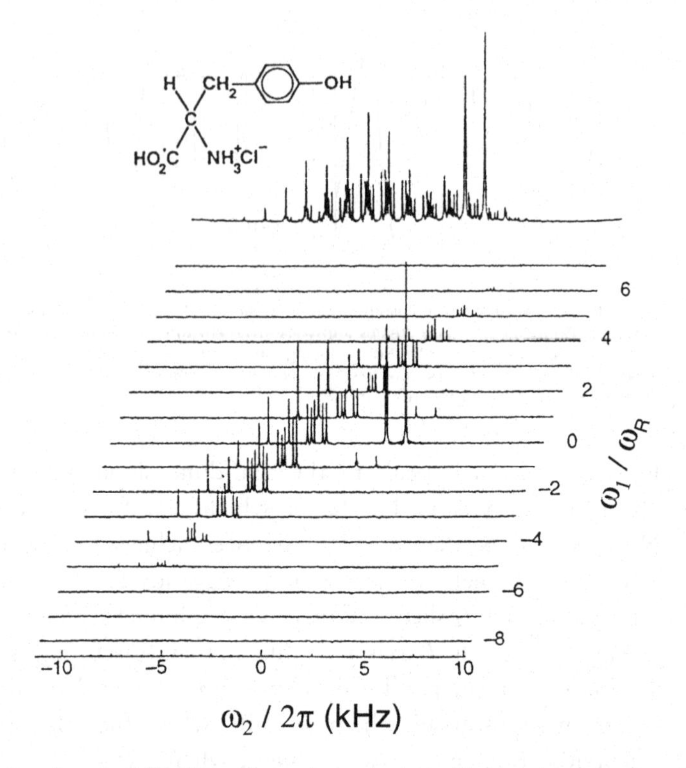

Fig. 3.11 Two-dimensional ^{13}C PASS spectrum of L-tyrosine hydrochloride (solid powder) [20]. The f_1 slices are labelled with the order of the sideband. The ^{13}C magic-angle spinning spectrum for this sample at the same spinning speed (1030 Hz) is shown at the top. The overlap of spinning sidebands in this one-dimensional spectrum prevents their quantitative analysis. Once separated in two dimensions by sideband order, the spectra are comparatively easy to analyse. (Taken from reference [20].)

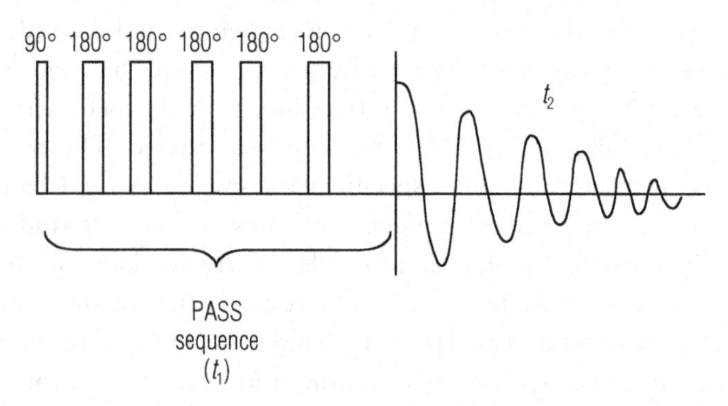

Fig. 3.12 The pulse sequence used in the 2D-PASS experiment. The positions of the five 180° pulses of the PASS part of the sequence are changed between successive t_1 increments of the two-dimensional experiment.

dimension in the final two-dimensional spectrum needs only a sufficient number of points to represent each order of sideband visible in the complete NMR spectrum, i.e. the number of t_1 points needs only to be of the order of the number of sidebands in the most extensive sideband pattern in the spectrum.

A recent modification of the 2D-PASS experiment is the DROSS experiment [21]. In this experiment, the apparent spinning rate governing the sideband intensites in ω_1 is scaled by a factor (less than 1) which is under the control of the experimenter. This has obvious advantages in that it allows a large number of spinning sideband intensities in ω_1 while using rapid magic-angle spinning which helps in the removal of dipolar coupling effects.

3.4 Measuring the orientation of chemical shielding tensors in the molecular frame for structure determination

There is clearly merit in knowing the orientation of the chemical shielding tensor in the molecular frame for the purposes of structure determination, if one is performing *ab-initio* calculations to model the chemical shielding tensor in terms of molecular geometry. However, in many cases, the orientation of the shielding tensor can be well predicted. For instance, for many functional groups in organic chemistry, the ^{13}C chemical shielding tensor orientation is relatively constant between different molecules. Then a knowledge of the shielding tensor orientation in the molecular frame is tantamount to a knowledge of the orientation of the particular functional group in the molecular frame, which can be very valuable in structure determination of proteins, for instance.

So the question is: how can we measure the orientation of a shielding tensor? The answer is to measure the shielding tensor orientation relative to some other interaction tensor orientation, whose orientation in the molecular frame we do know. The following is not intended to be an exhaustive review of this area, but rather serves to illustrate what is possible.

The obvious example here is to measure the orientation of the chemical shielding tensor relative to a particular dipole-coupling interaction, as the dipolar-coupling tensor is necessarily axial with its unique principal axis lying along the internuclear vector of the coupled spins (Fig. 3.13). The general philosophy for doing this is to measure the chemical shift anisotropy powder pattern or spinning sideband pattern (for instance by one of the methods in Section 3.3), but arranging for the powder pattern to be modulated in some way according to the dipolar-coupling interaction.

In the case of a ^{13}C spin with one bonded ^{1}H this can be done simply by recording a two-dimensional isotropic/anisotropic chemical shift separation experiment, such as that described in Section 3.3.1 [15, 16] using cross-polarization to generate the initial ^{13}C transverse magnetization [22]. If the cross-polarization step is kept very short (around 10 μs for a directly-bonded ^{13}C–^{1}H pair), then the only cross-polarization transfer will be from the closest ^{1}H, i.e. the directly bonded ^{1}H. In turn, the efficiency of the cross-polarization transfer depends on the strength of the ^{13}C–^{1}H dipolar coupling, which in turn depends on the orientation of the ^{13}C–^{1}H internuclear

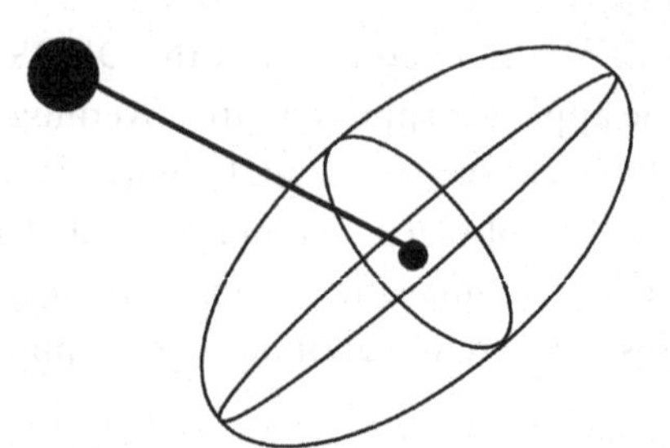

Fig. 3.13 Illustrating how the orientation of the chemical shielding tensor (represented here by an ellipsoid whose principal axes are coincident with the principal axis frame of the chemical shielding tensor) can be determined in the molecular frame by determining its orientation with respect to the dipolar coupling interaction indicated. The black circles represent nuclei within a molecule.

vector with respect to the applied field. This orientation is effectively constant during the cross-polarization transfer, even under magic-angle spinning, if the cross-polarization step is kept short. Hence, for a fixed cross-polarization time, the intensity of the transverse magnetization component arising from each molecular orientation in a powder sample will depend on the molecular orientation itself.

The chemical shift anisotropy powder patterns arising in the f_1 dimension of the two-dimensional spectrum consist of a continuum of sharp lines from each molecular orientation in the powder sample. These will now have their normal powder pattern intensities modulated according to the relative orientation of the shielding tensor and dipolar-coupling internuclear vector for each sharp line component of the powder pattern [22]. Simulation of the resultant lineshapes then reveals the orientation of the chemical shielding tensor relative to the internuclear vector.

An alternative is to measure the relative orientation of two shielding tensors (Fig. 3.14), where the orientation of each in the molecule is known or can be predicted. A particular example here is the correlation of the carbonyl ^{13}C shielding tensors in the backbone of a peptide or protein. The orientation of these shielding tensors with respect to the local CO moiety is known, so determining their relative orientation allows the determination of the torsion angles in the peptide backbone between the two CO groups in question. Clearly there is a need for specific isotope labelling here.

The correlation between the two shielding tensors can be determined by a simple, static, two-dimensional exchange experiment using the pulse sequence shown in Fig. 3.15. Initial transverse magnetization is generated by cross-polarization or a 90° pulse (as in Fig. 3.15) and allowed to evolve under the chemical shift anisotropy in t_1. A 90° pulse then transfers the remaining magnetization to z and any ^1H decoupling is switched off. ^{13}C–^{13}C dipolar coupling between the labelled spins assisted by ^{13}C–^1H dipolar coupling and ^1H spin diffusion during the subsequent period τ_{mix} then allows exchange of ^{13}C magnetization between the two ^{13}C sites of interest. A final 90° pulse to reinstate transverse magnetization and a Δ–180°–Δ echo step allows the acquisition of undistorted chemical shift anisotropy powder

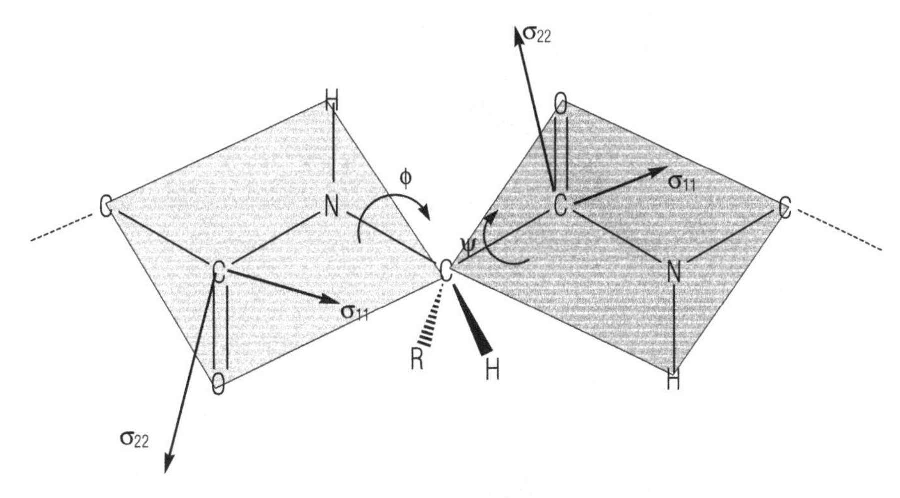

Fig. 3.14 Illustrating how the relative orientation of two shielding tensors can allow the determination of molecular torsion angles. The example used is that of a peptide fragment. The orientations of the backbone carbonyl shielding tensors are well known; components are labelled σ_{11} in the figure. σ_{22} is approximately 10° from the C–O double bond and σ_{11} is approximately 40° from the carbonyl–N bond, both components being in the peptide plane shown. Determining the relative orientation of two neighbouring carbonyl ^{13}C shielding tensors then allows the information about the intervening torsion angles ϕ and ψ to be found.

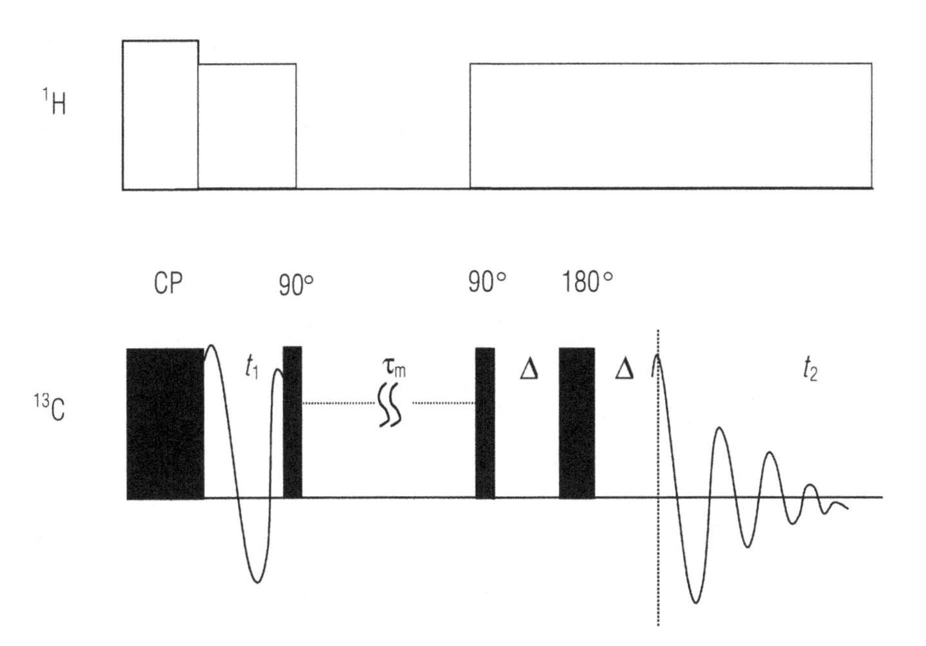

Fig. 3.15 The two-dimensional exchange pulse sequence for correlating two shielding tensors in static samples.

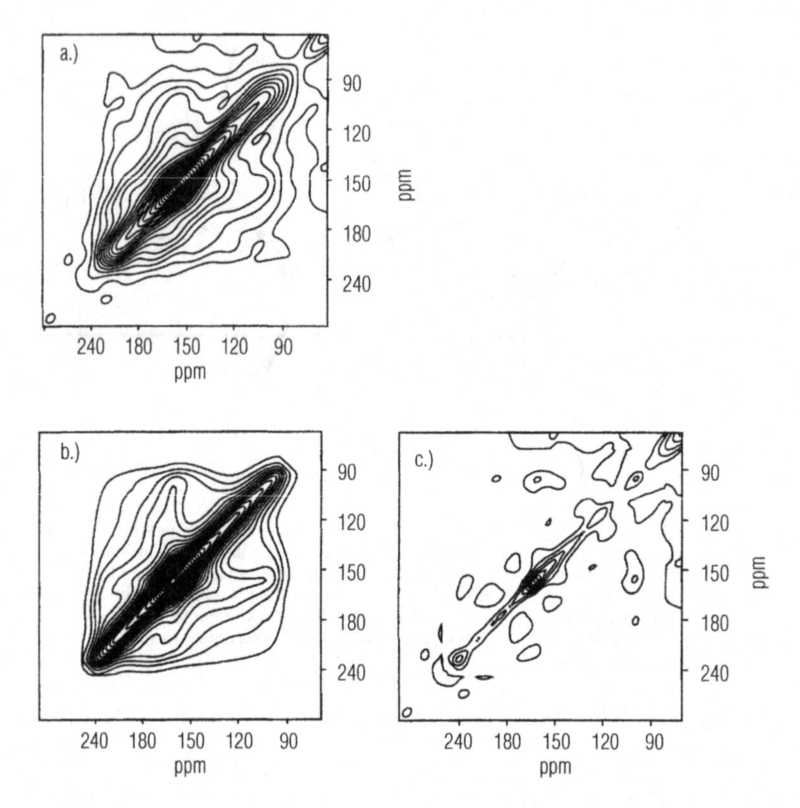

Fig. 3.16 (a) Two-dimensional ^{13}C experimental proton-driven spin-diffusion spectrum of glycine-1–^{13}C-labelled *N. madagascariensis* dragline silk at T = 150 K. A mixing time of 10 s was used. (b) Best fit of the experimental spectrum to a two-site exchange model with an amorphous background added (14% signal intensity from an isotropic exchange pattern). The conformation represents a $(Gly–Gly–X)_m–3_1$–helical structure with only two symmetrically inequivalent Gly–1–^{13}C sites. In contrast to the α-helical structures which are stabilized by intrachain hydrogen bonds, 3_1-helices are stabilized by interchain hydrogen bonds. (c) Difference between the experimental spectrum (a) and fit (b). Adapted from reference [23].

lineshapes in t_2/f_2. The resulting form of the spectrum is exemplified in Fig. 3.16 [23], which shows the two-dimensional correlation of ^{13}C chemical shift anisotropy powder patterns. Simulation of these allows the relative orientation of the two ^{13}C shielding tensors to be determined.

The same type of experiment can be performed under magic-angle spinning [24] with the addition of a dipolar recoupling sequence during the polarization transfer step, τ_{mix}, and ensuring that τ_{mix} is an integer multiple of rotor periods. Dipolar recoupling sequences are discussed in Section 4.3.1; essentially they reintroduce the dipolar coupling, which is otherwise averaged to zero by magic-angle spinning.

There are many variants of this latter experiment, for instance those using ^{13}C double-quantum coherence to correlate ^{13}C chemical shift tensors [25–27].

Notes

1. This is only true when the Zeeman interaction is significantly larger than any other interaction on the spin system, and so the applied field $\mathbf{B}_0$ provides the quantization axis for the spin system.
2. The antisymmetric part of the shielding tensor gives non-zero contributions to the higher-order terms in the average hamiltonian, and so only becomes important if the shielding interaction is large enough relative to the Zeeman term for these terms to become important.
3. Equation (3.13) implicitly uses the symmetric part of the shielding tensor as σ^{lab}. If the antisymmetric part of the shielding tensor were used as σ^{lab} in Equation (3.13), the result for $\sigma_{zz}^{\mathrm{lab}}$ would be zero. Thus the antisymmetric part of the shielding tensor makes no contribution to the chemical shift frequency of Equation (3.12) as previously stated. This will be the case as long as the chemical shift frequency depends only on the $\sigma_{zz}^{\mathrm{lab}}$ component of the laboratory frame shielding tensor, which it does if the Zeeman interaction provided by $\mathbf{B}_0$ is the dominant interaction on the spin system.
4. Note that here we are dealing with molecular *spin* orbitals, i.e. each molecular orbital has associated with it an electronic spin function (α or β). Thus each molecular orbital in this discussion can only contain one electron.
5. The explanation of the magic-angle turning experiment in this figure is slightly simplistic as the chemical shielding interaction which determines the chemical shift and its anisotropy is described by a second-rank tensor and not a single vector as implied by the diagram. We should look at what a second-rank tensor is averaged to when it is switched between these three orientations. A second-rank tensor can be represented by an ellipsoid whose principal axes (x, y, z) represent the principal axes of the tensor, and whose radius along each of the principal axes represents the tensor's associated principal values. The diagram represents how the principal z-axis only reorientates under magic-angle turning. In fact, a little thought shows that the (x, y) plane of the second-rank tensor becomes isotropic under this three- (or higher) site hopping.
6. For this reason, it is usually better to arrange for the isotropic signal to be collected in the directly observed dimension, t_2, in any isotropic/anisotropic separation experiment.

References

1. Jameson, C.J. & Mason, J. (1987) *The Chemical Shift in Multinuclear NMR* (chapter 3) (ed. J. Mason), pp. 51–83. Plenum Press, New York.
2. Strub, H., Beeler, A.J., Grant, D.M., Michel, J., Cutts, P.W. & Zilm, K.W. (1983) *J. Am. Chem. Soc.* **105**, 3333.
3. Huang, Y., Gilson, D.F.R. & Butler, I.S. (1992) *J. Chem. Soc. Dalton*, 2881.
4. deDios, A.C. & Oldfield, E. (1996) *Solid-State NMR* **6**, 101.
5. Antzutkin, O.N., Lee, Y.K. & Levitt, M.H. (1998) *J. Magn. Reson.* **135**, 144; Rich, J.E., Manolo, M.N. & deDios, A.C. (2000) *J. Phys. Chem.* **104A**, 5837.
6. Oldfield, E. (1995) *J. Biomol NMR* **5**, 217.
7. deDios, A.C. & Oldfield, E. (1994) *J. Am. Chem. Soc.* **116**, 11485.
8. deDios, A.C., Laws, D.D., & Oldfield, E. (1994) *J. Am. Chem. Soc.* **116**, 7784.
9. deDios, A.C. & Oldfield, E. (1996) Chemical shifts in biochemical systems. In: *Encyclopedia of Nuclear Magnetic Resonance*, p. 1330. Wiley, New York.
10. Laws, D.D., Le, H.-B., de Dios, A.C., Havlin, R.H. & Oldfield, E. (1995) *J. Am. Chem. Soc.* **117**, 9542.
11. Pearson, J.G., Wang, J.-F., Markley, J.L., Le, H.-B. & Oldfield, E. (1995) *J. Am. Chem. Soc.* **117**, 8823.
12. Le, H.-B. & Oldfield, E. (1996) *J. Phys. Chem.* 100, 16423.

13. Havlin, R.H., Le, H.-B., Laws, D.D., deDios, A.C. & Oldfield, E. (1997) *J. Am. Chem. Soc.* **119**, 11951.
14. Heller, J., Laws, D.D., Tomaselli, M., King, D.S., Wemmer, D.E., Pines, A., Havlin, R.H. & Oldfield, E. (1997) *J. Am. Chem. Soc.* **119**, 7827.
15. Tycko, R., Dabbagh, G. & Mirau, P.A. (1989) *J. Magn. Reson.* **85**, 265.
16. Liu, S.F., Mao, J.D. & Schmidt-Rohr, K. (2002) *J. Magn. Reson.* **155**, 15.
17. Frydman, L., Chingas, G.C., Lee, Y.K., Grandinetti, P.J., Eastman, M.A., Barrall, G.A. & Pines, A. (1992) *J. Chem. Phys.* **97**, 4800.
18. Gan, Z. (1992) *J. Am. Chem. Soc.* **114**, 8307.
19. Kolbert, A.C. & Griffin, R.G. (1990) *Chem. Phys. Lett.* **166**, 87.
20. Antzutkin, O.N., Shekar, S.C. & Levitt, M.H. (1995) *J. Magn. Reson. A.* **115**, 7.
21. Crockford, C., Geen, H. & Titman, J. (2001) *Chem. Phys. Lett.* **344**, 367.
22. Yao, X.L., Yamaguchi, S. & Hong, M. (2002) *J. Biomol. NMR* **24**, 51; Yao, X. & Hong, M. (2002) *J. Am. Chem. Soc.* **124**, 2730.
23. Kümmerlen, J., van Beek, J.D., Vollrath, F. & Meier, B.H. (1996) *Macromolecules* **29**, 2920.
24. Tycko, R., Weliky, D.P. & Berger, A.E. (1996) *J. Chem. Phys.* **105**, 7915.
25. Gregory, D.M., Mehta, M.A., Shiels, J.C. & Drobny, G.P. (1997) *J. Chem. Phys.* **107**, 28.
26. Bower, P.V., Oyler, N., Mehta, M.A., Long, J.R., Stayton, P.S. & Drobny. G.P. (1999) *J. Am. Chem. Soc.* **121**, 8373.
27. Blanco, F.J. & Tycko, R. (2001) *J. Magn. Reson.* **149**, 131.

Dipolar Coupling: Theory and Uses 4

4.1 Theory

Nuclear spin possesses a magnetic moment and, in a collection of spins, these interact through space. This is dipole–dipole or dipolar coupling. Classically, this interaction is akin to the interaction between pairs of bar magnets. In effect, each spin acts like a bar magnet and creates a local magnetic field which the surrounding spins feel and interact with. Note that this is distinct from scalar (J) coupling, which is an indirect coupling of the nuclear spins mediated by electrons. In solution, the dipole–dipole interaction is averaged to its isotropic value, zero, by molecular tumbling. This is not the case in solids, where this interaction is a major cause of linebroadening.

The strength of the interaction depends on the internuclear distance, in fact on $1/r^3$ where r is the internuclear distance. It also depends on molecular orientation, as shown schematically in Fig. 4.1. Each of the nuclear magnetic dipoles is orientated by the applied magnetic field in the NMR experiment. Thus, the field that each presents to the other depends on the relative disposition of the two spins in space, i.e. on the orientation of the molecular fragment they are held within relative to $\mathbf{B}_0$, as illustrated in Fig. 4.1.[1]

Classically, the energy of interaction between two point magnetic dipoles $\boldsymbol{\mu}_1$ and $\boldsymbol{\mu}_2$ separated by a distance r is:

$$U = \left\{ \frac{\boldsymbol{\mu}_1 \cdot \boldsymbol{\mu}_2}{r^3} - 3\frac{(\boldsymbol{\mu}_1 \cdot \mathbf{r})(\boldsymbol{\mu}_2 \cdot \mathbf{r})}{r^5} \right\} \frac{\mu_0}{4\pi} \tag{4.1}$$

where $\mathbf{r}$ is the vector between point magnetic dipoles.

Quantum mechanically, the magnetic moment operator $\hat{\boldsymbol{\mu}}$, is given by:

$$\hat{\boldsymbol{\mu}} = \gamma \hbar \hat{\mathbf{I}} \tag{4.2}$$

for a spin I.

Substituting this operator in the classical expression for energy of interaction, we obtain the interaction hamiltonian for dipolar coupling between two spins I and S in angular frequency units (rad s^{-1}):

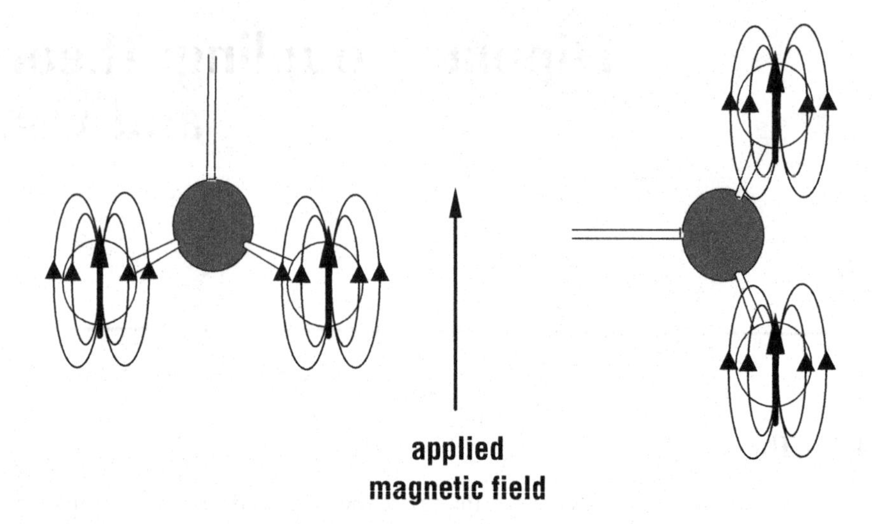

Fig. 4.1 Illustrating the dipolar interaction between two spins. Each spin creates a field (as a result of its spin angular momentum and concomitant magnetic dipole moment) which the other spin feels. The strength of the interaction depends on the internuclear distance ($1/r^3$) and on the molecular orientation, as shown. The dependence on molecular orientation effectively arises from the fact that the applied field in the NMR experiment ($\mathbf{B_0}$) orientates each spin's magnetic dipole moment, and hence orientates the field it presents to neighbouring spins. As illustrated, the strength of field the neighbouring spin feels depends on the orientation of the internuclear vector with respect to the applied field.

$$\hat{H}_{\mathrm{dd}} = -\left(\frac{\mu_0}{4\pi}\right)\gamma_I\gamma_S\hbar\left(\frac{\mathbf{I}\cdot\mathbf{S}}{r^3} - 3\frac{(\mathbf{I}\cdot\mathbf{r})(\mathbf{S}\cdot\mathbf{r})}{r^5}\right) \tag{4.3}$$

Expressing Equation (4.3) in spherical polar coordinates and expanding the scalar products, we obtain after some considerable rearrangement (and again, in angular frequency units)

$$\hat{H}_{\mathrm{dd}} = -\left(\frac{\mu_0}{4\pi}\right)\frac{\gamma_I\gamma_S\hbar}{r^3}[A+B+C+D+E+F] \tag{4.4}$$

where:

$$\begin{aligned}
A &= \hat{I}_z\hat{S}_z(3\cos^2\theta - 1)\\
B &= -\frac{1}{4}[\hat{I}_+\hat{S}_- + \hat{I}_-\hat{S}_+](3\cos^2\theta - 1)\\
C &= \frac{3}{2}[\hat{I}_z\hat{S}_+ + \hat{I}_+\hat{S}_z]\sin\theta\cos\theta e^{-i\phi}\\
D &= \frac{3}{2}[\hat{I}_z\hat{S}_- + \hat{I}_-\hat{S}_z]\sin\theta\cos\theta e^{+i\phi}\\
E &= \frac{3}{4}[\hat{I}_+\hat{S}_+]\sin^2\theta e^{-2i\phi}\\
F &= \frac{3}{4}[\hat{I}_-\hat{S}_-]\sin^2\theta e^{+2i\phi}
\end{aligned} \tag{4.5}$$

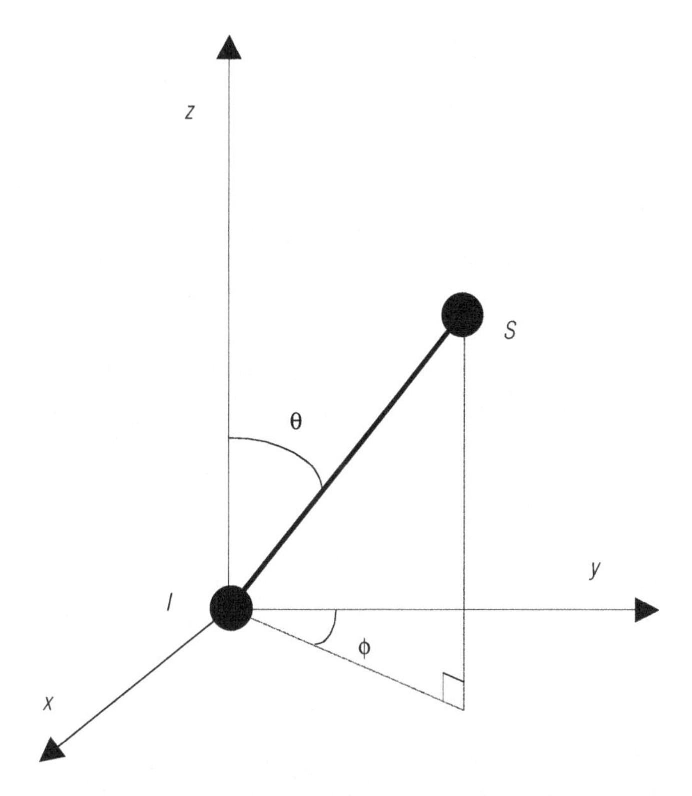

Fig. 4.2 Definition of the polar angles θ and φ specifying the orientation of the *I–S* internuclear vector with respect to the $\mathbf{B}_0$ field which is along the laboratory frame *z*-axis.

$\hat{I}_+$, $\hat{S}_+$ and $\hat{I}_-$, $\hat{S}_-$ are the raising and lowering operators respectively acting on spins *I* and *S*, defined in Box 1.1, and the polar angles θ and φ are defined in Fig. 4.2.

Alternatively, we may express the dipolar hamiltonian (in angular frequency units) in the Cartesian tensorial form

$$\hat{H}_{\mathrm{dd}} = -2\hat{\mathbf{I}}\cdot\mathbf{D}\cdot\hat{\mathbf{S}} \tag{4.6}$$

The spin angular momentum **S** is the equivalent of vector **A** in Equation (1.75) in the Chapter 1 discussion of nuclear spin interactions and it is the spin *S* which is the ultimate source of local field at spin *I*. The term **D** is the dipole-coupling tensor, with principal values of $-d/2$, $-d/2$, $+d$, where *d* is given by

$$d = \hbar\left(\frac{\mu_0}{4\pi}\right)\frac{1}{r^3}\gamma_I\gamma_S \tag{4.7}$$

and is known as the *dipolar-coupling constant* (in units of rad s^{-1}). The dipolar coupling tensor **D** describes how the magnetic field due to spin *S*, and felt at spin I, varies with the orientation of the I–S internuclear vector in the applied field. The spin angular momenta of I and S determine the local magnetic field they each give rise to. In turn, these angular momenta are quantized, with $\mathbf{B}_0$ defining the quantization axis for both. Thus, the local field that each spin gives rise to depends on the direction of $\mathbf{B}_0$ (and the spin state, i.e. component of spin parallel to $\mathbf{B}_0$). The strength of field at spin I due to spin S thus depends on the relative orientation of the I–S vector in the applied field, $\mathbf{B}_0$, and this is what is described in the dipole coupling tensor $\mathbf{D}$ along with the $1/r^3$ internuclear distance dependence. The dipolar hamiltonian can also be expressed in *spherical tensor form*; this is described in Box 4.2 later.

It should be noted that the dipole-coupling tensor is *traceless*, i.e. there is no isotropic component. Thus, when the dipole coupling is averaged by molecular motion in liquid samples, it is averaged to zero and there is no direct effect of the dipolar coupling on the NMR spectrum. The dipolar-coupling tensor $\mathbf{D}$ is always axially symmetric in its principal axis frame, with the unique axis of the principal axis frame lying along the I–S vector – see Section 1.4.1 for further discussion of tensors and their principal axis frames.

We now consider two possible cases of dipolar coupling, *homonuclear dipolar coupling*, where spins I and S are the same species, and *heteronuclear dipolar coupling*, where spins I and S are different.

4.1.1 *Homonuclear dipolar coupling*

If we want to understand the effect of dipolar coupling on an NMR spectrum, we need to transform the hamiltonians of Equations (4.4) and (4.6) to a rotating frame rotating at the Larmor frequency of the observed spin, because, as explained in Section 3.1.2, that is the frame from which we effectively observe the spins in an FT NMR experiment.

This frame transformation is easily done using the procedures in Box 2.1. The total hamiltonian governing a two-spin system, I–S, in this case is the sum of the Zeeman and dipolar coupling terms

$$\hat{H} = \hat{H}_0 + \hat{H}_{\mathrm{dd}} \tag{4.8}$$

where $\hat{H}_0 = \omega_0 \hat{I}_Z + \omega_0 \hat{S}_Z$, the Zeeman hamiltonian for the I–S spin system, where ω_0 is the Larmor frequency (the same for both spins in a homonuclear spin pair). The rotating frame from which we view the spin system is one which rotates about the laboratory z-axis ($\mathbf{B}_0$) at the Larmor frequency for both spins. The hamiltonian governing the spin system in this rotating frame is denoted $\hat{H}^*$ and is given by (see Box 2.1 for details)

$$\begin{aligned}\hat{H}^* &= \hat{R}_z^{-1}(\omega_0 t)\hat{H}\hat{R}_z(\omega_0 t) - \omega_0\hat{I}_z - \omega_0\hat{S}_z \\ &= \hat{R}_z^{-1}(\omega_0 t)\hat{H}_0\hat{R}_z(\omega_0 t) + \hat{R}_z^{-1}(\omega_0 t)\hat{H}_{\mathrm{dd}}\hat{R}_z(\omega_0 t) + \omega_0\hat{I}_z + \omega_0\hat{S}_z \qquad (4.9)\end{aligned}$$

where the rotation operator $\hat{R}_Z(\omega_0 t)$ describes the rotation of an object in spin space by an angle $\omega_0 t$ about the laboratory z-axis and acts on the spin coordinates $\hat{I}$ and $\hat{S}$ – it acts on both spins here as we are dealing with a homonuclear spin system. This z-rotation has no effect on the spin operators of $\hat{H}_0$, so the first and third terms of Equation (4.9) cancel, leaving just the second term, which, substituting from Equations (4.4) and (4.5) for the dipolar hamiltonian, yields

$$\hat{H}^*(t) = -d\{[A+B] + \hat{R}_z^{-1}(\omega_0 t)[C+D+E+F]\hat{R}_z(\omega_0 t)\} \qquad (4.10)$$

where d is the dipolar-coupling constant defined previously (Equation (4.7)), and terms A–F are defined in Equation (4.5). The terms A and B in the dipolar coupling hamiltonian are unaffected by the frame transformation. However, the rotating frame transformation does affect terms C, D, E and F (see Equation (4.5) and redefine the raising and lowering operators $\hat{I}_\pm$, $\hat{S}_\pm$ in terms of the Cartesian spin operators $\hat{I}_x$, $\hat{S}_x$ and $\hat{I}_y$, $\hat{S}_y$ to see the effects of a z-rotation); in particular, these terms now acquire a periodic time dependence (of frequency ω_0), which means that the frame transformation has led to a time dependence in the hamiltonian describing the spin system through the terms C–F. The best way to deal with this time dependence is to use average hamiltonian theory as in Box 2.1. Average hamiltonian theory approximates a periodically time-dependent hamiltonian such as that in Equation (4.10) as a sum of successively higher-order terms. The lowest-order term, confusingly denoted $\overline{H}^{(0)}$, but correctly called the first-order average hamiltonian (as it is first order in the interaction hamiltonian), is simply given by (see Box 2.1):

$$\overline{H}^{(0)} = \frac{1}{t_\mathrm{p}}\int_0^{t_\mathrm{p}} \hat{H}(t)\,\mathrm{d}t$$

and is a sufficiently good approximation to the full hamiltonian for the purposes of understanding the effect of the dipolar-coupling hamiltonian in NMR spectra, providing the magnitude of the dipolar-coupling interaction in frequency units is small relative to the frequency of the frame rotation, the Larmor frequency ω_0. Using Equation (4.10), we find the dipolar coupling contribution to the first-order average hamiltonian in the rotating frame to be:

$$\overline{H}_{\mathrm{dd}}^{(0)} = -d(3\cos^2\theta - 1)\Big[\underbrace{\hat{I}_z\hat{S}_z}_{\text{Term } A} - \underbrace{\tfrac{1}{2}(\hat{I}_x\hat{S}_x + \hat{I}_y\hat{S}_y)}_{\text{Term } B}\Big] \qquad (4.11)$$

where the time-dependent parts of the rotating frame hamiltonian of Equation (4.10) integrate to zero in the construction of this average hamiltonian.

Equation (4.11) is often written more succinctly as

$$\hat{H}_{\mathrm{dd}}^{\mathrm{homo}} = -d \cdot \frac{1}{2}(3\cos^2\theta - 1)[3\hat{I}_z\hat{S}_z - \hat{\mathbf{I}}\cdot\hat{\mathbf{S}}] \qquad (4.12)$$

where we have used the fact that $\hat{\mathbf{I}}\cdot\hat{\mathbf{S}} = \hat{I}_x\hat{S}_x + \hat{I}_y\hat{S}_y + \hat{I}_z\hat{S}_z$. Note that in Equation (4.12) we have dropped the bar and (0) superscript for ease of notation. Henceforth, we will use the nomenclature of Equation (4.12) and when we refer to the dipolar-coupling hamiltonian, we mean the dipolar contribution to the first-order rotating frame average hamiltonian. The form of the homonuclear dipolar hamiltonian in Equations (4.11) and (4.12) is often referred to as the *secular* (as it commutes with the Zeeman hamiltonian for the spin system) or the *truncated* form.

Box 4.1: Basis sets for multispin systems

Consider a spin system with N uncoupled spins in the static magnetic field of an NMR experiment. The hamiltonian describing this spin system is simply the sum of the operators for the energy of each individual spin in the field, i.e.

$$\hat{H}_0 = \omega_0 \sum_j^N \hat{I}_z^j \qquad \text{(i)}$$

where j denotes the nucleus and the operator $\hat{I}_z^j$ only acts on the spin coordinates of the spin j. The eigenfunctions of the $\hat{I}_z^j$ operators are $|I_j m_j\rangle$ or $\psi_{I_j m_j}$ depending on the notation being used (see Box 1.1). The eigenfunctions Ψ of the $\hat{H}_0$ hamiltonian, i.e. the Zeeman states for this spin system, must then be products of the $\psi_{I_j m_j}$ functions:

$$\Psi = \psi_{I_1 m_1}\psi_{I_2 m_2}\cdots\psi_{I_N m_N} \qquad \text{(ii)}$$

so that

$$\begin{aligned}\hat{H}_0\Psi &= \omega_0 \sum_j^N \hat{I}_z^j |I_1 m_1; I_2 m_2; \ldots I_N m_N\rangle \\ &= \omega_0(m_1 + m_2 + \ldots m_N)|I_1 m_1; I_2 m_2; \ldots I_N m_N\rangle \qquad \text{(iii)}\end{aligned}$$

which has the general form of the eigenfunction/eigenvalue equation given in Box 1.1. The eigenvalue, i.e. the energy of the spin system, E_0, is by inspection of (iii)

$$E_0/\hbar = \omega_0(m_1 + m_2 + \dots m_N) \qquad \text{(iv)}$$

The product wavefunctions/eigenfunctions of Equation (ii) constitute a complete set for an N spin system, from which any other function describing the system can be formed. These product wavefunctions are thus a good basis set with which to describe the spin system under more complex interactions on the spin system, such as dipole–dipole coupling.

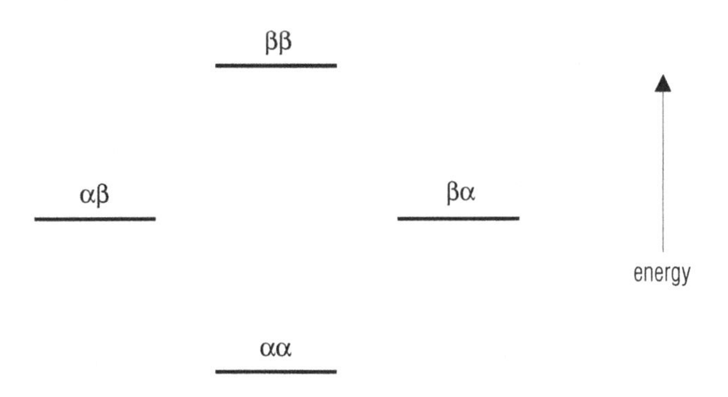

Fig. 4.3 The energy levels and wavefunctions in a homonuclear two spin-$\frac{1}{2}$ system, before dipolar coupling is considered, i.e. the Zeeman states. The αβ and βα levels are degenerate in a homonuclear spin system and are mixed by dipolar coupling.

4.1.2 *The effect of homonuclear dipolar coupling on a spin system*

In order to see what effect the dipole–dipole coupling operator has on a homonuclear spin system, consider the simple two-spin system which, before dipolar coupling effects are taken into account, may be represented by the product or Zeeman spin states illustrated in Fig. 4.3 for a two spin-$\frac{1}{2}$ system.

We now need to consider what effect the first-order average hamiltonian $\hat{H}_{\mathrm{dd}}^{\mathrm{homo}}$ has on the energies and wavefunctions of these states. Consider the effects of each term in the dipole–dipole operator in turn.

1. Term A: this term contains the spin operator $\hat{I}_z\hat{S}_z$. This has the effect of giving a first-order change in energy to all the states. For example, it causes an energy shift of $-\left(\frac{\mu_0}{4\pi}\right)\frac{\gamma_I\gamma_S\hbar^2}{r^3}(3\cos^2\theta - 1)\langle\alpha\alpha|\hat{I}_z\hat{S}_z|\alpha\alpha\rangle$ for the αα spin state. The eigenfunctions of this operator are simply the product Zeeman states, αα, αβ, βα, ββ. This can be verified by

operating with the $\hat{I}_z\hat{S}_z$ operator on each of these functions in turn, and seeing that the result is the original function, multiplied by a constant, i.e. $\hat{I}_z\hat{S}_z|\alpha\alpha\rangle = \frac{1}{4}|\alpha\alpha\rangle$.

2. Term *B*: this term contains the $|\hat{I}_+\hat{S}_- + \hat{I}_-\hat{S}_+|$ spin operator term. Therefore, in the matrix representation of this operator in the Zeeman basis (for a two-spin-$\frac{1}{2}$ system), there will be non-zero elements between the $\alpha\beta$ and $\beta\alpha$ basis functions, i.e.

$$\mathbf{B} = -\frac{1}{2}(3\cos^2\theta - 1)\begin{pmatrix} 0 & 0 & 0 & 0 \\ 0 & 0 & 1 & 0 \\ 0 & 1 & 0 & 0 \\ 0 & 0 & 0 & 0 \end{pmatrix} \tag{4.13}$$

where the basis functions in this representation are in the order $\alpha\alpha$, $\alpha\beta$, $\beta\alpha$, $\beta\beta$. We say that *B mixes* the $\alpha\beta$ and $\beta\alpha$ functions. Hence, the eigenfunctions of *B* include two linear combinations of the $\alpha\beta$ and $\beta\alpha$ functions. The other eigenfunctions are simply $\alpha\alpha$ and $\beta\beta$, since these are not mixed with any other functions by the *B* term.

Now consider a three-spin system. The dipolar hamiltonian for such a system is

$$\begin{aligned}\hat{H}_{\text{dd}}^{\text{homo}} = &-d_{12}\cdot\frac{1}{2}(3\cos^2\theta_{12} - 1)\left[2\hat{I}_{1z}\hat{I}_{2z} - \frac{1}{2}(\hat{I}_{1+}\hat{I}_{2-} + \hat{I}_{1-}\hat{I}_{2+})\right] \\ &-d_{23}\cdot\frac{1}{2}(3\cos^2\theta_{23} - 1)\left[2\hat{I}_{2z}\hat{I}_{3z} - \frac{1}{2}(\hat{I}_{2+}\hat{I}_{3-} + \hat{I}_{2-}\hat{I}_{3+})\right] \\ &-d_{13}\cdot\frac{1}{2}(3\cos^2\theta_{13} - 1)\left[2\hat{I}_{1z}\hat{I}_{3z} - \frac{1}{2}(\hat{I}_{1+}\hat{I}_{3-} + \hat{I}_{1-}\hat{I}_{3+})\right]\end{aligned} \tag{4.14}$$

where 1, 2, 3 label the three spins, and d_{ij} is the dipolar coupling constant associated with the coupling between spins *i* and *j*. The angle θ_{ij} is the angle between the *i*–*j* internuclear vector and the applied field $\mathbf{B}_0$. It is easy to show that the Zeeman states $\alpha\alpha\alpha$ and $\beta\beta\beta$ are eigenfunctions of this hamiltonian. However, the remaining possible states of the system (which will be linear combinations of the remaining Zeeman functions, among which there are several degeneracies) are less easy to find.

As for the two-spin system, the *B*-type terms in the dipolar hamiltonian mix degenerate Zeeman states. Thus, the eigenfunctions of the systems will consist of linear combinations of degenerate Zeeman levels, such as the $\alpha\alpha\beta$, $\alpha\beta\alpha$ and $\beta\alpha\alpha$ levels.

In general, for a many-spin system, the *B*-type terms in the homonuclear dipolar coupling hamiltonian will mix degenerate Zeeman levels, resulting in eigenfunctions for the multispin system which are linear combinations of

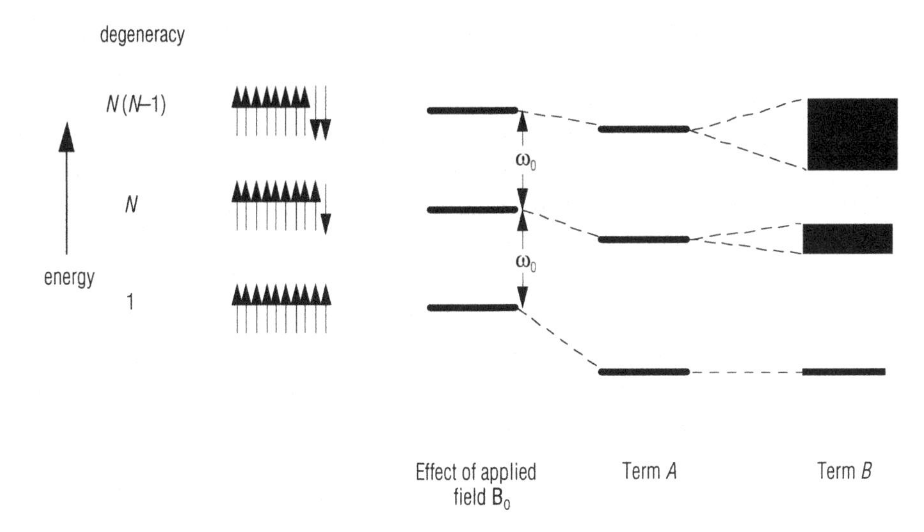

Fig. 4.4 In an N-spin, homonuclear system, there may be many Zeeman levels with the same M quantum number ($M = m_1 + m_2 + \ldots + m_N$); these are degenerate in the absence of dipolar coupling. Dipolar coupling has the effect of mixing these degenerate states and splitting their energy.

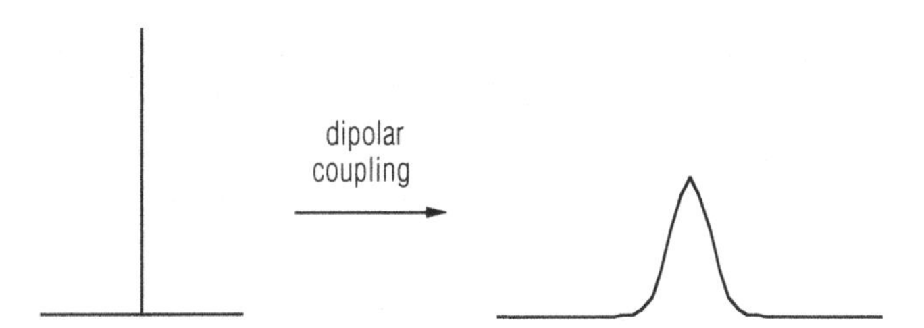

Fig. 4.5 The effect of homonuclear dipolar coupling in a multispin system is to broaden the resonance line, the lineshape tending to gaussian for a sufficiently large number of spins. For this reason, the ^{1}H linewidth in organic solids is frequently several tens of kHz wide.

the degenerate Zeeman levels. This has one immediate consequence for the NMR spectrum of such a system. As shown in Fig. 4.4, the energies of previously degenerate Zeeman levels are split, resulting in many different transition frequencies with $\Delta M = \pm 1$, where M is the total z-component of spin for the multispin system. Thus, even without the molecular orientation dependence of the transition frequencies, there is a huge range of transition frequencies in the NMR spectrum of such a spin system and this results in very broad lines (Fig. 4.5). For a large enough spin system, the lineshape approximates to gaussian (ignoring the usually much smaller orientation dependence of the transition frequencies).

However, there is a further consequence of the eigenfunctions of the dipolar-coupled spin system consisting of mixtures of degenerate Zeeman levels. Consider again the three-spin system. One of the eigenfunctions or states of this system consists of some linear combination of the ααβ, αβα and βαα Zeeman levels for instance (the combination coefficients depending on the geometry of arrangement of the three spins, and thus the dipolar coupling constants governing the interactions between them). If we make an observation of a spin system in such a state, the eigenfunction collapses to just one of the constituent Zeeman functions, and so we observe the spin system as if it were in one of the ααβ, αβα and βαα levels. The probability of observing it in any one Zeeman level is simply the square modulus of the combination coefficient for that particular Zeeman level in the system eigenfunction. If we make repeated observations of the spin system, we will see it in each of the three Zeeman levels at different times. So now consider how we view just one of the spins through these observations, say spin 1. Sometimes it will appear to have a spin state α (when the spin system collapses to ααβ or αβα on observation) and sometimes spin state β (when the spin system collapses to βαα). The fluctuation of the spin 1 spin state over repeated observations makes it look as though the spin is constantly changing its spin function, $\alpha \leftrightarrow \beta$. This is tantamount to saying that there is continual exchange of longitudinal magnetization between the spins of the dipolar-coupled system. Only the total z-component of spin, M, for the whole spin system is conserved, where M is given by

$$M = m_1 + m_2 + m_3 + \ldots + m_N \tag{4.15}$$

for an N-spin system, where m_i is the z-component of spin for spin i in the absence of dipolar coupling. This continual exchange of longitudinal magnetization between spins on observation of the spin system has several consequences for the spin system, particularly when we try to remove the effects of the dipolar interaction by magic-angle spinning (see Section 2.2.5).

The time dependence of the state of the multispin system under homonuclear dipolar coupling is the reason for the huge linebroadening seen in the NMR spectra of dipolar-coupled spin systems (Fig. 4.5).

4.1.3 Heteronuclear dipolar coupling

As for homonuclear dipolar coupling, we must recognize that in an NMR experiment, we observe the spin systems in a rotating frame of reference, and so we need to transform the spin coordinates of the hamiltonian describing the spin system into this rotating frame of reference to understand the outcome of our experiments. The rotating frame concerned rotates about the laboratory z-axis at the Larmor frequency of the observed spin, in this case taken to be I. To perform this frame transformation, we follow through exactly the same procedure as for homonuclear dipolar coupling in Section 4.1.1, except that the rotation operator we use to perform the rotating frame transformation is $\hat{R}_z(\omega_0 t)$ which acts *only* on the I spin coordinates, i.e. only acts on the observed spin, I. This results in a rotating frame hamiltonian of

$$\hat{H}^*(t) = -d\{A + \hat{R}_z^{-1}(\omega_0 t)[B + C + D + E + F]\hat{R}_z(\omega_0 t)\} \tag{4.16}$$

where terms A–F are defined in Equation (4.5). Term A in the heteronuclear dipolar-coupling hamiltonian of Equation (4.4) is the only term unaffected by the rotating frame transformation; all the remainder are, including term B which was unaffected by the transformation in the homonuclear case by virtue of this term containing terms of the form $\hat{I}_\alpha \hat{S}_\alpha$, $\alpha = x, y$, and both spins being acted on by the rotation operator in the rotating frame transformation. As for the homonuclear case, we find the first-order average hamiltonian as that which (principally) governs the NMR spectra we observe. Forming the first-order average hamiltonian has the effect of removing all the time-dependent terms, leaving just:

$$\hat{H}_{\text{dd}}^{\text{hetero}} = -d(3\cos^2\theta - 1)\hat{I}_z\hat{S}_z \tag{4.17}$$

which is to be compared with the homonuclear case:

$$\hat{H}_{\text{dd}}^{\text{homo}} = -d\cdot\frac{1}{2}(3\cos^2\theta - 1)[3\hat{I}_z\hat{S}_z - \hat{\mathbf{I}}\cdot\hat{\mathbf{S}}] \tag{4.18}$$

in angular frequency units, where d is the dipolar-coupling constant.

Thus, in the heteronuclear case, the dipolar-coupling hamiltonian is truncated further, and term B, important in homonuclear dipolar coupling, is unimportant to first order for heteronuclear dipolar coupling.

Another way to understand this is to appreciate that the term B in the dipolar-coupling hamiltonian represents the effects of the transverse components of the local field due to spin S, i.e. those in the plane perpendicular to $\mathbf{B}_0$, the applied field; hence the dependence of the B term on $\gamma_S \hat{S}_\alpha$, $\alpha = x, y$, which is the α-component of the magnetic dipole moment associated with spin S. Thus the term B is only significant if the transverse component of the magnetic field associated with spin S precesses at or near the resonance frequency of spin I, when it can cause transitions and so disturb the spin system. The transverse components of the local magnetic field due to a spin precess at the Larmor frequency of that spin, so this is only the case if spins I and S are the same nuclide (and with similar chemical shifts).

4.1.4 The effect of heteronuclear dipolar coupling on the spin system

The effect of heteronuclear dipolar coupling on a spin system has the same form as the more familiar heteronuclear *J* or scalar coupling. The line-broadening effects of term *B* are now absent. The effect on the spectrum is much simpler than for homonuclear dipolar coupling. As seen for homonuclear coupling, the term *A* simply shifts the energies of the nuclear spin levels, much in the same way as (first-order) scalar coupling. The form of the spin states is unchanged from the original Zeeman basis, i.e. there is no mixing of the Zeeman states under heteronuclear dipolar coupling, so the simple product basis functions

$$\psi_{I,m_I}\psi_{S,m_S} = |I,m_I\rangle|S,m_S\rangle = |I,m_I;S,m_S\rangle \tag{4.19}$$

are the eigenfunctions of the heteronuclear dipolar hamiltonian. The resulting dipolar contributions to the energy levels of the *IS* spin system are therefore

$$\begin{aligned} E_{\mathrm{dd}}^{\mathrm{hetero}}/\hbar &= \langle Im_I;Sm_S|\hat{H}_{\mathrm{dd}}^{\mathrm{hetero}}|Im_I;Sm_S\rangle \\ &= -d(3\cos^2\theta - 1)m_I m_S \end{aligned} \tag{4.20}$$

where m_I and m_S are the *z*-components of spin for spins *I* and *S*. Using the selection rule $\Delta m_I = \pm 1$ for the *I* spin spectrum and $\Delta m_S = \pm 1$ for the *S* spin spectrum, we can now establish the transition frequencies for the *I* and *S* spins in the dipolar-coupled *IS* spin system.

For the *I* spin they are:

$$\omega_{\mathrm{dd}}^{I}(\theta) = \omega_0^I \pm \frac{1}{2}d(3\cos^2\theta - 1) \tag{4.21}$$

where ω_0^I is the transition frequency in the absence of dipolar coupling between *I* and *S*.

If *I* and *S* are both spin-$\frac{1}{2}$ nuclei, there are two *I* spin (and two *S* spin) transitions (as shown in Fig. 4.6), each with a $(3\cos^2\theta - 1)$ dependence, and so for powder samples each gives rise to the typical powder pattern seen for axial chemical shift anisotropy. However, one transition has a $+(3\cos^2\theta - 1)$ dependence, while the other has a $-(3\cos^2\theta - 1)$ dependence, so the powder pattern arising from one transition is the mirror image of that arising from the other transition. Both transitions have the same isotropic frequency, ν_{iso}^I, however. The general form of the lineshape is found to be that shown in Fig. 4.7.

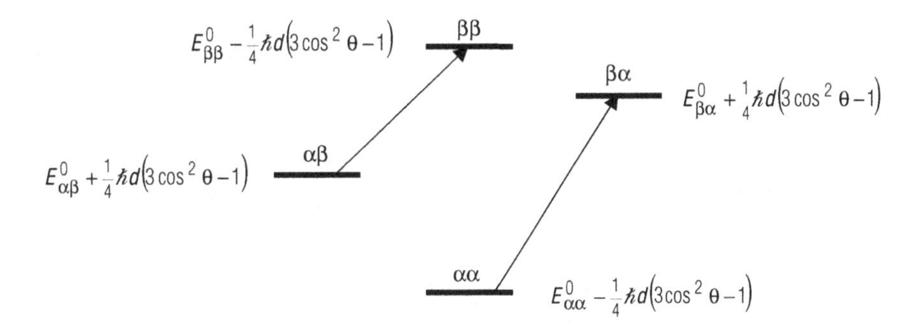

Fig. 4.6 The I spin transitions in a two-spin heteronuclear spin system (IS). The form of the spectrum for spin I in a powder sample is shown in Fig. 4.7. The $E^0_{\alpha\alpha}$, etc., are the energies of the spin levels in the absence of dipolar coupling between I and S.

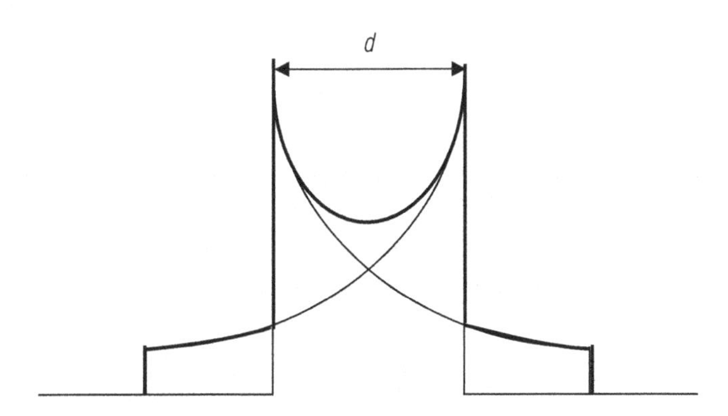

Fig. 4.7 The powder lineshape for the I spin in a heteronuclear two-spin system. The splitting of the 'horns' is equal to the dipolar coupling constant d. The horns are created by crystallites in which the I–S internuclear vector is perpendicular to the applied field $\mathbf{B_0}$.

4.1.5 Heteronuclear spin dipolar-coupled to a homonuclear network of spins

A special case of heteronuclear dipolar coupling occurs when a spin I is dipolar-coupled to a heteronuclear spin S, which is itself part of a homonuclear dipolar-coupled network of S spins. Suppose that the S spin which is closest, and therefore most strongly coupled to the I spin, is S_1. First consider just the S spins in the absence of I. As we have seen in Section 4.1.2, the spin state, i.e. α or β of the S_1 spin, varies as we make repeated observations of the S spin system. If we now introduce the I spin, it is clear the dipolar coupling between I and the S_1 spin will also vary as we make repeated observations of the I spin and its dipolar coupling, as the I–S dipolar coupling depends on the S_1 spin state. In such a case, we will observe large, homogeneous linebroadenings in the spectrum of I, reminiscent of those we might expect in the S spin spectrum, and for similar reasons. Any consideration of the IS_n spin system requires the entire network of S spins to be taken into account.

4.1.6 The spherical tensor form of the dipolar hamiltonian

So far, all our discussions of dipolar coupling have used the dipolar hamiltonian in Cartesian form, i.e. in terms of Cartesian spin operators and using Cartesian second-rank tensors to describe the spatial orientation dependence of the interaction. This approach is completely correct, but not always convenient. An alternative is to express the hamiltonian in terms of *spherical tensor operators* and *spherical tensors* describing the spatial orientation dependence of the interaction. This approach is fully described in Box 4.2 below. This form of the operator is particularly useful if we wish to perform rotations on the axis frame within which the dipolar hamiltonian is described. Thus, it is the natural form from which to describe the effects of magic-angle spinning. It is also the most useful form to start with if we wish to do any calculations of average hamiltonians which require transformations to a toggling frame, of which there are several examples later. Many of the techniques described in the remainder of this chapter, which use dipolar coupling, are analysed in terms of average hamiltonian theory and the dipolar coupling hamiltonian in spherical tensor form.

Box 4.2: The dipolar hamiltonian in terms of spherical tensor operators

The dipolar hamiltonian consists of a so-called *spin part*, consisting of spin operators, and a *spatial part*, which consists of geometrical functions describing the molecular-orientation dependence of the strength of the interaction. All nuclear spin interactions can be broken down in this way. In this Box, we show how the spin part of the dipolar hamiltonian can be written in terms of entities called *spherical tensor operators* [1, 2] and the space part in terms of *spherical tensor functions*. We then use this representation of the dipolar-coupling hamiltonian to derive the form of the hamiltonian under magic-angle spinning.

Spherical tensor operators

In this book, we have for the most part expressed hamiltonians and density operators in terms of the Cartesian spin operators, $\hat{I}_x$, $\hat{I}_y$, $\hat{I}_z$, etc. and products of these. Conceptually, these certainly provide a convenient way of describing spin interactions. However, we do not *have* to express spin hamiltonians and spin density operators in terms of these; there are other sets of operators that we might equally well use. Spherical tensor operators are one such set. Where we use these and where we use Cartesian operators is largely a matter of choice, but it is certainly true to say that for every situation, one set of operators will prove more convenient than another. Spherical tensor operators have specific symmetry properties with respect to rotations in three-dimensional space, and this makes their manipulation under such rotations much more straightforward than, for instance, that of products of Cartesian spin operators [1, 2]. Furthermore, we have seen elsewhere in this book how analysis of pulse sequences involves the use of a toggling frame, which in turn necessitates the rotation of the hamiltonian operator currently appropriate to the spin system. Clearly, under these circumstances, spherical tensor operators are likely to be a more convenient basis set for writing the hamiltonian than are other possible basis sets.

So-called *irreducible spherical tensor operators* transform according to the irreducible representations of the three-dimensional rotation point group [1, 2], also called the full rotation point group. An irreducible tensor operator is labelled in the following by $\hat{T}_{kq}$, where k is the *rank* of the operator. For a given k, there are $2k + 1$ different irreducible tensor operators, each with a different *order*, q; the possible values of q are $k, k - 1, \ldots, -k$. The set of $\{\hat{T}_{kk}, \hat{T}_{kk-1}, \ldots, \hat{T}_{k-k}\}$, i.e. irreducible tensor operators for a given k, transform under rotations like the kth irreducible representation of the full rotation point group. The q can therefore be thought of as designating different components of a set. It is this feature which makes these operators *tensor* operators. A useful analogy can be drawn between irreducible spherical tensor operators and atomic orbitals. Atomic orbitals also transform like the irreducible representations of the full rotation group, although clearly atomic orbitals are functions rather than operators. An *s*-orbital, for instance, has the same symmetry as the $\hat{T}_{00}$ irreducible operator, i.e. operator with $k = 0$ and (therefore) $q = 0$. A set of *p*-orbitals, i.e. three of them, transforms like the set of operators with rank 1, i.e. $\hat{T}_{11}$, $\hat{T}_{10}$, $\hat{T}_{1-1}$.

A specific symmetry property of irreducible tensor operators is that, under a rotation, their rank does not change. This is a consequence of their transforming like an irreducible representation of the spherical point group. It can be seen in the analogy with atomic orbitals; if you rotate a *p*-orbital, the result is another *p*-orbital, or more generally a linear combination of *p*-orbitals. Under a rotation of

Continued on p. 166

Box 4.2 *Cont.*

the axis frame in which the tensor is defined, from $(x^{old}, y^{old}, z^{old})$ to $(x^{new}, y^{new}, z^{new})$, the $\hat{T}_{kq}$ tensor operators transform as follows:

$$\begin{aligned}\hat{T}_{kq}(new) &= \hat{R}^{-1}(\alpha,\beta,\gamma)\hat{T}_{kq}(old)R(\alpha,\beta,\gamma) \\ &= \sum_{q'} \hat{T}_{kq'}(old)D^{k}_{q'q}(\alpha,\beta,\gamma)\end{aligned} \quad \text{(i)}$$

where $\hat{T}_{kq}(new)$ is the tensor element expressed in the new frame and $\hat{T}_{kq}(old)$ the tensor element expressed in the old frame. $\hat{R}(\alpha, \beta, \gamma)$ is the rotation operator described in Box 1.2, Chapter 1, and (α, β, γ) are the Euler angles which rotate the old frame $(x^{old}, y^{old}, z^{old})$, into the new frame $(x^{new}, y^{new}, z^{new})$, i.e. rotation of a frame initially coincident with $(x^{old}, y^{old}, z^{old})$ by γ about z^{old}, followed by rotation of β about y^{old}, then rotation by α about z^{old} takes this frame into $(x^{new}, y^{new}, z^{new})$. The $D^{k}_{q'q}(\alpha, \beta, \gamma)$ are elements of a matrix, a *Wigner rotation matrix* [1, 2], which can be decomposed into

$$D^{k}_{q'q}(\alpha,\beta,\gamma) = \exp(-i\alpha q')\exp(-i\gamma q)d^{k}_{q'q}(\beta) \quad \text{(ii)}$$

where $d^{k}_{q'q}(\beta)$ is a *reduced Wigner function* depending only on the Euler angle β. These functions are tabulated in many places.

So, in essence, rotating the axis frame in which an irreducible tensor operator is defined about any axis simply transforms the irreducible tensor operator into a linear combination of irreducible tensor operators of the same rank.

Using this rotation property, we can find linear combinations of Cartesian spin operators and products of spin operators which behave as irreducible tensor operators of different ranks, k, and orders, q, i.e. transform like the kth irreducible representation of the full rotation group (see references [1] and [2] for detailed discussions on how to construct these). For instance, if we use only single spin operators, then

$$\begin{aligned}&\hat{T}_{00} = \frac{1}{\sqrt{2}} && \text{i.e. a scalar} && \\ &\hat{T}_{10} = \sqrt{2}\hat{I}_z && \hat{T}_{1\pm1} = \mp\sqrt{\frac{1}{2}}(\hat{I}_x \pm i\hat{I}_y) && \\ &\hat{T}_{20} = \sqrt{\frac{1}{6}}(3\hat{I}_z^2 - \hat{\mathbf{I}}\cdot\hat{\mathbf{I}}) && \hat{T}_{2\pm1} = \mp(\hat{I}_\pm\hat{I}_z + \hat{I}_z\hat{I}_\pm) && \hat{T}_{2\pm2} = \frac{1}{2}\hat{I}_\pm^2\end{aligned} \quad \text{(iii)}$$

are the appropriate combinations of spin operators. If we have spin operators associated with two spins available, i.e. $\hat{I}_x$, $\hat{I}_y$, $\hat{I}_z$ and $\hat{S}_x$, $\hat{S}_y$, $\hat{S}_z$, then we can generate more irreducible tensor operators:

$$\hat{T}_{00}^{IS} = -\frac{2}{\sqrt{3}}\left(\hat{I}_z\hat{S}_z + \frac{1}{2}(\hat{I}_+\hat{S}_- + \hat{I}_-\hat{S}_+)\right)$$

$$\hat{T}_{10}^{IS} = \frac{1}{\sqrt{2}}(\hat{I}_-\hat{S}_+ - \hat{I}_+\hat{S}_-) \quad \hat{T}_{1\pm1}^{IS} = (-\hat{I}_\pm\hat{S}_z + \hat{I}_z\hat{S}_\pm) \tag{iv}$$

$$\hat{T}_{20}^{IS} = \sqrt{\frac{1}{6}}(3\hat{I}_z\hat{S}_z - \hat{\mathbf{I}}\cdot\hat{\mathbf{S}}) \quad \hat{T}_{2\pm1}^{IS} = \mp\frac{1}{2}(\hat{I}_\pm\hat{S}_z + \hat{I}_z\hat{S}_\pm) \quad \hat{T}_{2\pm2}^{IS} = \frac{1}{2}\hat{I}_\pm\hat{S}_\pm$$

where the *IS* superscripts on the tensor operators simply show which spins are involved in each.

Interaction tensors

In Chapter 1, we introduced the idea that nuclear spin interactions, such as dipolar coupling, could be expressed in terms of second-rank Cartesian tensors, **D** in the case of dipolar coupling, which expresses the orientation dependence of the interaction. We can equally well describe the orientation dependence of spin interactions in terms of *spherical tensors*. Note here that we are talking about tensors whose components are numbers or functions rather than operators as above. There is a link between these two, however, which is that both spherical tensor operators and the spherical tensors describing the orientation dependence of spin interactions transform as irreducible representations of the spherical point group. Before we can usefully write the dipolar hamiltonian in terms of spherical tensor operators, we will need to express the spatial orientation dependence of the dipolar interaction in terms of spherical tensor components also.

The transformations of spherical tensors under axis frame rotations are, not surprisingly, described in the same way as for tensor operators above, i.e. under a rotation of the axis frame described by Euler angles (α, b, γ), a numerical spherical tensor component Λ_{lm}, becomes

$$\hat{R}^{-1}(\alpha,\beta,\gamma)\Lambda_{lm}\hat{R}(\alpha,\beta,\gamma) = \sum_{m'}\Lambda_{lm'}D_{m'm}^{l}(\alpha,\beta,\gamma) \tag{v}$$

where the $D_{m'm}^{l}(\alpha, \beta, \gamma)$ are once again elements of a Wigner rotation matrix. The effect of rotating the defining axis frame on a Cartesian tensor, such as the dipolar tensor **D**, is described in Box 1.2, Chapter1.

The homonuclear dipolar hamiltonian under static and MAS conditions

In Chapter 1, we stated that the hamiltonians describing a spin interaction, *A*, could always be written in the form

Continued on p. 168

Box 4.2 Cont.

$$\hat{H}_A = \hat{\mathbf{I}}\cdot\mathbf{A}_{\text{loc}}\cdot\hat{\mathbf{J}} \tag{vi}$$

where $\mathbf{A}_{\text{loc}}$ is a second-rank Cartesian tensor which describes the strength and orientation dependence of the local spin interaction, and $\hat{\mathbf{J}}$ is a vector operator, whose exact nature depends on the particular spin interaction. For dipolar coupling between two spins, *I* and *S*, the hamiltonian expressed in this form is (in angular frequency units)

$$\hat{H}_{\text{dd}} = -2\hat{\mathbf{I}}\cdot\mathbf{D}\cdot\hat{\mathbf{S}} \tag{vii}$$

The second-rank Cartesian tensor **D** contains the information on the spatial dependence of the interaction. In Appendix B, we show how to transform equations of this form into spherical tensor form [3]. The result is an equation of the form:

$$\hat{H}_A = \hat{\mathbf{I}}\cdot\mathbf{A}_{\text{loc}}\cdot\hat{\mathbf{J}} \equiv \sum_{k=0}^{2}\sum_{q=-k}^{+k}(-1)^q \Lambda^A_{k-q}\hat{T}^A_{kq} \tag{viii}$$

where Λ^A_{k-q} is a component of a spherical tensor, Λ^A_k, which describes the spatial orientation dependence of the interaction *A* and $\hat{T}^A_{kq}$ is a component of a spherical tensor operator appropriate for interaction *A*. As shown in Appendix B, the $\hat{T}^A_{kq}$ are derived from coupling $\hat{\mathbf{I}}$ and $\hat{\mathbf{J}}$ in Equation (vi) together, and so the form of the $\hat{T}^A_{kq}$ operators depends on the interaction *A*. Appendix B lists the $\hat{T}^A_{kq}$ for dipolar coupling, chemical shielding and quadrupole coupling interactions. Here, we simply make a few salient observations. The combination of a $+q$ component of $\mathbf{\Lambda}_k$ and $-q$ component of $\hat{\mathbf{T}}_k$ in Equation (viii) ensures that the result of the product in that equation is a scalar, as it must be since the hamiltonian operator is a scalar operator (as it is the operator for energy, which is a scalar quantity). The summation over *k* in Equation (viii) goes up to $k = 2$, which takes account of the fact that Equations (vi) and (vii) are *bilinear* in Cartesian vector operators; a glance at Equation (iv) shows that the $k = 2$ $\hat{T}_{kq}$ operators also have this form. We now seek to find an expression of the form of Equation (viii) for the homonuclear dipolar hamiltonian.

The first-order average homonuclear dipolar hamiltonian in Cartesian form in the rotating frame of reference (*z* parallel to $\mathbf{B}_0$) for dipolar coupling between two spins, *I* and S, is

$$\hat{H}^{\text{homo}}_{\text{dd}} = -C_{IS}[3\hat{I}_z\hat{S}_z - \hat{\mathbf{I}}\cdot\hat{\mathbf{S}}] \tag{ix}$$

for a static, i.e. non-spinning, sample, where the constant C_{IS} is given by

$$C_{IS} = d\frac{1}{2}(3\cos^2\theta - 1)$$

where d is the dipolar-coupling constant for the IS spin pair and θ is the angle between the I–S internuclear vector and $\mathbf{B}_0$, as defined in Fig. 4.2.

The spin part of this hamiltonian can clearly be identified with the spherical tensor operator $\hat{T}_{20}^{IS}$ in Equation (iv). Thus, the summation in Equation (viii) reduces to a single term with $k = 2$ and $q = 0$

$$\hat{H}_{\mathrm{dd}}^{\mathrm{homo}} = -\Lambda_{20}^{IS}\hat{T}_{20}^{IS} \tag{x}$$

where Λ_{20}^{IS} can be found, by comparing Equations (ix) and (x), to be

$$\Lambda_{20}^{IS} = -\sqrt{6}C_{IS} = -\sqrt{\frac{3}{2}}d(3\cos^2\theta - 1) \tag{xi}$$

in which θ is the angle between the dipolar tensor principal z-axis (z^{PAF}) and the laboratory frame z-axis (parallel to $\mathbf{B}_0$). The IS superscipt on the spherical tensor component Λ_{20} simply designates the particular spin pair; d is the dipolar-coupling constant for the IS spin pair. The tensor operator $\hat{T}_{20}^{IS}$ is defined in Equation (iv) in terms of Cartesian spin operators defined with respect to the laboratory frame.

We can equally well derive Λ_{20}^{IS} from the dipolar-coupling tensor in spherical tensor form expressed in its principal axis frame. The dipolar-coupling tensor in its principal axis frame (PAF) for a spin pair IS is denoted $\boldsymbol{\lambda}^{IS}$ in spherical tensor form. The PAF is simply the axis frame in which the dipolar-coupling tensor is diagonal; this frame depends on the molecular structure of course. As shown in Appendix B, the only non-zero component of $\boldsymbol{\lambda}^{IS}$ is in fact $\lambda_{20}^{IS} = -\sqrt{6}d$. Thus the dipolar-coupling hamiltonian when the molecular orientation is such that the dipolar-coupling principal axis frame coincides with the laboratory frame, i.e. internuclear axis of the dipolar-coupled spin pair (z^{PAF}) is parallel to $\mathbf{B}_0$ (z in the laboratory frame), is

$$\hat{H}_{\mathrm{dd}}^{\mathrm{PAF}} = \lambda_{20}^{IS}\hat{T}_{20}^{IS} = -d(3\hat{I}_z\hat{S}_z - \hat{\mathbf{I}}\cdot\hat{\mathbf{S}}) \tag{xii}$$

as it should be. Of course, in general, the PAF does not coincide with the laboratory frame. It is fixed to the molecule and so varies with molecular orientation. Different PAF orientations with respect to the laboratory frame (see Fig. B4.2.1) reflect different strengths of dipolar coupling for different molecular orientations. More generally, the PAF will be orientated by the Euler angles (α, β, γ) with respect to the laboratory frame,[N2] where (α, β, γ) are the Euler angles which rotate the PAF into the laboratory frame. To find the strength of dipolar coupling for some arbitrary orientation of the PAF (i.e. arbitrary molecular orientation), we need to express the dipolar coupling tensor $\boldsymbol{\lambda}^{IS}$ in the laboratory frame, where we label it $\boldsymbol{\Lambda}^{IS}$.

The dipolar-coupling tensor in the laboratory and (PAF) frames is related via (see Equation (v))

Continued on p. 170

Box 4.2 *Cont.*

$$\Lambda_{2m}^{IS} = \sum_{m'=-2}^{+2} D_{m'm}^{2}(\alpha,\beta,\gamma)\lambda_{2m'}^{IS}$$
$$= D_{0m}^{2}(\alpha,\beta,\gamma)\lambda_{20}^{IS} \qquad \text{(xiii)}$$

where we have used the fact that λ_{20}^{IS} is the only non-zero component of $\boldsymbol{\lambda}^{IS}$ and where $D_{0m}^{2}(\alpha, \beta, \gamma)$ is a Wigner rotation matrix element, defined in Equation (ii). For the truncated dipolar hamiltonian of Equation (x), we are only interested in the Λ_{20}^{IS} component. The other components of $\boldsymbol{\Lambda}^{IS}$ are only relevant for the higher-order terms in the dipolar-coupling hamiltonian (see Section 4.1.1 for details of these other terms which arise from higher-order terms in the average hamiltonian). So we have for Λ_{20}^{IS} from Equation (xiii)

$$\Lambda_{20}^{IS} = D_{00}^{2}(\alpha,\beta,\gamma)\lambda_{20}^{IS}$$
$$= \frac{1}{2}(3\cos^2\beta - 1)\lambda_{20}^{IS} \qquad \text{(xiv)}$$

where we have substituted for the Wigner rotation matrix element $D_{00}^{2}(\alpha, \beta, \gamma)$ in the last line of Equation (xiv).

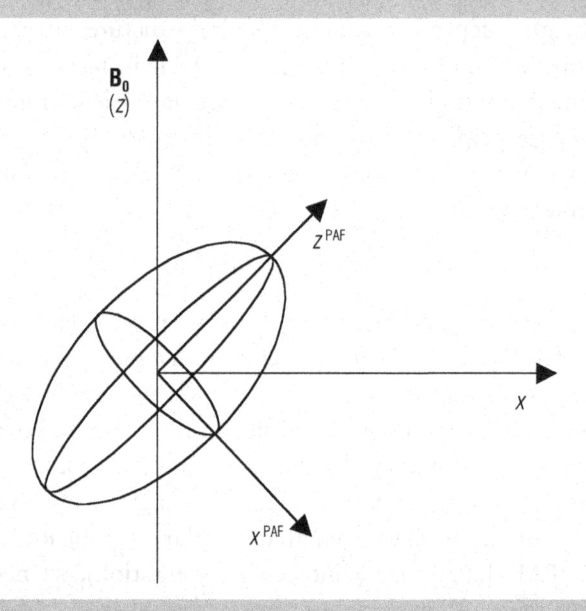

Fig. B4.2.1 The relationship between the laboratory frame (*x*, *y*, *z*) and dipolar-coupling tensor principal axis frame (PAF). The applied field $\mathbf{B}_0$ defines the *z*-axis of the laboratory frame. The dipolar-coupling tensor PAF *z*-axis is defined by the internuclear axis of the dipolar-coupled spin pair. The tensor is represented in this diagram by an ellipsoid whose principal axes are in the directions of the PAF; the radius of the ellipsoid in the direction of $\mathbf{B}_0$ is proportional to the strength of dipolar coupling for that orientation of PAF.

We now need to consider the effects of magic-angle spinning. To do this, we need to describe an extra axis frame, the rotor frame, which we designate (x^R, y^R, z^R). This frame is fixed to the rotor. A molecule inside the rotor changes its orientation as the rotor spins, and so the orientation of the PAFs associated with all spins in that molecule also changes. As in the static case, we need to express the dipolar-coupling tensor in the laboratory frame to find the strength of dipolar coupling for different PAF orientations. In the case of magic-angle spinning, we do this in two steps, first finding the dipolar-coupling tensor with respect to the rotor axis frame and then relating the tensor in that frame to the laboratory frame. The PAF remains fixed in orientation relative to the rotor-fixed frame; the Euler angles (α, β, γ) now relate the PAF to the rotor frame, R. The dipolar-coupling tensor in the rotor frame, $\mathbf{\Lambda}^{IS}$(R), is then given by an equation of the same form as Equation (xiii) previously, i.e.

$$\Lambda_{2m}^{IS}(R) = D_{0m}^{2}(\alpha,\beta,\gamma)\lambda_{20}^{IS} \qquad \text{(xv)}$$

where λ^{IS} is the dipolar-coupling tensor in its PAF as before.

The dipolar tensor in the laboratory frame $\mathbf{\Lambda}^{IS}$(lab) is then found from $\mathbf{\Lambda}^{IS}$ (R) in a similar manner:

$$\begin{aligned}\Lambda_{2m}^{IS}(\text{lab}) &= \sum_{m'=-2}^{+2} D_{m'm}^{2}(-\omega_R t,\theta_R,0)\Lambda_{2m'}^{IS}(\text{R}) \\ &= \sum_{m'=-2}^{+2} D_{m'm}^{2}(-\omega_R t,\theta_R,0)D_{0m'}^{2}(\alpha,\beta,\gamma)(\lambda_{20}^{IS})\end{aligned} \qquad \text{(xvi)}$$

where the rotor frame is related to the laboratory frame by the Euler angles ($-\omega_R t$, θ_R, 0) when the rotor has been spinning at angle θ_R, and rate ω_R for time t and where the y-axes of the rotor and laboratory frame are parallel at time $t = 0$ (see Fig. B4.2.2).

We require the Γ_{20}^{IS} (lab) component of the laboratory frame dipolar coupling tensor for the first-order average dipolar hamiltonian of Equation (x). From Equation (xvi), this is given by

$$\Lambda_{2m}^{IS}(\text{lab}) = \sum_{m'=-2}^{+2} D_{m0}^{2}(-\omega_R t,\theta_R,0)D_{0m}^{2}(\alpha,\beta,\gamma)\lambda_{20}^{IS} \qquad \text{(xvii)}$$

Expanding the Wigner rotation matrix elements (Equation (ii)) relating the rotor to the laboratory frame in this expression then gives

$$\begin{aligned}\Lambda_{2m}^{IS}(\text{lab}) &= \sum_{m'=-2}^{+2} \exp(im\omega_R t)d_{m0}^{2}(\theta_R)D_{0m}^{2}(\alpha,\beta,\gamma)\lambda_{20}^{IS} \\ &= \sum_{m'=-2}^{+2} \exp(im\omega_R t)\omega_{IS}^{(m)}\end{aligned} \qquad \text{(xviii)}$$

Continued on p. 172

Box 4.2 Cont.

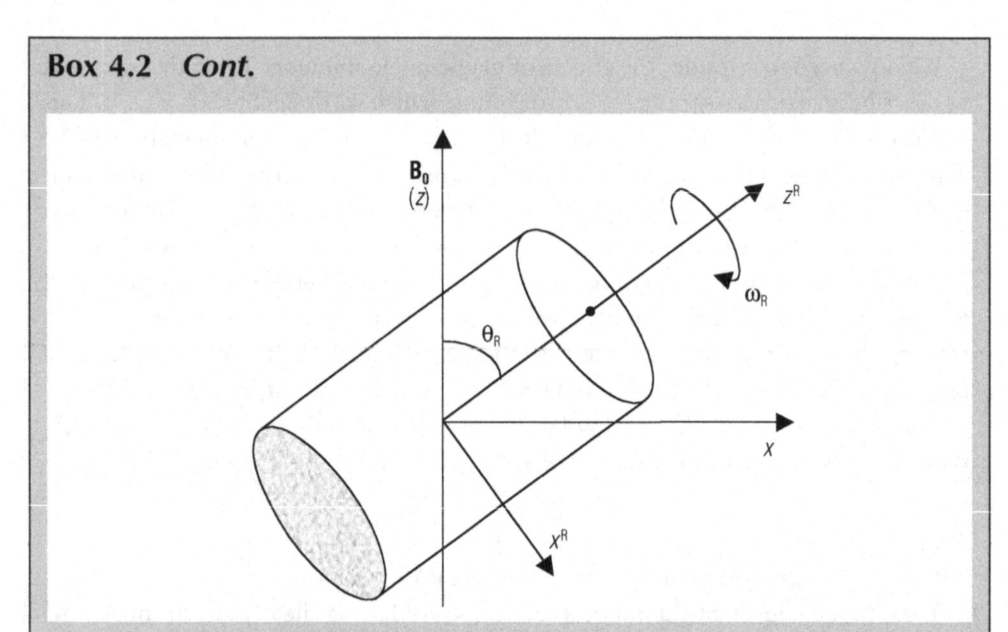

Fig. B4.2.2 The relationship between the laboratory frame and the rotor frame. The applied field $\mathbf{B}_0$ defines the *z*-axis of the laboratory frame. The rotor frame is fixed to the rotor and so spins with it under magic-angle spinning. θ_R is the angle between the rotor spinning axis and $\mathbf{B}_0$; ω_R is the spinning rate. Thus the rotor turns through an angle $\omega_R t$ after a time *t*. The rotor frame is shown at time $t = 0$.

where

$$\omega_{IS}^{(m)} = \sum_{m=-2}^{+2} d_{m0}^{2}(\theta_R) D_{0m}^{2}(\alpha,\beta,\gamma)\lambda_{20}^{IS} \tag{xix}$$

Thus the first-order average homonuclear dipolar hamiltonian under magic-angle spinning in spherical tensor form (for a two-spin system) is

$$\hat{H}_{dd}(t) = \hat{T}_{20}^{IS} \sum_{m=-2}^{+2} \exp(im\omega_R t)\omega_{IS}^{(m)} \tag{xx}$$

4.2 Introduction to the uses of dipolar coupling

The dipolar-coupling interaction can cause huge linebroadening as described previously; even where it is not huge, it still causes significant loss of resolution. Thus to the spectroscopist, dipolar coupling is often more of a problem than anything else. Chapter 2 deals with schemes for removing the effects of both homonuclear and heteronuclear dipolar coupling from NMR spectra. However, to the chemist and materials scientist, dipolar coupling probably represents the single most important spin interaction, and one which we would very much like to be able to measure. This is because of the direct dependence of its magnitude on the distance, r_{IS}, between the two spins I and S. The dipolar coupling is then the one spin interaction which can provide a means of determining internuclear distances and, from these, the geometry and conformation of molecules. Of course, this sort of structure determination is the traditional domain of diffraction techniques. However, there are many types of materials which do not lend themselves readily to diffraction techniques. Many compounds of interest, for instance biopolymers such as proteins, have an inherent degree of disorder and so do not form single crystals. Thus, any diffraction study can only produce rather limited information. This is not to say that there are not clear structural features in such materials; most have well defined secondary and tertiary structures, albeit along with some molecular degrees of freedom which confer a dynamic disorder on the overall structure. Other important materials, such as zeolites, are highly crystalline, but only form microcrystals, too small for single crystal diffraction studies. So-called amorphous materials always have some structure, despite their name, but the lack of regular repeating units (translational symmetry) means that, once again, diffraction techniques have only limited use. Nevertheless, it is possible to pick out common structural units in most such materials. Moreover, a knowledge of these structural units is essential to our understanding of the material's physical properties. Solid-state NMR gives the hope of structural information on all these types of material, via measurement of dipolar couplings.

A more qualitative use for dipolar coupling is in correlation spectroscopy (COSY) (Fig. 4.8). Such techniques are common in solution-state NMR, where the scalar or J coupling is used to determine which spins are linked by chemical bonds. A two-dimensional COSY spectrum showing a signal at (f_1, f_2) in the two-dimensional plane indicates that the spins giving rise to the signals at f_1 and f_2 are close together in the bonding network of the molecule under study. In solid-state NMR, dipolar coupling between spins can be used in a similar way to indicate which spins are close in space.

Another use of dipolar coupling is in spin counting, which is the determination of the number of spins in close proximity (Fig. 4.9). As shown in Box 4.1, in a homonuclear, many-spin system, the eigenstates (or coupled spin states) of the system are described by the total nuclear spin in the laboratory z-direction, M

$$M = m_1 + m_2 + m_3 + \ldots + m_N \tag{4.22}$$

where N is the number of coupled spins and m_i is the spin state of spin i. Remembering that the strength of dipolar coupling drops off as $1/r^3$, only those spins in close proximity will contribute to the coupled spin state. N

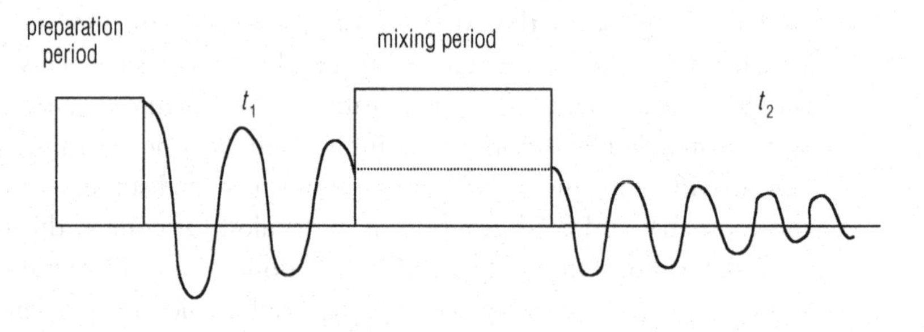

Frequency spectrum:

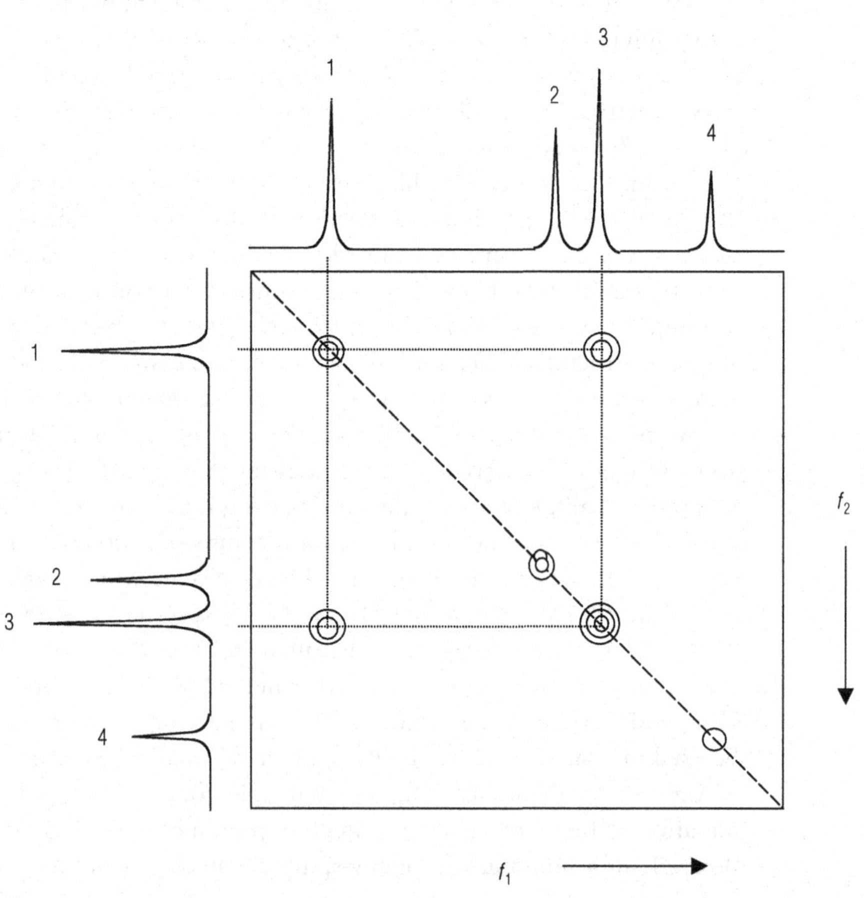

Fig. 4.8 The form of a homonuclear correlation experiment and spectrum. In the experiment, spins are allowed to evolve at their characteristic frequency during t_1 and t_2. During the mixing period between t_1 and t_2 magnetization is transferred between spins via the dipolar coupling or scalar coupling between them. A two-dimensional dataset is collected as a function of t_1 and t_2. Two-dimensional Fourier transformation of this time domain dataset then produces a two-dimensional frequency spectrum, with frequency axes labelled f_1 and f_2 (corresponding to the t_1 and t_2 time domain axes respectively). Signals along the $f_1 = f_2$ diagonal are *autocorrelation* peaks and arise from magnetization that did not transfer between spins during the mixing period (as well as possibly magnetization that transferred between like spins). This particular two-dimensional spectrum shows an off-diagonal peak between signals from spins 1 and 3, which in turn shows that magnetization has transferred between spins 1 and 3 during the mixing period. If the mixing period relied on dipolar-coupling effects for magnetization transfer, then this demonstrates that these spins are close in space. If the mixing period relied on the scalar or *J* coupling, then 1 and 3 are close together in the bonding network.

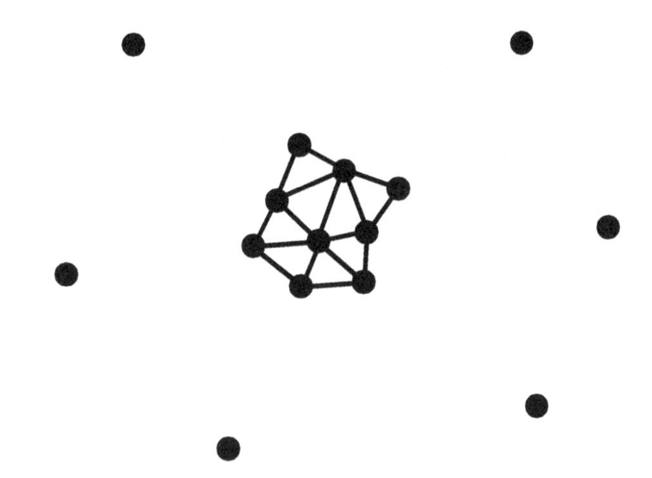

Fig. 4.9 Illustrating the concept of a *spin cluster*. Spins in close proximity are all dipole-coupled together. The states describing such a spin system are product spin states involving all N spins in the cluster and are characterized by a quantum number representing the total nuclear spin in the laboratory z-direction, M, where M is given by $M = m_1 + m_2 + \ldots + m_N$, i.e. the sum of the components provided by the N individual spins.

is thus the number of spins in close proximity, or in a cluster. There are many M states for a given cluster of N spins, and multiple-quantum coherences of order ΔM can be excited between them, where $\Delta M = M_a - M_b$, with a and b denoting the M levels involved in the coherence. The maximum order of coherence which can be excited for a given spin cluster is N for a spin-$\frac{1}{2}$ system (since the minimum M is $-\frac{1}{2} + -\frac{1}{2} + \ldots + -\frac{1}{2} = -\frac{1}{2}N$ and maximum M is $\frac{1}{2} + \frac{1}{2} + \ldots + \frac{1}{2} = \frac{1}{2}N$) and this gives a means of measuring N by investigating the orders of coherence which can be excited and their relative amplitudes.

In the remainder of this chapter, we first discuss how to measure dipolar couplings for both homonuclear and heteronuclear spin-$\frac{1}{2}$ and quadrupolar spin systems. In any NMR experiment, one of the first requirements is that of resolution. In a system where there are dipolar couplings, the requirements of resolution generally mean removing the effects of these couplings, which otherwise cause a high degree of linebroadening. However, this defeats the object of an experiment designed to measure those same couplings. The ingenuity of spectroscopists in devising experiments to measure dipolar couplings has been in dealing with these two somewhat contrary features.

4.3 Techniques for measuring homonuclear dipolar couplings

4.3.1 *Recoupling pulse sequences*

As already highlighted, a primary requirement for any NMR experiment is to produce sufficient resolution to separate signals from different chemical

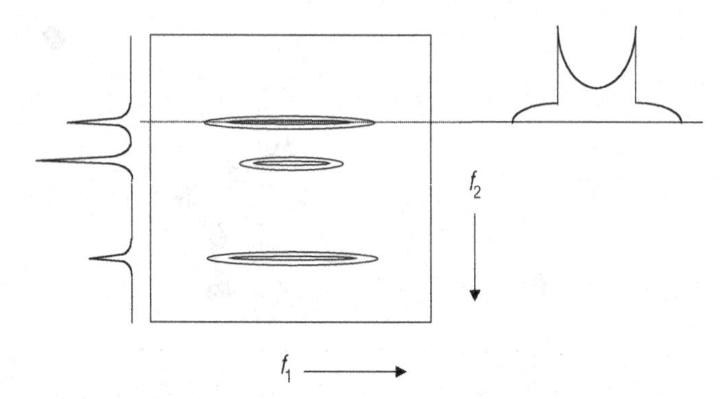

Fig. 4.10 A schematic illustration of the two-dimensional experiments used to measure homonuclear dipolar couplings. The experiment separates a high-resolution spectrum of isotropic chemical shifts in the f_2 spectral dimension from powder patterns representing the dipolar coupling interaction for each resolved site, in the f_1 spectral dimension.

sites. The linebroadening caused by dipolar coupling generally prevents this, and so its effects need to be removed from any part of the NMR experiment requiring high resolution. Magic-angle spinning is one of the simplest and most effective ways of removing the effects of dipolar coupling (see Chapter 2), and has the added bonus of removing the effects of chemical shift anisotropy also. Having effectively removed the dipolar coupling with magic-angle spinning, it can then be reintroduced for selected periods of the experiment by carefully designed pulse sequences, known as *recoupling sequences*.

In practice, dipolar couplings are measured using such pulse sequences as part of a two-dimensional experiment (Fig. 4.10). After initial excitation of transverse magnetization, a recoupling sequence is applied synchronously with the sample spinning in t_1. Thus, t_1 must (usually) be an integral number of rotor periods. An FID is then recorded as usual in t_2. Appropriate processing of the two-dimensional time domain data (recorded as a function of t_1 and t_2) then results in a two-dimensional frequency spectrum, with a high-resolution, magic-angle spinning spectrum in f_2 correlated in f_1 with spectra which represent the homonuclear dipolar coupling associated with each of the sites represented in the f_2 spectrum. For powder samples, the spectra in f_1 are generally some sort of powder pattern, although usually different from the powder patterns that would be expected in a conventional one-dimensional spectrum for a static sample.

The dipolar coupling constants ($d = (\mu_0/4\pi)(\gamma_I\gamma_S/r_{IS}^3)\hbar$ associated with each f_1 lineshape are extracted by simulating the f_1 lineshapes for different d until good agreement with the experimental lineshapes in achieved. Often, the t_1 time domain datasets are simulated rather than the frequency domain powder patterns; the two processes are equivalent. Recoupling pulse

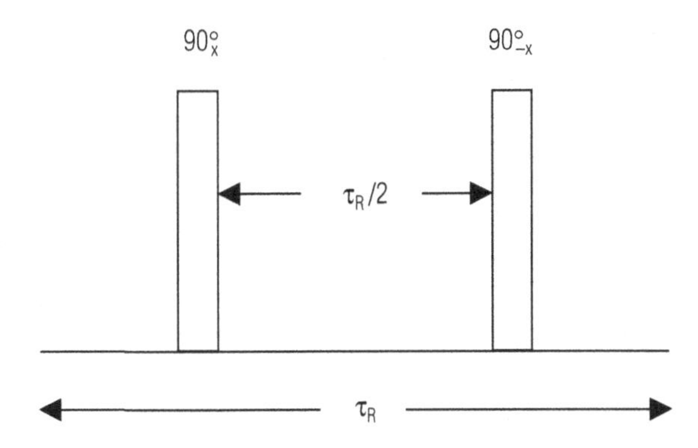

Fig. 4.11 The DRAMA (Dipolar Recovery at the Magic Angle) pulse sequence [4]. This pulse sequence has the effect of preventing the averaging to zero of homonuclear dipolar coupling interactions under magic-angle spinning. The two 90° pulses are placed symmetrically within each rotor period (τ_R) for which dipolar evolution is required.

sequences generally reintroduce *all* the homonuclear dipolar couplings associated with the irradiated spin type; clearly if more than two or three spins are coupled together, the corresponding spectra in f_1 are likely to be rather broad and featureless lineshapes which cannot be simulated unambiguously. Accordingly, these types of experiment are generally carried out on samples in which there are discrete spin pairs, often arranged through site-specific isotopic labelling of, for example, ^{13}C.

The first dipolar-recoupling pulse sequence was the DRAMA (Dipolar Recovery at the Magic Angle) sequence, invented by Tycko and co-workers [4] and although now superseded by other sequences, this pulse sequence still retains the distinction of having caught the imagination of the NMR community as to the possibilities of this type of experiment. This sequence is shown in Fig. 4.11 and consists of two 90° pulses symmetrically placed in each rotor period. Analysis of this pulse sequence requires the use of average hamiltonian theory and a toggling frame of reference as described in Box 4.3, below. This pulse sequence, like most other dipolar recoupling sequences, has the effect of also reintroducing any chemical shift anisotropy associated with the observed spins. This would give rise to powder lineshapes in f_1 which were determined by both the chemical shift anisotropy and dipolar coupling, and would complicate their analysis. Accordingly, the DRAMA sequence is usually used in the modified form [4] in Fig. 4.12, where the addition of 180° pulses in every second rotor period has the effect of refocusing chemical shift anisotropy and isotropic chemical shift offsets during the period of the pulse sequence in t_1. However, since the 180° pulses are rather sparse, high spinning frequencies (>5 kHz) are recommended for a better performance of DRAMA. Even so, DRAMA does not perform par-

(a)

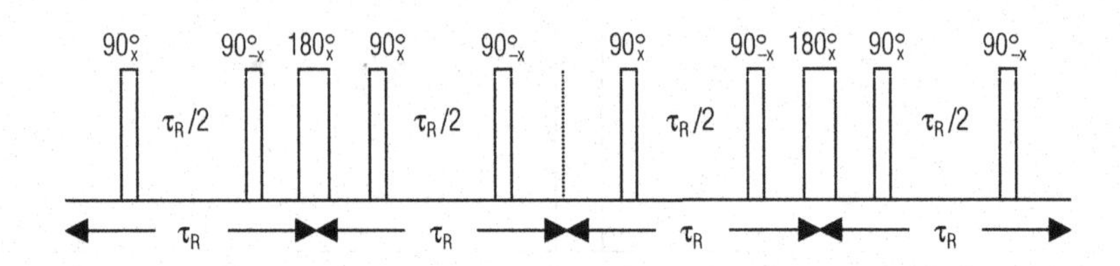

(b)

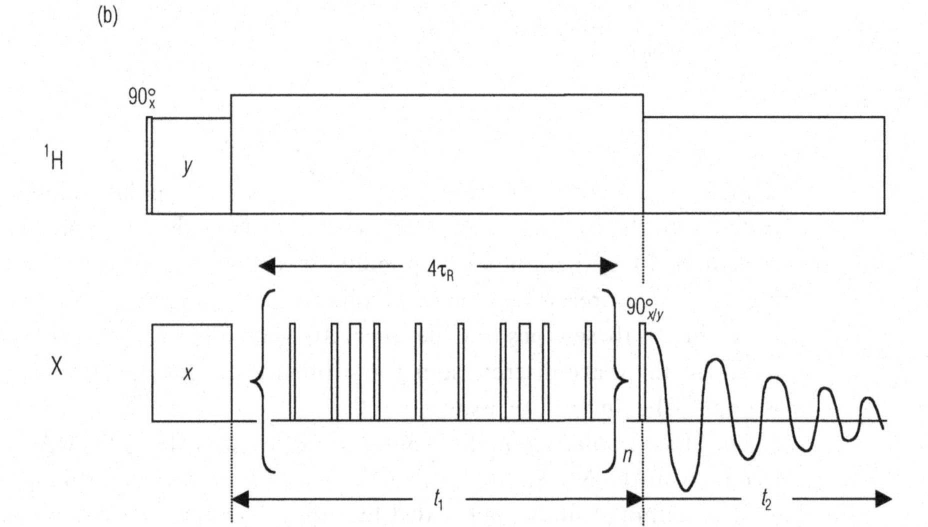

Fig. 4.12 (a) The modified form of the DRAMA pulse sequence which refocuses chemical shift anisotropies, while still preventing the averaging to zero of homonuclear dipolar couplings. (b) The two-dimensional NMR experiment which is used with the DRAMA sequence to obtain dipolar powder patterns for the X nucleus in f_1, correlated with a high-resolution magic-angle spinning spectrum in f_2. In this particular pulse sequence, cross-polarization from ^{1}H is used to generate the initial X transverse magnetization. The sequence in (a) then runs during t_1 of the experiment for $4n$ rotor cycles (τ_R). The ^{1}H decoupling power is increased during t_1 so that the ^{1}H and X rf powers no longer satisfy the Hartman–Hahn condition; this ensures that no cross-polarization inadvertently takes place during the pulse train on X during t_1.

ticularly well for ^{31}P systems for instance, because of the large chemical shift anisotropies associated with ^{31}P sites, which have a substantial effect on the spin evolution between the DRAMA 90° pulses.

In practice, two datasets are recorded, one using the pulse sequence in Fig. 4.12 with a 90°_x pulse at the end, and the other with a 90°_y pulse at the end. Dataset 1 then is modulated by the x-component of the transverse magnetization present in t_1, while dataset 2 is modulated by the y-component, i.e. the x- and y-components of the t_1 magnetization are effectively recorded

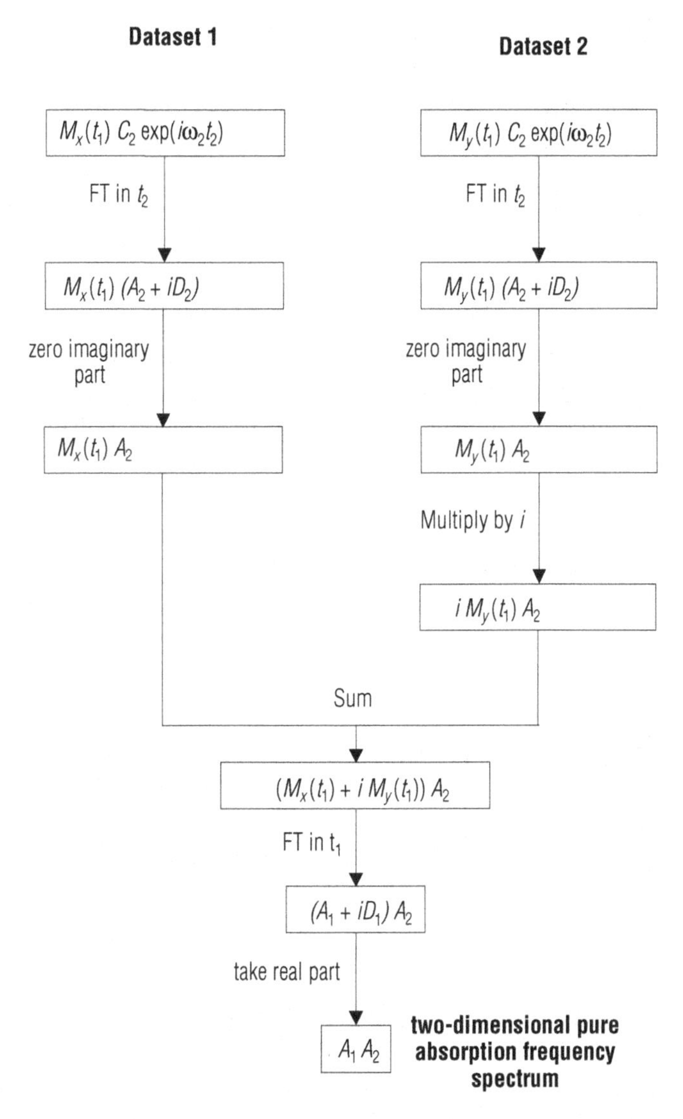

Fig. 4.13 The processing scheme for the two-dimensional time domain datasets arising from the experiment of Fig. 4.11, which uses a recoupling pulse sequence, such as the DRAMA pulse sequence (Figs 4.11 and 4.12) during t_1 of the experiment. The two datasets are produced in two separate experiments and have the forms indicated at the top of the diagram (see text for details). $M_x(t_1)$ and $M_y(t_1)$ are the *x*- and *y*-components of the X spin transverse magnetization in the t_1 period of the experiment. $C_2 \exp(i\omega_2 t_2)$ is the form of the signal recorded in t_2. The C_2 coefficient is the amplitude of the signal during the t_2 period of the experiment; it normally contains a decay term which is a function of time (i.e. accounts for transverse relaxation of the transverse magnetization in t_2). A_1 and A_2 are the real parts of the frequency spectra arising from Fourier transformation (FT) of the t_1 and t_2 time domain signals and D_1 and D_2 are their imaginary counterparts. The *A* terms correspond to absorption mode lineshapes, while the *D* terms correspond to dispersive mode lineshapes.

in separate experiments. This then allows pure absorption lineshapes to be obtained in the final two-dimensional frequency spectrum, if the processing scheme in Fig. 4.13 is followed. Similar processing can be applied to experiments employing other recoupling pulse sequences, but note that the final

Fourier transformation in t_1 is often omitted, and the t_1 data are simulated rather than the Fourier transformed f_1 spectra. This is because to obtain an undistorted f_1 frequency spectrum after Fourier transformation, a relatively large number of points are required in t_1, i.e. the t_1 dataset must not be truncated. If the t_1 data are simulated directly, fewer t_1 points can be used and still retain an accurate analysis.

Recoupling pulse sequences need to be robust to large chemical shift anisotropies, isotropic chemical shift offsets, chemical shift differences between recoupled spins and rf inhomogeneities. Large chemical shift anisotropies and offsets mean that the spins to be recoupled can have quite different frequencies, which is a problem in itself. However, an equally important problem arises in this situation if polycrystalline samples are being examined, as it means that the full chemical shift range of the sample is large, and so the recoupling sequence must be effective over this whole range. This is often referred to as the recoupling sequence needing a large bandwidth. Since the DRAMA experiment first appeared in the literature, many other sequences have been designed to perform better in one respect or another. The DRAMA pulse sequence is not robust to large chemical shift anisotropies and offsets; however, both features have been addressed in new pulse sequences based on the DRAMA experiment [5, 6]. The MELODRAMA experiment [7] is a reasonably robust sequence that uses rotor-synchronized spin locking to recouple homonuclear dipolar coupling. The RFDR (Radio-Frequency Driven Recoupling) experiment [8] uses one 180° pulse per rotor period, and shows efficient recoupling for a wide range of chemical shift offsets.

Box 4.3: Analysis of the DRAMA pulse sequence

This analysis of the DRAMA pulse sequence aims to show that this pulse sequence does indeed prevent the averaging to zero of the dipolar coupling by magic-angle spinning. It employs average hamiltonian theory and use of a toggling frame of reference for the hamiltonian describing the spin system and accordingly follows a similar path to the analysis of the WAHUHA pulse sequence in Box 2.1, Chapter 2. The reader is referred to this earlier section for details concerning the operation of average hamiltonian theory and the toggling frame.

We refer to the basic DRAMA pulse sequence [4] in Fig. 4.11 and assume that the rf pulse amplitudes are sufficiently large that all other interactions may be ignored during the pulses. We further assume that chemical shift offsets are zero in order to simplify the analysis, but still retain its salient features. The hamiltonian in the rotating frame during the pulses is then simply that due to the pulse,

$$\hat{H}_{\text{pulse}} = \omega_1 \hat{I}_x \tag{i}$$

for an x-pulse for instance, where ω_1 is the rf amplitude. During periods of free precession between the pulses, we assume the only spin interaction present is homonuclear dipolar coupling. We restrict ourselves to considering a spin system consisting of a single pair of spins as this leads to no loss of generality. The hamiltonian during the periods of free precession between the rf pulses (in the rotating frame) is then simply

$$\hat{H}_{\text{dd}} = -C_{IS}(t)(3\hat{I}_z\hat{S}_z - \hat{\mathbf{I}}\cdot\hat{\mathbf{S}}) \equiv \hat{H}_{zz}(t) \tag{ii}$$

where $C_{IS}(t)$ is the strength of the dipolar-coupling interaction, time dependent due to magic-angle spinning. The form of $C_{IS}(t)$ is discussed later.

We then form the average hamiltonian over one rotor period, all subsequent rotor periods in the pulse train in Fig. 4.11 being effectively identical. As discussed in Box 2.1, Chapter 2, the average hamiltonian over a period of time t_p is given by a sum of contributions

$$\overline{H}(t_p) = \overline{H}^{(0)} + \overline{H}^{(1)} + H^{(2)} + \ldots \tag{iii}$$

where the successive terms are given by

$$\begin{aligned}
\overline{H}^{(0)} &= \frac{1}{t_p}\{\hat{H}_1 t_1 + \hat{H}_2 t_2 + \ldots + \hat{H}_n t_n\} \\
\overline{H}^{(1)} &= -\frac{i}{2t_p}\{[\hat{H}_2 t_2, \hat{H}_1 t_1] + [\hat{H}_3 t_3, \hat{H}_1 t_1] + [\hat{H}_2 t_2, \hat{H}_3 t_3] + \ldots\} \\
\overline{H}^{(2)} &= -\frac{1}{6t_p}\{[\hat{H}_3 t_3,[\hat{H}_2 t_2, \hat{H}_1 t_1]] + [[\hat{H}_3 t_3, \hat{H}_2 t_2], \hat{H}_1 t_1] \\
&\quad + \frac{1}{2}[\hat{H}_2 t_2,[\hat{H}_2 t_2, \hat{H}_1 t_1]] + \frac{1}{2}[[\hat{H}_2 t_2, \hat{H}_1 t_1], \hat{H}_1 t_1] + \ldots\}
\end{aligned} \tag{iv}$$

in which $\hat{H}_i$ is the hamiltonian which describes the interactions in the spin system during the time period t_i. Now, as explained in Box 2.1, the pulse hamiltonian (Equation (i)) and dipolar hamiltonian (Equation (ii)) do not commute with each other. In other words, hamiltonians $\hat{H}_i$ at different time periods t_i in the rotor period in the DRAMA pulse sequence do not commute with each other, so the higher-order terms in the average hamiltonian, $\overline{H}^{(1)}$, $\overline{H}^{(2)}$, etc., are decidedly non-zero and may contribute significantly to the average hamiltonian. This makes the average hamiltonian tricky to calculate. To get around this, at the point of each pulse in the sequence, we transform the hamiltonian to a new frame, the so-called *toggling frame*, in which the pulse term in the hamiltonian vanishes. Then the hamiltonians describing the spin system throughout the rotor period all commute

Continued on p. 182

Box 4.3 *Cont.*

with one another, so that the $\overline{H}^{(0)}$ term of Equation (iii) alone is a good approximation to the average hamiltonian.[N3] A hamiltonian described in the toggling frame is said to be in the *interaction representation*. The act of performing these frame transformations takes account of the rf pulse effects as far as the density operator is concerned, i.e. if we calculate the density operator in the toggling frame at any point, we obtain the density operator with the effects of the rf pulses included, as shown in Box 2.1.

During the period of free precession which follows a pulse, the spin system simply evolves under the dipolar-coupling hamiltonian expressed in the current toggling frame. Thus, during the first $\tau_R/4$ period of the pulse sequence, the system evolves according to the usual rotating frame[N4] first-order average hamiltonian for dipolar coupling between spins I and S

$$\hat{H}(0 \rightarrow \tau_R/4) = -C_{IS}(t)(3\hat{I}_z\hat{S}_z - \hat{\mathbf{I}}\cdot\hat{\mathbf{S}}) \equiv \hat{H}_{zz}(t) \qquad \text{(v)}$$

The first 90°_x pulse of the DRAMA sequence then requires us to transform to a new toggling frame which is one rotated by $-90°$ about the rotating frame x-axis from the original (rotating) frame. This frame transformation has taken account of the rf pulse as far as the density operator is concerned. The dipolar hamiltonian in this new frame is equivalent to the original dipolar hamiltonian rotated by $+90°$ about the same axis and leaving the axis frame unchanged (see Box 1.2 in Chapter 1 for a detailed discussion of this). The advantage of rotating the hamiltonian rather than the axis frame in which it is defined is that the transformed hamiltonian is expressed in terms of rotating frame spin operators. When we come to sum hamiltonians expressed in different toggling frames, it is much easier if they are all in terms of spin operators defined in a single frame. The rotated dipolar hamiltonian, i.e. toggling frame hamiltonian, in this case is:

$$\hat{H}(\tau_R/4 \rightarrow 3\tau_R/4) = C_{IS}(t)(3\hat{I}_y\hat{S}_y - \hat{\mathbf{I}}\cdot\hat{\mathbf{S}}) \equiv \hat{H}_{yy}(t) \qquad \text{(vi)}$$

where y refers to the rotating frame spin operators. At the next 90°_{-x} pulse, we transform to another toggling frame, which is rotated from the previous toggling frame by $-90°$, this time about the rotating frame $-x$-axis. We now need to express the dipolar hamiltonian in the new toggling frame. This is most easily done as before by leaving the toggling frame where it is and rotating the objects in it, i.e. the dipolar hamiltonian, instead. So we need to rotate the dipolar hamiltonian in its previous toggling frame by $-90°$ about the rotating frame $-x$-axis. The rotating frame $-x$-axis is equivalent to the $-x$-axis of the toggling frame concerned, so this produces

$$\hat{H}(3\tau_R/4 \rightarrow \tau_R) = C_{IS}(t)(3\hat{I}_z\hat{S}_z - \hat{\mathbf{I}}\cdot\hat{\mathbf{S}}) \equiv \hat{H}_{zz}(t) \qquad \text{(vii)}$$

where again z refers to the rotating frame spin operators. The average hamiltonian $\overline{H}$ over the rotor period is then found by summing the hamiltonians which describe the spin system at the various points through the pulse sequence. The hamiltonians due to the pulses are zero with the toggling frames used, so we simply have to sum the hamiltonians from the periods of free precession:

$$\overline{H} = \frac{1}{\tau_R}\left(\int_0^{\tau_R/4} \hat{H}_{zz}(t)\mathrm{d}t + \int_{\tau_R/4}^{3\tau_R/4} \hat{H}_{yy}(t)\mathrm{d}t + \int_{3\tau_R/4}^{\tau_R} \hat{H}_{zz}(t)\mathrm{d}t \right) \qquad \text{(viii)}$$

To proceed further, we now need to explore the form of the time dependence of the dipolar coupling under magic-angle spinning. This was done in Box 4.2 previously for a dipolar-coupling strength written in terms of spherical tensor functions. Here we perform an equivalent derivation for the dipolar-coupling strength written in terms of the Cartesian dipolar tensor, **D**, rather than the spherical tensors of Box 4.2, since we are using Cartesian spin operators in this analysis.

The time dependence of the dipolar-coupling interaction under magic-angle spinning arises, of course, due to the molecular orientation dependence of the dipolar interaction. For a static sample, the geometric part of the dipolar coupling is

$$C_{IS} = \left(\frac{\mu_0}{4\pi}\right)\frac{\gamma_I\gamma_J}{r_{\mathrm{IS}}^3}\hbar\frac{1}{2}(3\cos^2\theta - 1) \qquad \text{(ix)}$$

An alternative way of writing the dipolar hamiltonian is as (in angular frequency units)

$$\hat{H}_{\mathrm{dd}} = -2\hat{\mathbf{I}}\cdot\mathbf{D}\cdot\hat{\mathbf{S}} \qquad \text{(x)}$$

where the Cartesian dipolar-coupling tensor $\boldsymbol{D}$ is expressed in the laboratory frame. Expanding this and comparing the result with Equation (ii) for the dipolar hamiltonian leads to the identity:

$$C_{IS} = D_{zz} \qquad \text{(xi)}$$

where D_{zz} is a component of the dipolar coupling tensor **D**, z referring to the laboratory frame z-axis.

Under magic-angle spinning, the dipolar tensor, **D**, acquires a time dependence in exactly the same way as the shielding tensor described in Section 2.2.1. There, Equations (2.2)–(2.8) derive the time dependence of the chemical shift frequency, ω_{cs}, which is equal to $-\omega_0\,\sigma_{zz}$, where σ_{zz} is the shielding tensor equivalent of D_{zz}. We can therefore use these equations to derive an expression for D_{zz} by substituting components of **D** for those of σ in Equations (2.2) to (2.8). Doing this and using the fact that the dipolar tensor is always axial with principal values $-d/2$, $-d/2$, $+d$, where d is the dipolar coupling constant ($d = (\mu_0/4\pi)(\gamma_I\gamma_S/r_{IS}^3)\hbar$, we obtain

$$D_{zz}(t) = d\left(-\frac{1}{\sqrt{2}}\sin 2\beta\cos(\omega_R t + \gamma) + \frac{1}{2}\sin^2\beta\cos(2\omega_R t + 2\gamma)\right) \qquad \text{(xii)}$$

Continued on p. 184

Box 4.3 *Cont.*

where β and γ are Euler angles describing the rotation of the dipolar tensor principal axis frame into the rotor-fixed frame, R, and ω_R is the spinning rate. The integrals in Equation (viii) can now be evaluated:

$$\int_{\tau_1}^{\tau_2} \hat{H}_{\alpha\alpha}(t)\mathrm{d}t = \int_{\tau_1}^{\tau_2} -D_{zz}(t)[3\hat{I}_\alpha \hat{S}_\alpha - \hat{\mathbf{I}}\cdot\hat{\mathbf{S}}]\mathrm{d}t$$

$$= -d[3\hat{I}_\alpha \hat{S}_\alpha - \hat{\mathbf{I}}\cdot\hat{\mathbf{S}}]\left(-\frac{1}{\sqrt{2}}\sin 2\beta \int_{\tau_1}^{\tau_2} \cos(\omega_R t + \gamma)\mathrm{d}t \right.$$

$$\left. +\frac{1}{2}\sin^2\beta \int_{\tau_1}^{\tau_2} \cos(2\omega_R t + 2\gamma)\mathrm{d}t\right)$$

$$= -d\hat{T}_{\alpha\alpha}\left(-\frac{1}{\omega_R\sqrt{2}}\sin 2\beta[\sin(\omega_R\tau_2 + \gamma) - \sin(\omega_R\tau_1 + \gamma)]\right.$$

$$\left. +\frac{1}{4\omega_R}\sin^2\beta[\sin(2\omega_R\tau_2 + 2\gamma) - \sin(2\omega_R\tau_1 + 2\gamma)]\right) \tag{xiii}$$

where

$$\hat{T}_{\alpha\alpha} = 3\hat{I}_\alpha \hat{S}_\alpha - \hat{\mathbf{I}}\cdot\hat{\mathbf{S}} \tag{xiv}$$

and from Equation (xiii) we can easily derive the average hamiltonian over a rotor period by substituting this expression for the integral in Equation (viii):

$$\overline{H} = \frac{1}{\sqrt{2}\pi} d \sin 2\beta \cos\gamma(\hat{T}_{zz} - \hat{T}_{yy}) \tag{xv}$$

This is the effective hamiltonian that the spin system evolves under for one rotor period. Without the DRAMA pulse sequence, the average hamiltonian over one rotor period would be zero (in the absence of chemical shift offsets). This is clearly not the case here and, moreover, the spin-system evolution depends on the dipolar coupling, as required. The evolution is also dependent on the orientation of the crystallites in the rotor-fixed frame (through β and γ), and so we can expect some kind of powder pattern in the spectrum arising from using the DRAMA pulse sequence. However, the orientation dependence is different from that for a normal, static sample.

Simulating powder patterns from the DRAMA experiment

In order to simulate the dipolar powder patterns which arise from a DRAMA pulse sequence, we use Equation (xv) for the average hamiltonian over one rotor period of the DRAMA sequence, and use this to calculate the density operator at the end of the pulse sequence using

$$\hat{\rho}(t) = \exp(-i\overline{H}t)\hat{\rho}(0)\exp(+i\overline{H}t) \qquad \text{(xvi)}$$

where $\hat{\rho}(0)$ is the density operator at the start of the experiment (spin system at equilibrium, so $\hat{\rho}(0) = \hat{I}_z + \hat{S}_z$, for the two-spin system; t is the length of time for which the DRAMA sequence runs (an integral number of rotor periods). This is done in practice by forming the matrix representations of $\hat{\rho}(0)$ and $\overline{H}$ in the two-spin product basis (i.e. αα, αβ, βα, ββ). The exponentials in Equation (xvi) are then evaluated by performing an eigenvalue/eigenvector analysis on the matrix of $\overline{H}$. The exponentials are then given by

$$\exp(i\overline{\mathbf{H}}t) = \mathbf{V}^{-1}\exp(\mathbf{E})\mathbf{V} \qquad \text{(xvii)}$$

where **V** is the eigenvector matrix and **E** is the (diagonal) matrix of eigenvalues with elements (eigenvalues) E_k. Exp(**E**) is simply a diagonal matrix with elements $\exp(E_k)$.

Having calculated the matrix representation of $\hat{\rho}(t)$, we can use this to calculate the expectation value of the x-magnetization at time t using

$$\langle(\hat{I}_x + \hat{S}_x)\rangle_t = \mathrm{Tr}((\hat{I}_x + \hat{S}_x)\hat{\rho}(t)) \qquad \text{(xviii)}$$

The trace in Equation (xviii) is evaluated from the matrices of the $\hat{I}_x + \hat{S}_x$ and $\hat{\rho}(t)$ operators in the two-spin product basis.

$\langle(\hat{I}_x + \hat{S}_x)\rangle_t$ needs to be calculated as a function of t (the t_1 values used in the experiment) for a given value of d, the dipolar-coupling constant which appears in the expression for $\overline{H}$, and summed over a representative sample of crystallite orientations, (β, γ) (on which $\overline{H}$ also depends) to produce a simulation of the t_1 time-domain data for a powder sample. This same basic procedure can be used to simulate the data arising from any recoupling pulse sequence; all that changes is the form of the average hamiltonian, $\overline{H}$.

4.3.2 Double-quantum filtered experiments

In any system with strongly dipolar-coupled spin pairs, it is in principle possible to excite double-quantum coherences involving the pairs of coupled spins (see Fig. 4.14). This feature can and is used to measure the dipolar couplings between the spins in a spin-pair system. The experiments rely on the fact that excitation efficiency for the double-quantum coherence depends upon the strength of the dipolar coupling. Another way of saying this is that the nutation rate of the double-quantum coherence during its excitation

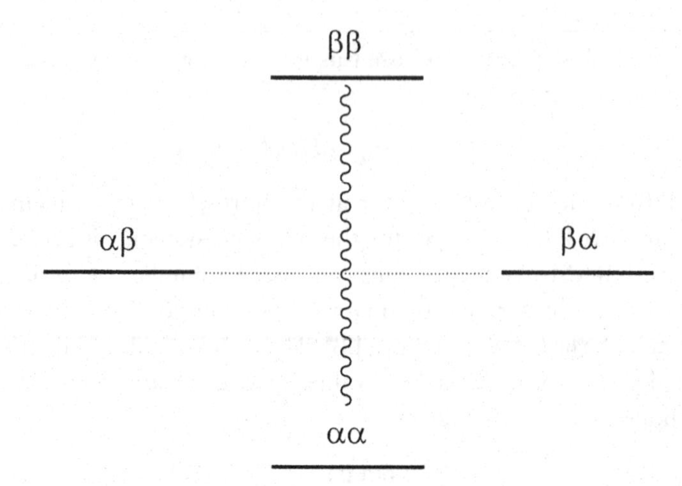

Fig. 4.14 The spin states in a dipolar-coupled spin-$\frac{1}{2}$ pair. Double-quantum coherence can be excited between the αα and ββ states ($\Delta M = \pm 2$), while zero-quantum coherence can be generated between αβ and βα ($\Delta M = 0$).

depends on the dipolar coupling between the spins involved in the double-quantum coherence. Thus, if we can follow the amplitude of double-quantum coherence produced as a function of excitation time, we should be able to analyse these data to extract the dipolar-coupling constant for the pairs of spins associated with the double-quantum coherence. The analysis will depend on the way in which the double-quantum coherence was excited. Now we cannot, of course, observe double-quantum coherences directly, so we must first convert any double-quantum coherence the experiment has excited into observable, single-quantum coherence and measure that, hence the expression double-quantum *filtered* experiments. In practice then we measure the intensity of the double-quantum filtered signal intensity as a function of double-quantum excitation time, and analyse the resulting data to extract dipolar-coupling constants. The format of the experiment is depicted in Fig. 4.15.

Double-quantum coherence can be easily excited with a simple 90°–τ–90° pulse sequence, where single-quantum coherence (corresponding to transverse magnetization) is excited by the first 90° pulse, allowed to evolve under the dipolar coupling during τ, creating terms in the density operator which the second 90° pulse then converts into double-quantum terms. However, most double-quantum filtered experiments need to be run under magic-angle spinning for the sake of resolution in the final spectrum. As magic-angle spinning averages all dipolar couplings to zero, if we wish to excite double-quantum coherences, first we have to reintroduce the dipolar coupling between the spins of interest. Section 4.3.1 discussed rf pulse sequences which reintroduce dipolar couplings under conditions of magic-angle spinning. Although these were first introduced with the aim of measuring dipolar

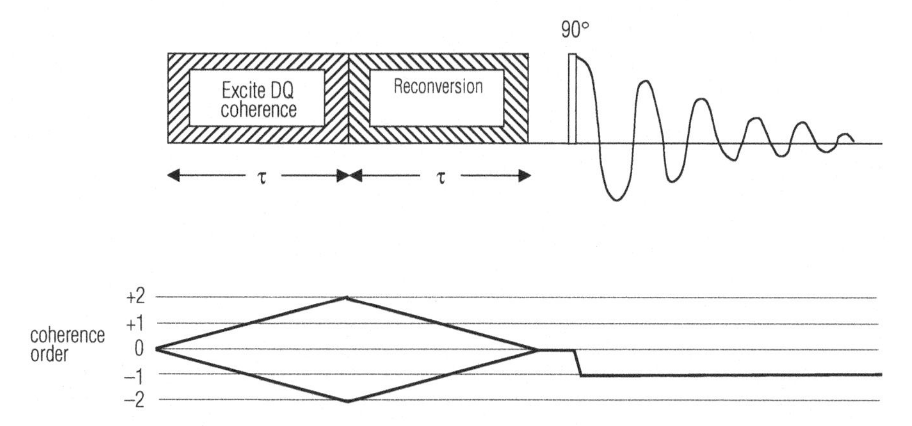

Fig. 4.15 The format of double-quantum filtered experiments for the measurement of dipolar couplings in homonuclear spin systems. Double-quantum coherence is excited then converted via longitudinal magnetization to observable single-quantum coherence. The amount of double-quantum coherence generated for a given spin pair depends on the strength of dipolar coupling between the pair. The experiment is repeated for different lengths of excitation time, τ, and the final signal intensity monitored as a function of τ.

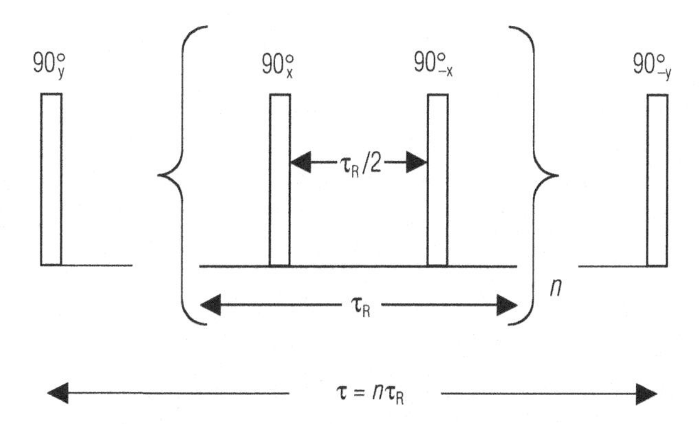

Fig. 4.16 The pulse sequence used to excite double-quantum coherence under magic-angle spinning for the analysis in Box 4.4. The sequence consists of a basic 90°–τ–90° pulse sequence, with a recoupling pulse sequence operating n times during the τ period to prevent the magic-angle spinning from averaging the dipolar coupling (which is ultimately responsible for the double-quantum coherence) to zero. The recoupling pulse sequence in this case is the DRAMA sequence (Fig. 4.11), but any dipolar recoupling pulse sequence can in principle be used.

couplings directly, these same pulse sequences can all be used to reintroduce dipolar couplings for the purpose of exciting multiple-quantum coherences in correlation experiments. Pulse sequences such as DRAMA can be readily used in this fashion [9]. Figure 4.16 shows a pulse sequence which uses DRAMA to excite double-quantum coherence. Box 4.4 describes the theory of their use in such applications. The conversion of multiple-quantum coherence to observable single-quantum coherence takes place in two steps. In the first step, the multiple-quantum coherence is reconverted to zero-

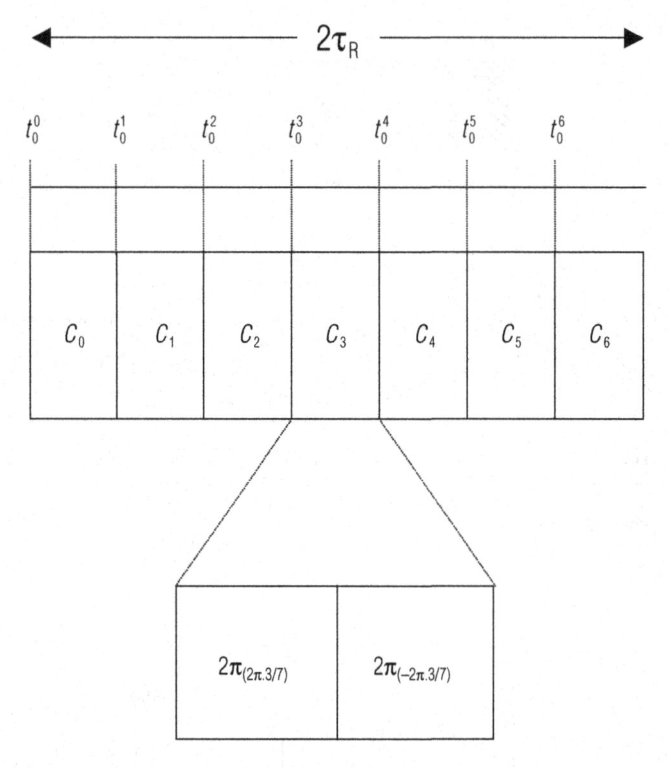

Fig. 4.17 The form of the C_n^N pulse sequences. The sequences consist of a series of n ($2\pi_\phi$ $2\pi_{-\phi}$) pulse cycles, each labelled C_p, timed so that they fit continuously into N rotor periods of sample spinning at the magic angle. The phase ϕ of the 2π pulses changes between each set so that for the pth cycle of pulses, $\phi = 2\pi\, p/n$. The whole sequence is generally run M times, where M is an integer. The particular pulse sequence shown here is the C_7^2 sequence. Also shown are the start timings of each cycle t_0^p.

quantum coherence (longitundinal magnetization). This is done by a reversal of the process by which multiple-quantum coherence was first excited from longitudinal magnetization [10]; this requires using the same pulse sequence as for the excitation, but with all the pulse phases shifted by 90°, as explained in Box 4.4. In the second step, single-quantum coherence is finally produced from the longitudinal magnetization by the simple application of a 90° pulse in the usual fashion.

More recently, the C7 sequence [11] and related R14 sequence [12] have been introduced by Levitt *et al.* for the explicit purpose of exciting double-quantum coherences under magic-angle spinning. These are members of general families of pulse sequences denoted C_n^N and R_n^N whose purpose is to recouple (under magic-angle spinning) any spin interaction of choice. Thus their applicability extends way beyond homonuclear dipolar recoupling, although it is that use that we discuss here. The C sequences [11] (Fig. 4.17) consist of a series of n ($2\pi_\phi$ $2\pi_{-\phi}$) pulse sets with the rf amplitude adjusted so that they fit continuously into N rotor periods of sample spinning at the magic angle. The phase ϕ of the 2π pulses changes between each set so that for the jth set of pulses, $\phi = 2\pi\,(j-1)/n$. The sequence is generally run for $M \times N$ rotor periods, where M is an integer, i.e. for several complete cycles of the pulse sequence (note, however, that the C7 pulse sequence works even with fractions of one rotor period, providing that complete sets of ($2\pi_\phi$ $2\pi_{-\phi}$) pulses are used). For the R sequences, the nomenclature is the same except that the $2\pi_\phi$ $2\pi_{-\phi}$ sets are replaced by π_ϕ $\pi_{-\phi}$. By adjusting N and n, the particular spin interaction which is recoupled can be changed [11]. The simplest sequence which recouples homonuclear dipolar coupling is the C_7^2 or, simply, C7 sequence [11]. The advantage of this sequence over those previously discussed is that it is much less sensitive to rf pulse amplitude errors and chemical shift offsets, in addition to having a higher recoupling efficiency.

The beauty of the C_n^N and R_n^N pulse sequences is the way they may be manipulated to remove those components of spin interactions we do not want and to keep those we do. The theory behind the operation of these sequences is well explained in reference [11], and outlined in Box 4.5 below for the C7 sequence.

Box 4.4: Excitation of double-quantum coherence under magic-angle spinning

As mentioned above, any of the pulse sequences discussed in Section 4.3.1 for the reintroduction of dipolar couplings under magic-angle spinning can be used (at least in principle) in the excitation of dipolar-coupling mediated, double-quantum coherence. Here, we shall use the example of the DRAMA pulse sequence [4, 9], as that was extensively analysed in Box 4.3. The sequence used to excite double-quantum coherence is the DRAMA sequence (or N cycles of it) sandwiched between two 90° pulses, whose phases are shifted by 90° from those of the DRAMA sequence (Fig. 4.16).

We use average hamiltonian theory to assess the effect of this complete sequence and, in doing so, follow much of the analysis of the DRAMA sequence in Box 4.3. We calculate the zeroth-order term in the average hamiltonian, $\overline{H}^{(0)}$, as providing we use the interaction representation, this is a good approximation to the total average hamiltonian. As in the analysis of the DRAMA sequence, we consider a two-spin system, I and S, and assume that we can neglect all other spin interactions other than that with the rf pulse field during a pulse. For simplicity here, we further assume that between pulses, the spin system evolves under the dipolar

Continued on p. 190

Box 4.4 *Cont.*

coupling between the two spins I and S only, i.e. we do not take account of chemical shift offsets etc. The hamiltonian describing this interaction is

$$\hat{H}_{\text{dd}} = -C_{IS}(t)(3\hat{I}_z\hat{S}_z - \hat{\mathbf{I}}\cdot\hat{\mathbf{S}}) \equiv \hat{H}_{zz}(t) \tag{i}$$

where the strength of the interaction is described by $C_{IS}(t)$ whose form is derived in Box 4.3.

In forming the zeroth-order average hamiltonian, we use the interaction representation; that is, we transform to a new toggling frame each time we get to a pulse in the pulse sequence, the new frame being such as to null the hamiltonian describing the effects of the pulse. We then have to transform the dipolar hamiltonian describing the evolution of the spin system after the pulse into the new frame. Equivalent to transforming the frame of reference is to rotate the dipolar hamiltonian instead; if the new frame is rotated by θ about axis α from the original rotating frame, then the equivalent rotation of the hamiltonian is by −θ about axis α.

So in the analysis of the excitation sequence of Fig. 4.16, we first transform to a toggling frame rotated by −90° about y of the rotating frame, to take account of the first 90°_y pulse. For the period of evolution (τ_1) which follows, we must transform the dipolar hamiltonian (Equation (i)) to this new frame, which is equivalent to rotating the dipolar hamiltonian +90° about y (of the rotating frame). This transforms $\hat{H}_{zz}(t)$ (Equation (i)) to $\hat{H}_{xx}(t)$, where x refers to the rotating frame x-axis and $\hat{H}_{\alpha\alpha}(t)$, $\alpha = x, y, z$, is given by

$$\hat{H}_{\alpha\alpha}(t) = -C_{IS}(t)(3\hat{I}_\alpha\hat{S}_\alpha - \hat{\mathbf{I}}\cdot\hat{\mathbf{S}}) \tag{ii}$$

The system then evolves under $\hat{H}_{xx}(t)$ for a time $\tau_R/4$. Then there is a 90°_x pulse, so we switch to a new toggling frame rotated by −90° about the rotating frame x-axis. The dipolar hamiltonian in this new frame is $\hat{H}_{yy}(t)$ (with y being the rotating frame y-axis) and the system evolves under this for a time $\tau_R/2$. Finally, the last 90°_{-x} pulse demands a toggling frame transformation in which the dipolar hamiltonian is transformed to $\hat{H}_{xx}(t)$, and evolution under this occurs for $\tau_R/4$. The zeroth-order average hamiltonian for one rotor period is then simply the time average of the hamiltonians occurring over one rotor period:

$$\overline{H}^{(0)}(\tau_R) = \frac{1}{\tau_R}\left(\int_0^{\tau_R/4} \hat{H}_{xx}(t)\mathrm{d}t + \int_{\tau_R/4}^{3\tau_R/4} \hat{H}_{yy}(t)\mathrm{d}t + \int_{3\tau_R/4}^{\tau_R} \hat{H}_{xx}(t)\mathrm{d}t\right) \tag{iii}$$

The symmetric nature of the time dependence of the dipolar hamiltonian with respect to the rotor period means that the first and last integrals in Equation (iii) are equal. The form of the integrals in Equation (iii) has been derived in Box 4.3

previously (Equation (xiii) in Box 4.3). Using this, we can simplify Equation (iii) to

$$\overline{H}^{(0)}(\tau_R) = \frac{3}{\sqrt{2}\pi} d \sin 2\beta \cos\gamma (\hat{I}_x\hat{S}_x - \hat{I}_y\hat{S}_y) \quad \text{(iv)}$$

where d is the dipolar-coupling constant and (β, γ) are two of the Euler angles which rotate the dipolar tensor PAF into a rotor-fixed frame (R) as in Box 4.3 and x and y refer to the usual rotating axis frame.

Note that we can achieve the same result by taking the average hamiltonian already derived for the DRAMA pulse sequence (Equation (xv) in Box 4.3) as the appropriate hamiltonian for the DRAMA sequence in the τ_R period after the first 90°_y pulse of the double-quantum excitation sequence. So, we only need the first toggling frame transformation of −90° about y to account for the first 90°_y pulse of the sequence. We then transform the average hamiltonian of Equation (xv) in Box 4.3 into this same frame by rotating this average hamiltonian by +90° about the rotating frame y-axis, which gives Equation (iv) above for the average hamiltonian over one rotor period of the double-quantum excitation sequence.

The form of the reconversion pulse sequence: the need for time-reversal symmetry

In this section we assess how the form of the reconversion part of the pulse sequence in Fig. 4.15 can affect the strength of the final signal in a two-dimensional double-quantum experiment. We will see how sensible choice of this sequence, namely a so-called *time-reversed* sequence [13] (relative to the excitation sequence), leads to good signal intensities.

Many different double-quantum coherences associated with different pairs of spins and/or different molecular orientations in the sample can be excited during the double-quantum excitation period of a double-quantum experiment. Each of these different double-quantum coherences is in general formed with a different *phase*. This is because the nutation rate for the double-quantum coherence depends upon the magnitude of the average hamiltonian which governs the excitation period. Now, Equation (iv) above shows that the excitation average hamiltonian depends on the dipolar-coupling constant between the pair of spins in question (d) and the particular molecular orientation (β, γ). In other words, the double-quantum coherence nutation angle, and so the phase of the double-quantum coherence formed, depends upon which spin pair the coherence is associated with and the orientation of the molecule. During the reconversion from double-quantum coherence back to longitudinal magnetization, a further phase shift is added to

Continued on p. 192

Box 4.4 *Cont.*

all the components of longitudinal magnetization arising from each component of double-quantum coherence. The net result is that the components of the longitudinal magnetization from different molecules in the sample, different spin pairs, etc., destructively interfere and partially cancel, so that the net signal at the end of the pulse sequence may be very small indeed. This can be expressed more formally as follows [13].

We take a general case where an average hamiltonian $\overline{H}_{ex}$ acting for a time τ governs the double-quantum excitation, and $\overline{H}_{recon}$ acting for a time τ' governs the reconversion (see Fig. B4.4.1).

The density operator at the end of the reconversion period, $\hat{\rho}(\tau + t_1 + \tau')$ can be derived from the equilibrium density operator at time 0 and is given by (see Chapter 1 for details of the time dependence of the density operator)

$$\begin{aligned}\hat{\rho}(\tau + t_1 + \tau') &= \exp(-i\overline{H}_{recon}\tau')\exp(-i\hat{H}_{int}t_1)\exp(-i\overline{H}_{ex}\tau)\times \\ &\quad (\hat{I}_z + \hat{S}_z)\exp(i\overline{H}_{ex}\tau)\exp(i\hat{H}_{int}t_1)\exp(i\overline{H}_{recon}\tau') \\ &= \hat{V}(\tau')\exp(-i\hat{H}_{int}t_1)\hat{U}(\tau)(\hat{I}_z + \hat{S}_z)\hat{U}^*(\tau)\exp(i\hat{H}_{int}t_1)\hat{V}^*(\tau') \qquad \text{(v)}\end{aligned}$$

where $(\hat{I}_z + \hat{S}_z)$ is the equilibrium density operator $(\hat{\rho}(0))$ at time 0 and $\hat{H}_{int}$ is the hamiltonian describing the internal spin interactions that govern the evolution of the density operator during the t_1 period of the experiment. For convenience, we have written the propagators $\exp(-i\overline{H}_{ex}\tau)$ and $\exp(-i\overline{H}_{recon}\tau')$ in Equation (v) as

$$\begin{aligned}\exp(-\overline{H}_{ex}\tau) &= \hat{U} \\ \exp(-iH_{recon}\tau') &= \hat{V}\end{aligned} \qquad \text{(vi)}$$

Note that if we wish to calculate the density matrix which arises after double-quantum coherence *only* has been excited in the excitation period, we should only retain those terms which correspond to double-quantum coherence in the density matrix at time τ, i.e. the end of the double-quantum excitation. The density operator at this point is

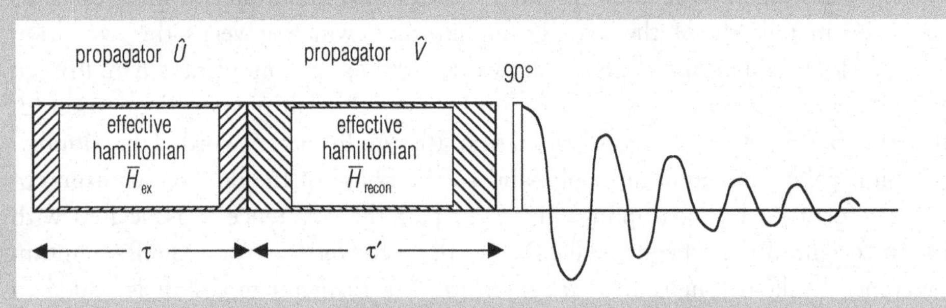

Fig. B4.4.1 The form of the general pulse sequence for multiple-quantum excitation – reconversion.

$$\hat{\rho}(\tau) = \hat{U}(\tau)(\hat{I}_z + \hat{S}_z)\hat{U}^*(\tau)$$

Now, we wish to know the longitudinal magnetization which remains at the end of the pulse sequence, i.e. at $t = \tau + t_1 + \tau'$. The expectation value of any observable A at time t can be found from the density operator at time t via (see Chapter 1 for details)

$$\langle A \rangle = \mathrm{Tr}(\hat{A}\hat{\rho}(t)) \tag{vii}$$

where $\hat{A}$ is the operator corresponding to the observable A. The expectation value of the longitudinal magnetization is proportional to $\langle \hat{I}_z + \hat{S}_z \rangle$, which, at the end of the sequence, is then given by

$$\begin{aligned}\langle(\hat{I}_z + \hat{S}_z)\rangle_{\tau+t_1+\tau'} &= \mathrm{Tr}((\hat{I}_z + \hat{S}_z)\hat{\rho}(\tau + t_1 + \tau')) \\ &= \mathrm{Tr}((\hat{I}_z + \hat{S}_z)\hat{V}(\tau')\exp(-i\hat{H}_{\mathrm{int}}t_1)\hat{U}(\tau)(\hat{I}_z + \hat{S}_z)\hat{U}^*(\tau)\exp(i\hat{H}_{\mathrm{int}}t_1)\hat{V}^*(\tau')) \\ &= \sum_{a,b}\langle a|\hat{U}(\hat{I}_z + \hat{S}_z)\hat{U}^*|b\rangle\langle b|\hat{V}^*(\hat{I}_z + \hat{S}_z)\hat{V}|a\rangle\exp(-i(\omega_a - \omega_b)t_1)\end{aligned} \tag{viii}$$

where we have substituted Equation (v) for $\hat{\rho}(\tau + t_1 + \tau')$. Calculating the trace in Equation (viii) requires us to form the matrices of $\hat{I}_z + \hat{S}_z$ and $\hat{\rho}(\tau + t_1 + \tau')$ in some basis. In Equation (viii) we have formed the matrices of these operators in terms of basis functions a, b, etc., which are eigenfunctions of the internal spin interaction hamiltonian $\hat{H}_{\mathrm{int}}$, with eigenvalues (energies) ω_a and ω_b respectively (see Box 1.1, Chapter 1, for more discussion of the matrix representation of operators). Having formed the matrices of these operators, it is then comparatively simple to evaluate the trace required in Equation (viii) for the expectation value of the longitudinal magnetization. It is the complex matrix elements, $\langle a|\hat{U}(\hat{I}_z + \hat{S}_z)\hat{U}^*|b\rangle$ and $\langle b|\hat{V}^*(\hat{I}_z + \hat{S}_z)\hat{V}|a\rangle$ which contain the phases of the components of longitudinal magnetization supplied by the excitation and reconversion periods respectively. The eigenfunctions a and b can be written as linear combinations of product Zeeman functions involving the two spins (I, S) in the spin system. The summation over these in Equation (viii) plus the further summation over different molecular orientations in a powder sample (not shown in Equation (viii), but $\hat{U}$, $\hat{V}$ and $\hat{H}_{\mathrm{int}}$ all depend on molecular orientation) leads to destructive interference between contributions from different pairs of functions a, b. This phenomenon has sometimes been referred to as dipolar dephasing.

The way to avoid this process and consequent loss of signal is to make the propagator $\hat{V}$, which describes the reconversion, equal to $\hat{U}^*$, where $\hat{U}$ is the excitation propagator. Then, the expectation value for the longitudinal magnetization reduces to:

$$\langle(\hat{I}_z + \hat{S}_z)\rangle_{\tau+t_1+\tau'} = \sum_{a,b}\left|\langle a|\hat{U}(\hat{I}_z + \hat{S}_z)\hat{U}^*|b\rangle\right|^2 \exp(-i(\omega_a - \omega_b)t_1) \tag{ix}$$

Continued on p. 194

Box 4.4 *Cont.*

So the expectation value for the longitudinal magnetization at the end of the pulse sequence in Fig. B4.4.1. now depends on the *squared modulus* of the complex matrix elements $\langle a|\hat{U}^*(\hat{I}_z + \hat{S}_z)\hat{U}|b\rangle$, so all the terms in the summation of Equation (ix) add constructively.

The only remaining question is how do we make $\hat{V} = \hat{U}^*$ in practice? Since $\exp(-i\overline{H}_{ex}\tau) = \hat{U}$, setting $\hat{V}$ equal to $\hat{U}^*$ gives

$$\hat{V} = \hat{U}^* = \exp(+i\overline{H}_{ex}\tau) \tag{x}$$

Now rewriting $\hat{V}$ in the form $\exp(-i\hat{H}t)$, the usual form for a propagator, where $\hat{H}$ is the hamiltonian acting during the period of the propagator, we have for $\hat{V}$

$$\hat{V} = \exp(-i(-\overline{H}_{ex})\tau) \tag{xi}$$

In other words, the average hamiltonian which must act during the reconversion period is *minus* the average hamiltonian we used in the excitation of double-quantum coherence, and it must act for the same length of time, τ, as the original excitation period acted for. Such an average hamiltonian is commonly called *time reversed*; this is somewhat of a misnomer because, of course, it is impossible to reverse time. However, it is the terminology in common usage.

So now we return to the use of the DRAMA sequence in double-quantum excitation/reconversion and examine how to achieve a time-reversed hamiltonian for the reconversion period. We want to use a pulse sequence which produces *minus* the average hamiltonian we used in the excitation of multiple-quantum coherence, $\overline{H}^{(0)}$, of Equation (iv) above. The pulse sequence which creates such an effective hamiltonian is the same excitation sequence but with all the pulses shifted in phase by 90°, i.e. an initial 90°_{-x} pulse followed by a 90°_y pulse at $\tau_R/4$, 90°_{-y} at $3\tau_R/4$ (and in each subsequent rotor period for which the reconversion is applied) with a final 90°_x pulse at the end of the sequence. Following the analysis of the excitation sequence above, this then leads to the operator $\hat{H}_{yy}$ acting during $t = 0$ to $t = \tau_R/4$, then $\hat{H}_{xx}$ for $\tau_R/4 < t \le 3\tau_R/4$ and finally $\hat{H}_{yy}$ for $3\tau_R/4 < t \le \tau_R$. Thus, we end up with an effective hamiltonian (equivalent to Equation (iii) above) of

$$\overline{H}^{(0)}(\tau_R) = \frac{1}{\tau_R}\left(\int_0^{\tau_R/4} \hat{H}_{yy}(t)dt + \int_{\tau_R/4}^{3\tau_R/4} \hat{H}_{xx}(t)dt + \int_{3\tau_R/4}^{\tau_R} \hat{H}_{yy}(t)dt\right) \tag{xii}$$

Evaluating the integrals in Equation (vii) using Equation (xiii) of Box 4.3 yields

$$\overline{H}^{(0)}(\tau_R) = -\frac{3}{\sqrt{2}\pi} d \sin 2\beta \cos\gamma(\hat{I}_x\hat{S}_x - \hat{I}_y\hat{S}_y) \tag{xiii}$$

which is indeed minus the effective hamiltonian for the excitation period, as required.

Analysis of the double-quantum filtered data

In a double-quantum filtered experiment, we measure the double-quantum filtered signal intensity as a function of double-quantum excitation time. We then wish to extract the dipolar coupling constant, d, for the dipolar coupling between the spins I and S giving rise to the double-quantum coherence. The double-quantum filtered signal intensity is proportional to the expectation value of the longitudinal magnetization at the end of the time-reversed reconversion period, assuming the subsequent conversion to transverse magnetization is done with a hard (non-selective) rf pulse. Equation (ix) above gives an expression for the expectation value of the longitudinal magnetization. It consists of a sum of matrix elements over all states for the spin system. The matrix elements involved in $\langle \hat{I}_z + \hat{S}_z \rangle$ further depend on the propagator $\hat{U}$, where

$$\begin{aligned}\hat{U} &= \exp(-i\overline{H}_{\text{ex}}\tau) \\ &= \exp(-i\overline{H}^{(0)}\tau) \\ &= \exp\left(-\frac{3i}{\sqrt{2}\pi} d \sin 2\beta \cos\gamma [\hat{I}_x\hat{S}_x - \hat{I}_y\hat{S}_y]\tau\right)\end{aligned} \qquad \text{(xiv)}$$

The matrix elements of Equation (ix) are then readily calculated (using Equation (xiv) for the propagator $\hat{U}$) for given values of d (the dipolar-coupling constant) and the molecular orientation angles β and γ, as the matrix elements $\langle a|\hat{U}(\hat{I}_z + \hat{S}_z)\hat{U}^*|b\rangle$ in Equation (ix) are simply the elements of the matrix **A**, where **A** is

$$\mathbf{A} = \mathbf{U}^*(\mathbf{I}_z + \mathbf{S}_z)\mathbf{U} \qquad \text{(xv)}$$

$\langle \hat{I}_z + \hat{S}_z \rangle$ in Equation (ix) also depends on the energies ω_a, etc., of the states of the spin system. These energies are the eigenvalues of $\hat{H}_{\text{int}}$, which describes all the spin interactions which are present during t_1 of the experiment. They may be evaluated by simply diagonalizing the corresponding matrix $\mathbf{H}_{\text{int}}$ of $\hat{H}_{\text{int}}$ in the Zeeman basis for the two-spin system.

With these components, it is then a relatively straightforward matter to calculate $\langle \hat{I}_z + \hat{S}_z \rangle$ for given values of d, β and γ. For powder samples, $\langle \hat{I}_z + \hat{S}_z \rangle$ needs to be summed over a distribution of β and γ angles representing the crystallite orientations of the powder. The result can then be compared with the experimentally measured signal intensity. The form of the propagator $\hat{U}$ (Equation (xiv)) contained in the expression for $\langle \hat{I}_z + \hat{S}_z \rangle$ (Equation (ix)) means that the signal intensity (which is proportional to $\langle \hat{I}_z + \hat{S}_z \rangle$) is expected to oscillate with frequency determined by d and the powder average of $\sin 2\beta \cos\gamma$.

Box 4.5: Analysis of the C7 pulse sequence for exciting double-quantum coherence in dipolar-coupled spin pairs

We consider a general pulse sequence of the C_n^N form [11] shown in Fig. 4.17 in which a sequence of n rf pulse cycles C is timed to fit into N periods of the sample rotation exactly. Each rf pulse cycle consists of $(2\pi_p\ 2\pi_{-p}$ where the phase ϕ_p is given by $\phi_p = 2\pi p/n$ ($p = 0, 1, 2, \ldots, n-1$) and the length of each cycle is $\tau_C = N\tau_R/n$. For simplicity, we consider that the rf pulses are on-resonance (no chemical shift offsets) and ignore chemical shift anisotropy, so that the only interactions are the dipolar coupling and interactions with the rf pulse field. Here we are *not* using hard rf pulses as in the DRAMA sequence, but relatively low amplitude rf pulses. Therefore at each point in the pulse sequence, the hamiltonian appropriate to the spin system is the sum of a term due to the rf pulse and a term due to the dipolar coupling; the size of the dipolar interaction is no longer negligible compared with the interaction with the rf pulse.

We use the form of the dipolar hamiltonian under magic-angle spinning derived in Box 4.2 in terms of spherical tensor operators and spherical tensor functions to describe the orientation dependence:

$$\hat{H}_{\mathrm{dd}}(t) = \sum_{\substack{IS\\ \text{spin}\\ \text{pairs}}} \hat{T}_{20}^{IS} \sum_{m=-2}^{+2} \exp(im\omega_{\mathrm{R}}t)\omega_{IS}^{(m)} \tag{i}$$

where the $\omega_{IS}^{(m)}$ describe the orientation dependence of the dipolar interaction (defined in Box 4.2) and the $\hat{T}_{20}^{IS}$ are spherical tensor operators (also defined in Box 4.2). As in the analysis of the DRAMA pulse sequence, we wish to form the average hamiltonian over the whole pulse sequence of n cycles. We begin by finding the average hamiltonian over one cycle C_p in which the rf phase ϕ_p is $2\pi p/n$.

If the pulse sequence starts at time 0, and the pulse cycle C_p starts at time t_p^0, then a time τ into the cycle, the dipolar hamiltonian is given by

$$\begin{aligned}\hat{H}_{\mathrm{dd}}(t_p^0+\tau) &= \sum_{\substack{IS\\ \text{pairs}}} \hat{T}_{20}^{IS} \sum_{m=-2}^{+2} \exp(im\omega_{\mathrm{R}}(t_p^0+\tau))\omega_{IS}^{(m)}\\ &= \sum_{\substack{IS\\ \text{pairs}}} \hat{T}_{20}^{ij} \sum_{m=-2}^{+2} \exp(im\omega_{\mathrm{R}}Np\tau_{\mathrm{R}}/n)\exp(im\omega_{\mathrm{R}}\tau)\omega_{IS}^{(m)}\\ &= \sum_{\substack{IS\\ \text{pairs}}} \hat{T}_{20}^{IS} \sum_{m=-2}^{+2} \exp(i2\pi mNp/n)\exp(im\omega_{\mathrm{R}}\tau)\omega_{IS}^{(m)} \end{aligned} \tag{ii}$$

where we have used $t_p^0 = Np\tau_R/n$, since each cycle is of length $N\tau_R/n$.

We now follow the same procedure as for the DRAMA sequence in Box 4.3. During the pulse cycle, we transform the spin coordinates in the hamiltonian describing the spin system to a toggling frame in which the effect of the rf pulse is null. This frame is one which is rotated from the normal rotating frame by $-\beta_{rf}(\tau)$ about the axis of the pulse, where β_{rf} is the nutation angle of the pulse, $\omega_1\tau$. The nutation angle of the rf pulse is the angle through which the rf pulse has turned the equilibrium magnetization in time t. The axis of the pulse is in the x–y plane of the rotating frame, an angle ϕ_p from the x-axis. The rotating frame is therefore rotated into the toggling frame via the Euler angles $(-(\pi/2 - \phi_p), -\beta_{rf}(\tau), (\pi/2 - \phi_p))$, i.e. rotate a frame coincident with the rotating frame by $(\pi/2 - \phi_p)$ about the rotating frame z-axis, then by $-\beta_{rf}(\tau)$ about the rotating frame y-axis, then by $-(\pi/2 - \phi_p)$ about the rotating frame z-axis (see Fig. B4.5.1).

Thus, we need to transform the spin operator, $\hat{T}_{20}^{IS} = \sqrt{\tfrac{1}{6}}(3\hat{I}_z\hat{S}_z - \hat{\mathbf{I}}\cdot\hat{\mathbf{S}})$, the spin part of the dipolar hamiltonian, into the spin space toggling frame.

The dipolar hamiltonian at time τ into the pulse cycle expressed in the new toggling frame is $\hat{H}^*(t_p^0 + \tau)$, given by

$$\begin{aligned}\hat{H}^*(t_p^0 + \tau) = \hat{R}^{-1}(-(\pi/2 - \phi_p), -\beta_{rf}(\tau), (\pi/2 - \phi_p))\hat{H}_{dd}(t_p^0 + \tau)\\ \times \hat{R}(-(\pi/2 - \phi_p), -\beta_{rf}(\tau), (\pi/2 - \phi_p)) \qquad \text{(iii)}\end{aligned}$$

in which the pulse part of the hamiltonian has vanished because of the frame rotation. The rotation operator $\hat{R}$ is the operator which rotates the spin coordinate

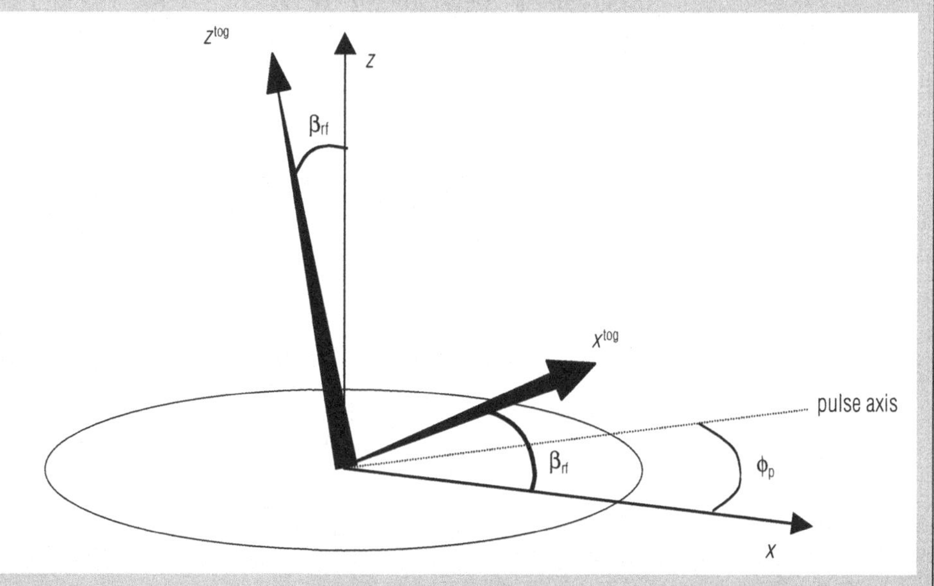

Fig. B4.5.1 The relationship between the rotating frame (x, y, z) and the toggling frame (x^{tog}, y^{tog}, z^{tog}) employed in the analysis of the C7 pulse sequence. A rotation of $-(\pi/2 - \phi_p)$ about z brings y into coincidence with the pulse axis in the rotating frame. We will call this intermediate frame (x_2, y_2, z_2). A rotation of $-\beta_{rf}$ about the y_2-axis (pulse axis) brings the z_2 into coincidence with the z^{tog} of the toggling frame. This new intermediate frame is (x_3, y_3, z_3). Finally, a rotation of $+(\pi/2 + \phi_p)$ about z_3 (z^{tog}) brings the intermediate frame x_3- and y_3-axes into coincidence with the x^{tog} and y^{tog} axes of the toggling frame.

Continued on p. 198

Box 4.5 *Cont.*

rotating frame into the toggling frame. It acts on the spin operators of the dipolar hamiltonian, i.e. $\hat{T}_{20}^{IS}$. Box 4.2 gives the transformation of spherical tensor operators under rotations; for this case

$$\begin{aligned}\hat{R}^{-1}\hat{T}_{20}^{IS}\hat{R} &= \sum_{q=-2}^{+2} D_{q0}^{2}(-(\pi/2-\phi_{\mathrm{p}}),-\beta_{\mathrm{rf}}(\tau),(\pi/2-\phi_{\mathrm{p}}))\hat{T}_{2q}^{IS} \\ &= \sum_{q=-2}^{+2} \exp(iq(\pi/2-\phi_{\mathrm{p}}))d_{q0}^{2}(-\beta_{\mathrm{rf}}(\tau))\hat{T}_{2q}^{IS} \\ &= \sum_{q=-2}^{+2} \exp(iq\,\pi/2)\exp(-iq2\pi p/n)d_{q0}^{2}(-\beta_{rf}(\tau))\hat{T}_{2q}^{IS} \\ &= \sum_{q=-2}^{+2} i^{q}\exp(-iq2\pi p/n)d_{q0}^{2}(-\beta_{\mathrm{rf}}(\tau))\hat{T}_{2q}^{IS} \qquad \text{(iv)}\end{aligned}$$

Thus, the hamiltonian in the toggling frame can be rewritten as

$$\begin{aligned}\hat{H}^{*}(t_{\mathrm{p}}^{0}+\tau) &= \sum_{\substack{IS\\ \text{pairs}}}\sum_{q=-2}^{+2}\sum_{m=-2}^{+2} i^{q}\exp(i2\pi mNp/n+im\omega_{\mathrm{R}}\tau) \\ &\quad \exp(-iq2\pi p/n)d_{q0}^{2}(-\beta_{\mathrm{rf}}(\tau))\omega_{IS}^{(m)}\hat{T}_{2q}^{IS} \\ &= \sum_{\substack{IS\\ \text{pairs}}}\sum_{q=-2}^{+2}\sum_{m=-2}^{+2} i^{q}\exp(i2\pi(Nm-q)p/n)\exp(im\omega_{\mathrm{R}}\tau) \\ &\quad d_{q0}^{2}(-\beta_{\mathrm{rf}}(\tau))\omega_{IS}^{(m)}\hat{T}_{2q}^{IS} \\ &= \sum_{\substack{IS\\ \text{pairs}}}\sum_{q=-2}^{+2}\sum_{m=-2}^{+2} \exp(i2\pi(Nm-q)p/n)\Omega_{qm}^{IS}(\tau)\hat{T}_{2q}^{IS} \qquad \text{(v)}\end{aligned}$$

where $\Omega_{qm}^{IS}(\tau)$ is given by

$$\Omega_{qm}^{IS}(\tau) = i^{q}\exp(im\omega_{\mathrm{R}}\tau)d_{q0}^{2}(-\beta_{\mathrm{rf}}(\tau))\omega_{IS}^{(m)} \qquad \text{(vi)}$$

The average hamiltonian to first order over the pth pulse cycle can now be formed by taking the average of $\hat{H}^{*}$ in Equation (v) over the period of the pulse cycle:

$$\hat{H}_{p}^{(0)} = \sum_{\substack{IS\\ \text{pairs}}}\sum_{q=-2}^{+2}\sum_{m=-2}^{+2} \exp(i2\pi(Nm-q)p/n)\overline{\Omega}_{qm}^{IS}\hat{T}_{2q}^{IS} \qquad \text{(vii)}$$

where

$$\overline{\Omega}_{qm}^{IS} = \frac{1}{\tau_{\mathrm{C}}}\int_{0}^{\tau_{\mathrm{C}}} \Omega_{qm}^{IS}(\tau)\mathrm{d}\tau \qquad \text{(viii)}$$

where τ_C is the length of the pulse cycle. The average hamiltonian over the whole pulse sequence is then found by summing Equation (vii) over all the n cycles in the pulse sequence:

$$\begin{aligned}\hat{H}^{(0)} &= \frac{1}{n}\sum_p \hat{H}_p^{(0)} \\ &= \frac{1}{n}\sum_p \sum_{\substack{IS \\ \text{pairs}}} \sum_{q=-2}^{+2} \sum_{m=-2}^{+2} \exp(i2\pi(Nm-q)p/n)\overline{\Omega}_{qm}^{IS}\hat{T}_{2q}^{IS}\end{aligned} \tag{ix}$$

where p labels the pulse cycles, C_p. The oscillating (exponential) term in this average hamiltonian means that, in general, the average hamiltonian sums to zero when the sum over p is performed. The only exception to this is if $(Nm - q)p/n$ = integer, when the exponential term simply becomes unity.

We wish to excite double-quantum coherence with the a C_n^N pulse sequence; for that we need to retain $\hat{T}_{2\pm2}^{IS}$ terms in the average hamiltonian of Equation (ix), i.e. terms with $q = \pm2$. If we also make a condition that we wish to suppress in the average hamiltonian, other terms arising from the dipolar hamiltonian i.e. $q = \pm1$ (and chemical shift terms which we have not considered here), then it turns out that the simplest sequence which satisfies the $(Nm - q)p/n$ = integer condition for $q = \pm2$ (but not $q = \pm1$) is that with $n = 7$ and $N = 2$. The C_7^2 sequence therefore generates double-quantum coherence in a dipolar-coupled spin pair.

4.3.3 Rotational resonance

We saw in Section 4.1 that in *degenerate* homonuclear spin systems, the B term in the dipolar hamiltonian (see Equation (4.11)) mixes degenerate Zeeman levels of the spin system and so causes exchange of longitudinal magnetization between the spins (see Section 4.1.2 for details). This is the ultimate cause of linebroadening in dipolar-coupled, homonuclear spin systems. In a homonuclear spin system where two spins I and S have a chemical shift difference, the operation of the B term in the dipolar hamiltonian is rendered ineffective because of the lack of degeneracy between the I and S spin states. However, under magic-angle spinning and under the condition known as *rotational resonance*, this situation can be altered [14]. The rotational resonance condition is that the chemical shift difference between the two spins, $\Delta\omega_I - \Delta\omega_S$, is equal to $n\omega_R$, where $n = \pm1, \pm2$ and ω_R is the sample spinning rate. Under this condition, the effect of the B term in the dipolar-coupling hamiltonian is reintroduced, but just between spins I and S for which the rotational resonance condition has been set. This feature can then be used to measure the dipolar coupling between spins I and S. This is usually done using one of the pulse sequences in Fig. 4.18 or similar [14]. Spin I (say) is selectively inverted. The

(a)

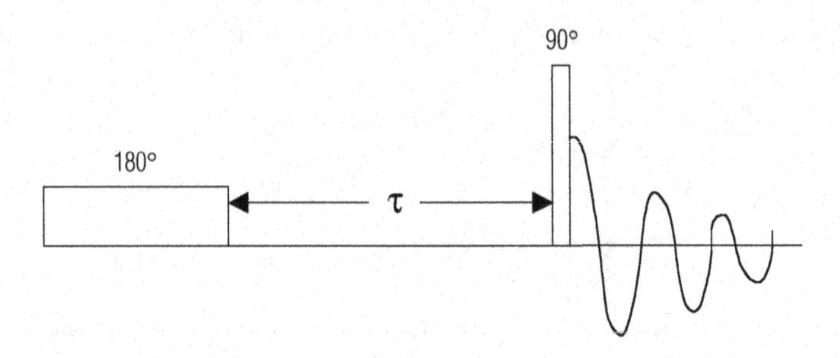

(b)

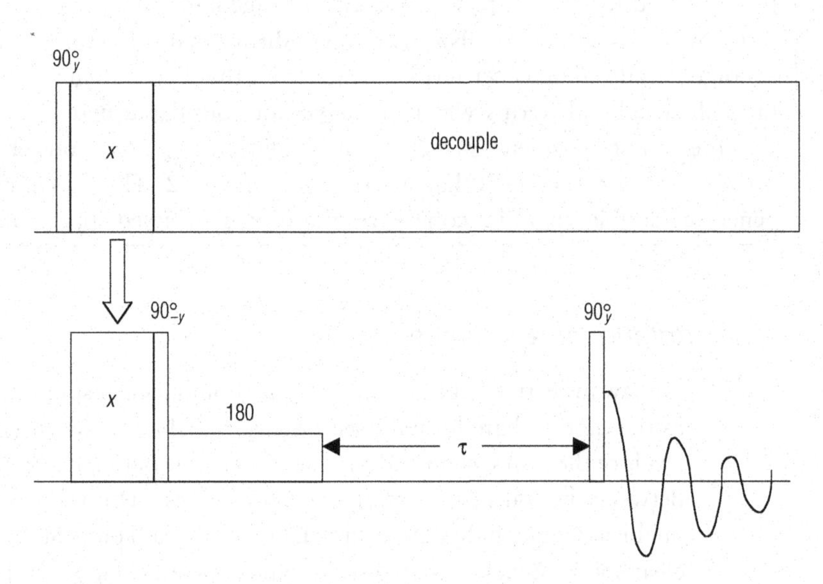

Fig. 4.18 (a) The pulse sequence used in the rotational resonance experiment described in the text. A rotational resonance experiment aims to measure the dipolar coupling between two non-degenerate homonuclear spins, labelled *I* and *S*. The pulse sequence selectively inverts one of the spins, using a low-power (and therefore long) 180° pulse centred on the resonance frequency of that spin. (The bandwidth of a pulse is inversely proportional to its length, so by using a long pulse, we ensure a small bandwidth of rf irradiation.) The rotational resonance condition (the sample spinning rate is set so that $n\omega_R$ is equal to the difference in isotropic chemical shift between spins *I* and *S*) is then applied for a period τ, during which time magnetization exchanges between *I* and *S*. Finally, a non-selective 90° pulse is applied to monitor the state of magnetization on both spins. (b) Cross-polarization can be used in a rotational resonance experiment to generate initial transverse magnetization on both spins *I* and *S*. A 90° pulse then converts this back to longitudinal magnetization, but of much greater magnitude than that which exists at equilibrium. The rotational resonance experiment then proceeds as for (a), with a selective inversion of one of the spins, a period τ under the rotational resonance condition followed by a non-selective 90° pulse to monitor both spins.

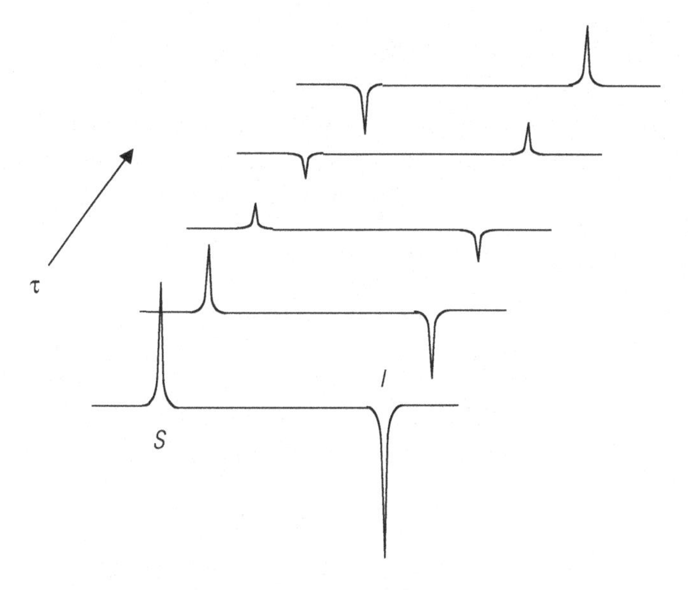

Fig. 4.19 The form of data resulting from a rotational resonance experiment. The experiment uses the pulse sequence of Fig. 4.18 to record NMR spectra of a two-spin system (*I* and *S*) as a function of τ. Spin *I* is the spin which is initially inverted in the experiment (see Fig. 4.18 for details). The signal intensities continue to oscillate as τ increases further (though with damping due to relaxation effects). In addition to changes of intensity with τ, the lineshapes of the *I* and *S* resonances are also affected by the rotational resonance condition, being split into asymmetric doublets [14].

rotational resonance condition is then applied for a time τ (equal to an integral number of rotor periods) to allow exchange of magnetization between spins *I* and *S*, after which a spectrum of *I* and *S* is recorded. The experiment is repeated for many τ; the form of spectra which arise as a function of τ is depicted in Fig. 4.19. For very short τ, the *I* spin signal appears at maximum intensity, but inverted (because of the initial selective inversion of the *I* spins), and the *S* spin signal is close to its maximum, positive intensity. As τ increases, so the *I* signal becomes less negative, and the *S* signal smaller as longitudinal magnetization is transferred between the *I* and *S* spins during the period of rotational resonance in the experiment. At still longer τ, the *I* signal becomes positive and the *S* signal negative and the signal intensities begin to oscillate at a frequency characteristic of the dipolar coupling between the *I* and *S* spins. For quantitative analysis, a plot of the *I* signal intensity minus the *S* signal intensity as a function of τ is constructed. Simulation of this plot allows determination of the dipolar-coupling constant for coupling between *I* and *S* and, hence, the *I*–*S* spin internuclear distance. One complication is that the rate of magnetization exchange between the two spins also depends on the *I* and *S* spin chemical shift anisotropies and the relative orientation of their shielding tensors and so a knowledge of these is required for the analysis of the data (see Box 4.6).

Box 4.6: Theory of rotational resonance

Consider a two-spin system in a sample undergoing magic-angle spinning. The hamiltonian governing this system, expressed in the rotating frame (rotating at the I, S spin Larmor frequency) and ignoring for the moment chemical shift anisotropy effects, is

$$\hat{H}(t) = \hat{H}_{\Delta} + \hat{H}_{\text{dd}}(t) \tag{i}$$

where

$$\begin{aligned} \hat{H}_{\Delta} &= \omega_{\text{iso}}^{I}\hat{I}_z + \omega_{\text{iso}}^{S}\hat{S}_z \\ \hat{H}_{\text{dd}}(t) &= -2C_{IS}(t)\Big[\underbrace{\hat{I}_z\hat{S}_{z-}}_{\text{Term A}} - \underbrace{\frac{1}{4}(\hat{I}_+\hat{S}_- + \hat{I}_-\hat{S}_+)}_{\text{Term B}}\Big] \end{aligned} \tag{ii}$$

The term $\hat{H}_{\Delta}$ accounts for the I and S spin isotropic chemical shift offsets and $\hat{H}_{\text{dd}}(t)$ describes the dipolar coupling between I and S (first-order average in the rotating frame). This latter term is time dependent because of the sample spinning at the magic angle.

We want to form an average hamiltonian over one rotor period, much as we did for the DRAMA pulse sequence in Box 4.3 previously, so that we can see the effect of the rotational resonance condition. The average hamiltonian consists of a sum of terms (see Box 2.1, Chapter 2 for details)

$$\overline{H}(t_{\text{p}}) = \overline{H}^{(0)} + \overline{H}^{(1)} + \overline{H}^{(2)} + \ldots \tag{iii}$$

where

$$\begin{aligned} \overline{H}^{(0)} &= \frac{1}{t_{\text{p}}}\{\hat{H}_1 t_1 + \hat{H}_2 t_2 + \ldots + \hat{H}_n t_n\} \\ \overline{H}^{(1)} &= -\frac{i}{2t_{\text{p}}}\{[\hat{H}_2 t_2, \hat{H}_1 t_1] + [\hat{H}_3 t_3, \hat{H}_1 t_1] + [\hat{H}_2 t_2, \hat{H}_3 t_3] + \ldots\} \\ \overline{H}^{(2)} &= -\frac{1}{6t_{\text{p}}}\Big\{[\hat{H}_3 t_3, [\hat{H}_2 t_2, \hat{H}_1 t_1]] + [[\hat{H}_3 t_3, \hat{H}_2 t_2], \hat{H}_1 t_1] \\ &\quad + \frac{1}{2}[\hat{H}_2 t_2, [\hat{H}_2 t_2, \hat{H}_1 t_1]] + \frac{1}{2}[[\hat{H}_2 t_2, \hat{H}_1 t_1], \hat{H}_1 t_1] + \ldots\Big\} \end{aligned} \tag{iv}$$

in which $\hat{H}_i$ is the hamiltonian during time period t_i. Now, the hamiltonian in Equation (i) does not commute with itself at different times t during the rotor period. That this is so can be seen by forming the matrix representation of the hamiltonian in the coupled spin basis, i.e. $m_I m_S$, and calculating the commutator

$$[\mathbf{H}_1 t_1, \mathbf{H}_2 t_2] = (\mathbf{H}_1\mathbf{H}_2 - \mathbf{H}_2\mathbf{H}_1)t_1 t_2$$

for two arbitrary times t_1 and t_2, where $\mathbf{H}_1$ and $\mathbf{H}_2$ are the matrix representations of the hamiltonian in Equation (i) at times t_1 and t_2. Clearly then, the higher-order terms, $\overline{H}^{(1)}$, $\overline{H}^{(2)}$, etc., in the average hamiltonian in Equation (iii) are non-zero, as the commutators they contain are non-zero. This makes the average hamiltonian very difficult to calculate, as we must calculate many higher-order terms to obtain a reasonable approximation to the average hamiltonian. It is the combination of the $\hat{H}_\Delta$ and $\hat{H}_{dd}(t)$ terms which prevents the hamiltonian in Equation (i) from commuting with itself at different times; $\hat{H}_{dd}(t)$ alone commutes with itself at different times. So, in order to construct the average hamiltonian, we transform to a new frame of reference, a frame in which the $\hat{H}_\Delta$ term in the total hamiltonian is effectively removed and in which the remaining hamiltonian commutes with itself at all times. The average hamiltonian (in the new frame) is then simply the first-order term of Equation (iii), all higher-order terms being zero (within the approximations made in the analysis). This first-order term is easily found by integrating the hamiltonian in the new frame over a rotor period.

The new or toggling frame is found as follows (and this argument follows exactly the same form as that discussion surrounding the toggling frame in Box 2.1, Chapter 2). The state of the spin system is described at any point in time by a density operator, which can in general be expressed as some linear combination of I and S spin operators. The density operator a further time t later is given by

$$\hat{\rho}(t) = \exp(-i\hat{H}t)\hat{\rho}(0)\exp(+i\hat{H}t) \tag{v}$$

where $\hat{\rho}(0)$ is the initial density operator and $\hat{H}$ is the hamiltonian governing the evolution of the system during t. In our case, $\hat{H}$ is given by Equation (i) and is a sum of $\hat{H}_\Delta$ and $\hat{H}_{dd}$. We want to change to a new frame of reference in which the effect of $\hat{H}_\Delta$ (on the density operator) is removed. As we shall see shortly, the effect of $\hat{H}_\Delta$ on the density operator is to rotate the I and S spin terms in the density operator about z, where z refers to the rotating frame z-axis. Thus if we transform to a frame of reference in which the I and S spin operators are rotated by the appropriate amounts about z, the frame transformation itself has taken on the effect of the $\hat{H}_\Delta$ term. $\hat{H}_\Delta$ can then be ignored in the new frame.

Effect of $\hat{H}_\Delta$ term on the density operator

The $\hat{H}_\Delta$ term acting for a time t gives a new density operator of

$$\begin{aligned}\hat{\rho}(t) &= \exp(-i\hat{H}_\Delta t)\hat{\rho}(0)\exp(+i\hat{H}_\Delta) \\ &= \exp(-i[\omega_{iso}^I \hat{I}_z + \omega_{iso}^S \hat{S}_z]t)\hat{\rho}(0)\exp(i[\omega_{iso}^I \hat{I}_z + \omega_{iso}^S \hat{S}_z]t) \\ &= \exp(-i\omega_{iso}^I \hat{I}_z t)\exp(-i\omega_{iso}^S \hat{S}_z t)\hat{\rho}(0)\exp(i\omega_{iso}^S \hat{S}_z t)\exp(i\omega_{iso}^I \hat{I}_z t)\end{aligned} \tag{vi}$$

Continued on p. 204

Box 4.6 *Cont.*

where we have used the general Equation (v) for $\hat{\rho}(t)$. The term

$$\exp(-i\omega_{\text{iso}}^{S}\hat{S}_z t)\hat{\rho}(0)\exp(i\omega_{\text{iso}}^{S}\hat{S}_z t)$$

in Equation (vi) for the density operator at time t has the effect of rotating any S spin operators in the initial density operator, $\hat{\rho}(0)$, by $\omega_{\text{iso}}^{S}t$ about z and the equivalent term for the I spin rotates any I spin operators by $\omega_{\text{iso}}^{I}t$ about z. This is equivalent to transforming to a new frame of reference where the I spin axis frame is rotated by $-\omega_{\text{iso}}^{I}t$ about its z-axis and the S spin axis frame by $-\omega_{\text{iso}}^{S}t$ about its z-axis. By the I spin axis frame, we mean the frame in which the I spin operators are defined and, throughout, z refers to the Larmor rotating frame z-axis.

The hamiltonian in the new rotated frame

The hamiltonian in this new frame of reference is denoted $\hat{H}^*(t)$ and is given by (see Box 2.1, Chapter 2 for details)

$$\hat{H}^*(t) = \hat{R}^{-1}\hat{H}(t)\hat{R} - \omega_{\text{iso}}^{I}\hat{I}_z - \omega_{\text{iso}}^{S}\hat{S}_z \tag{vii}$$

where the rotation operator $\hat{R}$ is

$$\hat{R} = \exp(-i(\omega_{\text{iso}}^{I}\hat{I}_z + \omega_{\text{iso}}^{S}\hat{S}_z)t) \tag{viii}$$

Now substituting Equation (i) for $\hat{H}(t)$ in Equation (vii), we see immediately that the $\hat{H}_\Delta$ term of $\hat{H}(t)$ is unaffected by the $\hat{R}$ rotation, as $\hat{R}$ describes a rotation about the I and S spin z-axes and the $\hat{H}_\Delta$ term contains only the operators $\hat{I}_z$ and $\hat{S}_z$, which are unaffected by z-rotations. The effect of the frame rotation on $\hat{H}_{\text{dd}}(t)$ can be found by considering this operator in the Cartesian form in Equation (4.11) and remembering that a frame rotation by angle θ about an axis is completely equivalent instead to rotating the operator $\hat{H}_{\text{dd}}(t)$ by $-\theta$ about the same axis within the original frame, i.e. rotating $\hat{I}_{x,y}$ by $\omega_{\text{iso}}^{I}t$ about z and rotating $\hat{S}_{x,y}$ by $\omega_{\text{iso}}^{S}t$ also about z, where z refers to the Larmor rotating frame z-axis. Doing this and after some rearrangement, we obtain

$$\hat{H}^*(t) = -2C_{IS}(t)\Big[\hat{I}_z\hat{S}_z - \frac{1}{4}(\hat{I}_+\hat{S}_- \exp(-i[\omega_{\text{iso}}^{I} - \omega_{\text{iso}}^{S}]t) + \hat{I}_-\hat{S}_+ \exp(+i[\omega_{\text{iso}}^{I} - \omega_{\text{iso}}^{S}]t))\Big] \tag{ix}$$

Note that the $\hat{H}_\Delta$ term has vanished in this toggling hamiltonian, cancelling as it does with the last two terms of Equation (vii). Thus, as promised at the outset, the effect of the $\hat{H}_\Delta$ term has been removed by an appropriate choice of axis frame.

The average hamiltonian

We can now form the average hamiltonian over a rotor period in the toggling frame by integrating Equation (ix) over one rotor period so as to form the $\overline{H}^{(0)}$ term of the average hamiltonian. The time dependence of the term $C_{IS}(t)$ under magic-angle spinning was described in Box 4.2 and was shown to be:

$$C_{IS}(t) = -\sqrt{\frac{1}{6}} \sum_{m=-2}^{+2} \omega_{IS}^{(m)} \exp(im\omega_R t) \qquad \text{(x)}$$

The oscillating terms in $C_{IS}(t)$ means that it integrates over a rotor period to zero and so, therefore, does the part of the average hamiltonian deriving from the $\hat{I}_z\hat{S}_z$ term in Equation (ix), which has no other time dependence. If we now impose the rotational resonance condition, $\omega_{iso}^I - \omega_{iso}^S = n\omega_R$, then Equation (ix) takes on the form

$$\hat{H}^*(t) = -\sqrt{\frac{1}{6}} \sum_{m=-2}^{+2} \omega_{IS}^{(m)} \exp(im\omega_R t)[\hat{I}_+\hat{S}_- \exp(-in\omega_R t) + \hat{I}_-\hat{S}_+ \exp(+in\omega_R t)] \qquad \text{(xi)}$$

ignoring that part we know to integrate to zero over a rotor period. When $n = 1, 2$, this hamiltonian contains time-*independent* terms as the $\exp(im\rho_R t)$ terms cancel with the $\exp(\pm in\omega_R t)$ terms. Thus, for $n = 1, 2$, $\hat{H}^*(t)$ integrates to a non-zero value over one rotor period, i.e. the average hamiltonian is non-zero. The final average hamiltonian over one rotor period therefore contains terms in $\hat{I}_+\hat{S}_-$ and $\hat{I}_-\hat{S}_+$, which ultimately arise from the B term in the dipolar-coupling operator. Like the B term in the normal dipolar-coupling hamiltonian, these terms in the average hamiltonian drive the exchange of longitudinal magnetization between the I and S spins.

In summary, setting the rotational resonance condition prevents averaging of the dipolar coupling to zero under magic-angle spinning and, moreover, reintroduces a part of the dipolar coupling, term B, which otherwise would be ineffective in a non-degenerate, homonuclear spin system.

In practice, the I and S spins probably both have chemical shift anisotropies associated with them. This means that the isotropic chemical shift offsets ω_{iso}^I and ω_{iso}^S terms in the original hamiltonian (Equation (i)) should be replaced by chemical shift offsets $\omega_{cs}^I(t)$ and $\omega_{cs}^S(t)$, time dependent due to the effect of magic-angle spinning on the (anisotropic) chemical shifts of spin I and S. In turn, this means that the $\exp(\pm i[\omega_{iso}^I - \omega_{iso}^S]t)$ terms of Equation (ix) for the toggling frame hamiltonian should be replaced by

Continued on p. 206

Box 4.6 *Cont.*

$$\exp\left(\pm i\int_0^t [\omega_{cs}^I(t') - \omega_{cs}^S(t')]dt'\right)$$

This term is simply the time-domain difference spectrum between spins I and S, in effect the 'FID' for the spin I signal minus that for the spin S signal. Both the I and S spin signals, under conditions of magic-angle spinning and non-zero chemical shift anisotropies, consist of a set of spinning sidebands, set at the spinning speed apart and radiating out from their respective isotropic chemical shift. Thus the term $\exp\left(\pm i\int_0^t [\omega_{cs}^I(t') - \omega_{cs}^S(t')]dt'\right)$ takes the form

$$\exp\left(\pm i\int_0^t [\omega_{cs}^I(t') - \omega_{cs}^S(t')]dt'\right) = \sum_{k=-\infty}^{\infty} a_\Delta^{(k)}(\exp\pm i(k\omega_R + \Delta\omega_{iso})t) \qquad \text{(xii)}$$

where $a_\Delta^{(k)}$ is the difference in intensity of the kth spinning sidebands for each spin, which occur at frequency $k\omega_R + \omega_{iso}$, and $\Delta\omega_{iso}$ is the difference between the I and S spin isotropic shifts. Setting the rotational resonance condition $\Delta\omega_{iso} = n\omega_R$ then gives

$$\exp\left(\pm i\int_0^t [\omega_{cs}^I(t') - \omega_{cs}^S(t')]dt'\right) = \sum_{k=-\infty}^{\infty} a_\Delta^{(k)} \exp(\pm i(k+n)\omega_R t) \qquad \text{(xiii)}$$

and substituting this in Equation (ix) in place of the $\exp(\pm i[\omega_{iso}^I - \omega_{iso}^S]t)$ terms gives a toggling frame hamiltonian of

$$\hat{H}^*(t) = -\sqrt{\frac{1}{6}}\sum_{k=-\infty}^{+\infty}\sum_{m=-2}^{+2} a_\Delta^{(k)}\omega_{IS}^{(m)}\exp(im\omega_R t)$$
$$\times[\hat{I}_+\hat{S}_- \exp(-i(k+n)\omega_R t) + \hat{I}_-\hat{S}_+ \exp(+i(k+n)\omega_R t)] \qquad \text{(xiv)}$$

again ignoring that term in $\hat{H}^*(t)$ known to average to zero when integrated over a rotor period. Now terms in this hamiltonian with $m = \pm(k + n)$ are time independent and so give non-zero contributions to the average hamiltonian over one rotor period. This is clearly a more wide-ranging condition than that previously derived. Equation (xiv) also shows that the average hamiltonian depends on the chemical shift anisotropies of the two spins through the terms $a_\Delta^{(k)}$. This dependence is stronger as n increases, i.e. for higher order rotational resonance conditions.

4.4 Techniques for measuring heteronuclear dipolar couplings

All of the techniques for measuring heteronuclear dipolar couplings have a similar form. All are based on some sort of *echo* experiment, in which the heteronuclear dipolar couplings would normally be refocused. However, complete refocusing can be prevented, for instance by applying rf pulses to one of the two coupled spins. The aim of the experiments is then to examine to what extent the perturbing rf pulses prevent refocusing, as this is a measure of the dipolar-coupling strength. Many such experiments have appeared and continue to appear in the literature; however, all are based on the Spin-Echo Double Resonance (SEDOR) and Rotational-Echo Double Resonance (REDOR) experiments which are described in detail here.

4.4.1 Spin-echo double resonance (SEDOR)

Spin-echo double resonance [15] typifies the general scheme of experiments designed to measure heteronuclear dipolar couplings. It is performed on static samples and so does not provide the resolution so often required and, consequently, is not often used nowadays. It is, however, still instructive to examine its operation (Fig. 4.20). After initial excitation of transverse magnetization on one spin, spin I in Fig. 4.20, a series of 180° pulses, with the

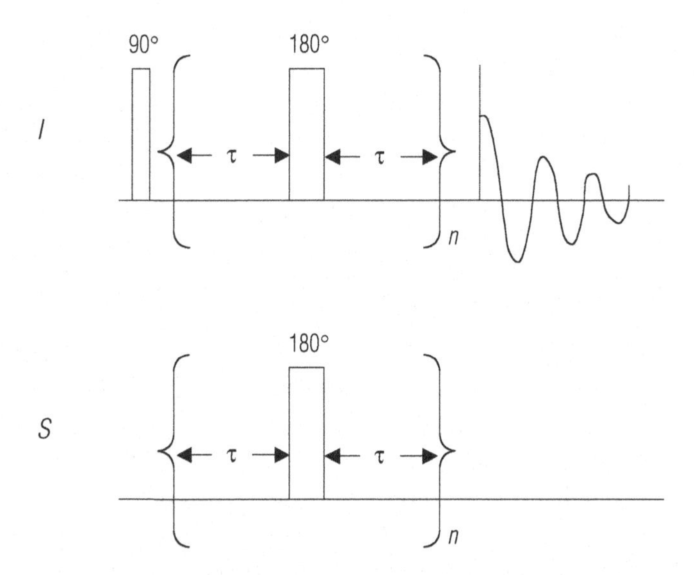

Fig. 4.20 The Spin-Echo Double Resonance (SEDOR) pulse sequence. This sequence is used to measure heteronuclear dipolar couplings in non-spinning samples. Two experiments have to be performed, a reference experiment in which refocusing pulses (refocusing the heteronuclear dipolar coupling) only are applied, and only to one spin, I in this case. In a second experiment, the refocusing of the I spin magnetization is interrupted by pulses applied to the S spin. The degree to which the S spin pulses prevent the refocusing of the I spin magnetization is a measure of the dipolar coupling between spins I and S. I spin spectra are collected as a function of n, the number of I spin echoes, from $n = 1$ until the I spin signal is lost through relaxation.

spacings shown are applied to the I spin [15]. As explained in Section 2.6, these 180° pulses have the effect of refocusing any inhomogeneous interactions, such as heteronuclear dipolar coupling. To recap the results of Section 2.6 briefly, if transverse magnetization dephases under whatever inhomogeneous interactions are present for a time τ, then a 180° pulse applied, the dephased transverse magnetization is refocused a further time τ after the 180° pulse. The series of 180° pulses in Fig. 4.20 causes the dephasing due to heteronuclear dipolar coupling to be refocused at times 2τ, 4τ, 6τ, etc., after the initial excitation of the I transverse magnetization. The echo intensities at these times are recorded as a reference data set. The only loss of intensity between the echoes is due to transverse relaxation. A second experiment is then performed in which 180° pulses are also applied to the S spin at times τ, 3τ, 5τ, etc. These have the effect of inverting the S spins at these points in time and thus changing the sign of the strength of the dipolar-coupling interaction between I and S, as the hamiltonian describing the interaction is

$$\hat{H}_{\mathrm{dd}}^{\mathrm{hetero}} = -2C_{IS}\hat{I}_z\hat{S}_z \tag{4.23}$$

where the constant C_{IS} is defined as in Box 4.2:

$$C_{IS} = d\frac{1}{2}(3\cos^2\theta - 1) \tag{4.24}$$

This in turn prevents the complete refocusing of the I–S dipolar coupling at the echo points of 2τ, 4τ, 6τ, etc. The difference in intensity of the echo maxima between the two experiments depends on the dipolar coupling strength, and the dipolar-coupling constant is then relatively easily extracted from the plots of echo intensity versus $n\tau$, $n = 2, 4, 6, \ldots$, for the case where I and S constitute an isolated spin pair.

4.4.2 Rotational-echo double resonance (REDOR)

This experiment is very similar in its operation to the SEDOR experiment described in the previous section, except that the I spin echoes are now provided by magic-angle spinning [16]. Under magic-angle spinning, any I spin transverse magnetization dephases under the I–S dipolar coupling during the first half of the rotor period and is then refocused during the second half of the rotor period. This is explained in more detail in Section 2.2.2, albeit for the case of dephasing under chemical shift anisotropy rather than heteronuclear dipolar coupling; however, the principles are exactly the same. The pulse sequence used in the REDOR experiment is shown in Fig. 4.21.

If 180° pulses are applied during the rotor period to the S spin, the I–S dipolar coupling is only partially refocused at the end of the rotor period

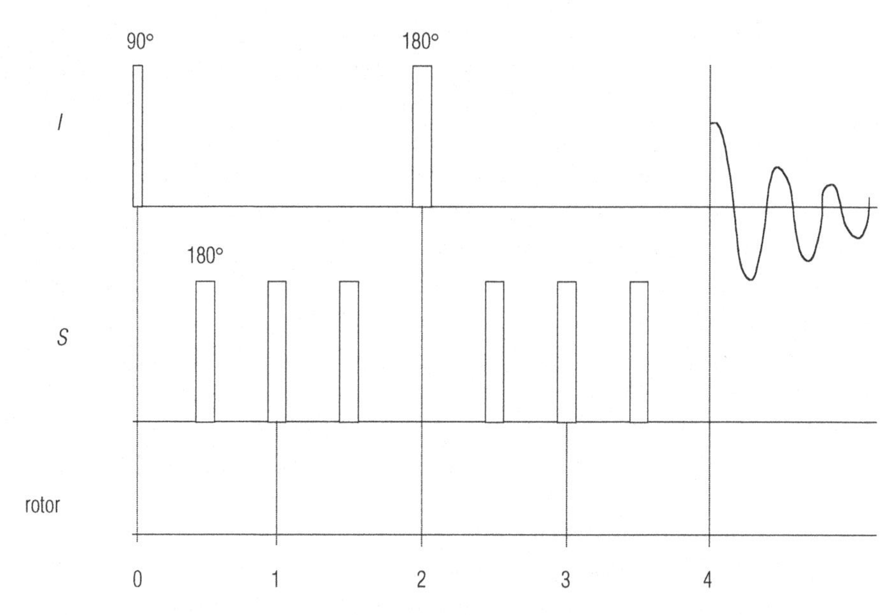

Fig. 4.21 The Rotational-Echo Double Resonance (REDOR) pulse sequence. As with the SEDOR sequence of Fig. 4.20, this pulse sequence is used to measure heteronuclear dipolar couplings, but this time on magic-angle spinning samples. Again, two experiments have to be performed with *I* as the observed spin in both cases. In the reference experiment, the dipolar coupling between the *IS* spin pair is refocused at the end of every rotor period, as normal under magic-angle spinning. In the second experiment, this refocusing of the dipolar coupling is prevented by a series of rotor-synchronized 180° pulses applied to one of the spins (*S* in this case). As with the SEDOR experiment, *I* spin spectra are collected for several different lengths of dephasing time, each of which is necessarily a multiple of $2\tau_R$, where τ_R is the rotor period. The initial *I* spin transverse magnetization is often generated via cross-polarization (from ^{1}H, for example), rather than with a 90° pulse as shown here.

for the same reason as the 180° pulses in the SEDOR experiment perturb the *I* spin echoes. The difference between the *I* spin intensity at the end of the rotor period in the experiment where no pulses are applied to the *S* spin (the normal, reference experiment, with intensity $I_0^{expt}(\tau_R)$) and the *I* spin intensity in the experiment where 180° pulses are applied to the *S* spin, ($I_f^{expt}(\tau_R)$), depends quantitatively on the dipolar-coupling strength.

In practice, the extent of *I* spin dephasing resulting from incomplete refocusing of the rotational echoes is monitored as a function of the number of rotor periods for which the dephasing is allowed to occur; as the experiment extends to larger numbers of rotor periods, so the net *I* spin dephasing under *S* spin 180° pulses increases. A 180° pulse is also applied to the *I* spins in each experiment halfway through the dephasing period to refocus any *I* spin chemical shift offset, which would otherwise contribute to the net *I* spin dephasing. The *S* spin 180° pulse which would otherwise have occurred at this halfway point is omitted, as explained in more detail in Box 4.7 below. A plot is then formed of $I_0^{expt}(N\tau_R) - I_f^{expt}(N\tau_R)$ as a function of number of rotor periods, *N*, and this can be quantitatively analysed to determine the IS dipolar-coupling constant. Such an analysis is generally unambigous if I–S is an isolated spin pair. However, in the case where the I spin is coupled to several S spins, the data can often be fitted by many different sets of dipolar-coupling constants and so are less useful.

This method is ideal for measuring dipolar-coupling constants of a few hundred to a few thousand Hertz. For very large dipolar couplings, the I spin transverse magnetization is completely dephased after only one or two rotor periods when 180° pulses are applied to the S spin. Under these circumstances, the plot of $I_0^{expt}(N\tau_R) - I_f^{expt}(N\tau_R)$ versus number of rotor periods does not contain enough data points for a good quantitative analysis. For very small dipolar couplings, the net dephasing when S spin 180° pulses are applied is just too small to be accurately measured by difference with the reference experiment. The net I spin dephasing depends on the rotor period of course, and is larger for slower spinning rates, so it is worthwhile considering what spinning rate might be most suitable for the particular dipolar-coupling constants to be measured.

Since the introduction of the REDOR experiment several variants of it have appeared in the literature; however, the principles of these are largely the same as that just described.

Box 4.7: Analysis of the REDOR experiment

This analysis uses average hamiltonian theory to assess the effect of the pulse sequence in Fig. 4.21 [17], and as such follows much the same course as that for the DRAMA experiment in Box 4.3.

We deal with the case of a heteronuclear spin pair, IS. The first-order average hamiltonian in the rotating frame describing the dipolar interaction between these two spins is

$$\hat{H}_{\text{dd}}^{\text{hetero}}(t) = -2C_{IS}(\beta, \gamma, t)\hat{I}_z\hat{S}_z \qquad \text{(i)}$$

where $C_{IS}(\beta, \gamma, t)$ is given by Equations (xi) and (xii) in Box 4.3 and describes the strength of the dipolar interaction for a given molecular orientation (β, γ) relative to the applied magnetic field and is time dependent here due to the magic-angle spinning which takes place during the experiment. We assume that the 180° pulses in the experiment are of sufficiently high amplitude that, during the period of a pulse, the interaction with the pulse rf field is much stronger than any other interactions. We may then neglect all other interactions during the pulses; the hamiltonian during the pulses is then simply that due to the pulse, namely

$$\hat{H}_{\text{pulse}} = \omega_1 \hat{I}_x \tag{ii}$$

for an x-pulse for instance, where ω_1 is the rf pulse amplitude.

We then form the average hamiltonian over one rotor period, all subsequent rotor periods in the S 180° pulse train in Fig. 4.21 being effectively identical. The rotor period in which a 180° pulse is applied to the I spins rather than S is discussed later, and is shown to be equivalent to other rotor periods in the sequence. As discussed in Box 2.1 (Chapter 2), the average hamiltonian over a period of time t_p is given by

$$\overline{H}(t_p) = \overline{H}^{(0)} + \overline{H}^{(1)} + \overline{H}^{(2)} + \ldots \tag{iii}$$

where the higher-order terms, $\overline{H}^{(1)}$, $\overline{H}^{(2)}$, depend on the commutators of the hamiltonians operating at different times in the t_p time period. Now, as explained in Box 4.3, the pulse hamiltonian (Equation (ii)) and dipolar hamiltonian (Equation (i)) do not commute with each other, so the higher-order terms, $\overline{H}^{(1)}$, $\overline{H}^{(2)}$, etc., are non-zero and contribute significantly to the total average hamiltonian. This makes the average hamiltonian difficult to calculate, as many higher-order terms may have to be considered. To get around this, at the point of each pulse in the sequence, we transform the hamiltonian to a new frame, the so-called *toggling frame*, in which the pulse hamiltonian vanishes. Then only the dipolar hamiltonian remains, which commutes with itself at all times, so that the $\overline{H}^{(0)}$ term of Equation (iii) alone is a good approximation to the average hamiltonian. In turn, $\overline{H}^{(0)}$ is simply the average of the hamiltonians which operate during the rotor period, i.e. the sum of the hamiltonians which operate multiplied by the fraction of the rotor period for which they operate.

So, we can now proceed to calculate the average hamiltonian over a rotor period [17]. In the first half of the rotor period, the system simply evolves under the IS dipolar coupling, described by the hamiltonian, $\hat{H}_{\text{dd}}^{\text{hetero}}(t)$, as in Equation (i) above. At the first S spin 180° pulse (half way through the rotor period), we transform the whole description of the spin system into a spin coordinate frame in which the pulse hamiltonian vanishes. The effect of an S spin 180° pulse is to rotate the parts of the density operator describing the S spin system by 180° about x. Thus, if we transform to a new S spin axis frame in which the S spin axis frame is rotated by −180° about the rotating frame x-axis, we have emulated the effect of the pulse; the hamiltonian describing the effect of the pulse in this new representation must then be zero.

For the second half of the rotor period, the system again evolves under the IS dipolar coupling, but we must now transform the hamiltonian operator describing this interaction into the new toggling frame, in which the S spin axis frame is rotated by −180° about the rotating frame x-axis. This is equivalent to rotating

Continued on p. 212

Box 4.7 *Cont.*

the dipolar hamiltonian itself by +180° about x and leaving the S spin axis frame alone. Rotating the S spin operators in the hamiltonian of Equation (i) above gives the new toggling frame hamiltonian to be

$$\hat{H}^*(t) = -2C_{IS}(\beta,\gamma,t)\hat{I}_z(-\hat{S}_z) = -2C_{IS}(\beta,\gamma,t)\hat{I}_z\hat{S}_z = -\hat{H}_{\text{dd}}^{\text{hetero}} \qquad \text{(iv)}$$

where z refers to the usual Larmor rotating frame z-axis. In the same manner, the S spin 180° pulse at the end of the rotor period then simply has the effect of transforming $-\hat{H}_{\text{dd}} \rightarrow \hat{H}_{\text{dd}}$, so that the hamiltonian describing the spin system at the beginning of the next rotor period is $\hat{H}_{\text{dd}}$ once again. The average hamiltonian over the rotor period is then simply

$$\overline{H}^{(0)} = \frac{1}{\tau_{\text{R}}}\left(\int_0^{\tau_{\text{R}}/2} \hat{H}_{\text{dd}}(t)\text{d}t - \int_{\tau_{\text{R}}/2}^{\tau_{\text{R}}} \hat{H}_{\text{dd}}(t)\text{d}t\right) \qquad \text{(v)}$$

The time dependence of $\hat{H}_{\text{dd}}$ is contained in the C_{IS} (β, γ, t) term (Equation (i)) which is given by Equations (xi) and (xii) in Box 4.3 to be:

$$C_{IS}(\beta,\gamma,t) = d\left(-\frac{1}{\sqrt{2}}\sin 2\beta\cos(\omega_{\text{R}}t+\gamma) + \frac{1}{2}\sin^2\beta\cos(2\omega_{\text{R}}t+2\gamma)\right) \qquad \text{(vi)}$$

where d is the dipolar coupling constant, $d = (\mu_0/4\pi)(\gamma_I\gamma_S/r_{IS}^3)\,\hbar$ for the IS spin pair and (β, γ) are two of the Euler angles describing the rotation of the dipolar PAF of the current molecular orientation into a rotor-fixed frame; ω_{R} is the rotor spinning rate. Using this equation in Equation (v), it is easy to see that $\overline{H}^{(0)}$ is non-zero and, moreover, that $\overline{H}^{(0)}$ depends on the dipolar coupling between I and S. It is the S spin 180° pulse midway through the rotor period that prevents the average hamiltonian from being zero (and, therefore, prevents the I–S dipolar coupling from being refocused). In contrast, consider the situation without any S spin π pulses. The average hamiltonian over a rotor period then is

$$\overline{H}^{(0)} = \frac{1}{\tau_{\text{R}}}\left(\int_0^{\tau_{\text{R}}} \hat{H}_{\text{dd}}(t)\text{d}t\right) \qquad \text{(vii)}$$

which (using the periodic time dependence in Equation (vi)) is zero, as it should be under magic-angle spinning. Thus the application of an S spin 180° pulse reverses the sign of the dipolar hamiltonian under which the spin system evolves and, in doing so, prevents the IS dipolar coupling from being refocused under magic-angle spinning, as it otherwise would be.

There is one rotor period in the pulse sequence of Fig. 4.21 in which a 180° pulse is applied to the I spins at the $\tau_{\text{R}}/2$ point (to refocus chemical shift offsets) and that which would normally be applied to the S spins is omitted. Using a similar analysis to that above, it is easy to see that a single 180° pulse applied to either spin has the required effect of reversing the sign of the dipolar hamiltonian in the toggling frame, and so prevents the dipolar coupling from being refocused. 180° pulses applied to both spins simultaneously, however, leave the dipolar hamiltonian unaltered, which is not what is required in this pulse sequence. Hence, the rotor period in which an I spin 180° pulse, but no S spin pulse, is applied is completely equivalent (in terms of average hamiltonian theory) to those rotor periods in which only S spin 180° pulses are applied.

We can then determine a quantitative expression for the loss of I spin signal intensity at the end of one rotor period resulting from the incomplete refocusing of the I–S dipolar coupling. The initial I spin transverse magnetization, prior to the S spin π pulse train, lies along a particular axis of the rotating frame. Any I spin magnetization at the end of the pulse sequence lying along this axis is effectively 'refocused' magnetization. Any I spin magnetization lying along an orthogonal axis arises from dephasing and does not contribute to the I spin signal intensity. We need to determine the amount of I spin magnetization lying along the initial axis at the end of the rotor period, as this is proportional to the I spin signal intensity. We do this by calculating the phase angle $\Delta\Phi$, acquired by the I spin transverse magnetization over the rotor period (see Fig. B4.7.1).

The I spin precession frequency in the rotating frame in the absence of chemical shift offsets is $\omega_{\mathrm{dd}}^{I}(\beta, \gamma, t)$ for a given molecular orientation (time dependent due to magic-angle spinning) and is given by

$$\omega_{\mathrm{dd}}^{I}(\beta,\gamma,t) = \pm C_{IS}(\beta,\gamma,t) \tag{viii}$$

for a spin-$\frac{1}{2}$ nucleus, where we have used Equation (i) to determine the I spin precession frequency (which corresponds to the dipolar contribution to the I spin transition frequency for $m_I = +\frac{1}{2} \rightarrow m_I = -\frac{1}{2}$) assuming spins I and S are both spin-$\frac{1}{2}$

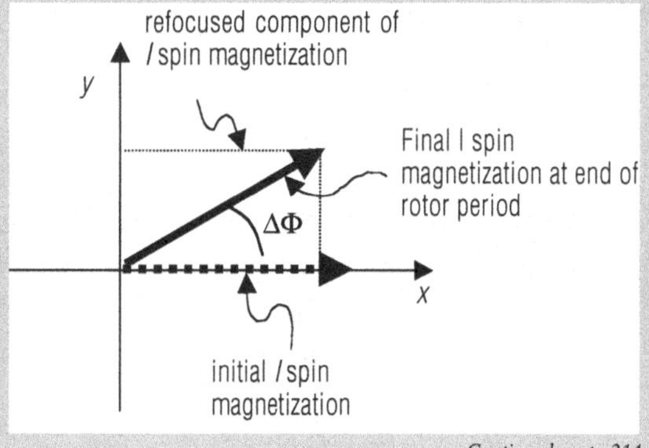

Fig. B4.7.1 Definition of the phase angle $\Delta\Phi$ acquired by I spin magnetization in the REDOR experiment (assuming no relaxation effect).

Continued on p. 214

Box 4.7 *Cont.*

nuclei. Thus the I spin phase angle after one rotor period and the S spin 180° pulse sequence in Fig. 4.21 is given by

$$\Delta\Phi(\beta,\gamma,\tau_R) = \left(\int_0^{\tau_R/2} \omega_{dd}^I(\beta,\gamma,t)dt - \int_{\tau_R/2}^{\tau_R} \omega_{dd}^I(\beta,\gamma,t)dt \right) \quad \text{(ix)}$$

The I spin signal intensity is proportional to cos $\Delta\Phi(\beta, \gamma, \tau_R)$, i.e. the x-component of the I spin transverse magnetization. Substituting for $\omega_{dd}^I(\beta, \gamma, t)$ in Equation (ix) from Equations (viii) and (vi), we obtain

$$\Delta\Phi(\beta,\gamma,\tau_R) = \pm 2\sqrt{2}\frac{d}{\omega_R}\sin 2\beta \sin\gamma \quad \text{(x)}$$

as the phase angle acquired by I spin magnetization for the molecular orientation described by Euler angles (β, γ). The I spin signal intensity for a powder sample at the end of the rotor period, $I_f(\tau_R)$, is then found by summing cos $\Delta\Phi(\beta, \gamma, \tau_R)$ over all possible crystallite orientations. Overall then, we have for the I spin signal intensity:

$$I_f(\tau_R) = \frac{I_0(0)}{2\pi} \int_0^{\pi/2} \int_0^{2\pi} \cos(\Delta\Phi(\beta,\gamma,\tau_R))\sin\beta d\beta d\gamma \quad \text{(xi)}$$

where $I_0(0)$ is the signal intensity at time 0, i.e. immediately after the initial 90° pulse creating transverse magnetization.

The phase angle acquired after N rotor periods is simply $N\,\Delta\Phi(\beta, \gamma, \tau_R)$. Equation (xi) takes no account of relaxation effects. In practice, there is always a loss of signal intensity during the dephasing period due to relaxation in addition to any dephasing. As already mentioned, a reference experiment is recorded without any S spin 180° pulses during the dephasing period in order to be able to take account of this. The difference between $I_f^{expt}(N\tau_R)$ and the reference spectrum intensity, $I_0^{expt}(N\tau_R)$ then corresponds to

$$\frac{I_0^{expt}(N\tau_R) - I_f^{expt}(N\tau_R)}{I_0^{expt}(0)} = 1 - \frac{1}{2\pi} \int_0^{\pi/2} \int_0^{2\pi} \cos(\Delta\Phi(\beta,\gamma,\tau_R))\sin\beta d\beta d\gamma \quad \text{(xii)}$$

from Equation (xi). $[I_0(N\tau_R) - I_f(N\tau_R)]/I_0(0)$ can be simulated for different values of the dipolar coupling constant d until good agreement is reached with the experimental plot. Analytical solutions for REDOR curves have been found by Mueller [18], and these can simplify the data analysis.

4.5 Techniques for dipolar-coupled quadrupolar–spin-$\frac{1}{2}$ pairs

Many materials of interest contain quadrupolar ($I > \frac{1}{2}$) nuclei, so it is almost inevitable that, at times, we will want to measure dipolar couplings between quadrupolar and spin-$\frac{1}{2}$ nuclei, and indeed, between quadrupolar nuclei. Quadrupolar nuclei are discussed in detail in Chapter 5; however, it is appropriate to discuss dipolar-coupling measurements pertaining to them here.

Quadrupolar nuclei often suffer quadrupolar linebroadening of the order of MHz. This feature means that the techniques discussed in Section 4.4 for the measurement of heteronuclear dipolar couplings cannot in general be applied to quadrupolar–spin-$\frac{1}{2}$ pairs, as techniques like REDOR require hard 180° pulses (i.e. rf amplitude much greater than any other spin interaction) to be applied at some stage to both spins. Rf amplitudes of a few hundred kHz can be achieved, but not much more, and clearly this is a lot less than a quadrupole interaction strength of a few MHz. Consequently, the REDOR experiment described in Section 4.4 simply will not work if the S spin to which the train of 180° pulses is applied is a quadrupolar nucleus suffering moderate (or larger) quadrupole coupling). The 180° pulses in the REDOR experiment are required to invert the S spin populations, but with quadrupolar linebroadening of a few MHz, no achievable rf pulse can actually irradiate all the spins of a powder sample in order to do this.

Having said that, REDOR has been successfully applied in a number of quadrupolar–spin-$\frac{1}{2}$ pair cases, when the S spin to which the 180° pulse train is applied is the spin-$\frac{1}{2}$ nucleus rather than the quadrupolar one [19]. In these cases, the quadrupolar spin is always a half-integer spin and it is only the central transition ($+\frac{1}{2} \rightarrow -\frac{1}{2}$) which is observed. As mentioned in Section 4.4, it is important to refocus any dephasing of the observed spin that occurs by any means other than dipolar coupling to the S spin, for example by chemical shift offsets or, in the case of a quadrupolar spin, by quadrupole coupling. In the case where the I spin is a spin-$\frac{1}{2}$, chemical shift offsets are refocused by a 180° pulse on the I spin in the middle of the S spin pulse train. Fortunately, it is relatively easy to refocus the dephasing of the central transition magnetization associated with the quadrupolar nucleus via a selective 180° pulse applied to the central transition of the quadrupolar spin in the middle of the S spin 180° pulse train. The dephasing here occurs due to quadrupole coupling as well as chemical shift effects. REDOR has also been used with MQMAS [20] (see Chapter 5) to improve the resolution in the experiment. However, these implementations of the REDOR experiment require that the spin-$\frac{1}{2}$ nucleus is abundant, otherwise the extent of dephasing by dipolar coupling on the quadrupolar spin magnetization is tiny. Moreover, they are only suitable for half-integer quadrupolar spins, and there are two potentially important integer quadrupolar spins, ^{14}N ($I = 1$) and ^{2}H ($I = 1$).

To this end, two experiments have been devised specifically to measure dipolar couplings between quadrupolar and spin-$\frac{1}{2}$ nuclei and which can be used with half-integer or integer quadrupolar spins alike. These are the Transfer of Population in Double Resonance (TRAPDOR) [21] and the Rotational-Echo, Adiabatic Passage, Double Resonance (REAPDOR) [22] experiments.

4.5.1 *Transfer of population in double resonance (TRAPDOR)*

The TRAPDOR experiment [21] is run under magic-angle spinning and is applied to quadrupolar–spin-$\frac{1}{2}$ pairs. The principle of the experiment is much the same as the REDOR experiment. One nucleus, I (the spin-$\frac{1}{2}$), is observed both with and without a perturbation being applied to the coupled nucleus, S (the quadrupolar one), the perturbation being designed to prevent the complete refocusing of the dipolar coupling between the pair under the magic-angle spinning. The experiment without any perturbation provides a reference dataset of signal intensities in the absence of any dipolar-coupling effects (just relaxation effects). The extent of the dephasing of the spin-$\frac{1}{2}$ transverse magnetization as a result of the incomplete refocusing in the second experiment is then a measure of the dipolar coupling between the spins. The experiment is repeated with the perturbation being applied for different lengths of time (the dephasing period) and a plot of the difference in signal intensity between the two experiments as a function of dephasing time is constructed. Analysis of this plot allows the determination of the dipolar coupling between the two spins.

In the REDOR experiment, the perturbation which prevents the complete refocusing of dipolar coupling is a series of rotor-synchronized 180° pulses applied to one of the spins. 180° pulses have the effect of inverting the spin populations among the levels of the particular spin type the pulses are applied to. In turn, this has the effect of reversing the sign of the dipolar-coupling interaction (which depends on the operator $\hat{I}_z\hat{S}_z$) part way through the rotor period, so that the average dipolar coupling over a rotor period is no longer zero.

The TRAPDOR experiment prevents complete refocusing of dipolar coupling between a spin pair in a similar way, namely by altering the strength of the dipolar coupling throughout the rotor period by changing the populations of the Zeeman spin levels of one of the spins. In the TRAPDOR experiment, it is the spin-$\frac{1}{2}$ nucleus which is observed and the populations of the quadrupolar spin levels which are changed through the rotor period. However, the method of changing the quadrupolar spin state populations is

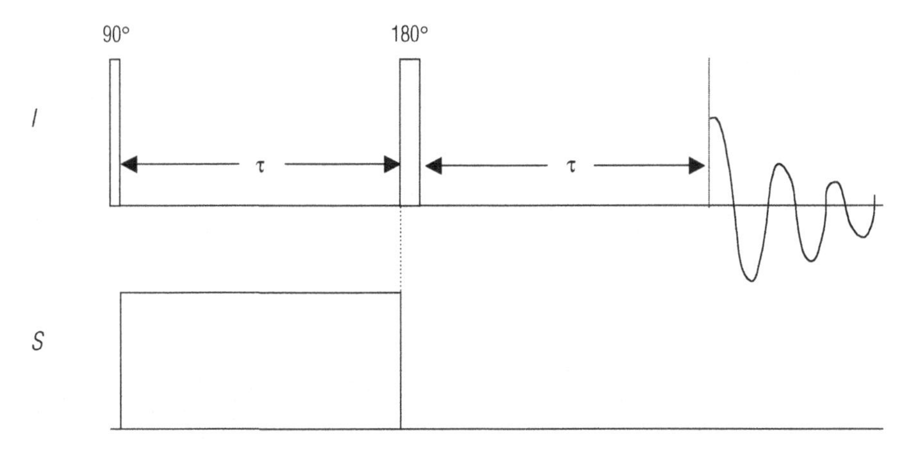

Fig. 4.22 The TRAPDOR pulse sequence, used for measuring dipolar couplings between quadrupolar (particularly $I = 1$) spins and spin-$\frac{1}{2}$ nuclei. The I spin sequence is run under magic-angle spinning both with and without (reference spectrum) the S spin irradiation. I spin spectra are collected for different values of τ from zero upwards, with the proviso that τ is always an integral number of rotor periods. In the absence of the S spin irradiation, the I spin magnetization is fully refocused at 2τ (except for any loss of signal due to relaxation). The S spin irradiation during the first τ period affects the S spin state, and so alters the dipolar coupling between I and S in this period. Switching off the S spin irradiation in the second τ period then ensures that the I–S dipolar coupling has a different time dependence in the second τ period from the first, so that the I spin magnetization is not fully refocused now at the end of the period. Cross-polarization from, e.g., ^{1}H, is often used to generate the initial I spin transverse magnetization, rather than the 90° pulse shown here.

quite different to the method used in the REDOR experiment, and rather neat. Instead of a series of 180° pulses as in the REDOR experiment, a continuous rf pulse is applied to the quadrupolar spin for half the required dephasing period and then nothing is done in the second half of the dephasing period (Fig. 4.22). The effect of an rf pulse on quadrupolar spin states is discussed in detail in Chapter 5. Here, suffice to say that both the interaction with the rf field and the quadrupole coupling must be considered during the pulse, as the quadrupole coupling is generally large compared to the rf pulse amplitude, so cannot be ignored.

Different molecular orientations in a powder sample have different strengths of quadrupole coupling and so give rise to different energies of spin eigenstates during the pulse (and at other times too of course). A plot of the energies of the three spin levels for a spin-1 nucleus as a function of quadrupole coupling strength is shown in Fig. 4.23. Also shown on this plot are the forms of the spin levels at extreme values of the quadrupole coupling.

Now, under magic-angle spinning, the molecular orientations are changing continuously, and so therefore does the quadrupole coupling strength associated with any one crystallite. In effect then, the nature and energy of the spin-1 eigenstates under magic-angle spinning oscillate between the two extremes on the horizontal axis during a rotor period. This means that the

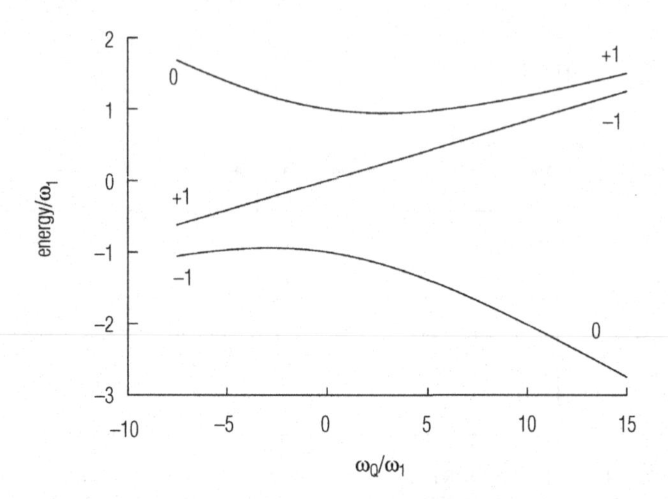

Fig. 4.23 The energies of the three spin levels of an $I = 1$ spin under rf irradiation, as a function of ω_Q/ω_1, where ω_Q is the quadrupole splitting (see Equation 5.10) and ω_1 is the rf irradiation amplitude. Shown on the plot are the nature of the spin states at the extreme values of quadrupole coupling strength.

Zeeman level |+1> for instance, which exists for large negative ω_Q, becomes the Zeeman |–1> level as ω_Q becomes large and positive as a result of sample spinning. Providing the change between the two ω_Q values is done adiabatically, the result is that the population of the |+1> Zeeman level becomes that of the |–1 > level, i.e. populations of these Zeeman levels are exchanged, as required in the TRAPDOR experiment. Further examination of Fig. 4.23 shows that, in addition, the |–1> Zeeman level becomes the |0> Zeeman level and |0> becomes |+1> as ω_Q goes from being large and negative to large and positive. Each time the rotor orientation changes such that ω_Q goes through zero (a so-called *zero crossing*) this change of spin levels and concomitant exchange of Zeeman level populations occurs. In turn, changing the nature of the quadrupolar spin eigenstates alters the dipolar coupling between this spin-1 nucleus and the dipolar-coupled spin-$\frac{1}{2}$ one. For instance, the change in eigenstate from |+1> to |–1> as ω_Q varies between large and positive and large and negative changes the $\hat{S}_z$ eigenvalue from +1 to –1, which has a clear effect on the dipolar coupling which depends on $\hat{I}_z\hat{S}_z$. This then prevents the complete averaging of the dipolar coupling by the magic-angle spinning. Each molecular orientation in a powder sample experiences two or four zero crossings per rotor period, depending on the initial crystallite orientation; each zero crossing contributes to the net dephasing of the spin-$\frac{1}{2}$ transverse magnetization.

As already mentioned, this experiment requires the changes in the spin system to be adiabatic. The parameter which determines whether or not this is the case is the adiabaticity parameter, α, which has been determined to be [21]

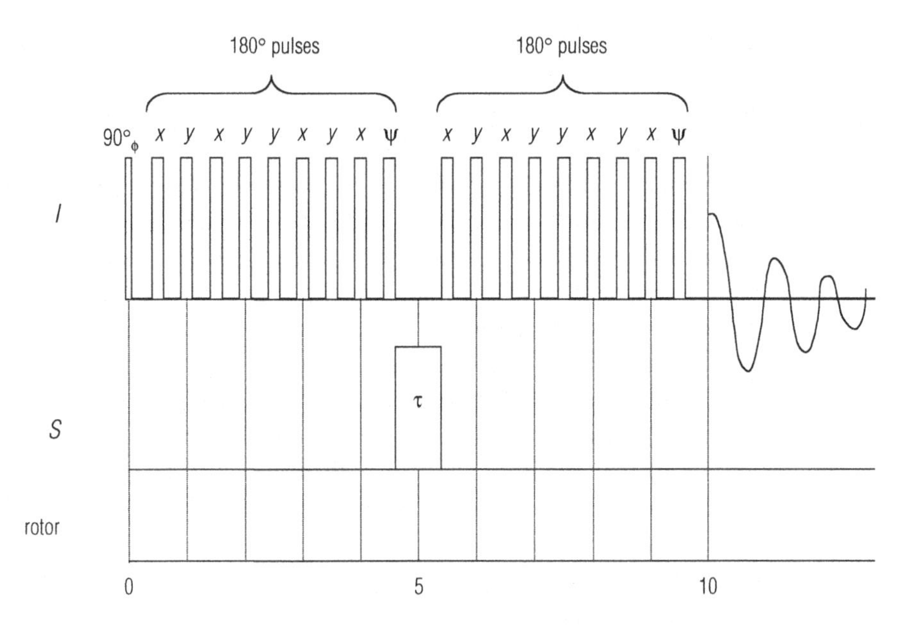

Fig. 4.24 The REAPDOR pulse sequence, used for measuring dipolar couplings between quadrupolar (particularly $I = 1$) spins and spin-$\frac{1}{2}$ nuclei. A reference experiment is recorded without any S spin irradiation. In this experiment, dipolar evolution occurs under the first series of I spin 180° pulses and is refocused (along with isotropic chemical shifts, chemical shift anisotropies) at the point at which acquisition starts. In the experiment with S spin irradiation, a pulse of length τ ($<\tau_R$) is applied to the S spin in the middle of the dipolar evolution period, which prevents the refocusing of the I–S dipolar coupling at the point of acquisition. The experiment is repeated for different dipolar evolution periods. The minimum dipolar evolution period is $8\tau_R$ (as in the illustrated sequence) and is incremented by $8\tau_R$ in successive experiments. As with the TRAPDOR experiment, the initial I spin magnetization can be derived from cross-polarization rather than a 90° pulse as shown here.

$$\alpha = \frac{\omega_1^2}{\chi \omega_R} \tag{4.25}$$

where ω_1 is the rf pulse amplitude, χ is the S spin quadrupole coupling constant (see Chapter 5 for more details) and ω_R is the sample spinning rate.

Analysis of the results from a TRAPDOR experiment involves calculating the number of zero crossings [21] which occur for each crystallite during the dephasing period of the experiment and assuming that completely efficient transfer of populations occurs at each zero crossing, i.e. perfectly adiabatic passage occurs.

4.5.2 Rotational-echo adiabatic-passage double-resonance (REAPDOR)

The pulse sequence for the REAPDOR experiment [22] is shown in Fig. 4.24. Its operation is similar in principle to the TRAPDOR and REDOR experiments previously described (Sections 4.5.1 and 4.4.2 respectively). As in the TRAPDOR experiment, the spin-$\frac{1}{2}$ nucleus is observed. The 180° pulse train applied to the spin-$\frac{1}{2}$ nucleus is a REDOR-like sequence (see Section 4.4) designed to prevent the complete refocusing of the *I–S* dipolar coupling under magic-angle spinning; the phases of the pulses shown in Fig. 4.24 ensure that chemical shift offsets, however, are refocused as required. The missing 180° pulse in the middle of the spin-$\frac{1}{2}$ pulse train means that in the absence of any pulses on the coupled quadrupolar spin, any evolution under the dipolar coupling which takes place in the first half of the pulse train is refocused in the second half. This then constitutes the reference experiment in REDOR terms. In a second experiment, a transfer of populations between the spin states of the quadrupolar nucleus is effected in a similar way to the TRAPDOR experiment, via a continuous rf pulse applied to that nucleus. However, whereas in the TRAPDOR experiment the quadrupole spin pulse is applied for half the total dephasing time required, in the REAPDOR experiment, it is always applied for a constant time, which is less than one rotor period.

The two experiments (reference and one with rf pulse on the quadrupolar nucleus) are repeated for different lengths of the spin-$\frac{1}{2}$ pulse train and the extent of dephasing monitored. The spin-$\frac{1}{2}$ pulse train must be incremented by blocks of 8 pulses being added to each half of the pulse train, so as to retain the property of refocusing chemical shift offsets.

The advantage of the REAPDOR experiment over TRAPDOR [22] is that its analysis is less susceptible to error, if the condition of perfect adiabatic passage (see Section 4.5.1) is not met. Also, a much shorter pulse is required for the transfer of populations on the quadrupolar nucleus, which is less demanding on the spectrometer probehead. However, the REAPDOR experiment may be more susceptible to small errors in the spin-$\frac{1}{2}$ pulse train (amplitude and timing).

4.6 Techniques for measuring dipolar couplings between quadrupolar nuclei

As discussed in Section 4.5, the large quadrupole coupling suffered by many quadrupolar nuclei renders many of the techniques for measuring dipolar couplings discussed in Sections 4.3–4.5 well nigh useless for purely quadrupolar spin systems. Very few techniques exist for measuring dipolar couplings between quadrupolar spin pairs. Rotational resonance has been shown to be useful in cases where the central transitions (for half-integer quadrupolar nuclei) of the dipolar-coupled spins are well separated [23] (see Section 4.3.3 for a discussion of rotational resonance). Dipolar couplings between quadrupolar nuclei can be used in correlation experiments (see

Section 4.7), which gives an indication of which spins are close in space, but does not give a quantitative measure of the dipolar-coupling constant.

4.7 Correlation experiments

Correlation spectroscopy is common in solution-state NMR, where the scalar or *J* coupling is most often used to correlate spins that are close together in the bonding network. Most techniques in the solid state utilize the dipolar coupling between spins to indicate which spins are close in space, although solution-like INADEQUATE-type experiments using the *J* or scalar coupling also exist [24]. All these experiments rely on the fact that multiple-quantum coherences can be excited in dipolar-coupled spin systems.

4.7.1 Homonuclear correlation experiments for spin-$\frac{1}{2}$ systems

There are two types of experiment that can be performed on homonuclear spin systems to establish spatial correlations of spins, both of which utilize mutliple-quantum coherences excited between dipolar-coupled spins. Most often, the experiments use the double-quantum coherence which can be excited between a spin-$\frac{1}{2}$ pair, but zero-quantum coherence is relevant in some experiments.

One type of two-dimensional correlation NMR experiment is shown in Fig. 4.25, in which double-quantum coherence is excited and then allowed

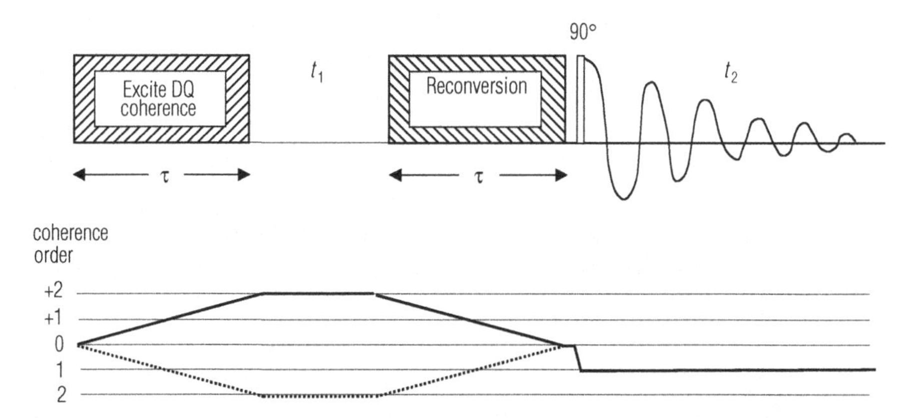

Fig. 4.25 Schematic illustration of a two-dimensional correlation experiment in which a double-quantum coherence spectrum, arising from dipolar-coupled spin pairs, is correlated with the normal single-quantum spectrum. The experiments are generally performed under rapid magic-angle spinning so that the spectra in both dimensions are high resolution. Double-quantum coherence is initially excited with the aid of some type of recoupling pulse sequence (to counter the effect of magic-angle spinning which removes the dipolar couplings necessary for the generation of double-quantum coherences). The double-quantum coherence is then allowed to evolve in t_1 before being transferred back into longitudinal magnetization. A 90° pulse then generates observable transverse magnetization from this, for monitoring in t_2. Two separate experiments, employing the $+2 \rightarrow 0 \rightarrow -1$ and $-2 \rightarrow 0 \rightarrow -1$ pathways respectively, are generally performed, in order to produce pure absorption two-dimensional spectra.

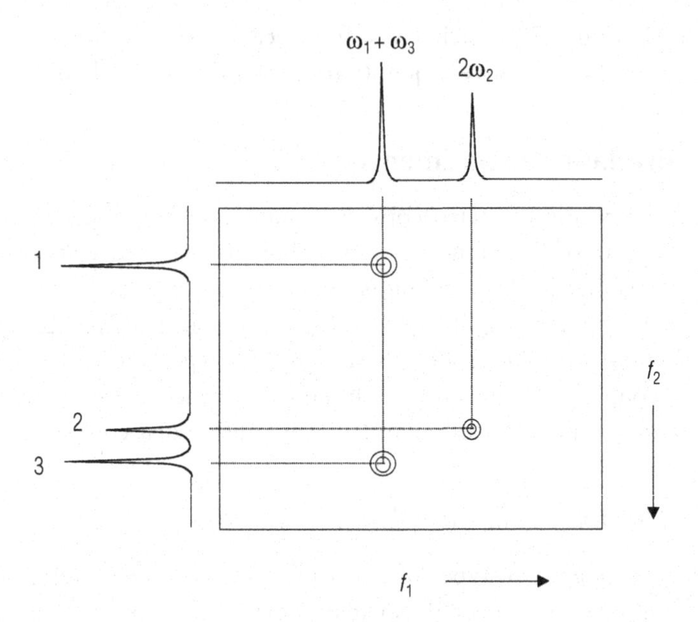

Fig. 4.26 The form of the two-dimensional frequency spectrum arising from the double/single-quantum correlation experiment of Fig. 4.25. The double-quantum spectrum appears in f_1 correlated with the single-quantum spectrum in f_2. A signal at ($\omega_I + \omega_S$) in the double-quantum spectrum (correlated with signals at ν_I and ν_S in the single-quantum spectrum) indicates that the spins with frequencies ω_I and ω_S have a sufficiently large dipolar coupling between them to generate a significant double-quantum coherence, i.e. these spins are close in space. The spectrum shown indicates that spins 1 and 3 are close together and spins labelled 2 are also close to each other.

to evolve in t_1 before being converted to single-quantum coherence to be observed in t_2 [10]. The resulting two-dimensional frequency spectrum correlates the double-quantum spectrum in f_1 with a double-quantum-filtered single-quantum spectrum in f_2. The advantage of this type of experiment is that the final two-dimensional spectrum only shows signals arising from spin pairs where the two spins are moderately close in space. The intensity of each signal is dependent on the proximity of the two spins; in general, the closer the spins, the stronger the signal. Moreover, because the double-quantum spectrum is correlated with the single-quantum spectrum in this two-dimensional spectrum, we can identify *which* two spins contribute to each signal, and so identify the network of coupled spins (Fig. 4.26). Methods for exciting double-quantum coherence between spin pairs are discussed in Section 4.3.2.

The second type of two-dimensional correlation experiment is the more usual one which correlates two normal, single-quantum spectra with cross-peaks occurring between signals from spins which are close in space. The general form of such experiments was shown in Fig. 4.8. After initial excitation, transverse magnetization is allowed to evolve during t_1 at its characteristic frequency. There then follows a mixing period during which magnetization is exchanged or transferred between spins, before the transverse magnetization is regenerated and observed in t_2. The mixing period contains a pulse sequence designed to excite double- and/or zero-quantum coherences between dipolar-coupled spin pairs, i.e. generate coherence involving spin pairs rather than the single spin, single-quantum coherence present previously in t_1. When multiple-quantum coherence involving multiple spins is converted back to single-spin, single-quantum coherence, some coherence ends up on each of the spins involved in the multiple-quantum coherence, i.e. magnetization has in effect been transferred between the spins. Thus, if single-quantum coherence associated with spin 1(say) in t_1 is converted to a double-quantum coherence between spin 1 and spin 2, the reconversion back to single-quantum coherence then results in single-quantum coherence on spins 1 *and* 2, which then evolves at their respective characteristic frequencies in t_2. A similar fate is suffered by single-quantum coherence associated with spin 2 in t_1. After processing, the resulting two-dimensional spectrum has signals from spins 1 and 2 in the f_1 and f_2 spectra (from the single-quantum coherences which evolve in t_1 and t_2). Moreover, it has cross-peaks between the spin 1 and spin 2 signals, because coherence which started on spin 1 in t_1 has ended up on spin 2 in t_2 (and vice versa).

For both types of experiment, magic-angle spinning is usually used for the sake of resolution in both dimensions of the experiment. This of course has the effect of averaging to zero any dipolar couplings on which the correlation experiments rely. Accordingly, the excitation of multiple-quantum coherences must involve some type of recoupling sequence, as discussed in detail in Section 4.3.2 on double-quantum filtering.

There is much interest in performing multiple-quantum correlation experiments in homonuclear spin systems, both in cases where the spin of interest is abundant and where there are relatively isolated spin pairs. Abundant ^1H spin systems in particular have been investigated [10]. Generally speaking, the dipolar coupling between ^1H spins in systems where protons are abundant causes massive linebroadening, so resolution is a problem in any solid-state ^1H NMR spectroscopy. However, very fast magic-angle spinning is able to remove this broadening. The double-quantum/single-quantum correlation type of technique has been successfully applied in ^1H NMR to examine the spatial proximity of spins. Very fast magic-angle spinning (often in excess of 30 kHz) is needed to produce high resolution in the single-quantum ^1H dimension of the experiment. An example of such an experiment is shown in Fig. 4.27 [10].

The more traditional type of single-quantum/single-quantum two-dimensional correlation experiment tends to be applied to systems with less abundant spins, particularly systems which have been partly isotopically labelled. An example of such use is shown in Fig. 4.28 on a uniformly-

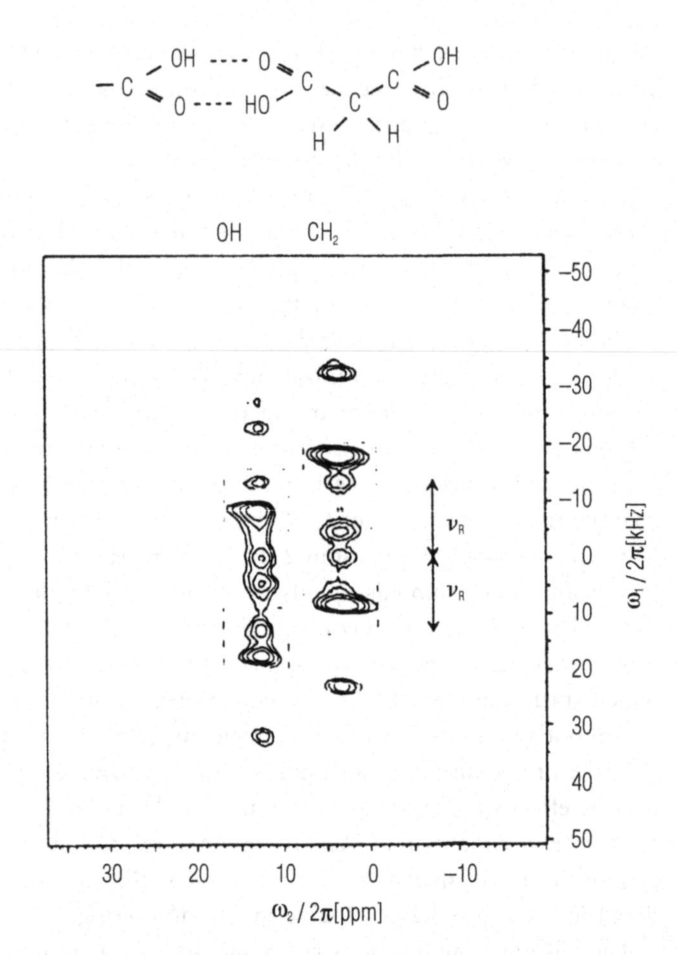

Fig. 4.27 A two-dimensional ^{1}H double-quantum/single-quantum correlation spectrum of malonic acid [10]. The experiment is conducted under very rapid magic-angle spinning to produce high-resolution spectra in both dimensions. Double-quantum coherences are generated between the two CH_2 protons and the two OH protons in close proximity and between each proton of the CH_2 and OH groups. Thus each double-quantum spectrum (in f_1) consists of four lines plus spinning sidebands.

labelled cytidine [25]. In this experiment, the RFDR recoupling pulse sequence [8] is used in the mixing period.

4.7.2 Homonuclear correlation experiments for quadrupolar spin systems

In principle, the same experiments that have been devised for spin-$\frac{1}{2}$ correlation experiments could be applied to quadrupolar spin systems, providing hard, non-selective pulses are used throughout (as for spin-$\frac{1}{2}$ systems). However, this latter proviso is rather harder to achieve for quadrupolar spins than for spin-$\frac{1}{2}$, because quadrupolar spins are generally subject to

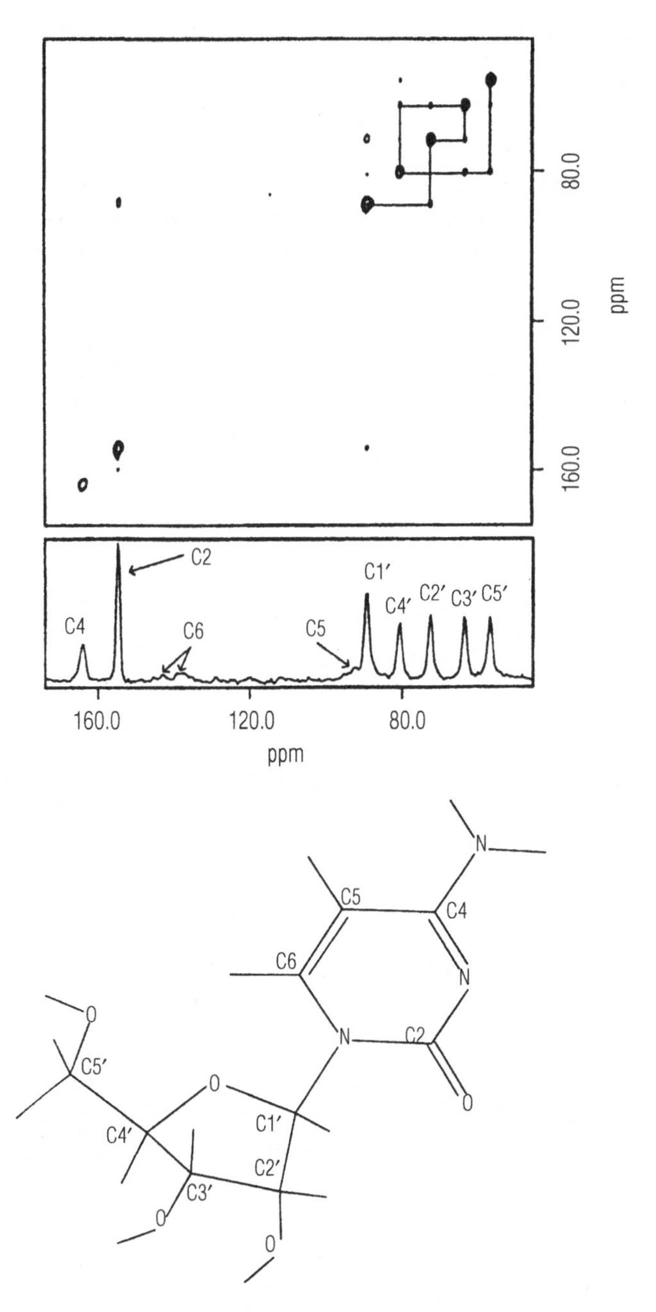

Fig. 4.28 Two-dimensional ^{13}C single-quantum/single-quantum correlation spectrum of cytidine–U–^{13}C–^{15}N [25]. The RFDR recoupling scheme [8] is used in the mixing period of the experiment to transfer polarization between ^{13}C spins which are close in space.

quadrupole-coupling interactions which can broaden their respective NMR spectra by MHz. Given that rf pulses generate interactions which are at best a few hundred kHz in magnitude, it is clear that any rf pulse can only irradiate a small portion of the total spectrum. Accordingly, alternative methods must be found for generating multiple-quantum coherences between quadrupolar spins, but only two methods have been reported in the literature so far. One experiment [26] generates $4I$-quantum coherences between dipolar-coupled spin I pairs, where I is a half-integer spin. The $4I$-quantum coherence is allowed to evolve in t_1 before being converted to observable single-quantum coherence for observation in t_2. The resulting $4I$-quantum/single-quantum two-dimensional correlation spectrum is equivalent to the double-quantum/single-quantum correlation spectrum for spin-$\frac{1}{2}$ nuclei discussed in Section 4.7.1. The second method is a single-quantum/single-quantum correlation technique [27], where a single, relatively long, but weak rf pulse is used in the mixing period to generate double-quantum coherence involving the central transitions of dipolar-coupled half-integer quadrupolar nuclei. By using a weak rf pulse, the averaging effect of magic-angle spinning on the dipolar coupling is disrupted [27]. An additional complication for quadrupolar spins is that it is quite possible to excite multiple-quantum coherences on a single spin, which do not of course depend on the dipolar coupling between spins at all. The advantage of the $4I$-quantum correlation experiment is that $4I$-quantum coherence can only arise from dipolar-coupled spins and not from a single spin [27].

4.7.3 Heteronuclear correlation experiments for spin-$\frac{1}{2}$

Single-quantum/single-quantum heteronuclear correlation experiments, similar to those described in Section 4.7.1 for homonuclear spin systems, are quite common in solid-state NMR. The form of the experiments is depicted in Fig. 4.29, and is very similar to that for homonuclear spin systems, although the pulse sequence in the mixing period is often much simpler than that used in homonuclear spin systems. Most often, the mixing sequence used in heteronuclear correlation experiments to promote transfer of magnetization between heteronuclei is a simple cross-polarization step, with a properly adjusted contact pulse being applied to each nucleus (Fig. 4.29). Cross-polarization is discussed in detail in Section 2.5.

Magic-angle spinning is usually applied in such experiments, and clearly this can affect the efficiency of the cross-polarization step. Several methods have been used to circumvent this problem, such as using ramped contact pulses, as discussed in Section 2.5.

One frequently used heteronuclear correlation experiment is the WISE (WIdeline SpEctroscopy) experiment, which is particularly used in studies

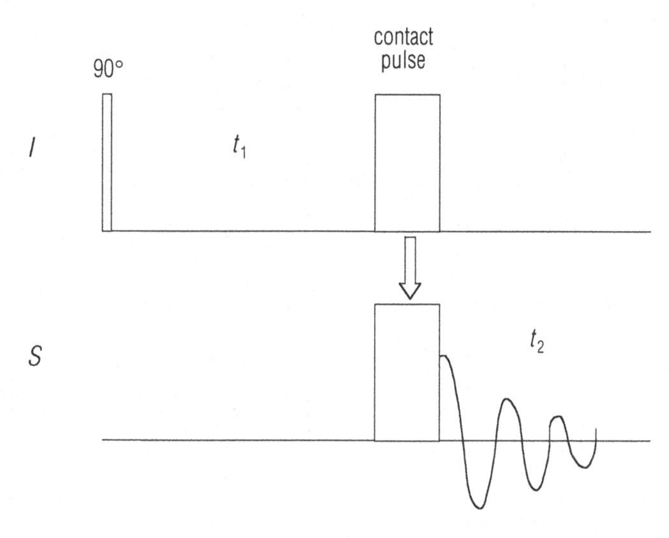

Fig. 4.29 The form of a heteronuclear correlation experiment. The experiment begins by exciting single-quantum coherence (transverse magnetization) on one of the spins (*I* in this case) and allows it to evolve at its characteristic frequency in t_1. Contact pulses applied to both spins at the end of t_1 result in cross-polarization from spin *I* to spin *S*, after which the resulting *S* spin transverse magnetization is observed in t_2. The resulting two-dimensional frequency spectrum displays a single-quantum *I* spin spectrum in f_1 and a single-quantum *S* spin spectrum in f_2, but only signals from *I* spins which are close to *S* spins appear, and vice versa. A peak in the two-dimensional spectrum at (ω_I, ω_S) indicates that the spins with spectral frequencies ω_I and ω_S are close in space.

of molecular motion. This experiment is discussed in Chapter 6. A specific example of heteronuclear correlation for structural study is shown in Fig. 4.30 [28].

4.8 Spin-counting experiments

Dipolar coupling in extensive homonuclear spin systems allows spin-counting experiments to be performed. By spin counting, we mean the determination of the number of spins in close proximity (see Fig. 4.9). As shown in Box 4.1, in a homonuclear, many-spin system, the eigenstates (or coupled spin states) of the system are described by the total nuclear spin in the laboratory *z*-direction, *M*

$$M = m_1 + m_2 + m_3 + \ldots + m_N \tag{4.22}$$

where N is the number of coupled spins. As mentioned in Section 4.2, multiple-quantum coherences of order ΔM can be excited between them, where $\Delta M = M_i - M_j$ with i and j being the M states involved in the coherence. By following the amplitude of different multiple-quantum coherences as a function of their excitation time, the value of N, the number of spins in the cluster, can be estimated.

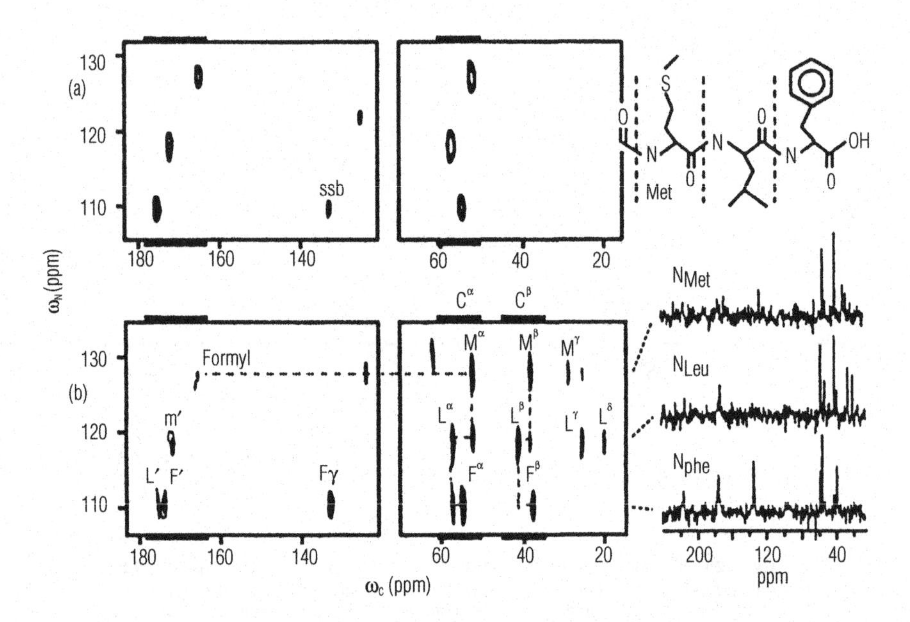

Fig. 4.30 Two-dimensional ^{15}N–^{13}C (single-quantum) correlation spectrum of the formyl–Met–Leu–Phe amino acid sequence shown in the figure. The experiment is conducted under magic-angle spinning and uses a REDOR sequence to recouple ^{13}C–^{15}N dipolar coupling during mixing periods. ^{13}C transverse magnetization is created initially by cross-polarization from ^1H. A REDOR-based polarization transfer from $^{13}C \rightarrow {}^{15}N$ then takes place during a mixing period τ_m, following which ^{15}N transverse magnetization is allowed to evolve for t_1. A similar REDOR mixing sequence is then applied for τ_m to transfer the remaining ^{15}N transverse magnetization back to ^{13}C. A ^{13}C FID is then collected in t_2. By this sequence of polarization transfer steps, only signals from ^{13}C and ^{15}N sites which are close in space are seen in the final spectrum. The longer the mixing time, the greater the distance between $^{13}C/^{15}N$ spins which contribute to the final spectrum. Two spectra are shown (a) mixing time $\tau_m = 614.4\,\mu s$ ($2\tau_R$) and (b) $\tau_m = 2.46$ ms ($8\tau_R$).

Most spin-counting experiments have been performed on non-spinning samples [29], although experiments under magic-angle spinning are also possible [30]. The aim of the experiment is two-fold: first, to separate coherences of different order, and then to examine their amplitudes as a function of the length of time for which the coherences are excited. By following the extent to which higher-order coherences are excited as a function of excitation time, detailed information on the size of spin clusters can be obtained.

4.8.1 The formation of multiple-quantum coherences

In Section 4.3.2, we showed how double-quantum coherences can be generated (albeit under magic-angle spinning). In fact, higher-order coherences can be excited with exactly the same pulse sequences that generate double-quantum coherences [13]. To investigate this, we examine the simplest possible pulse sequence known to create double-quantum coherence, namely 90°–τ–90°. The density operator is a convenient means of studying the state of a spin system from our point of view here, as the density operator formalism leads naturally to a description of what coherences have been generated.

The equilibrium density operator before the 90°–τ–90° sequence is applied is proportional to $\hat{I}_z$, where $\hat{I}_z$ is given here by

$$\hat{I}_z = \sum_i^N \hat{I}_z^i \tag{4.26}$$

where the sum is over all the spins in the cluster. The initial (hard) 90° pulse, of phase y for the sake of argument, transforms the density operator to $\hat{I}_x \left(\hat{I}_x = \sum_i^N \hat{I}_x^i \right)$. The density operator then evolves during τ under a hamiltonian, $\hat{H}_{\text{int}}$, which describes all the spin interactions which are present, namely chemical shifts and dipolar couplings between spins (in the rotating frame):

$$\hat{H}_{\text{int}} = \hat{H}_{\text{cs}} + \hat{H}_{\text{dd}}^{\text{homo}} \tag{4.27}$$

where

$$\begin{aligned} \hat{H}_{\text{cs}} &= \sum_i \omega_{\text{cs},i} \hat{I}_z^i \\ H_{\text{dd}}^{\text{homo}} &= -\sum_{i>j} C_{ij}(3\hat{I}_z^i \hat{I}_z^j - \hat{\mathbf{I}}^i \cdot \hat{\mathbf{I}}^j) \end{aligned} \tag{4.28}$$

The density operator after a time t of this evolution is given by the usual expression (see Chapter 1)

$$\hat{\rho}(t) = \exp(-i\hat{H}_{\text{int}}t)\hat{\rho}(0)\exp(i\hat{H}_{\text{int}}t) \tag{4.29}$$

where $\hat{\rho}(0)$ is the density operator at the start of the evolution period, i.e. $\hat{I}_x$. Equation (4.29) can be expanded by use of the Magnus expansion (see Box 1.2, Chapter 1):

$$\hat{\rho}(t) = \hat{\rho}(0) + it[\hat{\rho}(0), \hat{H}_{\text{int}}] - \frac{1}{2}t_2[[\hat{\rho}(0), \hat{H}_{\text{int}}], \hat{H}_{\text{int}}] + \ldots \tag{4.30}$$

Now, we examine what terms arise in the density operator at time t by virtue of Equation (4.30). The commutator $[\hat{\rho}(0), \hat{H}_{\text{int}}]$ (= $\hat{\rho}(0)\hat{H}_{\text{int}} - \hat{H}_{\text{int}}\hat{\rho}(0)$) in Equation (4.30) for the density operator gives rise to spin operator terms such as

$$\begin{aligned} \hat{I}_{1x}\hat{I}_{1z}\hat{I}_{2z} - \hat{I}_{1z}\hat{I}_{2z}\hat{I}_{1x} &= (\hat{I}_{1x}\hat{I}_{1z} - \hat{I}_{1z}\hat{I}_{1x})\hat{I}_{2z} \\ &= [\hat{I}_{1x}, \hat{I}_{1z}]\hat{I}_{2z} \\ &= -i\hbar\hat{I}_{1y}\hat{I}_{2z} \end{aligned} \tag{4.31}$$

in the density operator $\hat{\rho}(t)$ when we substitute for $\hat{\rho}(0)$ and $\hat{H}_{\mathrm{int}}$ and where 1 and 2 label particular spins in the spin cluster. In generating (4.30), we have used the fact that spin operators associated with different spins all commute with one another. The term $[[\hat{\rho}(0), \hat{H}_{\mathrm{int}}], \hat{H}_{\mathrm{int}}]$ in Equation (4.30) then yields spin operator terms such as

$$\begin{aligned} \hat{I}_{1y}\hat{I}_{2z}\cdot\hat{I}_{1z}\hat{I}_{3z} - \hat{I}_{1z}\hat{I}_{3z}\cdot\hat{I}_{1y}\hat{I}_{2z} &= (\hat{I}_{1y}\hat{I}_{1z} - \hat{I}_{1z}\hat{I}_{1y})\hat{I}_{2z}\hat{I}_{3z} \\ &= [\hat{I}_{1y}, \hat{I}_{1z}]\hat{I}_{2z}\hat{I}_{3z} \\ &= i\hbar\hat{I}_{1x}\hat{I}_{2z}\hat{I}_{3z} \end{aligned} \qquad (4.32)$$

in the density operator $\hat{\rho}(t)$, i.e. terms involving up to three spins. The higher-order terms (not shown explicitly in Equation (4.30)) have nested commutators which give rise to terms in the density operator involving progressively more and more spins, up to terms of the form $\hat{I}_{1x}\hat{I}_{2z}\hat{I}_{3z} \ldots \hat{I}_{(N-1)z}\hat{I}_{Nz}$ and all permutations of this.

A 90°_x pulse at the end of the τ period then converts a term $\hat{I}_{1x}\hat{I}_{2z}\hat{I}_{3z}$ (for instance) in the density operator into $\hat{I}_{1x}\hat{I}_{2y}\hat{I}_{3y}$ and terms like $\hat{I}_{1x}\hat{I}_{2z}\hat{I}_{3z} \ldots \hat{I}_{(N-1)z}\hat{I}_{Nz}$ into $\hat{I}_{1x}\hat{I}_{2y}\hat{I}_{3y} \ldots \hat{I}_{(N-1)y}\hat{I}_{Ny}$. If the $\hat{I}_x$ and $\hat{I}_y$ terms are rewritten in terms of the raising and lowering operators, $\hat{I}_+$ and $\hat{I}_-$ $\left(\hat{I}_x = \frac{1}{2}(\hat{I}_+ + \hat{I}_-); \hat{I}_y = \frac{1}{2i}(\hat{I}_+ - \hat{I}_-)\right)$ it then becomes apparent that $\hat{I}_{1x}\hat{I}_{2y}\hat{I}_{3y}$, for instance, consists of a sum of terms ranging from $\hat{I}_{1+}\hat{I}_{2+}\hat{I}_{3+}$, representing triple (+3) quantum coherence, through things like $\hat{I}_{1+}\hat{I}_{2+}\hat{I}_{3-}$ (+1 quantum coherence) to $\hat{I}_{1-}\hat{I}_{2-}\hat{I}_{3-}$ (−3 quantum coherence). In general, a term $\hat{I}_{1x}\hat{I}_{2y}\hat{I}_{3y} \ldots \hat{I}_{(N-1)y}\hat{I}_{Ny}$ represents all orders of coherence from the maximum $+N$ to the minimum $-N$ in steps of two.

It is clear then that high orders of coherence can be excited with this simple pulse sequence, the limit on the maximum order being which and how many pairs of spins appear in the dipolar-coupling hamiltonian, $\hat{H}^{\mathrm{homo}}_{\mathrm{dd}}$ in Equation (4.28).

The next question is what is the amplitude of each of these multiple-quantum coherences? The amplitude of each coherence is its coefficient in the expression for the density operator, $\hat{\rho}(t)$. The term $\hat{I}_{1x}\hat{I}_{2y}\hat{I}_{3y}$, for instance, arising from the $[[\hat{\rho}(0), \hat{H}_{\mathrm{int}}], \hat{H}_{\mathrm{int}}]$ term in Equation (4.30) for the density operator, is multiplied by $\tau^2 d_{12}d_{13}$, where d_{12} is the dipolar-coupling constant for the dipolar coupling between spins 1 and 2 and d_{13} is that for spins 1 and 3; τ is the length of the evolution period between the 90° pulses in the excitation period.

In a similar manner, higher-order terms in Equation (4.30) give rise to terms in the density operator involving more spins and are multiplied by $\tau^n d_{12} \ldots d_{1n}$, where n is the order of the term in Equation (4.30). The terms arising from the nth-order terms of Equation (4.30) give rise to a maximum coherence order of $+(n + 1)$ in the density operator. Clearly, the amplitude of these coherences depends on the size of $\tau^n d_{12} \ldots d_{1n}$. If $\tau\, d_{ij} << 1$ then the higher-order terms of Equation (4.30) for the density operator are negligible and so the amplitude of higher-order coherences is very small. As $\tau\, d_{ij}$ increases, however, the amplitude of the higher-order terms increases. Thus, as the τ delay in the pulse sequence is increased, so the amplitude of the higher-order coherences increases. The density operator, however, is always dominated by terms corresponding to the lower-order coherences, simply from the form of Equation (4.30) and the dependence of each term on a factor $\tau^n d_{12} \ldots d_{1n}$. It is worth noting that the amplitude of all coherences also depends on the various dipolar-coupling constants between the spins involved in the coherence.

Finally, it is worth noting that a 90°–τ–90° sequence where the two 90° pulses differ in phase by 90° results in odd orders of coherence being excited (as in the above analysis), whereas when the two 90° pulses have the same phase, even orders of coherence are excited. That this is so can be easily seen by examining the above analysis for a 90°_y–τ–90°_y sequence. The final 90°_y pulse (rather than 90°_x) transforms terms like $\hat{I}_{1x}\hat{I}_{2z}\hat{I}_{3z}$ into $-\hat{I}_{1z}\hat{I}_{2x}\hat{I}_{3x}$, which contains as its maximum quantum coherence $-\hat{I}_{1z}\hat{I}_{2+}\hat{I}_{3+}$ representing +2 coherence and $-\hat{I}_{1z}\hat{I}_{2-}\hat{I}_{3-}$ (−2-quantum coherence) as its minimum quantum coherence, plus two other terms representing zero quantum coherence.

4.8.2 Implementation of spin-counting experiments

Although multiple-quantum coherences can be excited with a 90°–τ–90° pulse sequence as illustrated in the previous section, it is only really effective at exciting the higher orders if the rf pulse power is very high, higher than can be achieved in practice. Accordingly, other pulse sequences have been developed which are effective at moderate rf pulse powers [13]. The one which is commonly used is shown in Fig. 4.31. The experiments are generally carried out on non-spinning samples, so there is no need to counter the effect of magic-angle spinning during the multiple-quantum coherence excitation. The multiple-quantum coherences are then allowed to evolve during t_1 before being converted back to longitudinal magnetization. The reconversion step is achieved by a 'time-reversed' version of the excitation sequence. The need for 'time reversal' to prevent dipolar dephasing is discussed in Box 4.4. It is even more important here than in double-quantum experiments, because a given order of coherence, n, can arise from the nth order term in Equation (4.30) for the density operator, *and from all higher-order terms greater than n* in Equation (4.30). Thus, there are potentially many contributions to a given order of coherence, which all in general have

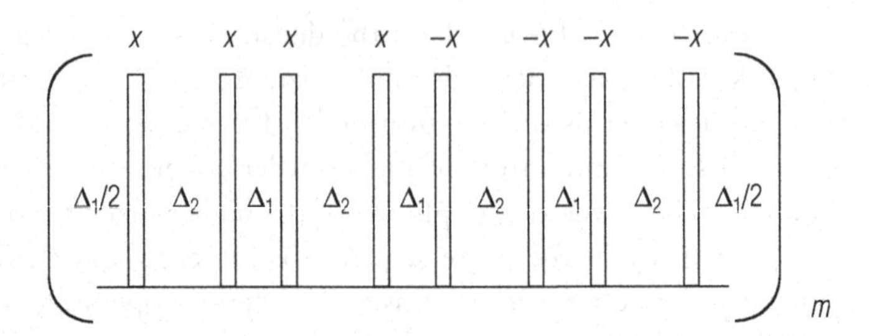

Fig. 4.31 The pulse sequence used for exciting multiple-quantum coherence orders in spin-counting experiments (static samples). All pulses are 90° pulses with the phases shown. The delays between pulses Δ_1 and Δ_2 are such that $\Delta_2 = 2\Delta_1 + t_p$ where t_p is the 90° pulse length. In general, the longer the excitation sequence (i.e. the higher the value of m), the higher the orders of multiple-quantum coherences excited.

different phases (and therefore destructively interfere) unless the whole pulse sequence has so-called time-reversal symmetry.

After the reconversion to longitudinal magnetization, a period of a few milliseconds is allowed for unwanted coherences to dephase before a 90° pulse is applied to create transverse magnetization which is monitored in the subsequent t_2 period in the usual way. Two-dimensional Fourier transformation would then produce the normal single-quantum spectrum in ν_2 and a spectrum in ν_1 containing signals from all the multiple-quantum coherence orders excited in the experiment. Although signals from different multiple-quantum coherence orders have different frequencies, they may not be very different, and so there is a tendency for signals from different coherence orders to overlap. To separate the different coherence orders, the phases of the rf pulses in the reconversion step are shifted in concert with t_1. This results in a phase shift of all the components of longitudinal magnetization resulting from the reconversion, the phase shift being proportional to the order of coherence each component arose from. Thus the data acquired in t_2 behave as though each coherence order in t_1 is phase shifted in proportion to the order of the coherence, i.e. appears to have evolved through an extra, order-dependent angle. Two-dimensional Fourier transformation now results in well-separated signals for different coherence orders in ν_1.

In general, the longer the excitation period in these experiments, the higher the maximum order of coherence excited, as discussed in the previous section. Much information about the size and geometry of spin clusters can be obtained by following the amplitude of different coherence orders as a function of excitation time [28].

Notes

1. Figure 4.1 only depicts the z-components of spin, i.e. those quantized along $\mathbf{B}_0$. There is also, of course, an x, y component, which precesses about $\mathbf{B}_0$ at the Larmor frequency of

the spin (in the laboratory frame). This compoonent is important for homonuclear spin pairs, as a magnetic field rotating at the Larmor frequency induces transitions on nearby spins with the same Larmor frequency, i.e. degenerate homonuclear spins. The x, y component is not, however, important for heteronuclear spin pairs, for which Fig. 4.1 is a complete picture.

2. Note that in an NMR experiment, we observe the molecular orientation as if we were in the laboratory frame. It is only the spin coordinates that we observe from a rotating frame.
3. The only remaining approximation is that the rf pulse amplitude is much larger than the dipolar coupling, i.e. the dipolar coupling can be ignored during a pulse.
4. The rotating frame here is the usual rotating frame that we first met in Chapter 1; the frame rotates about $\mathbf{B}_0$ at frequency ω_{rf}, the frequency of the rf pulse. In this frame, the time dependence of the magnetic field due to the pulse is removed. The first-order average dipolar hamiltonian in this rotating frame is the same as that derived in Section 4.1.2, where the first-order average hamiltonian in a rotating frame which rotates about $\mathbf{B}_0$ at the Larmor frequency ω_0 of the spin is derived. The only difference is an offset term if the rf irradiation is not on resonance, i.e. $\omega_{rf} \neq \omega_0$.

References

1. Brink, D.M. & Satchler, G.R. (1994) *Angular Momentum.* Clarendon Press, Oxford.
2. Zare, R.N. (1988) *Angular Momentum.* Wiley, New York.
3. Spiess, H.W. (1978) *Dynamic NMR Spectroscopy. NMR Basic Principles and Progress,* **15**, 55.
4. Tycko, R. & Dabbagh, G. (1990) *Chem. Phys. Lett.* **173**, 461.
5. Tycko, R. & Smith, S.O. (1993) *J. Chem. Phys.* **98**, 932.
6. Klug, C.A., Zhu, W., Merritt, M.E. & Schaefer, J. (1994) *J. Magn. Reson.* **109**, 134.
7. Sun, B.Q., Costa, P.R., Kocisko, D., Lansbury, R.T. & Griffin, R.G. (1995) *J. Chem. Phys.* **102**, 702.
8. Sodrickson, D.K., Levitt, M.H., Vega, S. & Griffin, R.G. (1993) *J. Chem. Phys.* **98**, 6742.
9. Tycko, R. & Dabbagh, G. (1991) *J. Am. Chem. Soc.* **113**, 9444.
10. Gottwald, J., Demco, D.E., Graf, R. & Spiess, H.W. (1995) *Chem. Phys. Lett.* **243**, 314.
11. Brinkman, A., Eden, M. & Levitt, M.H. (2000) *J. Chem. Phys.* **112**, 8539.
12. Carravetta, M., Eden, M., Zhao, X., Brinkmann, A. & Levitt, M.H. (2000) *Chem. Phys. Lett.* **321**, 205.
13. Baum, J., Munowitz, M., Garroway, A.N. & Pines, A. (1985) *J. Chem. Phys.* **83**, 2015.
14. Raleigh, D.P., Levitt, M.H. & Griffin, R.G. (1988) *Chem. Phys. Lett.* **146**, 71.
15. Shore, S.E., Ansermet, J.P., Slichter, C.P. & Sinfelt, J.H. (1987) *Phys. Rev. Lett.* **58**, 953.
16. Gullion, T. & Schaefer, J. (1989) *J. Magn. Reson.* **81**, 196.
17. Pan, Y., Gullion, T. & Schaefer, J. (1990) *J. Magn. Reson.* **90**, 330.
18. Mueller, K.T. (1995) *J. Magn. Reson. A* **113**, 81.
19. See, for instance, Jarvie, T.P., Wenslow, R.M. & Mueller, K.T. (1995) *J. Am. Chem. Soc.* **117**, 570.
20. Fernandez, C., Lang, D.P., Amoureux, J.-P. & Pruski, M. (1998) *J. Am. Chem. Soc.* **120**, 2672; Pruski, M., Bailly, A., Lang, D.P., Amoureux, J.-P. & Fernandez, C. (1999) *Chem. Phys. Lett.* **307**, 35.
21. Grey, C.P. & Veeman, W.S. (1992) *Chem. Phys. Lett.* **192**, 379.
22. Gullion, T. (1995) *Chem. Phys. Lett.* **246**, 325.
23. Nijman, M., Ernst, M., Kentgens, A.P.M. & Meier, B.H. (2000) *Molec. Phys.* **98**, 161.
24. Lesage, A., Auger, D.C., Calderelli, S. & Emsley, L. (1997) *J. Am. Chem. Soc.* **119**, 7867; Lesage, A., Sakellariou, D.D., Steuernagel, S. & Emsley, L. (1998) *J. Am. Chem. Soc.* **120**, 13194.
25. Kiihne, S.R., Geahigan, K.B., Oyler, N.A., Zebroski, H., Mehta, M.A. & Drobny, G.P. (1999) *J. Phys. Chem. A* **103**, 3890.

26. Duer, M.J. & Painter, A.J. (1999) *Chem. Phys. Lett.* **313**, 763.
27. Baldus, M., Rovnyak, D. & Griffin, R.G. (2000) *J. Chem. Phys.* **112**, 5902.
28. Hong, M. & Griffin, R.G. (1998) *J. Am. Chem. Soc.* **120**, 7113.
29. Baum, J. & Pines, A. (1986) *J. Am. Chem. Soc.* **108**, 7447.
30. Meier, B.H. & Earl, W.L. (1986) *J. Chem. Phys.* **85**, 4905.

5 Quadrupole Coupling: Theory and Uses

5.1 Introduction

Some 74% of NMR active nuclei have a spin greater than $\frac{1}{2}$ and are termed *quadrupolar*. That is to say, they possess a nuclear electric quadrupole moment, which in turn is able to interact with any electric field gradient at the nucleus. Such gradients arise naturally in solids for all nuclei not at a site of cubic symmetry by virtue of the distribution of other nuclei and electrons in their vicinity.

The electric quadrupole moment in the nucleus arises from the distribution of charge there. A distribution of charge, such as protons in a nucleus, cannot be adequately described by simply specifying the total charge. In general, a proper description requires expanding the charge distribution function as a series of *multipoles*. The total charge is the zeroth-order multipole; the electric dipole moment is the first-order multipole in the expansion. The next highest term is the electric *quadrupole moment*, which has the distribution of charge illustrated in Fig. 5.1.

All nuclei with a spin greater than $\frac{1}{2}$ necessarily possess an electric quadrupole moment in addition to the magnetic dipole moment possessed by spin-$\frac{1}{2}$ nuclei. Electric quadrupoles interact with electric field gradients. Thus, a nucleus with a spin greater than $\frac{1}{2}$ not only interacts with the applied magnetic field and all local magnetic fields, but also with any electric field gradients present at the nucleus. This interaction affects the nuclear spin energy levels in addition to the other magnetic interactions already described.

The strength of the interaction depends upon the magnitude of the nuclear quadrupole moment and the strength of the electric field gradient. The electric quadrupole moment of a nucleus is generally given as eQ, where e is the proton charge; eQ is constant for a given nuclear species and does not change with the chemical environment of the nucleus, for instance.

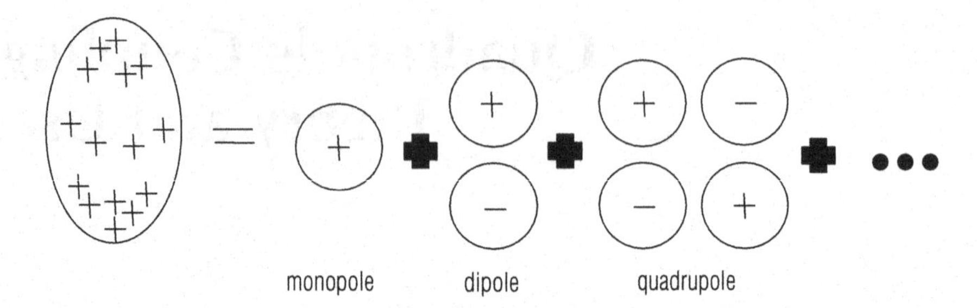

Fig. 5.1 Schematic illustration of the expansion of a charge distribution (such as that in a nucleus) as a series of multipoles. The electric dipole term is always zero for a nuclear charge distribution. If it were not there would be a net force acting on the charge distribution from the electric field arising from the nuclear charge itself. This would act so as to rearrange the charge distribution and remove the electric dipole moment.

It is important to realize that in a strong applied field $\mathbf{B}_0$ (such that the Zeeman interaction is of the order of a factor of 10 greater than any quadrupole interaction or greater), the applied field acts as the unitization axis for the nuclear spin levels, which are thus still able to be described by specifying the quantum numbers, I and m, the total spin and the component of spin directed along the applied field respectively. If, however, the strength of interaction of the nucleus with the applied field is of similar order of magnitude to the strength of its interaction with a local field gradient, then the applied field can no longer be considered as the quantization axis. This situation arises for nuclear species with large quadrupole moments, such as ^{87}Rb and ^{14}N at sites of low symmetry, which necessarily have large electric field gradients associated with them.

In the case where the applied magnetic field still acts as the quantization axis for the nuclear spins, the applied field $\mathbf{B}_0$ has the effect of orientating the nuclear magnetic dipole moment, i.e. tends to align a component of the nuclear magnetic moments of the spins with the applied field (for positive γ). Both the nuclear magnetic dipole moment and the nuclear electric quadrupole moment are fixed within the nucleus and although the latter does not interact with the applied field $\mathbf{B}_0$, the quadrupole moment is still orientated in some manner with respect to this field by virtue of the fact that the magnetic dipole moment tends to align with it. Meanwhile, the electric field gradient that the nuclear quadrupole moment does interact with depends on the immediate environment of the nuclues, through the whereabouts of electrons and other nuclei, i.e. on the geometry of bonds around the nucleus. Thus, the strength of the quadrupole interaction has a dependence on the orientation of the molecule (which determines the electric field gradient) with respect to the applied magnetic field $\mathbf{B}_0$ (which indirectly determines the orientation of the nuclear electric quadrupole moment), even though this is not a magnetic interaction.

5.2 Theory

5.2.1 The quadrupole hamiltonian

The basic form of the quadrupole hamiltonian describing the interaction between a nuclear electric quadrupole moment and an electric field gradient **V** is [1] (in angular frequency units):

$$\hat{H}_Q = \frac{eQ}{2I(2I-1)\hbar}\hat{\mathbf{I}} \cdot \mathbf{V} \cdot \hat{\mathbf{I}} \tag{5.1}$$

for a spin I, where **V** is the electric field gradient at the nucleus (a second-rank Cartesian tensor), $\hat{\mathbf{I}}$ is the nuclear spin vector and Q is the nuclear quadrupole moment. This can be simply rewritten in terms of the Cartesian components of $\hat{\mathbf{I}}$ and **V** by expanding the tensor product in Equation (5.1):

$$\hat{H}_Q = \frac{eQ}{6I(2I-1)\hbar}\sum_{\substack{\alpha,\beta= \\ x,y,z}} V_{\alpha\beta}\left[\frac{3}{2}(\hat{I}_\alpha\hat{I}_\beta + \hat{I}_\beta\hat{I}_\alpha) - \delta_{\alpha\beta}\hat{I}^2\right] \tag{5.2}$$

where α, β refer to x, y, z of whatever frame of reference the spin operators and electric field gradient tensor are defined in.

It is conventional to define the parameters

$$\begin{aligned} eq &= V_{zz}^{\mathrm{PAF}} \\ \eta_Q &= \frac{V_{xx}^{\mathrm{PAF}} - V_{yy}^{\mathrm{PAF}}}{V_{zz}^{\mathrm{PAF}}} \end{aligned} \tag{5.3}$$

where η_Q is the *(quadrupolar) asymmetry parameter* and e is the magnitude of the electron charge. The trace of the electric field gradient tensor **V** is zero, so it has no isotropic component. Thus eq (= V_{zz}^{PAF}) is the anisotropy of the electric field gradient tensor. The anisotropy and asymmetry of a second-rank Cartesian tensor were discussed in Section 1.4.1, which should be referred to for definitions of these parameters. The electric field gradient tensor in Equations (5.3) is expressed in its principal axis frame (PAF), i.e. the frame in which it is diagonal (see Section 1.4.1); the elements of the tensor in this frame are labelled $V_{\alpha\alpha}^{\mathrm{PAF}}$. The axes in this frame are labelled x^{PAF}, y^{PAF}, z^{PAF} to distinguish them from the laboratory frame (x, y, z). The orientation of the PAF is determined by the chemical structure around the nucleus and so is fixed relative to the molecule or crystallite. In the PAF frame, the quadrupole hamiltonian in (5.2) becomes simply (again, in angular frequency units):

$$\hat{H}_Q = \frac{e^2qQ}{4I(2I-1)\hbar}\left[3\hat{I}_{z^{\mathrm{PAF}}}^2 - \hat{I}^2 + \frac{1}{2}\eta_Q\left(\hat{I}_{x^{\mathrm{PAF}}}^2 - \hat{I}_{y^{\mathrm{PAF}}}^2\right)\right] \tag{5.4}$$

The derivation of Equation (5.4) from Equation (5.2) uses the fact that the electric field gradient tensor is traceless (in any frame), i.e. $V_{xx} + V_{yy} + V_{zz} = 0$. If Tr(**V**) were not zero, there would be a net force on the nucleus from the electric quadrupole interaction and thus the nucleus would not be in an equilibrium spatial position with respect to the surrounding electrons and nuclei which create the electric field gradient. The constant $e^2qQ/\hbar$ is known as the *quadrupole-coupling constant* and is given the symbol χ here (units rad s^{-1}). The quadrupole-coupling constant in units of Hz is often denoted C_Q (= $e^2qQ/\hbar$). The *quadrupole-coupling tensor* is also a term often used; this is given by $\boldsymbol{\chi} = (eQ/\hbar)\mathbf{V}$.

The formulation of the quadrupole hamiltonian in Equation (5.4) is completely correct, but not very useful for dealing with NMR experiments. In the NMR experiment, the Zeeman interaction between the applied magnetic field and nuclear spin is the dominant interaction. The hamiltonian describing this interaction is

$$\hat{H}_0 = \omega_0 \hat{I}_z \tag{5.5}$$

where ω_0 is the Larmor frequency and z is the direction of the applied magnetic field in the laboratory frame. Clearly we need to re-express Equation (5.4) in terms of the laboratory frame (x, y, z) in which the Zeeman interaction is naturally described, rather than in terms of (x^{PAF}, y^{PAF}, z^{PAF}), the principal axis frame for the quadrupole interaction. $\hat{I}_{x^{\text{PAF}}}$, $\hat{I}_{y^{\text{PAF}}}$, $\hat{I}_{z^{\text{PAF}}}$, in Equation (5.4) are related to $\hat{I}_x$, $\hat{I}_y$, $\hat{I}_z$ (laboratory frame) through

$$\begin{pmatrix} \hat{I}_x \\ \hat{I}_y \\ \hat{I}_z \end{pmatrix} = \mathbf{R}(\theta, \phi) \begin{pmatrix} \hat{I}_{x^{\text{PAF}}} \\ \hat{I}_{y^{\text{PAF}}} \\ \hat{I}_{z^{\text{PAF}}} \end{pmatrix} \tag{5.6}$$

where $\mathbf{R}(\theta, \phi)$ is the rotation matrix which rotates the principal axis frame (x^{PAF}, y^{PAF}, z^{PAF}) into the laboratory frame (x, y, z) (see Box 1.2 in Chapter 1 for further discussion of rotations). The angles (θ, ϕ) are spherical polar angles describing the orientation of the laboratory frame z-axis (as defined by $\mathbf{B}_0$) in the PAF; no third angle is needed to describe the relationship between the two frames, as the x- and y-axes of the laboratory frame are arbitrary. We can then write $\hat{H}_Q$ in the laboratory frame (x, y, z) by substituting appropriate linear combinations of $\hat{I}_x$, $\hat{I}_y$, $\hat{I}_z$ in Equation (5.4) for $\hat{I}_{x^{\text{PAF}}}$, $\hat{I}_{y^{\text{PAF}}}$, $\hat{I}_{z^{\text{PAF}}}$ with the aid of Equation (5.6) to find the linear combinations:

$$\begin{aligned} \hat{H}_Q = \frac{e^2qQ}{4I(2I-1)\hbar} \{ & \tfrac{1}{2}(3\cos^2\theta - 1)(3\hat{I}_z^2 - \hat{I}^2) \\ & + \tfrac{3}{2}\sin\theta\cos\theta[\hat{I}_z(\hat{I}_+ + \hat{I}_-) + (\hat{I}_+ + \hat{I}_-)\hat{I}_z] \\ & + \tfrac{3}{4}\sin^2\theta(\hat{I}_+^2 + \hat{I}_-^2)\} \end{aligned}$$

$$+\eta_Q \frac{e^2 qQ}{4I(2I-1)\hbar}\{\tfrac{1}{2}\cos 2\phi[(1-\cos^2\theta)(3\hat{I}_z^2 - \hat{I}^2)$$
$$+(\cos^2\theta + 1)(\hat{I}_+^2 + \hat{I}_-^2)]$$
$$+\tfrac{1}{2}\sin\theta[(\cos\theta\cos 2\phi - i\sin 2\phi)(\hat{I}_+\hat{I}_z + \hat{I}_z\hat{I}_+)$$
$$+(\cos\theta\cos 2\phi + i\sin 2\phi)(\hat{I}_-\hat{I}_z + \hat{I}_z\hat{I}_-)]$$
$$+(i/4)\sin 2\phi\cos\theta(\hat{I}_+^2 - \hat{I}_-^2)\} \quad (5.7)$$

where x, y, z refer to the laboratory frame. θ and ϕ are determined by the orientation of the chemical structure containing the nucleus with respect to the applied field (along the laboratory z-axis). If the nucleus is part of a molecule, θ and ϕ are determined by the molecular orientation in the applied field. The operators $\hat{I}_+$ and $\hat{I}_-$ are related to $\hat{I}_x$ and $\hat{I}_y$ and are defined in Box 1.1 in Chapter 1.

Clearly, in cases of an axially symmetric electric field gradient ($\eta = 0$), only the first term of Equation (5.7) is non-zero.

It is apparent from Equation (5.7) that the matrix of $\hat{H}_Q$ in the Zeeman basis contains off-diagonal terms as well as diagonal terms. Thus, to find the exact energies of the spin levels under the quadrupole coupling, one would need to diagonalize the matrix of $\hat{H}_Q$ in the Zeeman basis to obtain the eigenstates (the new states of the spin system under $\hat{H}_Q$; linear combinations of the Zeeman states) and corresponding eigenvalues (their energies). Clearly, this is a time-consuming process. If the Zeeman interaction is the dominant term in the hamiltonian for the spin system, then only the secular parts of $\hat{H}_Q$ affect the energies of the states of the spin system, i.e. those terms in $\hat{H}_Q$ which commute with the Zeeman hamiltonian of Equation (5.5) (see Section 1.4 for discussion of this point). The non-secular parts of the matrix of $\hat{H}_Q$ (in this case, the off-diagonal terms of the matrix in the Zeeman basis) are small compared with the diagonal terms provided by the Zeeman term in the total hamiltonian. Thus, they have little effect on the eigenvalues of the diagonalized matrix of the complete hamiltonian for the spin system $\hat{H}_0 + \hat{H}_Q$ (where $\hat{H}_0$ is the Zeeman hamiltonian (Equation (5.5)). Under these circumstances, the approximate energies of the spin system are found by recourse to perturbation theory.[N1] Corrections up to second-order are generally sufficient for quadrupole interactions which cause splittings less than ~1/10 of the Zeeman splitting in the particular field used in the NMR experiment. The first- and second-order energy corrections to the energies of the Zeeman levels from quadrupole coupling from perturbation theory are [2, 3]:

$$E_m^{(1)} = \frac{e^2 qQ}{4I(2I-1)}(3m^2 - I(I+1))\frac{1}{2}[(3\cos^2\theta - 1) + \eta\cos 2\phi\sin^2\theta] \quad (5.8)$$

$$
\begin{aligned}
E_m^{(2)} = {} & -\left(\frac{e^2qQ}{4I(2I-1)}\right)^2 \frac{m}{\omega_0} \times \left\{ -\frac{1}{5}(I(I+1)-3m^2)(3+\eta_Q^2) \right. \\
& + \frac{1}{28}(8I(I+1)-12m^2-3)[(\eta_Q^2-3)(3\cos^2\theta-1)+6\eta_Q\sin^2\theta\cos 2\phi] \\
& + \frac{1}{8}(18I(I+1)-34m^2-5)\left[\frac{1}{140}(18+\eta^2)(35\cos^4\theta-30\cos^2\theta+3) \right. \\
& \left.\left. + \frac{3}{7}\eta_Q\sin^2\theta(7\cos^2\theta-1)\cos 2\phi + \frac{1}{4}\eta_Q^2\sin^4\theta\cos 4\phi\right]\right\}
\end{aligned}
\tag{5.9}
$$

where m denotes the magnetic quantum number associated with the particular Zeeman level.

There are several points arising from these equations which are worth noting:

- For a half-integer quadrupolar nucleus, the first-order energy correction due to quadrupole coupling for the $+\frac{1}{2} \rightarrow -\frac{1}{2}$ transition, the so-called *central transition*, is zero. Thus, this transition is expected to be sharp when the quadrupole coupling is small (so that only the first-order energy correction to the Zeeman levels is important). However, this transition is affected by quadrupole coupling to second-order, so will be broadened for a powder sample when the quadrupole coupling is large. Figure 5.2 shows the effect of the quadrupole-coupling interaction on the Zeeman levels in the rotating frame, i.e. the frame in which NMR transition frequencies are measured.
- The first-order term (Equation (5.8)) has the same molecular orientation dependence as the chemical shielding and dipole–dipole interactions. Thus the spectrum of a quadrupolar nucleus with relatively small quadrupole coupling (so that only the first-order energy correction to the Zeeman levels is important) will (for a powder sample) contain powder patterns reminiscent of those arising from chemical shift anisotropy and heteronuclear dipole–dipole coupling (see Sections 3.1.3 and 4.1.4 respectively). Figure 5.3 shows the form of the spectrum expected for a spin-1 and spin-$\frac{3}{2}$ nucleus in a powder sample under first-order quadrupole coupling of axial symmetry, i.e. $\eta_Q = 0$.
- The first term of Equation (5.9) for the second-order energy correction has no molecular/PAF orientation dependence; it is an isotropic term. This is very important as it means that the isotropic shift seen in any spectrum of a quadrupolar nucleus has a contribution from the quadrupole coupling as well as the usual chemical shift.
- The second-order term depends inversely on the Larmor frequency ($1/\omega_0$); thus the importance of this term diminishes with increasing field strength.

For many applications, the quadrupole hamiltonian is best rewritten in terms of spherical tensor operators. This is detailed in Box 5.1.

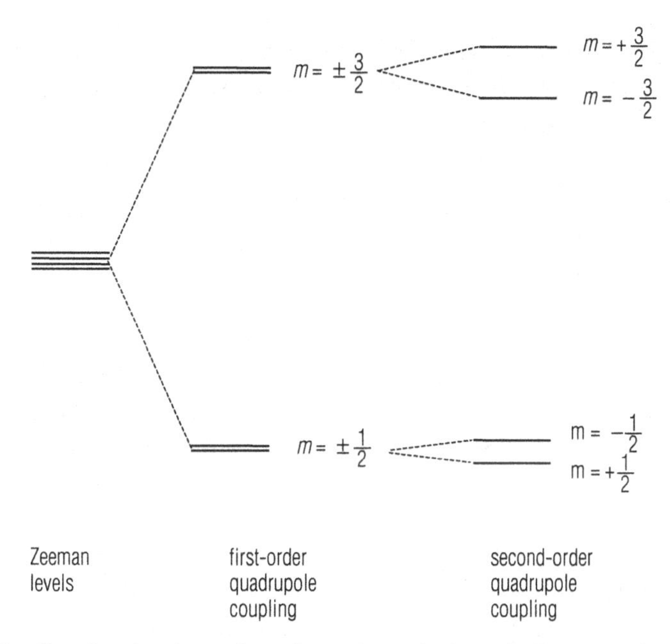

Fig. 5.2 The effect of quadrupole coupling to first- and second-order on the Zeeman spin levels for a spin-$\frac{3}{2}$ nucleus in the rotating frame, i.e. in a frame in which all transition frequencies are measured with respect to the Larmor frequency ω_0 of the spin. This is the relevant frame from which to view the situation as it is the frame (in effect) in which we measure the resonance frequencies of a spin system in an NMR experiment. See Section 1.5 for details.

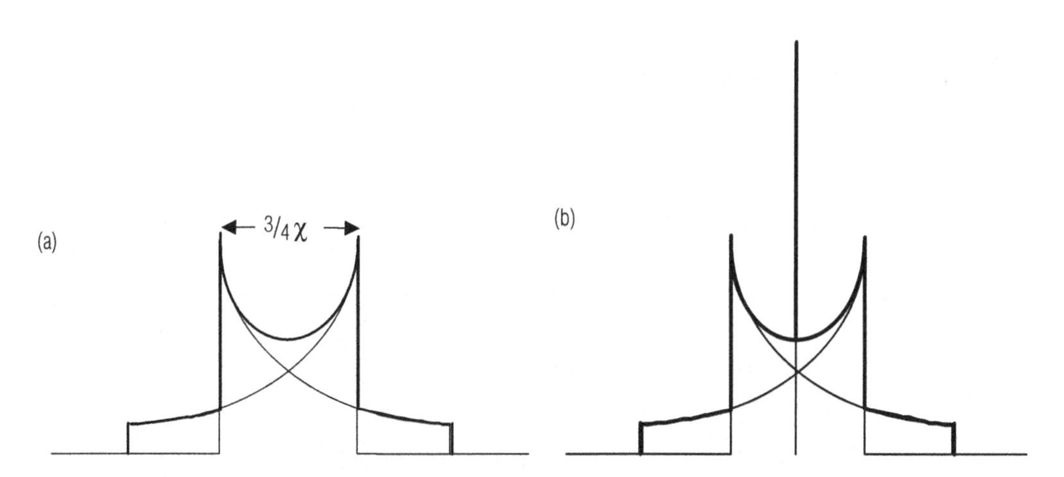

Fig. 5.3 (a) The first-order quadrupolar powder lineshape for a spin-1 nucleus at a site of axial symmetry. There are two possible transitions for a spin-1 nucleus ($m = +1$ to $m = 0$ and $m = 0$ to $m = -1$), which give rise to mirror image lineshapes. The splitting of the 'horns' in the pattern is equal to $\frac{3}{4}\chi$, where χ is the quadrupole-coupling constant. The horns are created by crystallites in which the principal z-axis of the quadrupolar coupling tensor is perpendicular to the applied field $\mathbf{B}_0$. (b) The first-order quadrupolar powder lineshape for a spin-$\frac{3}{2}$ nucleus at a site of axial symmetry. There are three possible $\Delta m = \pm 1$ transitions for a spin-$\frac{3}{2}$ nucleus. The so-called *satellite transitions*, $m = +\frac{3}{2}$ to $m = +\frac{1}{2}$ and $m = -\frac{1}{2}$ to $m = -\frac{3}{2}$, give rise to mirror image lineshapes. The central transition, $m = +\frac{1}{2}$ to $m = -\frac{1}{2}$, has no orientation dependence (to first order in the applied field $\mathbf{B}_0$) and so gives a sharp signal.

Box 5.1: The quadrupole hamiltonian in terms of spherical tensor operators: the effect of the rotating frame and magic-angle spinning

The quadrupole hamiltonian in terms of spherical tensor operators

The quadrupole hamiltonian of Equation (5.7) can be rewritten in terms of spherical tensor operators. This is not merely an academic exercise. As we saw in Chapter 4, spherical tensor operators have well-defined transformation properties under rotations and so are relatively easy to deal with under such circumstances. In NMR, we usually work within the rotating frame (see Chapter 1), i.e. a frame rotating at the Larmor frequency about the laboratory *z*-axis, so we will ultimately need to perform this rotation on the quadrupole hamiltonian (and any other components of the total hamiltonian describing the spin system). This is most easily carried out when the hamiltonian describing the spin system is expressed in terms of spherical tensor operators.

As discussed in Appendix B, any nuclear spin interaction hamiltonian can be written in the form:

$$\hat{H} = \sum_{q=-2}^{+2} (-1)^q \hat{T}_{2q}^A \Lambda_{2-q}^A \qquad \text{(i)}$$

where $\hat{T}_{2q}^A$ is the spherical tensor operator of rank 2 and order q appropriate for interaction A as defined in Appendix B, and the Λ_{2-q}^A is the order $-q$ component of a second-rank spherical tensor which describes the strength and orientation dependence of the interaction. Thus, the laboratory frame quadrupole hamiltonian of Equation (5.7) can be rewritten as

$$\hat{H}_Q = \sum_{q=-2}^{+2} (-1)^q \hat{T}_{2q}^Q \Lambda_{2-q}^Q \qquad \text{(ii)}$$

where the Λ_{2q}^Q is a component of a spherical tensor (in the laboratory frame) which describes the strength and orientation dependence of the quadrupole coupling. The Λ_{2q}^Q are equivalent to the Λ_{2q}^{IS} in the dipolar hamiltonian in Box 4.2. The $\hat{T}_{2q}$ appropriate for a single spin interaction (such as the quadrupole interaction) may be formed from the Cartesian components of $\hat{\mathbf{I}}$. These are derived in Appendix B and are repeated here for convenience:

$$\begin{aligned} \hat{T}_{20}^Q &= \sqrt{\frac{1}{6}}(3\hat{I}_z^2 - \hat{\mathbf{I}}\,.\,\hat{\mathbf{I}}) \\ \hat{T}_{2\pm1}^Q &= \mp(\hat{I}_\pm \hat{I}_z + \hat{I}_z \hat{I}_\pm) \qquad \text{(iii)} \\ \hat{T}_{2\pm2}^Q &= \frac{1}{2}\hat{I}_\pm^2 \end{aligned}$$

We now need to determine the Λ_{2q}^{Q} terms. To do this, we follow the procedure used in Box 4.2 to find the Λ_{2q}^{IS} for the dipolar-coupling interaction, and start with the quadrupole hamiltonian in its principal axis frame (PAF). In the principal axis frame, the quadrupole hamiltonian is

$$\hat{H}_Q^{\mathrm{PAF}} = \sum_{q=-2}^{+2} (-1)^q \hat{T}_{2q}^{Q} \lambda_{2-q}^{Q} \tag{iv}$$

where the λ is the quadrupole-coupling tensor in the principal axis frame, with components in units of rad s^{-1} (see Appendix B for derivation of these terms):

$$\begin{aligned} \lambda_{20}^{Q} &= \sqrt{6}\,\frac{e^2qQ}{4I(2I-1)} \\ \lambda_{2\pm1}^{Q} &= 0 \\ \lambda_{2\pm2}^{Q} &= \frac{e^2qQ}{4I(2I-1)}\eta_Q \end{aligned} \tag{v}$$

The quadrupole-coupling tensor in the laboratory frame and principal axis frames is related via (see Box 4.2, Equation (v)):

$$\Lambda_{2q}^{Q} = \sum_{q'=-2}^{+2} D_{q'q}^{2}(\alpha, \beta, \gamma)\lambda_{2q'}^{Q} \tag{vi}$$

where α, β, γ are the Euler angles which rotate the principal axis frame into the laboratory frame and $D_{q'q}^{2}(\alpha, \beta, \gamma)$ is a Wigner rotation matrix element (see Box 4.2):

$$D_{q'q}^{2}(\alpha, \beta, \gamma) = \exp(-i\alpha q')\exp(-i\gamma q)d_{q'q}^{2}(\beta) \tag{vii}$$

where $d_{q'q}^{2}(\beta)$ is a reduced Wigner rotation matrix element. Thus, we can rewrite the full quadrupole hamiltonian of Equation (5.7) as

$$\hat{H}_Q = \sum_{q=-2}^{+2} (-1)^q \hat{T}_{2q}^{Q} \sum_{q'=-2}^{+2} D_{-q'q}^{2}(\alpha, \beta, \gamma)\lambda_{2-q'}^{Q} \tag{viii}$$

We will use this later.

The effect of the rotating frame: first- and second-order average hamiltonians for the quadrupole interaction

The full hamiltonian for a spin I in an NMR experiment must, of course, also include a term for the Zeeman interaction:

$$\hat{H}_0 = \omega_0 \hat{I}_z \tag{ix}$$

Continued on p. 244

Box 5.1 *Cont.*

where ω_0 is the Larmor frequency, so that the total hamiltonian is

$$\hat{H} = \hat{H}_0 + \hat{H}_Q \tag{x}$$

Now, providing the Zeeman interaction is much larger than the quadrupole interaction, i.e. $(\omega_0/2\pi) >> (e^2qQ/2I(2I - 1)\hbar)$, the effects of the Zeeman term can be removed by transforming to the rotating frame, a frame which rotates about the laboratory z-axis at rate ω_0, the Larmor frequency. A similar transformation was used in Section 1.2.2 to remove the time dependence of the term in the hamiltonian due to an rf pulse (although there the rotating frame rotated about z at the frequency of the rf irradiation, which is only ω_0 if the rf pulse is on resonance). The rotating frame transformation is used here in part simply to remove the effects of the large Zeeman term and allow us to see more clearly the effects of the quadrupole term. It is also used because our NMR experiments are in effect recorded in such a rotating frame, as we measure the NMR signal relative to some carrier signal with frequency $\omega_0 + \Delta\omega$, where $\Delta\omega$ is the offset of the rf pulse used in the NMR experiment from the Larmor frequency. Thus, to determine the outcome of an NMR experiment, we must work within this same rotating frame.

In the rotating frame, the total hamiltonian $\hat{H}^*$ is given by (see Box 2.1)

$$\begin{aligned} \hat{H}^*(t) &= \hat{R}_z^{-1}(\omega_0 t)\hat{H}\hat{R}_z(\omega_0 t) - \omega_0 \hat{I}_z \\ &= \hat{R}_z^{-1}(\omega_0 t)\hat{H}_0\hat{R}_z(\omega_0 t) + \hat{R}_z^{-1}(\omega_0 t)\hat{H}_Q\hat{R}_z(\omega_0 t) - \omega_0 \hat{I}_z \end{aligned} \tag{xi}$$

where $\hat{R}_z(\omega_0 t)$ is the rotation operator which rotates an object by an angle $\omega_0 t$ about the laboratory z-axis. The Zeeman term $\hat{H}_0 = \omega_0\hat{I}_z$ is unaffected by a z-rotation, so the first term of Equation (xi) cancels with the third term, $\omega_0\hat{I}_z$. Thus the rotating frame hamiltonian becomes

$$\begin{aligned} \hat{H}^*(t) &= \hat{R}_z^{-1}(\omega_0 t)\hat{H}_Q\hat{R}_z(\omega_0 t) \\ &= \sum_{q=-2}^{+2} (-1)^q \hat{R}_z^{-1}(\omega_0 t)\hat{T}_{2q}^Q\hat{R}_z(\omega_0 t)\Lambda_{2-q}^Q \end{aligned} \tag{xii}$$

The $\hat{T}_{20}^Q$ term in Equation (xii) is unaffected by a z-rotation (see Equation (iii); $\hat{T}_{20}^Q$ consists only of $\hat{I}_z$ terms and scalar terms), but the remaining $\hat{T}_{2q}^Q$ terms are all affected and this imparts a time dependence to these terms, and hence to the rotating frame hamiltonian. The rotation of the $\hat{T}_{2q}^Q$ operators as in Equation (xii) can be described in terms of Wigner rotation matrices (see Box 4.2):

$$\begin{aligned} \hat{R}_z^{-1}(\omega_0 t)\hat{T}_{2q}^Q\hat{R}_z(\omega_0 t) &= \sum_{q'=-2}^{+2} D_{q'q}^2(0, 0, \omega_0 t)\hat{T}_{2q}^Q \\ &= \exp(+i\omega_0 tq)\hat{T}_{2q}^Q \end{aligned} \tag{xiii}$$

Thus the hamiltonian in the rotating frame is

$$\hat{H}^*(t) = \sum_{q=-2}^{+2} (-1)^q \exp(+i\omega_0 tq)\hat{T}^Q_{2q}\Lambda^Q_{2-q} \quad \text{(xiv)}$$

The time dependence of this hamiltonian is clearly unhelpful. However, we saw in Chapter 2, Box 2.1, how to deal with a periodically time-dependent hamiltonian such as that in Equation (xiv), by forming an average hamiltonian over one cycle of the periodic time dependence. The first-order term in the average hamiltonian (see Box 2.1) for a continuously time-varying hamiltonian is simply

$$\overline{H}^{(0)} = \frac{1}{t_p}\int_0^{t_p} \hat{H}(t)\mathrm{d}t \quad \text{(xv)}$$

For the rotating frame hamiltonian of Equation (xiv), the first-order average hamiltonian (first order in the hamiltonian) is then:

$$\begin{aligned}\overline{H}^{(0)} &= \frac{\omega_0}{2\pi}\sum_{q=-2}^{+2}(-1)^q \hat{T}^Q_{2q}\Lambda^Q_{2-q}\int_0^{2\pi/\omega_0}\exp(i\omega_0 tq)\mathrm{d}t \\ &= \hat{T}^Q_{20}\Lambda^Q_{20} - \frac{i}{2\pi}\sum_{q=\pm1,\pm2}\frac{(-1)^q}{q}[\exp(i2q\pi) - 1]\hat{T}^Q_{2q}\Lambda^Q_{2-q} \\ &= \hat{T}^Q_{20}\Lambda^Q_{20} \end{aligned} \quad \text{(xvi)}$$

It is worthwhile here expanding the Λ^Q_{20} term of Equation (xvi). The Λ^Q_{20} term is found using Equation (vi) previously, which expresses it in terms of the principal axis frame (PAF) quadrupole interaction spherical tensor λ^Q:

$$\Lambda^Q_{20} = D^2_{00}(\alpha,\beta,\gamma)\lambda^Q_{20} + D^2_{20}(\alpha,\beta,\gamma)\lambda^Q_{22} + D^2_{-20}(\alpha,\beta,\gamma)\lambda^Q_{2-2} \quad \text{(xvii)}$$

where α, β, γ are the Euler angles which rotate the principal axis frame into the laboratory frame. Substituting for the Wigner rotation matrix elements and the spherical tensor components λ^Q_{2q} we obtain

$$\Lambda^Q_{20} = \sqrt{6}\cdot\frac{e^2qQ}{4I(2I-1)}\frac{1}{2}[(3\cos^2\beta - 1) + \eta_Q\sin^2\beta\cos 2\alpha] \quad \text{(xviii)}$$

Hence, the first-order rotating frame quadrupole hamiltonian using Equations (xvi) and (xix) is

$$\hat{H}^{(0)}_Q = \sqrt{6}\cdot\frac{e^2qQ}{4I(2I-1)}\frac{1}{2}[(3\cos^2\beta - 1) + \eta_Q\sin^2\beta\cos 2\alpha]\hat{T}^Q_{20} \quad \text{(xix)}$$

This will cause an energy change to the Zeeman spin levels equivalent to that found from perturbation theory in Equation (5.8).

We now need to find the second-order contribution to the average hamiltonian.

Continued on p. 246

Box 5.1 *Cont.*

The general form of the second-order average hamiltonian for a continuously time varying hamiltonian is (see Box 2.1)

$$\overline{H}^{(1)} = -\frac{i}{2t_{\text{p}}}\int_0^{t_{\text{p}}} \text{d}t \int_0^t [\hat{H}(t), \hat{H}(t')]\text{d}t' \qquad \text{(xx)}$$

The second-order term in the rotating frame average quadrupole hamiltonian is thus

$$\begin{aligned}\overline{H}_Q^{(1)} &= -\frac{i\omega_0}{4\pi}\int_0^{2\pi/\omega_0} \text{d}t \int_0^t [\hat{H}^*(t), \hat{H}^*(t')]\text{d}t' \\ &= -\frac{i\omega_0}{4\pi}\sum_{q=-2}^{+2}\sum_{q'=-2}^{+2}(-1)^{q+q'}\Lambda_{2-q}^{Q}\Lambda_{2-q'}^{Q}[\hat{T}_{2q}^{Q}, \hat{T}_{2q'}^{Q}]\int_0^{2\pi/\omega_0} \text{d}t \int_0^t \exp(i\omega_0(q+q')t')\text{d}t'\end{aligned} \qquad \text{(xxi)}$$

where the rotating frame hamiltonian $\hat{H}^*$ is given by Equation (xiv). In Equation (xxi) only those terms with $q = -q'$ give rise to non-zero contributions to $\overline{H}_Q^{(1)}$, as only in this case does the time-dependent, oscillating exponential term vanish. Otherwise, such oscillating terms will integrate over the hamiltonian cycle to give zero. Thus, the second-order average hamiltonian is of the form

$$\overline{H}_Q^{(1)} = -\frac{i\pi}{2\omega_0}\sum_{q=-2}^{+2}\Lambda_{2-q}^{Q}\Lambda_{2q}^{Q}[\hat{T}_{2q}^{Q}, \hat{T}_{2-q}^{Q}] \qquad \text{(xxii)}$$

where the Λ_{2q}^{Q} components are given by Equation (vi) and the $\hat{T}_{2q}^{Q}$ by Equation (iii). The products of $\Lambda_{2-q}^{Q}\Lambda_{2q}^{Q}$ can be written as linear combinations of new spherical tensors V_{k0}^{Q}, where k runs from $2 + 2 = 4$ to $2 - 2 = 0$ in steps of 2 [4]. The commutators of Equation (xxii) can be evaluated (in terms of other spherical tensor operators) [4]. Once all this is done (a lengthy process which we will not bore the reader with here) the result is:

$$\begin{aligned}\hat{H}_Q^{(1)} = &-\left(\frac{e^2qQ}{4I(2I-1)}\right)^2\frac{1}{\omega_0}\frac{2}{5}\{[-3\sqrt{10}\hat{T}_{30}^{Q} + \hat{T}_{10}^{Q}(3-4I(I+1))]V_{00}^{Q} \\ &+[-12\sqrt{10}\hat{T}_{30}^{Q} - \hat{T}_{10}^{Q}(3-4I(I+1))]V_{20}^{Q} \\ &+[-34\sqrt{10}\hat{T}_{30}^{Q} + 3\hat{T}_{10}^{Q}(3-4I(I+1))]V_{40}^{Q}\}\end{aligned} \qquad \text{(xxiii)}$$

where the spherical tensor operators are defined as follows:

$$\begin{aligned}\hat{T}_{10}^{Q} &= \hat{I}_z \\ \hat{T}_{20}^{Q} &= \sqrt{\frac{1}{6}}(3\hat{I}_z^2 - I(I+1)) \\ \hat{T}_{30}^{Q} &= \sqrt{\frac{1}{10}}(5\hat{I}_z^2 - 3I(I+1) + 1)\hat{I}_z\end{aligned} \qquad \text{(xxiv)}$$

and the V_{k0}^{Q} parameters, which contain all the geometrical terms, are [5]:

$$V_{k0}^{Q} = \sum_{n} D_{n0}^{k}(\alpha, \beta, \gamma)A_{kn} \qquad \text{(xxv)}$$

with

$$A_{00} = -\frac{1}{5}(3+\eta_Q^2)$$

$$A_{20} = \frac{1}{14}(\eta_Q^2 - 3) \qquad A_{2\pm2} = \frac{1}{7}\sqrt{\frac{3}{2}}\eta_Q$$

$$A_{40} = \frac{1}{140}(18+\eta_Q^2) \qquad A_{4\pm2} = \frac{3}{70}\sqrt{\frac{5}{2}}\eta_Q \qquad A_{4\pm4} = \frac{1}{4\sqrt{70}}\eta_Q^2$$

The first-order average hamiltonian from (xix) can be expressed in a similar manner as

$$\hat{H}_Q^{(0)} = \frac{e^2qQ}{4I(2I-1)}\sqrt{6}W_{20}^{Q}\hat{T}_{20}^{Q} \qquad \text{(xxvii)}$$

where W_{20}^{Q} is given by $W_{20}^{Q} = \sum_{q=0,\pm2} D_{q0}^{2}(\alpha, \beta, \gamma)B_{2q}$ with $B_{20} = 1$ and $B_{2\pm2} = \sqrt{\frac{1}{6}}\eta_Q$. There is one final point to discuss with respect to the rotating frame transformation: how many terms in the average hamiltonian do we need to take to get a good approximation to the full rotating frame hamiltonian? The average hamiltonian is a sum of terms:

$$\overline{H}(t_p) = \overline{H}^{(0)} + \overline{H}^{(1)} + \overline{H}^{(2)} + \ldots \qquad \text{(xxviii)}$$

where $\overline{H}^{(i)}$ is the $(i-1)$th-order term (in the interaction hamiltonian). The $(i-1)$th-order term for the quadrupole-coupling average hamiltonian in the rotating frame contains a coefficient $(e^2qQ/4I(2I-1))\cdot(1/(\omega_0)^i)$. Thus the number of terms we need to take in the expansion of the average hamiltonian depends on the magnitude of the quadrupole coupling $(e^2qQ/4I(2I-1))$ relative to the Larmor frequency ω_0. In high field NMR experiments, $(e^2qQ/4I(2I-1))$ is often less than $\omega_0/10$. Under these circumstances, the expansion of the average hamiltonian to second order is generally sufficient for most purposes. If, however, $(e^2qQ/4I(2I-1))$ is of the order of ω_0 or larger, then clearly the series expansion of $\overline{H}$ will never converge. Under these circumstances, it is not useful to use the rotating frame transformation and average hamiltonian approach; one must use the full laboratory frame quadrupole plus Zeeman hamiltonian instead. Clearly, this has implications for the manner in which experiments are recorded also.

Continued on p. 248

Box 5.1 *Cont.*

The energy levels under quadrupole coupling

We can use Equations (xxiii) and (xxvii) to derive equations for the first- and second-order (with respect to the quadrupole coupling constant) quadrupolar contributions to the rotating frame Zeeman spin levels. These are derived to be (in frequency units):

$$\omega_Q^{(1)} = \frac{e^2qQ}{4I(2I-1)}(3m^2 - I(I+1))W_{20}^Q \qquad \text{(xxix)}$$

(xxvi)

$$\omega_Q^{(2)} = -\left(\frac{e^2qQ}{4I(2I-1)}\right)^2 \frac{2}{\omega_0} m \times \{[I(I+1) - 3m^2]V_{00}^Q + [8I(I+1) - 12m^2 - 3]V_{20}^Q + [18I(I+1) - 34m^2 - 5]V_{40}^Q\} \qquad \text{(xxx)}$$

This formulation is useful in order to see the geometric dependence of the energies in terms of the V_{k0}^Q defined in Equation (xxv). Much discussion in the rest of this chapter will use this formulation.

The effect of magic-angle spinning

The effect of magic-angle spinning on the average hamiltonians in Equations (xxvii) and (xxiii) is simple to deduce. As for the discussion of magic-angle spinning on the dipolar coupling hamiltonian in Box 4.2, we consider the principal axis frame (PAF) of the quadrupole-coupling tensor as being orientated via Euler angles (α, β, γ) in a rotor-fixed axis frame (R), which is then orientated in the laboratory frame via Euler angles ($-\omega_R t$, θ_R, 0). The transformation of the quadrupole-coupling tensor expressed in spherical tensor form under rotations of the axis frame in which it is expressed from the PAF to the rotor (R) frame to the laboratory frame is easily accomplished using equations of the form of Equation (vi). The average hamiltonian components for quadrupole coupling under magic-angle spinning are then those of Equations (xxiii) and (xxvii), but with the V_{k0}^Q (which ultimately contain the components of the transformed quadrupole-coupling tensor) replaced by

$$V_{k0}^Q = \sum_m D_{m0}^k(-\omega_R t, \theta_R, 0) \sum_n D_{nm}^k(\alpha, \beta, \gamma) A_{kn} \qquad \text{(xxxi)}$$

where the $D_{nm}^k(\alpha, \beta, \gamma)$ are Wigner rotation matrix elements which take account of the rotation of the axis frame for the quadrupole coupling tensor from the PAF to R and the $D_{m0}^k(-\omega_R t, \theta_R, 0)$ take account of the frame rotation from R to the laboratory frame.

It is worth noting at this point that magic-angle spinning will average the W_{20}^Q terms of Equation (xxix) to zero.

5.2.2 *The effect of rf pulses*

For quadrupolar nuclei in many samples, the strength of the quadrupolar interaction can be larger than the strength of interaction with an rf pulse. Modern NMR spectrometers typically produce rf pulses with amplitudes of the order of 100 kHz, while quadrupole-coupling constants are often of the order of MHz. Thus the quadrupole interaction cannot be ignored during an rf pulse. It is not simply that the quadrupole coupling has some small effect which need only concern purists; in the majority of cases the quadrupole coupling has a drastic effect that dominates the whole experiment.

A large quadrupole coupling has a further effect if we are considering irradiation of the sample by rf pulses. If a low amplitude rf pulse is applied at the Larmor frequency, only the central transition is effectively irradiated. The satellite transitions (i.e. the other single-quantum transitions) are removed from the Larmor frequency by $(2m - 1)\omega_Q$ as a result of the quadrupole coupling where m is the initial (Zeeman) level involved in the transition and ω_Q, the *quadrupole splitting*, is given by:

$$\omega_Q = \frac{3e^2qQ}{4I(2I-1)}\frac{1}{2}[(3\cos^2\theta - 1) + \eta_Q \cos 2\phi \sin^2\theta] \qquad (5.10)$$

Thus for most molecular (or PAF) orientations (defined by the spherical polar angles (θ, ϕ) relative to the laboratory frame), the satellite transitions are well off resonance.

During a pulse, the effective hamiltonian describing the spin system in the rotating frame (here the rotating frame is rotating about the laboratory z-axis at the frequency of the rf pulse, ω_{rf}) is $\hat{H}_{pulse}$, where

$$\hat{H}_{pulse} = \hat{H}_{cs} + \hat{H}_Q^{eff} + \hat{H}_{rf} + \hat{H}_\Delta \qquad (5.11)$$

where $\hat{H}_{cs}$ is the chemical shift hamiltonian and $\hat{H}_{rf}$ describes the effects of the rf pulse:

$$\hat{H}_{cs} = \omega_{cs} I_z \qquad (5.12)$$

$$\begin{aligned} \hat{H}_{rf} &= \omega_1(\hat{I}_x \cos\phi + \hat{I}_y \sin\phi) \\ &= \frac{\omega_1}{\sqrt{2}}(\hat{T}_{1-1}\exp(+i\phi) - \hat{T}_{11}\exp(-i\phi)) \end{aligned} \qquad (5.13)$$

ω_{cs} is the chemical shift (see Section 3.1); ω_1 is the amplitude of the rf pulse and ϕ its phase. The tensor operators in $\hat{H}_{rf}$ are defined as

$$\hat{T}_{1\pm1} = \mp\sqrt{\frac{1}{2}}\hat{I}_\pm \qquad (5.14)$$

and $\hat{I}_{\pm}$ are defined in Box 1.1. The pulse may be applied at any frequency, ω_{rf}, within the quadrupolar powder pattern. The rotating frame in which $\hat{H}_{pulse}$ is defined is one which rotates at frequency ω_{rf} about $\mathbf{B}_0$, so that the time dependence of the rf magnetic field vanishes (see Section 1.2). The Zeeman term of the hamiltonian (Equation (5.5)) also vanishes in the rotating frame if the pulse is applied at the Larmor frequency. The term $\hat{H}_{\Delta}$ accounts for off-resonance effects [6, 7] if the pulse is applied at a frequency ω_{rf} other than the Larmor frequency, ω_0, when the Zeeman term of the hamiltonian becomes, in the frame rotating at ω_{rf} about $\mathbf{B}_0$:

$$\hat{H}_{\Delta} = (\omega_0 - \omega_{rf})\hat{I}_z \tag{5.15}$$

In order to determine the effects of the rf pulse on the spin system, we need to use $\hat{H}_{pulse}$ to calculate the density matrix describing the spin system a time τ after the start of the pulse.

As described in Chapter 1, the density matrix ρ is usually calculated within the Zeeman basis. At any given time τ may be calculated from that at time 0 using the solution to the Liouville–von Neumann equation:

$$\rho(\tau) = \mathbf{U}^{-1}(\tau)\rho(0)\mathbf{U}(\tau) \tag{5.16}$$

where the propagator matrix $\mathbf{U}(\tau)$ is

$$\mathbf{U}(\tau) = \hat{T}\exp\left(-i\int_0^{\tau}\mathbf{H}(t)\mathrm{d}t\right) \tag{5.17}$$

$\mathbf{H}(t)$ is the hamiltonian matrix in the Zeeman basis which describes the spin system at time and $\hat{T}$ is the Dyson time ordering operator. The hamiltonian matrix does not commute with itself at different times t, so $\mathbf{U}(\tau)$ is calculated recursively using [8]:

$$\underset{\delta\tau\to 0}{\mathbf{U}}(\tau + \delta\tau) = \mathbf{V}^{-1}(\tau)\exp(\mathbf{E}(\tau)\,\delta\tau)\mathbf{V}(\tau)\mathbf{U}(\tau) \tag{5.18}$$

$\mathbf{E}(\tau)$ is a diagonal matrix of eigenvalues of $\mathbf{H}(\tau)$ and $\mathbf{V}(\tau)$ is the matrix of eigenvectors of $\mathbf{H}(\tau)$. Clearly, such a calculation is a time-consuming business.

However, we can get a general picture of the effects of the rf pulses by considering two limiting cases [7]. In the first, $\omega_1 >> \omega_Q$, so that we have non-selective excitation. All transitions are effectively on resonance, so that we can ignore $\hat{H}_{\Delta}$ as well as the quadrupole and chemical shift terms in Equation (5.11) for the hamiltonian describing the spin system. Thus we are left with just $\hat{H}_{pulse} \approx \hat{H}_{rf} = \omega_1\hat{I}_y$ for the case of a y-pulse. This is just the same as for an isolated spin-$\frac{1}{2}$ system and was dealt with in detail in Section 1.3; as deduced there, a density matrix element $\rho_{m,m+1}(\tau)$ a time τ after the start of the pulse is simply given by

$$\frac{\rho_{m,m+1}(\tau)}{\left(\dfrac{1}{Z}\dfrac{\hbar\omega_0}{kT}\right)} = \langle m|\hat{I}_y|m+1\rangle \sin(\omega_1\tau)$$

$$= \frac{i}{2}[I(I+1) - m(m+1)]^{\frac{1}{2}} \sin(\omega_1\tau) \tag{5.19}$$

where $\langle m|\hat{I}_y|m+1\rangle$ is a matrix element of the $\hat{I}_y$ operator and is defined in Box 1.1 in Chapter 1. Z is defined in Equation (1.59).

In the limiting case of the quadrupole coupling being much larger than the interaction with the rf pulse [7], i.e. $\omega_Q >> \omega_1$, we necessarily have selective excitation of the transition on resonance, i.e. that at or near frequency ω_{rf}. We assume no irradiation of any other transitions, so that we only need to consider the on-resonance transition, $m \rightarrow m+1$. Accordingly, we use the submatrix of $\hat{H}_{pulse}$ (Equation (5.11)) in the basis of $(m, m+1)$ in our calculations as though these were the only two spin levels which exist. The submatrix we require is then given by (again considering a y-pulse)

$$\begin{aligned}\mathbf{H}_{pulse}^{m,m+1} &= \mathbf{H}_{cs}^{m,m+1} + \mathbf{H}_{Q}^{m,m+1} + \mathbf{H}_{\Delta}^{m,m+1} + \mathbf{H}_{rf}^{m,m+1}\\ &= \mathbf{H}_{\Sigma}^{m,m+1} + \mathbf{H}_{rf}^{m,m+1}\\ &= \begin{pmatrix}\langle m|\hat{H}_\Sigma|m\rangle & 0\\ 0 & \langle m+1|\hat{H}_\Sigma|m+1\rangle\end{pmatrix} + \begin{pmatrix}0 & \dfrac{i}{2}\omega_1 W_m\\ -\dfrac{i}{2}\omega_1 W_m & 0\end{pmatrix}\end{aligned} \tag{5.20}$$

where $\mathbf{H}_{\Sigma}^{m,m+1} = \mathbf{H}_{cs}^{m,m+1} + \mathbf{H}_{Q}^{m,m+1} + \mathbf{H}_{\Delta}^{m,m+1}$ (and similarly $\hat{H}_\Sigma = \hat{H}_{cs} + H_Q^{eff} + \hat{H}_\Delta$). The factor W_m arises from evaluating the $\hat{I}_y$ matrix elements in $\hat{H}_{rf}$ within the $(m, m+1)$ basis and is

$$W_m = [I(I+1) - m(m+1)]^{\frac{1}{2}} \tag{5.21}$$

We are only interested in the off-diagonal elements of the density matrix, as only these give rise to NMR signal. The $\mathbf{H}_{\Sigma}^{m,m+1}$ term in Equation (5.20) is diagonal and so does not contribute to the off-diagonal elements of the density matrix. Only the second term of Equation (5.20), $\mathbf{H}_{rf}^{m,m+1}$, will contribute to off-diagonal elements of the density matrix. This term can then be used as the hamiltonian matrix for the purposes of calculating the $\rho_{m,m+1}$ term of the density matrix via Equation (5.16). This involves diagonalizing $\mathbf{H}_{rf}^{m,m+1}$, but this is straightforward. The final result is [6, 7]:

$$\frac{\rho_{m,m+1}(\tau)}{\left(\dfrac{1}{Z}\dfrac{\hbar\omega_0}{kT}\right)} = i\frac{1}{2}\sin(W_m\omega_1\tau) \tag{5.22}$$

where W_m is defined in Equation (5.21). Thus, the effective rf field strength in this selective excitation is a factor $W_m = [I(I+1) - m(m+1)]^{\frac{1}{2}}$ larger than that for non-selective excitation. Thus the length τ of rf pulse which gives maximum signal intensity, i.e. an effective selective 90° pulse, is in general shorter under selective excitation conditions. It is interesting to note that Equation (5.22) for the density matrix element under the condition of $\omega_Q >> \omega_1$ does not depend on the quadrupole coupling. The main effect of the very strong quadrupole coupling in this case is to ensure that only one transition, $m \rightarrow m + 1$, is excited, in other words to produce the selective excitation condition.

For values of ω_1 intermediate between the limits of selective and non-selective excitation, the density matrix elements, and thus the degree of excitation, in general depend upon both $\omega_1$and ω_Q. The dependence upon ω_Q, which itself depends on molecular/crystallite orientation (Equation (5.10)), means that not only does the degree of excitation depend upon the quadrupole-coupling constant, and so the chemical site of the quadrupolar nucleus, it also depends upon the molecular/crystallite orientation in the applied field. Thus in a powder sample, nuclei in equivalent chemical sites, but in different crystallites, experience (potentially very) different excitation. These two factors result in distorted powder patterns, and intensities which are no longer proportional to the number of spins present.

The ω_Q dependence of the rf excitation for quadrupolar nuclei is a perennial problem which has yet to be satisfactorily solved. The problem diminishes as the length of the rf pulse decreases and its amplitude increases. Short ($\omega_1\tau < \pi/6$ radians), hard pulses are therefore commonly used in NMR studies of quadrupolar nuclei. In situations where the quadrupole coupling is very large, there is little option other than to use continuous wave-type experiments [9].

5.2.3 *The effects of quadrupolar nuclei on the spectra of spin-$\frac{1}{2}$ nuclei*

Quadrupole coupling can have effects beyond those already described for the quadrupolar nucleus itself. A particular example of this which is quite prevalent is the case of observing a spin-$\frac{1}{2}$ nucleus which is dipolar coupled to a quadrupolar nucleus. The analysis of such a system will show the general effect which the quadrupole coupling can have beyond those already shown in the previous analyses. The hamiltonian for a spin system which consists of a spin-$\frac{1}{2}$ nucleus (*I*) dipolar coupled to a quadrupolar nucleus (*S*) in the laboratory frame is

$$\hat{H} = \hat{H}_{0,I} + \hat{H}_{0,S} + \hat{H}_{IS} + \hat{H}_Q \tag{5.23}$$

where $\hat{H}_{IS}$ describes the dipolar coupling between *I* and *S* and can be expressed as

$$\hat{H}_{IS} = \sum_{q=-2}^{+2} (-1)^q \hat{T}_{2q}^{IS} \Lambda_{2-q}^{IS} \tag{5.24}$$

where we have included both secular and non-secular terms. The Λ_{2q}^{IS} tensor components describe the strength and orientation dependence of the dipolar interaction (and are given in Box 4.2) and the $\hat{T}_{2q}^{IS}$ are the spherical tensor operators appropriate to dipolar coupling between spins I and S (see Box 4.2). $\hat{H}_Q$ describes the laboratory frame quadrupole coupling on spin S and is (see Box 5.1):

$$\hat{H}_Q = \sum_{q=-2}^{+2} (-1)^q \hat{T}_{2q}^{Q} \Lambda_{2-q}^{Q} \tag{5.25}$$

We have ignored any chemical shift effects on either spin for clarity.

We will be observing the I spin, and so we need to transform the total hamiltonian for the spin system (Equation (5.23)) into the frame from which we will be observing that spin, i.e. a rotating frame which rotates at the Larmor frequency of spin I for both the I and S spin coordinates. The rotating frame transformation is necessary as that is the frame within which we effectively perform our NMR experiments on the I spin, and so we must work within such a frame if we are to understand the outcome of our experiments. In Box 5.1, we examined the effect of putting a hamiltonian which was a simple sum of a Zeeman term plus a quadrupolar term into a rotating frame. This resulted in a time-dependent rotating frame hamiltonian which we used average hamiltonian theory to deal with.

Performing the required rotating frame transformation on the hamiltonian of Equation (5.23), we obtain as the rotating frame hamiltonian

$$\hat{H}^*(t) = (\omega_{0,S} - \omega_{0,I})\hat{S}_z + \hat{H}_{IS}^*(t) + \hat{H}_Q^*(t) \tag{5.26}$$

where the first term arises from the S spin Zeeman term, the I spin Zeeman term vanishes in the rotating frame transformation and the dipolar and quadrupolar terms become time dependent by virtue of the rotating frame transformation (see Box 5.1 for details). From Equation (5.26) for the rotating frame hamiltonian, it is straightforward to derive the first-order average hamiltonian, using the procedures in Box 5.1. As we will be observing the I spin, we are only interested in terms in the average hamiltonian which can affect the I spin, i.e. contain I spin operators. After some simple algebra, this component is found to be:

$$\overline{H}_I^{(0)} = \overline{H}_{IS}^{(0)} \tag{5.27}$$

where $\overline{H}_{IS}^{(0)}$ is given by the usual expression for the secular part of the heteronuclear dipolar coupling (see Box 4.2):

$$\hat{H}_{IS}^{(0)} = \Lambda_{20}^{IS}\hat{T}_{20}^{IS} \tag{5.28}$$

Thus, to first order, the I spin spectrum is affected only by the I–S dipolar coupling (as we have ignored chemical shift effects). This is perhaps only what we would expect. The interesting part happens when we consider the second-order average hamiltonian. The second-order average hamiltonian is given by

$$\overline{H}^{(1)} = -\frac{i}{2t_\mathrm{p}}\int_0^{t_\mathrm{p}} \mathrm{d}t \int_0^{t} [\hat{H}^*(t), \hat{H}^*(t')]\,\mathrm{d}t' \tag{5.29}$$

where $\hat{H}^*$ is the rotating frame hamiltonian. In Box 5.1, when we only had the quadrupolar term in the rotating frame hamiltonian, the commutators of Equation (5.29) necessarily consisted of products of the quadrupolar hamiltonian with itself. However, in the present case the rotating frame hamiltonian consists of a sum of a dipolar term and a quadrupolar term. Thus, in the second-order average hamiltonian, in addition to the terms arising purely from the quadrupole coupling, there will be cross-terms arising from commutators of the dipolar-coupling operator with the quadrupole-coupling operator. Again, we are only interested in terms in the second-order average hamiltonian which contain I spin operators and so affect the I spin spectrum. Following the same process as in Box 5.1, we find non-zero terms of the form

$$\overline{H}_I^{(1)} = -\frac{i\pi}{2\omega_{0,I}} \sum_{q=-2}^{+2} \Lambda_{2-q}^{\mathrm{Q}}\Lambda_{2q}^{\mathrm{dd}}[\hat{T}_{2q}^{\mathrm{Q}}, \hat{T}_{2-q}^{\mathrm{dd}}] \tag{5.30}$$

in this category. In the full second-order average hamiltonian there will be the usual second-order pure quadrupolar terms (which only affect the quadrupolar S spin of course) and second-order pure dipolar terms which, although in principle affect the I spin, are likely to be very small and so can be ignored, as usual, for the dipolar coupling. As in the case of pure quadrupole coupling in Box 5.1, the products $\Lambda_{2-q}^{\mathrm{Q}}\Lambda_{2q}^{\mathrm{dd}}$ in Equation (5.30) can be written as linear combinations of new spherical tensors of rank 4, 2 and 0 and order zero.

Thus, the spatial orientation dependence of the rotating frame hamiltonian for a spin-$\frac{1}{2}$ nucleus dipolar coupled to a quadrupolar nucleus to second order is expressed by spherical tensors of rank 0, 2 and 4, which have magnitudes of the order of $d \cdot (e^2qQ/4I(2I-1))/(\omega_{0,I})$ where d is the dipolar coupling constant for dipolar coupling between spins I and S. The dipolar coupling may be small, but if the quadrupole coupling is large, this dependence is distinctly non-negligible.

The relevance of this is that (a) there is an isotropic term (the rank 0 spherical tensor component) arising in the spin-$\frac{1}{2}$ spectrum by virtue of its dipolar coupling to the quadrupolar nucleus and (b) while the effects of the rank 2 spatial orientation dependence in the second-order average hamiltonian can be removed by magic-angle spinning, that of the rank 4 component cannot. Thus a molecular orientation dependence will remain in the spectrum even under magic-angle spinning. This will cause the formation of powder patterns with widths of the order of $d \cdot (e^2qQ/4I(2I - 1))/(\omega_{0,I})$ (albeit scaled by the magic-angle spinning averaging of the rank 4 term in the second-order average hamiltonian).

This is the cause of the apparent splitting of lines in ^{13}C spectra of ^{13}C spins close to ^{14}N spins. ^{14}N ($S = 1$) often suffers very large quadrupole couplings (e.g. χ is around 3.2 MHz for amide nitrogens in the main chains of peptides) and the dipolar coupling of such spins to ^{13}C can have a noticeable effect on the ^{13}C spectrum [10].

5.3 High-resolution NMR experiments for half-integer quadrupolar nuclei

Up until now in this chapter, we have been dealing with a general quadrupolar spin I. Now we make the distinction between integer and half-integer quadrupolar spins. Actually, the vast majority of quadrupolar spins are half integer. Of the integer quadrupolar spins, ^2H ($I = 1$) is the only one of major interest to chemists. This nucleus has a relatively small quadrupole moment, giving quadrupole coupling constants around 160–190 kHz for ^2H in organic compounds, for instance. It has been used extensively in studies of molecular dynamics and is discussed in detail in Chapter 6. ^{14}N ($I = 1$) is also potentially of interest to chemists. However, it has a sizeable electric quadrupole moment resulting in quadrupole-coupling constants of MHz in many cases. Clearly such huge linebroadening does not make this nucleus amenable to study and, as yet, no good high-resolution techniques are available for spin-1 nuclei. However, its effects are used indirectly in the TRAPDOR and REAPDOR experiments described in Section 4.5. ^6Li ($I = 1$) has a very small value of Q and consequently behaves more or less as a spin-$\frac{1}{2}$ because the quadrupole splittings are not resolvable. A high-resolution Li spectrum may thus be obtained using ^6Li rather than ^7Li, despite the very small chemical shift range of Li.

In contrast, there are now four feasible experiments for producing high-resolution spectra of half-integer quadrupolar nuclei. In addition to this, the central transition ($m = +\frac{1}{2} \rightarrow m = -\frac{1}{2}$) of half-integer quadrupolar nuclei is not broadened to first order by quadrupole coupling (although it is to second order). The satellite transitions are, however, and this can render them largely unobservable in NMR experiments at the Larmor frequency, as they are too far off resonance. The lack of first-order broadening on the central transition makes it significantly easier to observe than the satellites, and so in this section we focus exclusively on the central transition. Although not broadened to first order, there still remain second- and higher-order contributions to the frequency of this transition which depend on molecular orientation and therefore cause linebroadening in powder samples, albeit rather less than would arise from first-order coupling. We shall only be dealing with powder samples in the subsequent discussion.

5.3.1 *Magic-angle spinning (MAS)*

Magic-angle spinning (MAS) in an invaluable linenarrowing technique in solid-state NMR. As seen in Section 2.2, for inhomogeneous interactions whose molecular orientation dependence depends only on second-rank rotation matrix terms, the anisotropic parts of the interaction are averaged to zero, providing the rate of sample spinning is significantly larger than the anisotropic linewidth. Thus magic-angle spinning will average the first-order quadrupole broadening of satellite transitions, but, for most samples, the rate of spinning cannot be fast enough for complete averaging and a large array of spinning sidebands will result. The central transition is unaffected by quadrupole coupling to first order of course, but is still affected to second order. Magic-angle spinning has a linenarrowing effect on the central transition lineshape of half-integer quadrupolar nuclei; however, complete removal of the anisotropic quadrupolar effects by this technique is not possible. As Equation (xxx) in Box 5.1 shows, the second-order effects of quadrupole coupling depend on second- *and* fourth-rank rotation matrices. The former may be averaged to zero by sufficiently fast magic-angle spinning. The latter are averaged by magic-angle spinning too, but not to zero. Thus an anisotropic powder pattern remains, even under conditions of very rapid magic-angle spinning. Figure 5.4 shows the effect of magic-angle spinning on central transitions of half-integer spins for different asymmetry parameters, η_Q.

The frequency of the central transition to second order under magic-angle spinning is easily derived from Equation (xxx) in Box 5.1:

$$\omega_{\frac{1}{2},-\frac{1}{2}} = -\left(\frac{e^2qQ}{4I(2I-1)}\right)^2 \frac{1}{\omega_0}(3-4I(I+1))\left\{\frac{1}{2}V_{00}^Q + 4V_{20}^Q + 9V_{40}^Q\right\} \quad (5.31)$$

where the V_{k0}^Q parameters are given in Equation (xxv) in Box 5.1. The V_{k0}^Q parameters depend on rotation matrices of rank k. Under the conditions of

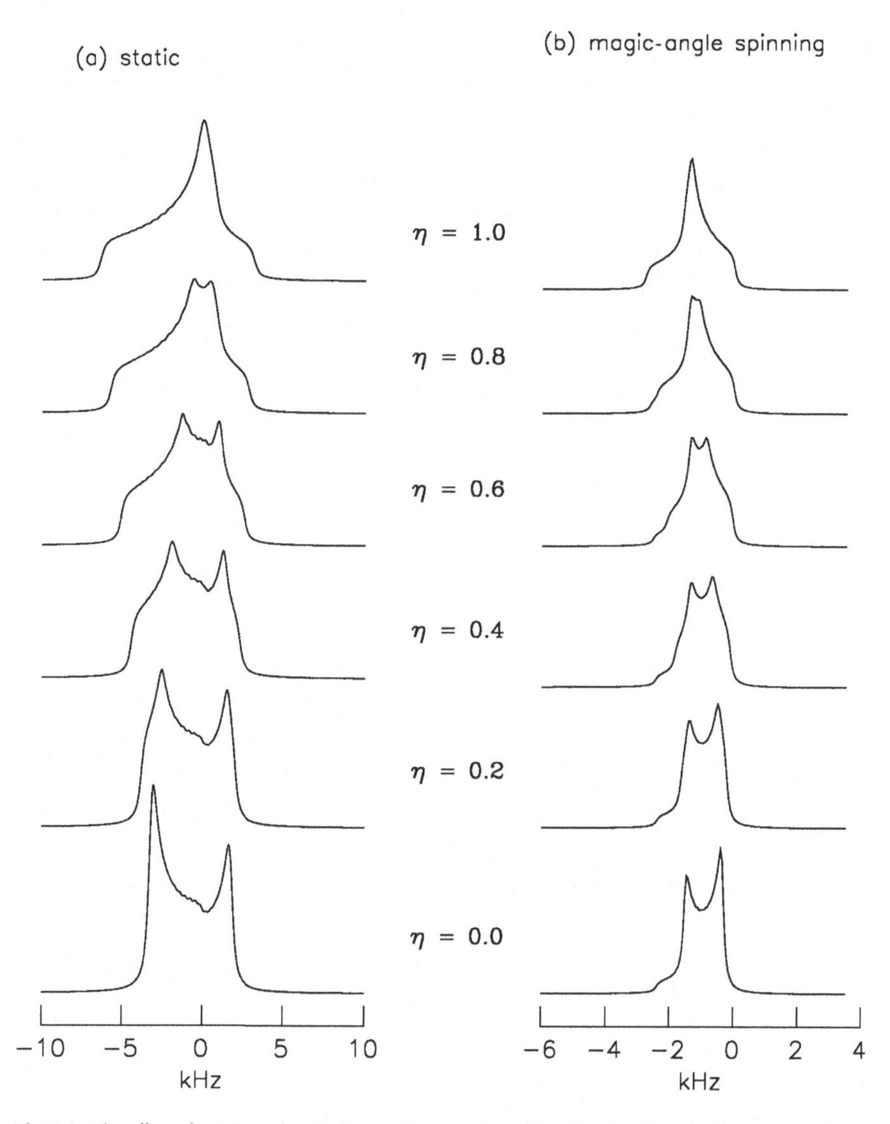

Fig. 5.4 The effect of magic-angle spinning on the central transition lineshape for a half-integer quadrupolar nucleus. (a) Central transition lineshape under static, i.e non-spinning conditions. (b) The same transition under magic-angle spinning. The spectra are plotted according to a normalizing parameter y where $y = \omega_{\frac{1}{2}}^{(2)} \Big/ (\omega_q^2/6\omega_0)\left(I(I+1) - \frac{3}{4}\right)$ and $\omega_q = (3e^2qQ)/2I(2I-1)$; $\omega_{\frac{1}{2}}^{(2)}$ is the central transition frequency to second order as calculated from Equation (5.9) for static conditions and from Equation (5.31) for MAS conditions. The asymmetry parameter is given with each spectrum. The excitation pulses producing each spectrum are assumed to be ideal, i.e. $\omega_1 >> \omega_q$. The spinning rate in the magic-angle spinning spectra is assumed to be much larger than ω_q.

very rapid sample spinning, only the D_{00}^{k} (α, β, γ) terms involved in the respective V_{k0}^{Q} parameters survive.

$$D_{00}^{0}(\alpha, \beta, \gamma) = 1 \tag{5.32}$$

$$D_{00}^{2}(\alpha, \beta, \gamma) = \frac{1}{2}(3\cos^2\beta - 1) \tag{5.33}$$

$$D^4_{00}(\alpha, \beta, \gamma) = \frac{1}{8}(35\cos^4\beta - 30\cos^2\beta + 3) \tag{5.34}$$

For magic-angle spinning, i.e. $\theta_R = 54.74°$, the D^2_{00} term averages to zero over one rotor cycle, removing the V^Q_{20} term in the transition frequency (Equation (5.31)) as expected for a second-rank term under magic-angle spinning. Figure 5.5 shows the variation with angle of the terms D^2_{00} and D^4_{00}, which are second- and fourth-order Legendre polynomials in $\cos\beta$, respectively. At the magic angle the D^4_{00} term is not zero but becomes a scaling factor (–0.389) and, thus, the anisotropic V^Q_{40} term, although reduced by magic-angle spinning, remains. In addition, there is an isotropic shift $\omega^{iso}_{\frac{1}{2},-\frac{1}{2}}$ (arising from the V^Q_{00} term) of:

$$\omega^{iso}_{\frac{1}{2},-\frac{1}{2}} = \left(\frac{e^2qQ}{4I(2I-1)}\right)^2 \frac{1}{10\omega_0}(3 - 4I(I+1))\{3 + \eta^2\} \tag{5.35}$$

This shift is in addition to any chemical shift offset there might be and has no dependence on spinning rate, i.e. it also exists for static samples. It is important to bear in mind that the centre of gravity of the central transition powder patterns is the chemical shift *plus* the isotropic quadrupolar shift as defined in Equation (5.35). The quadrupolar shift, coupled with the orientation-dependent linebroadening of the central transition, can make it very difficult to determine the true chemical shift accurately. One way around this is to use *satellite transition spectroscopy* [11]. This is particularly useful for higher spin numbers such as $\frac{5}{2}, \frac{7}{2}$, etc. In satellite transition

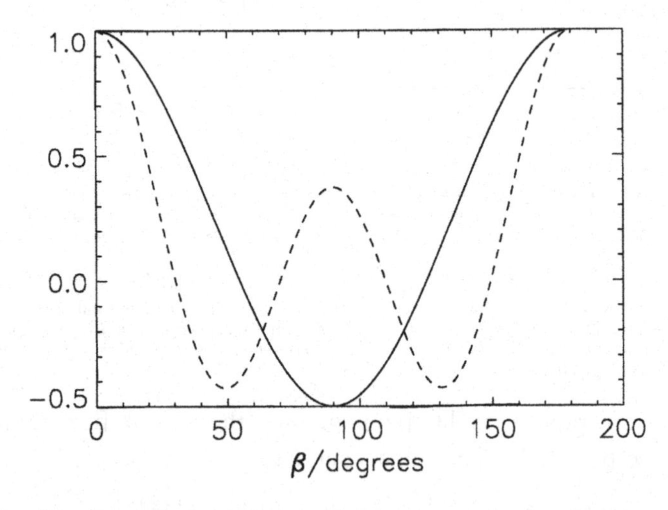

Fig. 5.5 Plots of the D^2_{00} (solid line) and D^4_{00} (dashed line) functions as a function of β. See Equations (5.33) and (5.34) for definitions of these functions.

spectroscopy [11], one simply observes the satellite transitions under magic-angle spinning. Under these conditions, satellite transition lineshapes (usually unobserved) are broken up into a series of spinning sidebands, albeit of relatively low intensity. However, it is much easier to find the centre of gravity of a satellite transition under magic-angle spinning than that of the central transition; it is simply the average of the positions of the +*N*- and −*N*-order spinning sidebands. Providing the quadrupole parameters are known, and the satellite transition correctly assigned, the isotropic quadrupolar shift associated with the transition can be calculated from the first term of Equation (5.9) and hence the chemical shift extracted from the satellite transition centre of gravity. When satellite transitions are observed, there is also a smaller V_{40}^{Q} associated with the $m = \pm\frac{1}{2} \rightarrow \pm\frac{3}{2}$ and other satellite transitions than with the central transition. This results in an improvement in the resolution of overlapping V_{40}^{Q} powder patterns in satellite bands [12] as shown in Fig. 5.6.

5.3.2 Double rotation (DOR)

As already explained, second-order quadrupole effects mean that the frequency of the central transition is molecular-orientation-dependent, being determined by both second- and fourth-rank rotation matrices, which express the molecular frame/PAF orientation with respect to the applied field. The double-rotation technique [13, 14] achieves high-resolution spectra of the central transitions by spinning the sample simultaneously at two angles. One angle is the magic angle, and so averages the second-rank rotation matrix terms in Equation (5.31) to zero, while the other angle (30.6° or 70.1°) averages the fourth-rank terms to zero. This latter point may be determined from Fig. 5.5 or by putting $\langle\beta\rangle = 30.6°$ or 70.1° in Equation (5.34) for $D_{00}^{4}(\alpha, \beta, \gamma)$.

This conceptually simple technique is far from simple to implement, however, and involves one rotor spinning inside another (Fig. 5.7) [13, 14]. In order to achieve spectra free from spinning sidebands, it is necessary to spin each rotor at rates of the order of the anisotropic linewidth. Unfortunately this is rarely achievable and, in some cases, spinning sidebands may obscure isotropic resonances. Nevertheless, the technique has been used with a great deal of success and is likely to find many further applications as NMR experiments are performed at higher fields. At higher fields, the anisotropic quadrupolar linewidths are reduced (due to the dependence of the transition frequency on $1/\omega_0$), and so spinning sidebands are reduced in intensity for the same spinning rates. An example of DOR at high field is shown in Fig. 5.8 [15].

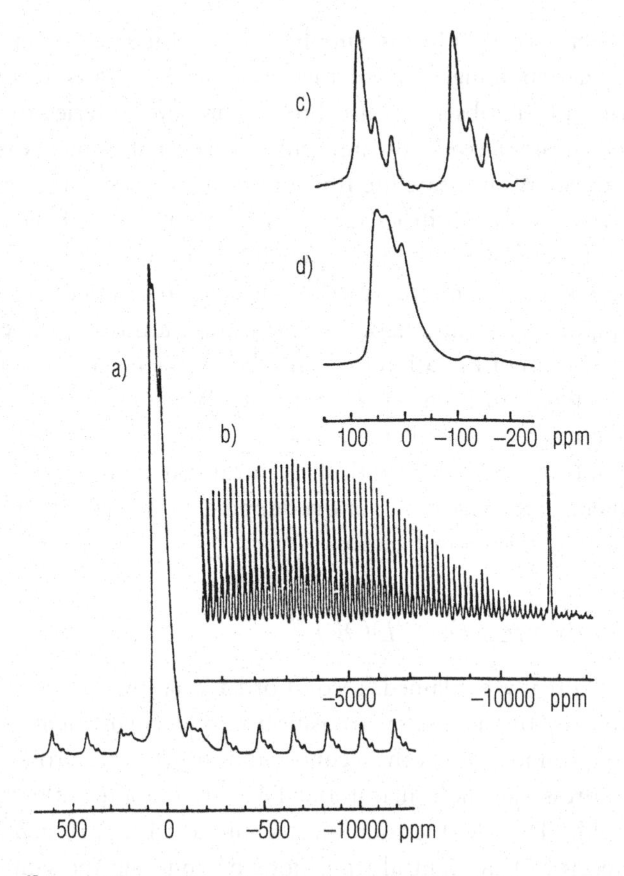

Fig. 5.6 (a) The ^{27}Al (I = 5/2) central transitions and part of the satellite transition manifold of a lead aluminoborate glass [12]. (b) The right-hand part of the total satellite transition spinning sideband spectrum. (c) The third and fourth satellite transition sidebands, showing resolution of three signals corresponding to the four-, five- and six-coordinate aluminium sites. (d) The central transition centrebands, showing poor resolution of the three aluminium sites. Second-order broadening on the $\pm 1/2 \leftrightarrow \pm 3/2$ transition for $I = \frac{5}{2}$ is $\frac{1}{8}$ that of the central $\frac{1}{2} \leftrightarrow -\frac{1}{2}$ transition. Therefore, there is much higher site resolution in the satellite transition sidebands than in the central transition centrebands.

5.3.3 Dynamic-angle spinning (DAS)

Dynamic-Angle Spinning (DAS) [16] is a two-dimensional NMR experiment which achieves isotropic signals in one dimension and quadrupolar-broadened, anisotropic powder patterns in the other.

The experiment takes advantage of the fact that the transition frequency (of the central transition and others) depends on the sample spinning angle, θ_R (Equation (5.31)), to perform a type of refocusing or echo experiment. Evolution under the quadrupolar broadening during a first period of free precession is refocused during a second period by changing the transition or evolution frequency between the two periods, via a change in sample spinning angle.

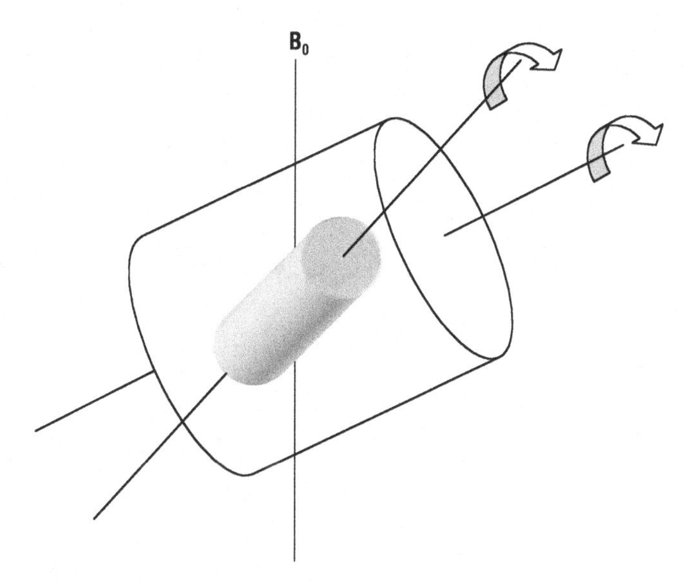

Fig. 5.7 Schematic diagram of the DOuble Rotation (DOR) experiment. There are two rotors, one spinning inside the other. The outer rotor spins about an axis inclined at 54.74° to the applied field $\mathbf{B}_0$, while the inner one spins about an axis inclined at 30.6° (or 70.1°).

In detail (Fig. 5.9), initial coherence corresponding to the central transition is generated by a selective 90° pulse while the sample is spinning at the first angle, θ_1. Evolution under the quadrupolar broadening, modified by spinning at θ_1, is then allowed to occur for a period t_1. At the end of t_1, the remaining coherence is stored along the applied magnetic field (using a selective 90° pulse) while the spinning angle is changed to θ_2. The central transition coherence is then regenerated (via another selective 90° pulse) and allowed to evolve for a further time kt_1 under the quadrupole coupling while spinning at the new angle. The constant k is set (by prior calculation) so that the quadrupolar evolution in the first period t_1 is exactly undone by the subsequent evolution in kt_1. At the end of kt_1, a free induction decay is recorded during t_2 in the normal way. A two-dimensional dataset is collected by incrementing t_1; subsequent Fourier transformation in both dimensions yields in f_1 (corresponding to the transformed t_1 dimension) an isotropic spectrum from which the anisotropic effects of the quadrupole coupling have been removed. As with DOR, the isotropic quadrupolar shift (Equation (5.35)) remains. Further discussion on recording and processing such spectra is given later in Section 5.3.6.

There are many pairs of spinning angles [16] which allow this refocusing effect; for the combination $\theta_1 = 37.38°$, $\theta_2 = 79.19°$, $k = 1$, so that the refocusing period is equal in length to the initial evolution period.

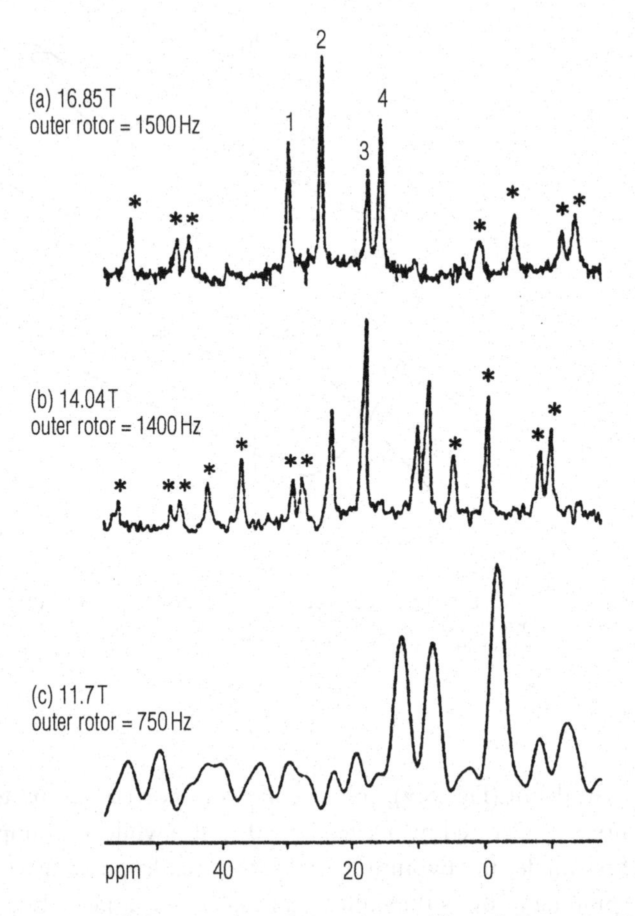

Fig. 5.8 ^{17}O DOR spectra of siliceous zeolite Y (faujasite) [15] at different field strengths. Four ^{17}O signals are clearly resolved at the higher field strengths. Spinning sidebands are marked with an asterisk. The spinning rate of the outer rotor in the experiments is given with each spectrum. The spectra are referenced with respect to ^{17}O in H_2O (at 0 ppm).

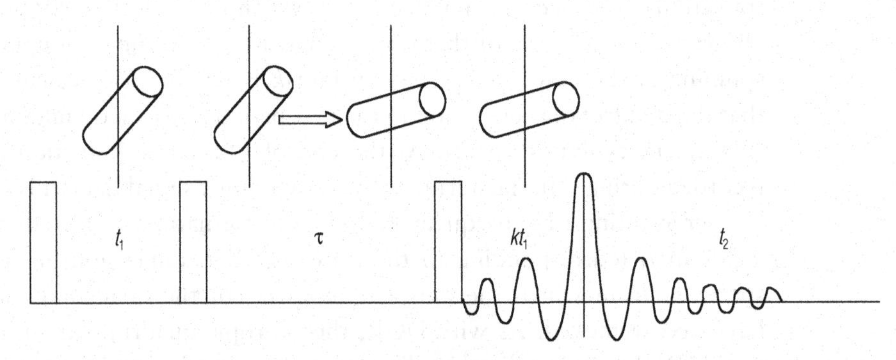

Fig. 5.9 Schematic illustration of the Dynamic Angle Spinning (DAS) experiment. All the pulses are selective 90° pulses for the central transition of the quadrupolar spin in question. The sample is spun at angle θ_1 during t_1. At the end of t_1 a selective 90° pulse is used to store remaining magnetization along $\mathbf{B_0}$. During the subsequent τ delay, the spinning angle is changed to θ_2. Another 90° pulse restores the magnetization to the transverse plane. The evolution under the (second-order) quadrupole coupling during t_1 is refocused after a period kt_1 at the second spinning angle and an FID is then recorded in t_2. When $\theta_1 = 37.4°$ and $\theta_2 = 79.2°$, the constant $k = 1$. In practice, the selective 90° pulses are calibrated at each spinning angle required.

The advantage of this experiment over DOR is that only one spinning angle is required at a time, and so is easier to implement in one sense. Furthermore, there is no difficulty in obtaining rapid spinning rates, so as to reduce the intensity of spinning sidebands. There is, however, the difficulty of changing spinning angle during the experiment; the change must be perfomed in a time which is much shorter than the spin-lattice relaxation time for the signal or the stored coherence relaxes back to equilibrium during the angle change. In addition, the storage step while the angle is changed results in the loss of half the signal compared with a single-pulse experiment, because only one component of the transverse magnetization (x or y) can be stored.

A further disadvantage compared to DOR is the fact that none of the DAS spinning angle pairs include the magic angle. Interactions such as the chemical shift anisotropy and dipole–dipole coupling are scaled at the DAS angles but *not* averaged to zero and so their effects remain in both dimensions of the two-dimensional spectrum. The 'best' angle pair in this respect is the $k = 5$ pair of 0°/63° with spectra acquired close to the magic angle at 63°. DAS spectra may be acquired without contributions from dipole–dipole coupling or chemical shift anisotropy by the addition of an extra 'hop' to the magic angle before acquisition [17]. However, this results in a further loss of signal and the experiment will take 8 times as long to achieve signal-to-noise equivalent to a single pulse experiment. Figure 5.10(a) shows the overlapping magic-angle spinning pattern obtained for the ^{17}O spectrum of the SiO_2 polymorph coesite (5 oxygen sites) [17]. The advantage of the separation of the isotropic and anisotropic contributions to the spectrum afforded by DAS (Fig. 5.10(b)) is very clear [17]. This two-dimensional spectrum was acquired with a final hop to the magic angle so that the anisotropic lineshapes had no chemical shift anisotropy or dipole–dipole contributions.

5.3.4 Multiple-quantum magic-angle spinning (MQMAS)

The MQMAS experiment was first proposed in the literature in 1995 [18] for achieving high-resolution spectra for half-integer quadrupolar nuclei while spinning the sample at just one angle, the magic angle, throughout the experiment. This is much easier to implement than DOR or DAS and is achievable on any modern NMR spectrometer. Since 1995, this experiment has become the primary method by which high-resolution spectra of quadrupolar nuclei are obtained. Indeed, the very existence of this experiment has encouraged the recording of NMR experiments involving quadrupolar nuclei, a previously somewhat neglected area. It is a two-dimensional experiment and achieves resolution in much the same way as the DAS experiment, via a refocusing of evolution during the t_1 period in a

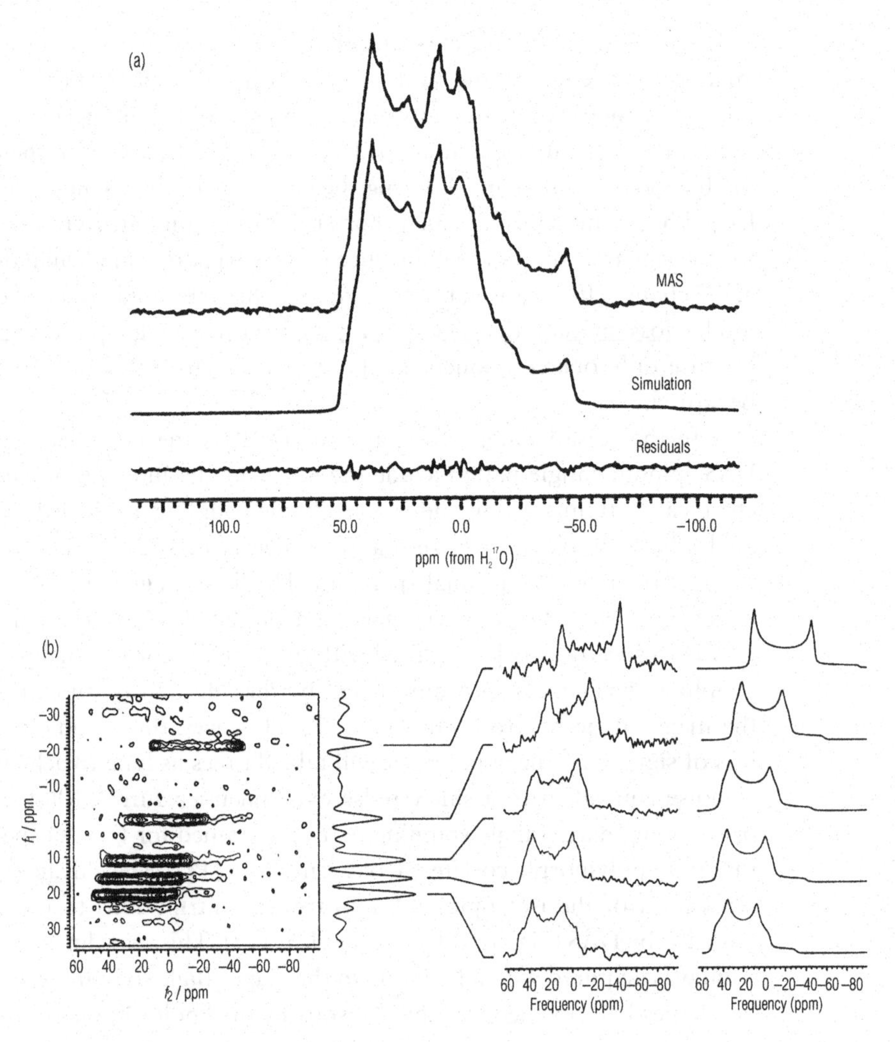

Fig. 5.10 (a) The ^{17}O magic-angle spinning spectrum of the silicate, coesite [17]. (b) The two-dimensional ^{17}O DAS spectrum of coesite [17]. The right-hand side shows f_2 slices through the isotropic signals of the f_1 dimension. All spectra are recorded at 11.7 T.

second period of free precession, kt_1. In order to do this, a change of transition frequency is needed between the two periods, which in the DAS experiment is done by changing the sample spinning angle. In the MQMAS experiment (Fig. 5.11), it is done by changing the *order* of the evolving coherence. In particular, a multiple-quantum coherence corresponding to a symmetric $+m \rightarrow -m$ transition evolves during t_1. This is then converted to (observable) single-quantum coherence, which evolves during kt_1.

The frequency of a $+m \rightarrow -m$ transition for a half-integer quadrupolar nucleus depends only on second-order terms and is given by (from Equation (xxi) in Box 5.1):

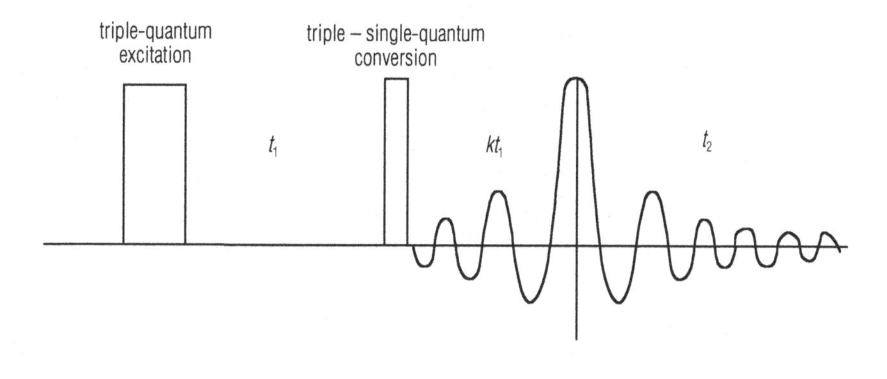

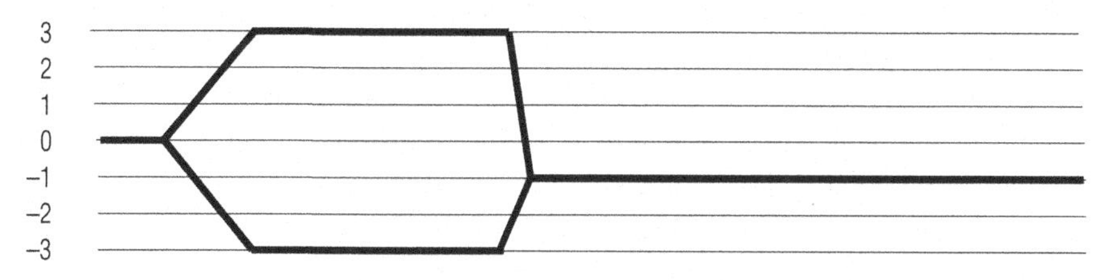

Fig. 5.11 Schematic illustration of the multiple-quantum magic-angle spinning (MQMAS) experiment for a spin-$\frac{3}{2}$. Both excitation and transfer pulses are optimized for each sample. The sample is spun at the magic angle throughout the experiment to remove the effects of second-rank terms of the second-order quadrupole coupling (and chemical shift anisotropy and dipolar coupling). Triple-quantum coherence is excited initially and allowed to evolve during t_1. The remaining coherence is then transferred to single-quantum coherence. After a period kt_1 ($k = \frac{7}{9}$ for spin-$\frac{3}{2}$), evolution under the fourth-rank quadrupolar terms in t_1 is refocused and an FID is recorded in t_2. For half-integer spins with $I > \frac{3}{2}$ other multiple-quantum coherences associated with $+m \rightarrow -m$ transitions are also available for use in this experiment.

$$\omega_{m,-m} = \left(\frac{e^2qQ}{4I(2I-1)}\right)^2 \frac{4}{\omega_0} m\{[I(I+1) - 3m^2]V_{00}^{Q} + [8I(I+1) - 12m^2 - 3]V_{20}^{Q} + [18I(I+1) - 34m^2 - 5]V_{40}^{Q}\} \quad (5.36)$$

where the kth rank V_{k0}^{Q} tensor components depend on kth rank rotation matrices (see Box 5.1). In the MQMAS experiment, multiple-quantum coherence of order $2m$ is first excited and allowed to evolve during t_1 at a rate given by Equation (5.36). The experiment is conducted under magic-angle spinning which may be assumed to average to zero the second-rank terms of V_{20}^{Q} in this equation throughout the experiment. Thus, the only anisotropy in the evolution during t_1 arises from the V_{40}^{Q} term. At the end of t_1, a second rf pulse transfers the remaining multiple-quantum coherence into single-quantum (–1) coherence associated with the $+\frac{1}{2} \rightarrow -\frac{1}{2}$ transition. The new coherence has a frequency given by Equation (5.36) with $m = \frac{1}{2}$ (with V_{20}^{Q} terms again averaged to zero by magic-angle spinning). The important point is that the evolution of the multiple-quantum coherence under the V_{40}^{Q} term during t_1 is now 'undone' by the evolution of the single-quantum coherence during kt_1 under the V_{40}^{Q} term associated with the central transition second-order quadrupole broadening. The value of k is given by the ratio of the coefficients of the V_{40}^{Q} terms in the second-order quadrupole-coupling term in the frequency of the multiple-quantum coherence ($+m \rightarrow -m$) and the central transition, $R_4(I, m)$, i.e. the ratio of $[18I(I + 1) - 34m^2 - 5]m$ with the appropriate m for the multiple-quantum transition, and $m = \frac{1}{2}$ (see Equation (5.36)). At the end of kt_1, a normal FID is collected. The values of $R_4(I, m)$ appropriate for different spin quantum numbers I and m are given in Table 5.1. A two-dimensional dataset is collected in the same way as for DAS; in the MQMAS data, an echo analogous to the DAS echo is seen to shift in time with increasing t_1 values. Fourier transformation results in a two-dimensional frequency spectrum in which the f_1 dimension exhibits an isotropic spectrum and the f_2 dimension, the anisotropic powder patterns associated with the central transitions (corresponding to the single-quantum coherence which evolves during t_2) of the different sites. The isotropic f_1 dimension exhibits signals at frequencies which are a combination of the single- and multiple-quantum isotropic shifts:

$$\omega_1^{\text{iso}} = \frac{1}{k+1}(2m\omega_{\text{iso}} + \omega_{m,-m}^{\text{iso}}) + \frac{k}{k+1}\left(\omega_{\text{iso}} + \omega_{\frac{1}{2},-\frac{1}{2}}^{\text{iso}}\right) \qquad (5.37)$$

where ω_{iso} is the isotropic chemical shift, $\omega_{\frac{1}{2},-\frac{1}{2}}^{\text{iso}}$ is given by Equation (5.35) and

$$\omega_{m,-m}^{\text{iso}} = -\left(\frac{e^2qQ}{4I(2I-1)}\right)^2 \frac{4}{5\omega_0} m(I(I+1) - 3m^2)(3 + \eta_Q^2) \qquad (5.38)$$

Table 5.1 The $R_4(I, n)$ values for the MQMAS experiment for different spin quantum numbers, I, and different multiple-quantum transitions, $n = 2m$, where m is the quantum number defining the initial level in the multiple-quantum transition.

I	n	$R_4(I, n)$
$\frac{3}{2}$	3	−7/9
$\frac{5}{2}$	3	12/19
	5	−12/25
$\frac{7}{2}$	3	101/45
	5	11/9
	7	−161/45
$\frac{9}{2}$	3	91/36
	5	95/36
	7	7/18
	9	−31/6

The multiple-quantum coherence that evolves during t_1 can be excited in the conventional way for exciting multiple-quantum coherence in solution-state experiments with a 90°–τ–90° sequence, or with a single pulse [18–21]. Not surprisingly, the amplitude of multiple-quantum coherence generated depends on the strength of the quadrupole coupling. Vega and Naor [21] have shown that the amplitude of triple-quantum coherence for a spin-$\frac{3}{2}$ nucleus generated by a single pulse of length τ_{rf}, whose amplitude ω_1 is much less than the quadrupole splitting ω_Q (defined in Equation (5.10)), is

$$c_{1,4} = \frac{3}{2}\frac{\omega_0}{kT}\sin\left(-\frac{3\omega_1^3}{8\omega_Q^2}\tau_{rf}\right) \tag{5.39}$$

where $c_{1,4}$ is the amplitude of the (1, 4) element of the density matrix (see Section 1.3), which corresponds to triple-quantum coherence for a spin-$\frac{3}{2}$ when the density matrix is expressed in the Zeeman basis. The phase of this coherence is –90° out of phase with the pulse phase.

Practical considerations

Since the first paper describing the MQMAS experiment, there have been numerous papers describing modifications of the experiment [22–35]. Most of these have concentrated on the optimal conditions for excitation of the multiple-quantum coherence and on its subsequent transformation to –1-quantum coherence.

The multiple-quantum coherence is selected via phase cycling as described in Section 1.5.2. To select $\pm p$ quantum coherence, we need to phase cycle the excitation sequence (whatever it is) in $|p|$ steps of $360/|p|$ degrees (see Section 1.5.2). Specific phase cycling schemes for MQMAS experiments are given in the next section.

- *Excitation conditions* If using a single excitation pulse, it has been shown from simulations that the optimal pulse length is around $0.8/\nu_1$ where ν_1 is the amplitude of the excitation pulse (in Hz) [19, 25]. However, multiples of 360° pulses are often used, as these generate very little of the unwanted single-quantum coherence. If using a 90°–τ–90° sequence for excitation, the optimal conditions are for the two 90° pulses to differ in phase by 90° and for the τ delay to be $n\tau_R$, $n = 0, 1, 2, \ldots$ with τ_R being the rotor period [20]. Whatever the scheme used, generally speaking, the higher the pulse power, the more efficient the excitation, at least until the pulse amplitude (in Hz) approaches the quadrupole coupling constant in order of magnitude [25]. It is desirable to optimize the pulse lengths for the excitation on each sample to generate the maximum signal. It is worth noting that the excitation efficiency is highly dependent on the strength of quadrupole coupling. This means that the excitation efficiency is likely to vary considerably for different crystallite orientations. Moreover, there will be very large differences in excitation efficiency between different chemical sites with different quadrupole coupling constants; sometimes these differences can be so profound that unless the excitation conditions are carefully chosen, some sites will not be represented at all in the final spectrum. In any case, the intensities will not be quantitative, and if information on site populations is required, it has to be obtained by simulation.

- *MQ → SQ transfer* It has been shown, again by simulation, that the optimal pulse length for the MQ → SQ transfer is around $0.2/\nu_1$ [19, 25]. As with the excitation pulse, the higher the pulse power, the more efficient the transfer, up to the limit of the pulse amplitude being of the order of the quadrupole-coupling constant. Fast amplitude modulated pulses have proved very successful for the MQ → SQ conversion also [35].

For both the excitation and transfer steps, adiabatic transfer methods have been used, with very good results [28]. These require a little more care in the set-up of the experiment, but have the advantage of giving quadrupolar powder patterns in the ω_2 dimension which closely approximate the lineshapes that would arise in a simple one-dimensional experiment. For other excitation and conversion schemes, the lineshapes arising in ω_2 can be very distorted relative to those from a one-dimensional experiment. Furthermore, adiabatic methods are less sensitive to the quadrupole-coupling strength, so while intensites arising in the experiments are still not exactly quantitative, they are considerably more so than with other excitation schemes.

Finally, it is clear from the foregoing discussion that both the excitation and transfer steps in the MQMAS experiment have efficiencies which depend on the quadrupole splitting, ω_Q and hence on molecular orientation. This feature can cause the appearance of spinning sidebands in the f_1 dimension of the final two-dimensional MQMAS spectrum [36]. The extent of the spinning sidebands depends on the orientation dependence of the excitation and transfer steps, but can be considerable. In addition, dipolar coupling and chemical shift anisotropy effects are magnified in this dimension [37] and exacerbate the effect. For these reasons, it is often desirable to record MQMAS spectra with t_1 incremented in integer rotor periods (or half rotor periods) to avoid spinning sidebands cluttering the f_1 dimension.

5.3.5 Satellite transition magic-angle spinning (STMAS)

Satellite transition MAS (STMAS) was invented by Gan [38] in 2000 and follows much the same principles as MQMAS, except that rather than correlating a multiple-quantum transition with the central transition, STMAS correlates a satellite transition, i.e. $m = \pm n - \frac{1}{2} \leftrightarrow m = \pm n + \frac{1}{2}$, $n = 1, 2, \ldots (I - \frac{1}{2})$ with the central transition in a two-dimensional experiment.

The simplest pulse sequences for STMAS are shown in Fig. 5.12. The desired satellite transition is excited initially and the coherence due to this evolves in t_1. Satellite transitions are broadened by first- and second-order quadrupole-broadening. The transition frequency for a satellite transition from $m \rightarrow m + 1$ can be found from Equations (xxix) and (xxx) in Box 5.1 to be:

$$\omega_{m,\,m+1} = \omega^{(1)}_{m,\,m+1} + \omega^{(2)}_{m,\,m+1} \tag{5.40}$$

where the first- and second-order contributions, $\omega^{(1)}_{m,m+1}$ and $\omega^{(2)}_{m,m+1}$ respectively are given by

$$\omega^{(1)}_{m,\,m+1} = \frac{3e^2qQ}{4I(2I-1)}(2m+1)W^{Q}_{20} \tag{5.41}$$

$$\begin{aligned}\omega^{(2)}_{m,\,m+1} = -\left(\frac{e^2qQ}{4I(2I-1)}\right)^2 \frac{2}{\omega_0}\{&[I(I+1) - 9m(m+1) - 3]\,V^{Q}_{00} \\ &+ [8I(I+1) - 36m(m+1) - 15]\,V^{Q}_{20} \\ &+ 3[6I(I+1) - 34m(m+1) - 13]\,V^{Q}_{40}\}\end{aligned} \tag{5.42}$$

where the kth rank V^{Q}_{k0} tensor components depend on kth rank rotation matrices (see Box 5.1). In STMAS, the first-order quadrupole broadening of the satellite transition coherence evolution is removed by magic-angle spinning. However, the first-order broadening is in general so large (of the order of MHz for most samples for which STMAS is likely to be applied), that it is necessary to synchronize the t_1 data collection with the sample spinning to avoid what would otherwise be huge numbers of spinning sidebands in the associated ω_1 dimension of the final two-dimensional experiment. Thus, the t_1 period must be incremented by the rotor period, τ_R, between successive experiments in the collection of the two-dimensional dataset.

The second-order quadrupolar broadening of the satellite transition (as for other transitions) is a sum of zeroth-, second- and fourth-rank terms (Equation (5.42)), depending respectively on zeroth-, second- and fourth-rank rotation matrices. The magic-angle spinning used in the experiment also has the effect of averaging to zero the second-rank terms governing the second-order quadrupole-coupling effects in the satellite transition, so that the evolution of the satellite transition coherence proceeds simply under the effects of the fourth-rank terms in the second-order quadrupole broadening and, of course, the isotropic terms associated with the second-order quadrupole coupling and chemical shift.

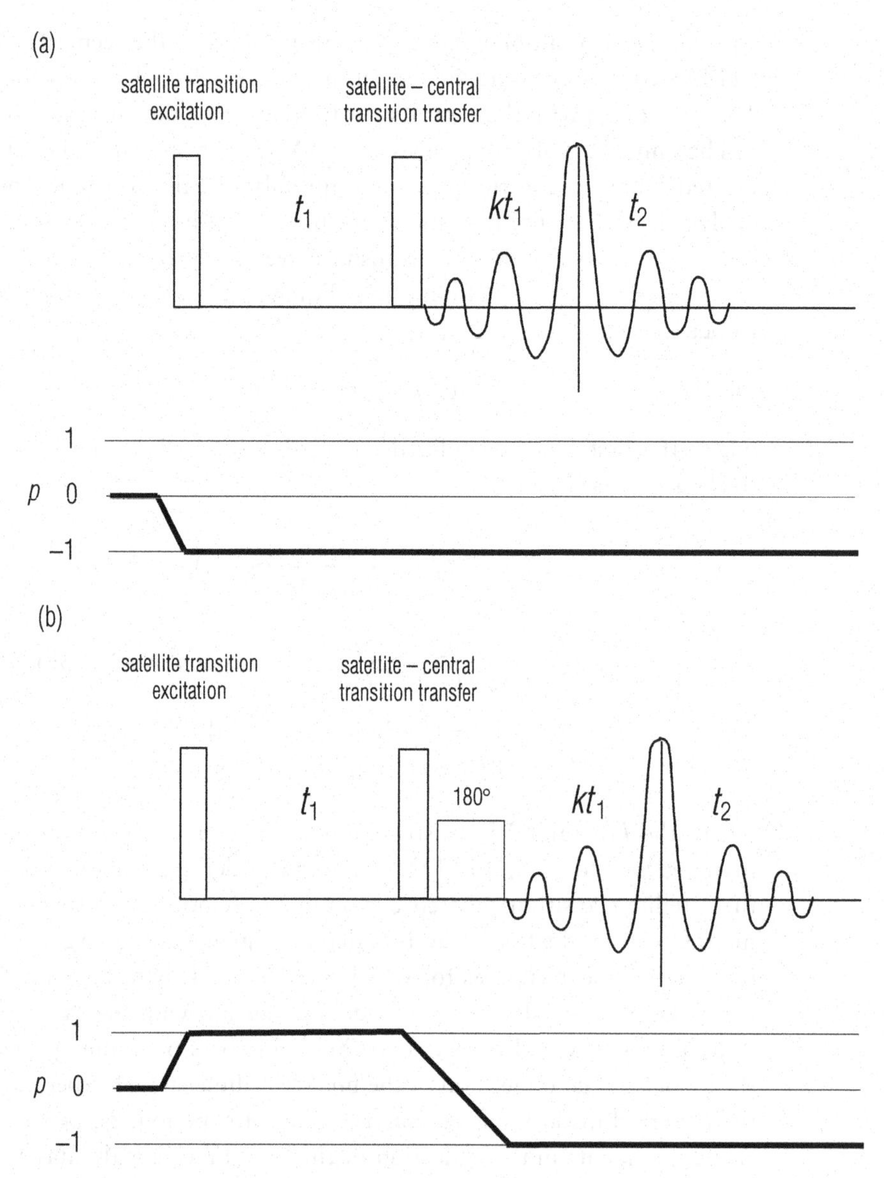

Fig. 5.12 The simplest pulse sequences which achieve satellite–central transition correlation – the STMAS experiment [38, 40]. Sequence (a) is used where the ratio between the fourth-rank terms governing the second-order quadrupole coupling of the satellite transition and central transition is negative. Sequence (b) is used where this ratio is positive. The third pulse in this sequence is a selective 180° pulse. As with MQMAS and DAS, two two-dimensional datasets must be collected to achieve pure absorption spectra. This involves changing the phase of pulse 1 by 90° between the two experiments. The transfer of coherence between the satellite and central transitions occurs with different efficiencies in the two experiments. Phase cycling of the second pulse with $+x$, $+y$, $-x$, $-y$ removes this difference. No other phase cycling is necessary in principle as the −1 coherence at the end of the pulse sequence is self-selecting. In practice, however, phase cycling of the selective 180° pulse in (b) is used to remove the effects of pulse imperfections. The t_1 period must be incremented by the rotor period between successive experiments in the two-dimensional dataset collection. More details are given in Section 5.3.5. In the experiment, pulses 1 and 2 are optimized on the sample to achieve maximum signal intensity for the satellite–central transition echo (see Fig. 5.13).

At the end of t_1, the satellite coherence is transferred to the central transition, $m = +\frac{1}{2} \leftrightarrow -\frac{1}{2}$, via an rf pulse (the transfer pulse). The coherence associated with the central transition then evolves for a period kt_1. The central transition coherence is unaffected by first-order quadrupole coupling, so evolves purely under the effects of second-order quadrupole coupling (Equation (5.31)) and any isotropic chemical shifts. Of the terms governing the second-order quadrupole coupling (Equation (5.31)), the second-rank term, V_{20}^{Q}, is averaged to zero by magic-angle spinning, once again leaving just the fourth-rank and isotropic terms (V_{40}^{Q} and V_{00}^{Q} respectively) where the kth rank V_{k0}^{Q} terms depend on kth order rotation matrices (see Box 5.1) to govern the coherence evolution during the kt_1 period. The magnitude of the fourth-rank terms in the second-order quadrupole-coupling frequency for the satellite transition and the central transition are simply related by a numerical factor, $R_4(I, n)$, where n indicates the particular satellite transition, as detailed above, and $R_4(I, n)$ is given by

$$R_4(I,n) = \frac{2\left(6I(I+1) - 34\left(n - \frac{1}{2}\right)\left(n + \frac{1}{2}\right) - 13\right)}{9(3 - 4I(I+1))} \tag{5.43}$$

i.e. the ratio of the coefficients of the V_{40}^{Q} terms in Equations (5.42) and (5.31). Thus, if k in the central transition evolution period is set equal to $R_4(I, n)$, at the end of the kt_1 period, the evolution of the satellite transition under its fourth-rank, second-order quadrupole-coupling frequency during the initial t_1 period will have been undone by the evolution of the central transition coherence under its fourth-rank, second-order quadrupole-coupling frequency. An FID of the subsequent evolution arising from the central transition in t_2 is recorded. Subsequent processing yields a two-dimensional spectrum with an isotropic projection in ω_1 and central transition powder patterns in ω_2, in much the same way as for MQMAS.

Some consideration needs to be given to the coherence order pathway needed to achieve the desired spectrum in STMAS. If the ratio $R_4(I, n)$ between the fourth-rank terms in the second-order quadrupole-coupling frequency of the satellite and central transitions is negative, then we need the same coherence order in t_1 and kt_1 to achieve the desired refocusing of the second-order quadrupole effects. Since we must ultimately observe -1 order coherence, the desired pathway is $0 \rightarrow -1$ (in t_1) $\rightarrow 1$ (in kt_1 and t_2). If the ratio $R_4(I, n)$ is positive, we must have opposite signs of coherence order in t_1 and kt_1 (and therefore t_2, as we just lose intensity if we add another change of coherence order between kt_1 and t_2). Thus, we need the coherence order pathway $0 \rightarrow +1$ (in t_1) $\rightarrow -1$ (in kt_1 and t_2). This can be arranged by phase cycling both pulses in the pulse sequence, but it is more efficient to generate the $+1 \rightarrow -1$ change by a selective 180° pulse on the central transition immediately after the transfer of coherence to this transition. This is shown in Fig. 5.12. The ratios $R_4(I, n)$ for different spins and their various satellite transitions are given in Table 5.2.

So far, we have implicitly assumed that we will only excite one satellite transition in the experiment and thus that the evolution in t_1 will be governed by just one second-order quadrupole-coupling dependence. Clearly, this can only be true for $I = \frac{3}{2}$. For half-integer spins with $I > \frac{3}{2}$, there is more than one satellite transition available. In general, all satellite transitions will be excited, at least to some extent, and thus affect the final spectrum [39]. Obviously, the value of k in the second t_1 period can only be set to refocus one of the satellite transition evolutions. The effect of the other satellite transitions will appear in the final two-dimensional STMAS spectrum as ridge-like correlations between the satellite transitions and central transitions, with slopes in the two-dimensional spectrum depending on the ratio of $R_4(I, n)$ for the particular satellite transition and the value of k chosen for the experiment. There is then the question of which satellite transition the experiment should be optimized for. Simulations suggest that the transfer of coherence from satellite to central transition is most efficient for the innermost satellite transitions which share an energy level with the central transition [40].

Like MQMAS, the STMAS experiment has the advantage of being carried out purely at the magic angle, making it a much more straightforward experiment than DAS or DOR. The advantage of STMAS over MQMAS is that as it involves only single-quantum coherences throughout, it affords potentially much greater signal-to-noise in a given acquisition time. The excita-

Table 5.2 The $R_4(I, n)$ values (ratio between the fourth-rank terms in the second-order quadrupolar broadening in a satellite transition and the central transition) for the STMAS experiment for different spin quantum numbers, I, and different satellite transitions, $\pm n - \frac{1}{2} \leftrightarrow \pm n + \frac{1}{2}$, where $n = 1, 2, \ldots, I - \frac{1}{2}$.

I	n	$R_4(I, n)$
$\frac{3}{2}$	1	−8/9
$\frac{5}{2}$	1	7/24
	2	−11/16
$\frac{7}{2}$	1	28/45
	2	−23/45
	3	−12/5
$\frac{9}{2}$	1	55/72
	2	1/18
	3	−9/8
	4	−50/18

tion of the satellite transition also has a much weaker dependence on the quadrupole coupling constant than does the excitation of multiple-quantum coherence in MQMAS experiments [40]. This means that in a sample with a wide range of quadrupole-coupling constants for the observed nucleus, STMAS is more likely to result in observation of all the sites in the spectrum than MQMAS, where excitation conditions are necessarily more selective, unless very high rf amplitudes are used in the excitation of the multiple-quantum coherence.

The down side of STMAS is that the experiment can potentially suffer from resolution problems if several satellite transitions are excited, as these may overlap with each other [39]. For a sample with several distinct chemical sites for the observed nucleus, this can lead to problems identifying all the resonances. Also, in the STMAS experiment, the central transition–central transition autocorrelation cannot be avoided as its production involves the same coherence pathway as for satellite–central transition correlations. Thus, the two-dimensional spectrum can become very crowded. There are, however, ways of minimizing or removing the central transition–central transition autocorrelation peaks (see below).

A further drawback with the STMAS experiment is its extreme sensitivity to the setting of the magic angle. Any slight mis-setting of the magic angle will reintroduce first-order quadrupole broadening and second-rank, second-order quadrupole broadening into the satellite transition coherences which evolve in t_1, with the consequence that each signal in the isotropic dimension of the two-dimensional spectrum becomes a powder pattern, rather than a sharp line. Not only does this have obvious implications for resolution in this dimension, but it can lead to apparent splittings of the signals, and so problems in interpreting the spectrum. All this makes it sound as though STMAS is a poor relation to MQMAS. This, however, is definitely not the case; some superb results have been obtained [39], and the intensity enhancement is not trivial. For samples where signal intensity is an issue, or where there are multiple sites with widely-varying quadrupole coupling constants, STMAS may become the experiment of choice.

Practical considerations

- *Optimizing the rf pulses* In the STMAS experiment, both the excitation and transfer pulse amplitudes and lengths are adjusted experimentally. Simulations and practice suggest that high-amplitude, small flip angle pulses (in the region of 18–36°) are optimal for both excitation and transfer [40].
- *Optimizing the magic angle* As the STMAS experiment is very sensitive to mis-settings of the magic angle, it is very important to set the angle very accurately (around ±0.004° is required for samples with

quadrupolar splittings of the order of 1 MHz). A suitable procedure for optimizing the magic angle is as follows [39]. Step 1 is to set the angle in the normal manner, as detailed in Section 2.2.4, by maximizing the number of rotor echoes seen in the normal FID of a simple pulse-acquire experiment. This can be carried out on the sample of interest by observing the rotor echoes arising from the satellite transitions. Step 2 is to run one slice (i.e. one finite t_1 ($= N\ \tau_R$) value) of the STMAS experiment. The form of the recorded FID seen in such an experiment is shown schematically in Fig. 5.13 for a spin-$\frac{3}{2}$ nucleus. The spinning angle should then be adjusted so as to maximize the satellite–central transition correlation echo. Wherever possible, the angle optimization should be carried out on the sample of interest, with the sample remaining untouched in the probehead after the adjustment until the STMAS experiment has been run. Changing samples or adjusting the spinning rate can all cause changes in the angle setting, which, while relatively unimportant for most experiments, can be critical for the STMAS experiment.

- *Setting the spinning rate* Clearly the spectral width in ω_1 of the two-dimensional STMAS spectrum is equal to the sample spinning rate, as the t_1 incrementation must be set to the rotor period (adjusted for the finite length of the pulses either side of t_1) in STMAS. Thus, the spinning rate needs to be fast enough to provide a sufficient spectral width in the isotropic dimension of the STMAS experiment.
- *Minimizing the central transition–central transition correlation peak* The central transition–central transition autocorrelation peak exacerbates spectral crowding and can be removed or reduced in the following ways:

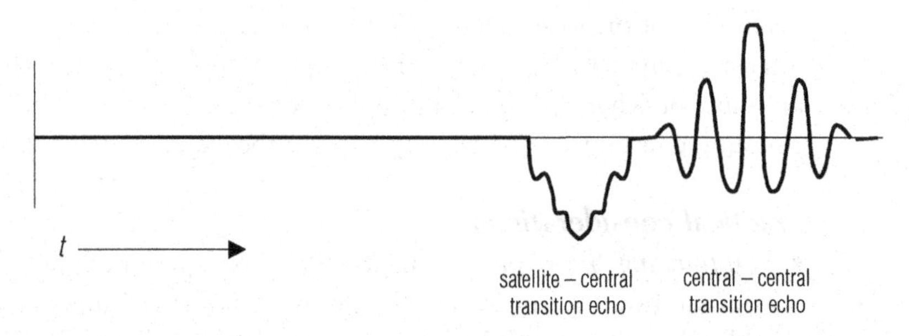

Fig. 5.13 Schematic illustration of the form of the recorded FID in an STMAS experiment for a spin-$\frac{3}{2}$ nucleus (see Fig. 5.12 for the pulse sequence). The refocusing of the satellite coherence evolution in t_1 by evolution of the central transition coherence in kt_1 leads to the formation of an echo at $t = kt_1$, where t is the time from the satellite–central transition transfer pulse. There is also unavoidable coherence in t_1 due to the central transition, which leads to an echo at $t = t_1$. For half-integer spins with $I > \frac{3}{2}$ there will be further satellite–central transition echoes due to all the other possible satellite transitions. These occur at $t = R_4(I, n)t_1$, where the ratios $R_4(I, n)$ are given in Table 5.2 for each spin I and each satellite transition n.

1. Use half-rotor synchronization in t_1 [41]. This can completely remove the unwanted autocorrelation peak, but is not successful if the central transition has strong spinning sidebands associated with it [41].
2. Presaturate the central transition prior to each run of the STMAS pulse sequence. In practice, however, true presaturation is difficult to achieve.
3. Optimize the choice of pulse lengths and amplitudes in the STMAS pulse sequence to minimize the central transition–central transition echo (see Fig. 5.13). This can be done in the same way as the magic-angle setting, by observing a single slice of a shifted-echo STMAS experiment.

5.3.6 Recording two-dimensional datasets for DAS, MQMAS and STMAS

General methods of producing pure absorption lineshapes in two-dimensional NMR spectra were discussed in Chapter 1. DAS, MQMAS and STMAS experiments all result in an echo being produced at kt_1 the decay of which in t_2 gives rise to the recorded signal. This echo formation allows different procedures to be used for achieving pure absorption lineshapes [24, 42] from those detailed elsewhere in this book. Fourier transformation of the t_2 FID gives a frequency spectrum with dispersion-mode components. This is because, in effect, the dataset is only defined from $t_2 = 0$ (the top of the echo) to $t_2 \rightarrow \infty$. However, Fourier transformation involves an integral from $t_2 \rightarrow -\infty$ to $+\infty$. Thus the dataset appears to the Fourier transformation procedure as a step function (zero from t_2, $-\infty$ to 0 and 1 thereafter) multiplied by the recorded data. The result in the frequency domain is therefore a convolution of the Fourier transform of the step function (a dispersive lineshape) and the Fourier transform of the recorded data (the required lineshape). In order to achieve pure absorption mode lineshapes, information is needed about the signal prior to $t_2 = 0$. This can be obtained indirectly as in (a) below, or directly as in (b). In both cases, the t_2 FID is generally recorded from immediately after the last pulse in the sequence rather than from the top of the echo maximum. With this definition of the t_2 dimension, the echo maximum appears at $t_2 = kt_1$. This shift of the time origin requires a phase correction as discussed below.

(a) Hypercomplex method

Record two two-dimensional datasets with a change in the rf phase in the excitation sequence between them such that the phase of the coherences in

t_1 is shifted by 90° between the two experiments. Phase cycling is applied in both experiments so that $+p$ and $-p$ of the appropriate coherence $|p|$ are selected in t_1 ($|1|$ for DAS and STMAS; $|2m|$ for MQMAS). This results in echo formation at kt_1 from one of the components and so-called *anti-echo* formation at $-kt_1$ from the other component [24]. It is the decay of the net sum of these two echo components which is recorded in t_2 (Fig. 5.14). Because data recording can only begin after the last pulse, the whole echo plus anti-echo cannot be recorded in one dataset. The true echo signal can, however, be reconstructed from the two datasets so obtained using the hypercomplex method [24]. The form of the signals in the two datasets was discussed in Section 1.5.3. From these, it can readily be seen that the echo and anti-echo signals are constructed from [24]:

$$\begin{aligned} S_E(t_1, t_2) &= S_X(t_1, t_2) - iS_Y(t_1, t_2) \\ S_A(t_1, t_2) &= S_X(t_1, t_2) + iS_Y(t_1, t_2) \end{aligned} \tag{5.44}$$

where S_X is the time-domain dataset generated from t_1 coherence of relative phase 0° and S_Y is the time-domain dataset generated from t_1 coherence of relative phase 90°. As illustrated in Fig. 5.14, the complete echo can then be reconstructed in the t_2 dimension by constructing a single new dataset with twice as many t_2 points as each individual dataset, and placing the anti-echo signal in the first half of the new dataset, and the echo signal in the latter half.

Fourier transformation in both dimensions results in a two-dimensional pure absorption mode spectrum which is symmetrical about $\omega_2 = 0$ (where

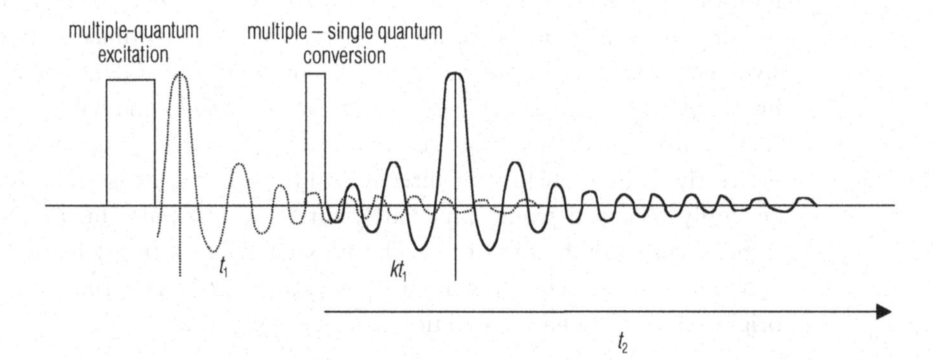

Fig. 5.14 Formation of the echo (solid line) and anti-echo (dashed line) in the MQMAS experiment. During t_1, both $+p$ and $-p$ multiple-quantum coherences are selected. One of these components leads to formation of the anti-echo at $-kt_1$ from the last pulse and the other, the formation of the echo at $+kt_1$ from the last pulse. For spin-$\frac{3}{2}$ and triple-quantum coherence, the −3 coherence leads to echo formation and +3 to anti-echo formation. The sum of the dashed and solid lines is recorded in the FID, which is generally recorded from the end of the last pulse.

ω_2 corresponds to the Fourier transformed t_2 domain); the two symmetric halves of the spectrum are identical and either half represents the required NMR spectrum. The two halves may be added to improve signal-to-noise.

An alternative and completely equivalent approach [24] is to Fourier transform S_E and S_A in t_2 and then in t_1. A first-order phase correction in the ω_2 dimension needs to be made after Fourier transformation in t_2. This is because the echo maximum in the experiment does not appear at the time origin, $t_2 = 0$, but is shifted in t_2 away from $t_2 = 0$ as a function of t_1. In fact, the echo maximum appears at $t_2 = kt_1$ in successive t_1 experiments, where k is a constant determined by the particular experiment. A Fourier transform theorem states that if a function $f(t)$ is displaced by Δt, i.e. $f(t) \rightarrow f(t + \Delta t)$, then its Fourier transform, $F(\omega)$, is multiplied by $\exp(-\omega\Delta t)$. This is simply a first-order phase shift; to undo this phase shift in our two-dimensional experiment, we must multiply the Fourier transformed ω_2 dimension of the echo dataset by $\exp(i\omega_2 kt_1)$ and of the anti-echo dataset by $\exp(-i\omega_2 kt_1)$ in each t_1 slice.

The final pure absorption mode spectrum $S(\omega_1, \omega_2)$ is then produced by combining the individual Fourier transformed and ω_2-phased datasets according to

$$S'(\omega_1, \omega_2) = S'_E(\omega_1, \omega_2) + S'_A(-\omega_1, \omega_2) \qquad (5.45)$$

where the primes denote datasets that have received the ω_2 phase correction. An example phase cycling scheme for this method for the triple-quantum MQMAS experiment for spin-$\frac{3}{2}$ is shown in Table 5.3 [24].

It is very important when using this technique that the two coherence pathways through $0 \rightarrow \pm p \rightarrow -1$ come through to the end of the sequence with the same amplitude. As illustrated in Fig. 5.11, the desired coherence pathways require simultaneous transformations $+p \rightarrow -1$ and $-p \rightarrow -1$; these transformations involve different changes in coherence order and, unfortunately, they generally occur with different efficiency. This often leads to large spectral distortions. To avoid this, a *zero-quantum filter* or *z-filter* can be applied [26]. As shown in Fig. 5.15, this involves an extra pulse in the sequence so that the coherence pathway becomes $\pm p \rightarrow 0 \rightarrow -1$. This now involves $+p \rightarrow 0$ and $-p \rightarrow 0$ transformations which have the same change in coherence order and so can be expected to have the same efficiencies. In principle, the time delay τ in the z-filter needs to be only long enough to change the phase and amplitude for the next rf pulse, the selective 90° pulse. In practice, better results are sometimes obtained if the delay is longer, tens or even hundreds of microseconds. This is presumably because the delay acts to remove other unwanted coherences which appear because of incomplete relaxation between cycles, and so inefficient removal of unwanted coherences by phase cycling.

Table 5.3 The phase-cycling scheme for the triple-quantum MQMAS experiment for $I = \frac{3}{2}$ using the hypercomplex method for producing pure absorption lineshapes in the two-dimensional frequency spectrum. (i) the X-dataset phase cycling; (ii) the Y-dataset phase cycling. Pulse phases are labelled according to the pulse sequence shown at the top of the tables; the coherence pathway is also shown.

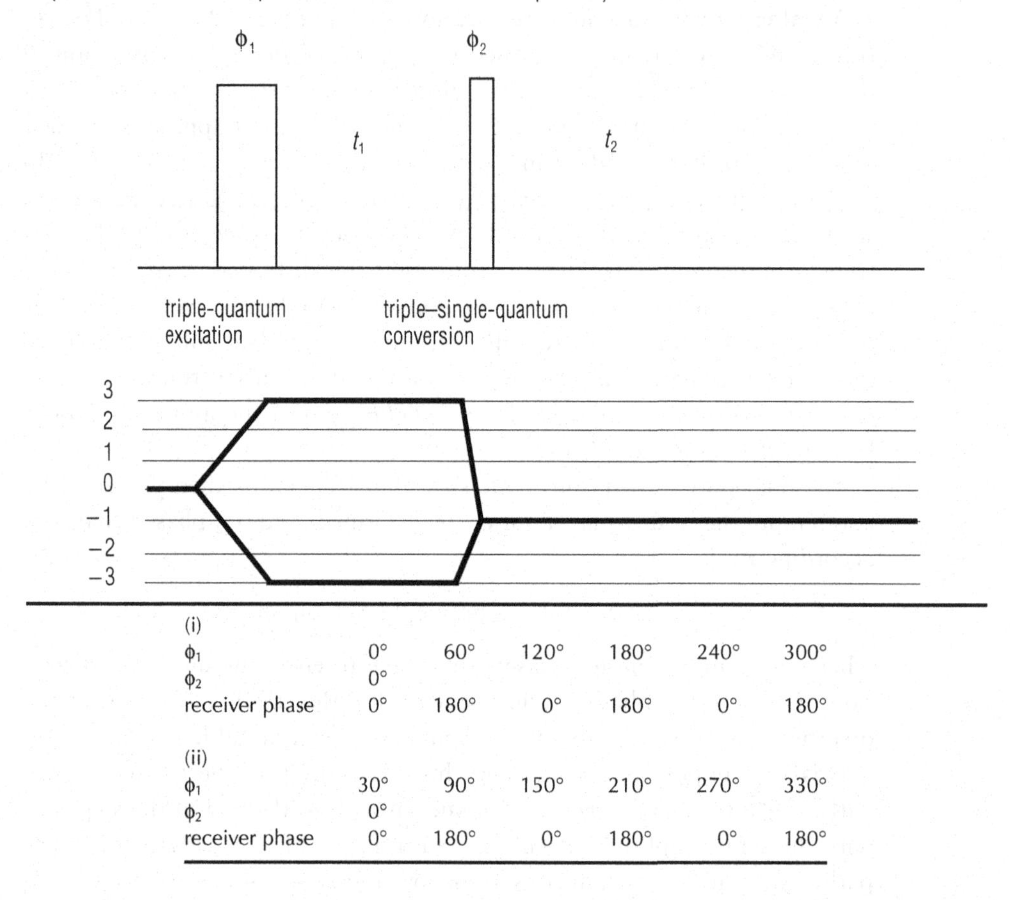

(i)						
ϕ_1	0°	60°	120°	180°	240°	300°
ϕ_2	0°					
receiver phase	0°	180°	0°	180°	0°	180°

(ii)						
ϕ_1	30°	90°	150°	210°	270°	330°
ϕ_2	0°					
receiver phase	0°	180°	0°	180°	0°	180°

(b) Acquisition of whole echoes

Record a single, two-dimensional dataset using a *shifted-echo sequence* [24, 30, 37]. In this approach, a refocusing selective 180° pulse (selective for the central transition) is added a time τ after the end of the pulse sequence for MQMAS and DAS. The positioning of the selective 180° pulse requires further consideration for the STMAS experiment, as explained in Fig. 5.16. This shifts the echo which arises from a DAS or MQMAS experiment a further time τ after the 180° pulse and allows the whole echo to be recorded. In this method, only one coherence ($+p$ or $-p$) is selected in t_1. The form of the signal is then $\exp(i\omega_1 t_1)\exp(i\omega_2 t_2)$ with t_2 defined from $-\infty \rightarrow +\infty$ so that, after Fourier transformation in t_2, the signal is $\exp(i\omega_1 t_1)A_2$ using the terminology in Section 1.5, after appropriate phasing as discussed below.

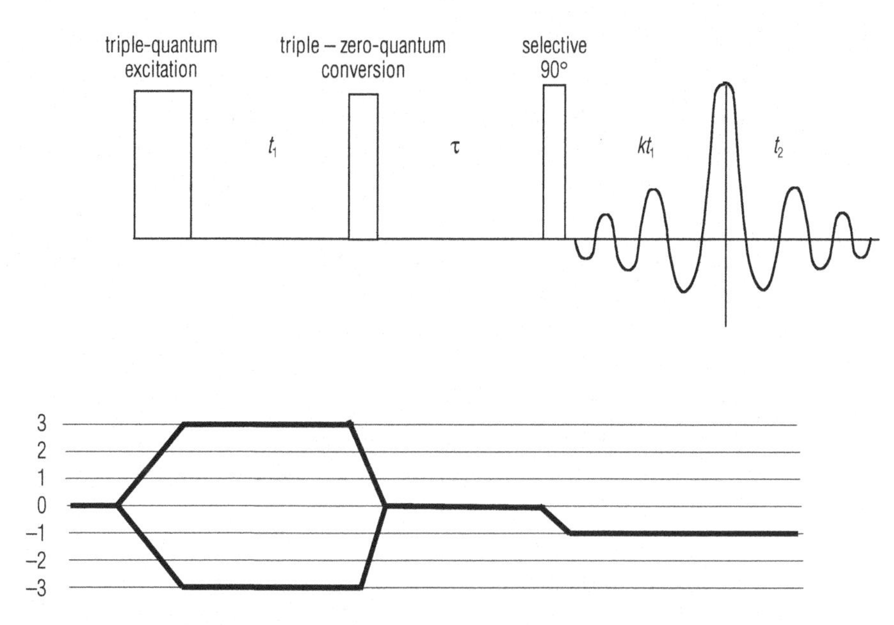

Fig. 5.15 The zero-quantum filtered MQMAS pulse sequence for a spin-$\frac{3}{2}$ nucleus. The two coherence pathways $0 \rightarrow -3 \rightarrow 0$ and $0 \rightarrow +3 \rightarrow 0$ have equal amplitudes because they are symmetric. In the basic MQMAS sequence the two pathways are $0 \rightarrow -3 \rightarrow -1$ and $0 \rightarrow +3 \rightarrow -1$. The final step in such an experiment involves coherence order changes of +2 and −4 respectively for the two coherence pathways. These different changes of coherence order generally have different efficiencies, which results in the two pathways giving different amplitudes to the final FID recorded in t_2. In turn, this leads to gross spectral distortions. The zero-quantum filtered sequence avoids this problem.

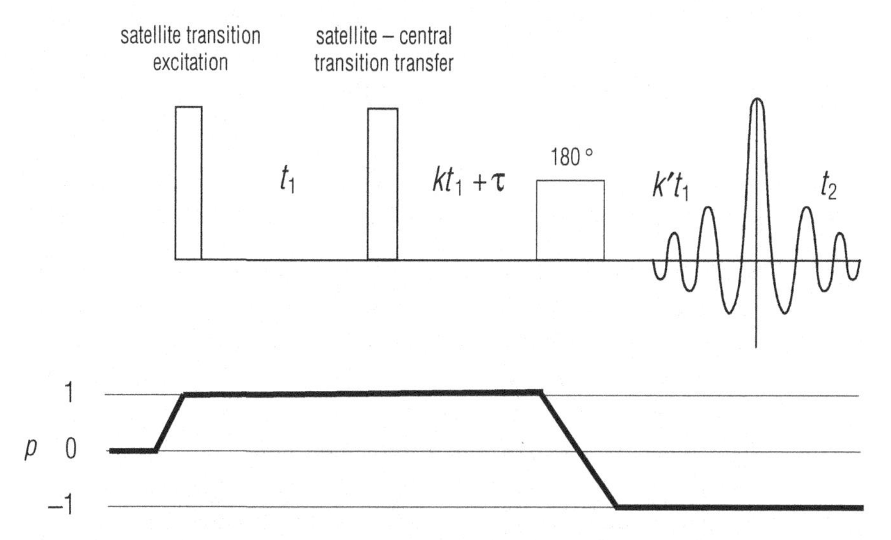

Fig. 5.16 The shifted-echo experiment for the STMAS experiment [39]. The experiment works on the same principles as the shifted-echo experiment for DAS and MQMAS. Here, however, the position of the selective 180° pulse depends on the sign of the ratio between the second-rank terms in the second-order quadrupolar broadening under MAS for the satellite and central transitions involved in the experiment. The ratios $R_4(I, n)$ for various spins I and satellite transitions n are given in Table 5.2. For negative ratios, $k = R_4(I, n)$, $k' = 0$; for positive ratios, $k = 0$, $k' = R_4(I, n)$. This ensures the correct refocusing of the evolution of the +1 coherence as explained in Section 5.3.5. In each case, τ is chosen to be long enough to collect the entire echo for all t_1 values being used in the two-dimensional experiment.

Fourier transformation in t_2 gives only an absorptive part because the dataset is, in effect, defined for $t_2 = -\infty \rightarrow +\infty$. A dispersive (imaginary) part appears when the dataset is truncated to values for $t_2 = 0 \rightarrow +\infty$. After Fourier transformation in t_1, the final signal is (again after appropriate phase corrections) $(A_1 + iD_1)A_2 = A_1A_2 + iD_1A_2$ so that the real part of the final spectrum is purely absorptive.

As discussed in (a), a first-order phase correction is needed in ω_2 before Fourier transformation in t_1 to 'correct' for the fact that the echo maximum shifts in successive t_1 experiments; we need to multiply the ω_2 dimension by $\exp(i\omega_2 k t_1)$ in all t_1 slices. In addition to this first-order phase correction, we must also perform a phase correction to correct for the fact that the echo maximum is shifted by a constant τ, the echo delay, in addition to the t_1-dependent shift in all t_1 experiments. This requires multiplication in the ω_2 dimension by $\exp(i\omega_2\tau)$ in all t_1 slices.

The shifted-echo method is not appropriate for samples in which there is a significant homogeneous linebroadening, due to the loss of signal imposed by the echo delay (homogeneous linebroadening cannot be refocused by an rf pulse). When using this method, it is desirable to set the receiver phase so that the shifted echo appears as a symmetric echo with maximum intensity in the real part of the time domain. The experiment can be arranged so that the echo moves to increasing t_2 times as t_1 increases, or in the reverse direction. The phase cycling schemes for these two alternatives are given in Table 5.4 for the triple-quantum MQMAS for spin-$\frac{3}{2}$ experiment [24].

Note that for both the DAS and MQMAS experiments, the spectral width in f_1 is $1/[(k+1)\delta t_1]$ where δt_1 is the t_1 dwell time, for the definition of t_1 used here.

5.4 Other techniques for half-integer quadrupolar nuclei

REDOR experiments employing spin-$\frac{1}{2}$ – quadrupolar nuclei (half-integer) spin pairs have been developed [43–45] for the purposes of studying internuclear distances between quadrupolar nuclei; REDOR is discussed further in Section 3.3.2. Some of these combine REDOR with MQMAS [43].

Very large quadrupole-coupling constants still cause problems, even for half-integer spins. For quadrupole-coupling constants which exceed the order of 10 MHz, even the central transition becomes excessively broad to observe. Under these circumstances, the techniques described in Section 5.3 become considerably less useful and one has to resort to quadrupole-echo detection of the central transition, usually under magic-angle spinning. More recently, a quadrupolar Carr–Purcell–Meiboom–Gill (CPMG) experiment has been developed which performs better than traditional quadrupole-echo experiments [46]. This is discussed further in Section 6.5.1.

Table 5.4 The phase-cycling scheme for the triple-quantum MQMAS experiment for $I = \frac{3}{2}$ using the shifted-echo method for producing pure absorption lineshapes in the two-dimensional frequency spectrum. Pulse phases are labelled according to the pulse sequence shown at the top of the tables. The coherence pathway is also shown; only one of the coherences, +3 or −3, is selected. The phase cycle for the +3 → +1 → −1 pathway is given.

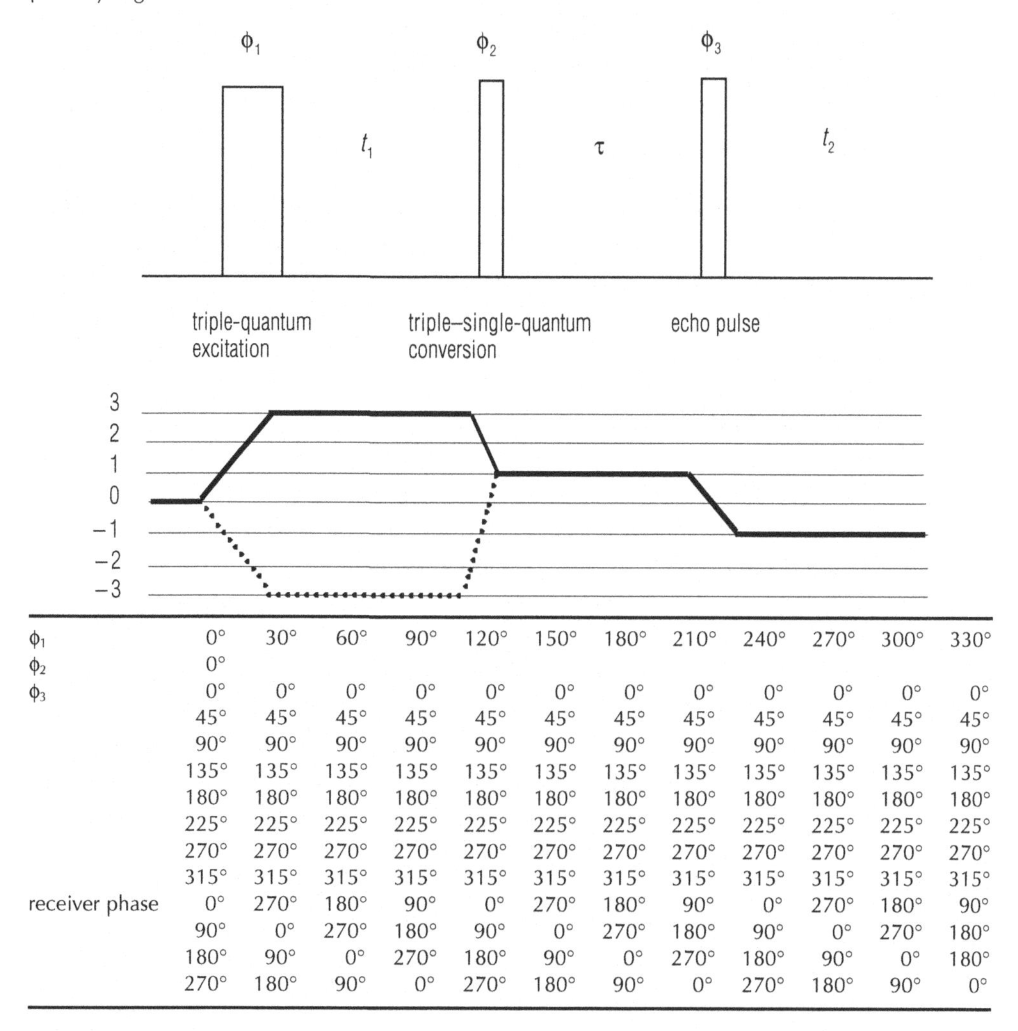

ϕ_1	0°	30°	60°	90°	120°	150°	180°	210°	240°	270°	300°	330°
ϕ_2	0°											
ϕ_3	0°	0°	0°	0°	0°	0°	0°	0°	0°	0°	0°	0°
	45°	45°	45°	45°	45°	45°	45°	45°	45°	45°	45°	45°
	90°	90°	90°	90°	90°	90°	90°	90°	90°	90°	90°	90°
	135°	135°	135°	135°	135°	135°	135°	135°	135°	135°	135°	135°
	180°	180°	180°	180°	180°	180°	180°	180°	180°	180°	180°	180°
	225°	225°	225°	225°	225°	225°	225°	225°	225°	225°	225°	225°
	270°	270°	270°	270°	270°	270°	270°	270°	270°	270°	270°	270°
	315°	315°	315°	315°	315°	315°	315°	315°	315°	315°	315°	315°
receiver phase	0°	270°	180°	90°	0°	270°	180°	90°	0°	270°	180°	90°
	90°	0°	270°	180°	90°	0°	270°	180°	90°	0°	270°	180°
	180°	90°	0°	270°	180°	90°	0°	270°	180°	90°	0°	180°
	270°	180°	90°	0°	270°	180°	90°	0°	270°	180°	90°	0°

There are, however, two further techniques applied to quadrupolar nuclei which should be mentioned, quadrupole nutation and cross-polarization. Quadrupole nutation is a simple experiment which has largely been superseded by the MQMAS experiment, which uses the same equipment and usually separates signals with different quadrupole-coupling constants more reliably. However, there is a great deal of beauty in the simplicity of the quadrupole nutation experiment, and it is included here because many papers in the literature in the pre-MQMAS era rely on it. Cross-polarization is not used extensively on quadrupolar nuclei, partly because it can be tricky to set up, but also because many quadrupolar nuclei of interest are highly abundant, and so cross-polarization is not needed to create signal intensity. However, it can be useful for judging which spins are close in space and, indeed, this has been its primary use for quadrupolar nuclei to date. It has also been used in the MQMAS experiment [47].

5.4.1 Quadrupole nutation

The pulse sequence of the quadrupole nutation experiment is shown in Fig. 5.17 [48–50]. It is a two-dimensional experiment in which an irradiating rf pulse is applied for a period t_1 with subsequent detection of an FID during t_2. The purpose of the experiment is to study the evolution of the spin system in the rotating frame (rotating about $\mathbf{B}_0$ at ω_{rf}, the frequency of the rf irradiation) in the presence of the radiofrequency field $\mathbf{B}_1$ provided by the rf pulse in t_1. In the rotating frame, the effect of the $\mathbf{B}_0$ field is only very small, providing the rf pulse is close to resonance, so the principal field which the spins feel is the $\mathbf{B}_1$ field. Thus, the evolution during t_1 yields a low frequency ($B_1 \sim 0.001\ B_0$) NMR spectrum, known as the *nutation spectrum*, while having the sensitivity of a high-field experiment.

Throughout this chapter, we have considered the case where the Zeeman interaction is much larger than the quadrupole interaction, so that the eigenstates of the spin system are simply small perturbations of the Zeeman states. However, during the t_1 period, the Zeeman term in the rotating frame is very small in comparison to the other terms in the rotating-frame hamiltonian describing the spin system in this period. The hamiltonian describing the spin system during the rf pulse irradiation in t_1 in the rotating frame is [50]

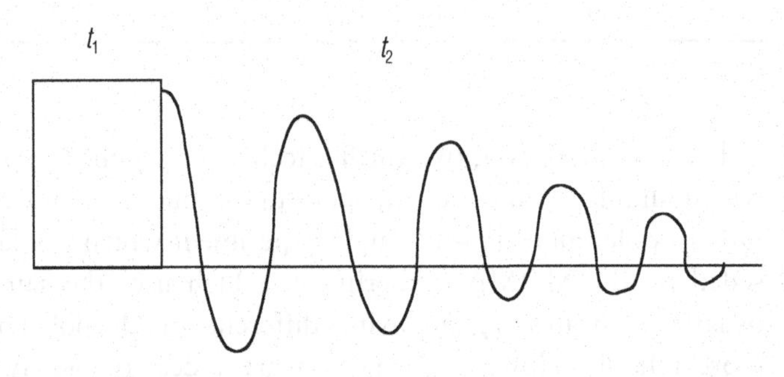

Fig. 5.17 The two-dimensional quadrupole nutation experiment. The length of the excitation pulse is increased step-wise in successive t_1 slices and the FID corresponding to the central transition of the half-integer quadrupolar nucleus under study is recorded in t_2.

$$\begin{aligned}\hat{H}_{\text{pulse}} &= \hat{H}_{\text{rf}} + \hat{H}_Q + \hat{H}_\Delta \\ &= \omega_1\hat{I}_y + \frac{1}{3}\omega_Q(3\hat{I}_z^2 - \hat{I}^2) + (\omega_0 - \omega)\hat{I}_z\end{aligned} \tag{5.46}$$

$\hat{H}_{\text{rf}}$ describes the effect of the rf pulse (taken to be a *y*-pulse here), with the pulse amplitude, $\omega_1 = \gamma B_1$; $\hat{H}_Q$ describes the effects of the quadrupole coupling to first order and $\hat{H}_\Delta$ describes off-resonance effects as detailed in Section 5.2.2. If we assume the rf pulse is applied close to the Larmor frequency ($\hat{H}_\Delta \sim 0$), the eigenstates of the system during t_1 depend on the relative sizes of the two remaining terms in $\hat{H}_{\text{pulse}}$, namely $\hat{H}_{\text{rf}}$ and $\hat{H}_Q$. In other words, the eigenstates of the system depend on ω_1/ω_Q (ω_Q is defined in Equation (5.10)), and while they may be described as linear combinations of the Zeeman spin functions (since these constitute a complete set for description of the state of the spin system), they are certainly not small perturbations of them.

During t_2, the $\mathbf{B}_1$ field, i.e. the rf pulse, is switched off and the hamiltonian governing the system becomes in the laboratory frame [50]

$$\hat{H}_2 = \hat{H}_0 + \hat{H}_{\text{cs}} + \hat{H}_Q \tag{5.47}$$

where $\hat{H}_0$ is the Zeeman term (Equation (5.5)) and $\hat{H}_{\text{cs}}$ describes the effects of chemical shift. The eigenstates of the system now return to the usual Zeeman spin states.[N2] The FID due to coherence between the $+\frac{1}{2}$ and $-\frac{1}{2}$ spin levels, i.e. the central transition, is collected during t_2 (this is the on-resonance transition for rf irradiation near the Larmor frequency); as we have already seen in Section 5.2.1, this transition is only affected by $\hat{H}_Q$ in second order.

A two-dimensional dataset is collected by incrementing t_1, the length of the rf pulse during t_1. Two-dimensional Fourier transformation of this time-domain dataset yields a two-dimensional frequency spectrum with the normal central transition lineshape in f_2 and the nutation spectrum in f_1. As the nutation spectrum is principally governed by ω_1 and ω_Q, chemical shift effects are largely absent from this dimension.

In the extreme case of $\omega_1 >> \omega_Q$, $\hat{H}_{\text{pulse}}$ reduces to

$$\hat{H}_{\text{pulse}} \approx \hat{H}_{\text{rf}} = \omega_1\hat{I}_y \tag{5.48}$$

so the evolution (nutation) frequency of the spin system is clearly just ω_1, the rf pulse amplitude, as for any other case of an isolated (uncoupled) spin in the presence of an on-resonance pulse. So, in this case, the nutation spectrum would just consist of a single line at frequency ω_1.

In the other extreme, $\omega_Q >> \omega_1$, $\hat{H}_{\text{pulse}}$ is simply

$$\hat{H}_{\text{pulse}} \approx \hat{H}_Q = \frac{1}{3}\omega_Q(3\hat{I}_z^2 - \hat{I}^2) \tag{5.49}$$

As shown previously in Section 5.2.2, such a case constitutes selective excitation of the transition which is on resonance. Assuming that the pulse frequency is on resonance for the central transition, the corresponding evolution or nutation frequency in this case is $(I + \frac{1}{2})\omega_1$. In this case, again a single line is seen in the nutation spectrum, but this time at frequency $(I + \frac{1}{2})\omega_1$.

For intermediate cases, the nutation spectrum is quite complex, dependent on ω_1/ω_Q (remembering that ω_Q depends on molecular orientation) and, in general, consisting of several frequencies. Simulation of experimental nutation spectra in this regime allows accurate determination of the quadrupole coupling constant and asymmetry. Figure 5.18 shows the nutation spectrum calculated for a spin-$\frac{5}{2}$ nucleus in a powder sample for different χ/ω_1, where χ is the quadrupole-coupling constant, including the limiting cases.

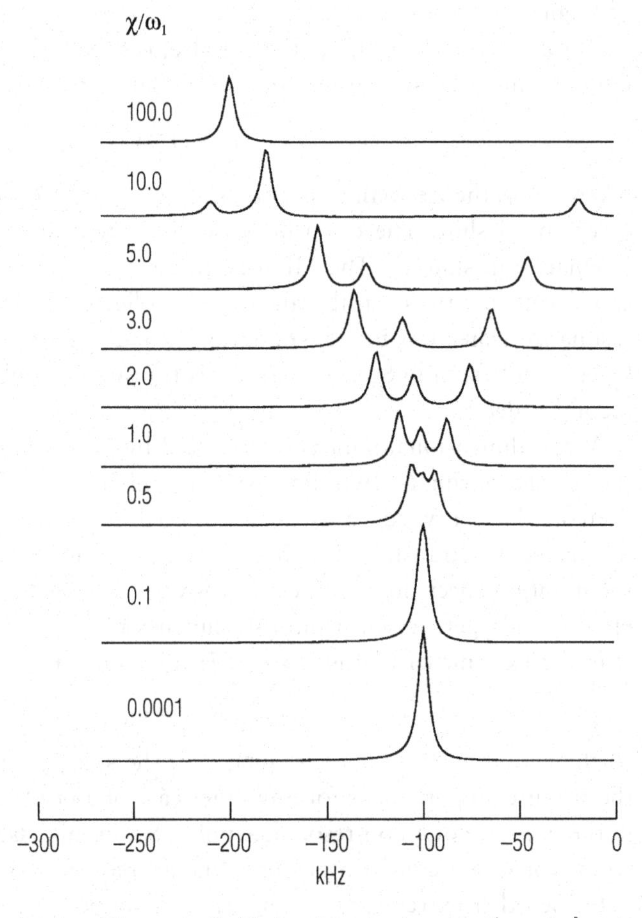

Fig. 5.18 Quadrupole nutation spectra (f_1 of the two-dimensional spectrum) for a spin-$\frac{5}{2}$ nucleus in a powder sample for various χ/ω_1 where ω_1 is the rf power used in the experiment and χ is the quadrupole-coupling constant. Note the limiting cases when $\chi/\omega_1 << 1$, which gives a single line at $\omega = \omega_1$ and when $\chi/\omega_1 >> 1$, which gives a single line at $\omega = (I + \frac{1}{2})\ \omega_1$. $\omega_1/2\pi$ is 100 kHz in all the simulations.

5.4.2 Cross-polarization

Many quadrupolar nuclei that are studied are near to 100% abundant, so there is little need for cross-polarization in order to improve signal-to-noise. However, cross-polarization from spin-$\frac{1}{2}$ nuclei to half-integer quadrupolar nuclei is used from time to time in solid-state NMR in order to ascertain which spins are close in space.

The cross-polarization experiment for polarization transfer from spin-$\frac{1}{2}$ to quadrupolar nuclei takes much the same form as from the conventional spin-$\frac{1}{2}$–spin-$\frac{1}{2}$ experiment discussed in Section 2.4.1. The pulse sequence is shown again in Fig. 5.19 for convenience, where the cross-polarization is from the I-spin-$\frac{1}{2}$ to the quadrupolar S spin.

The principal difference from the pure spin-$\frac{1}{2}$ case is in the Hartmann–Hahn match. For the spin-$\frac{1}{2}$–spin-$\frac{1}{2}$ cross-polarization case, the Hartmann–Hahn condition is simply $\omega_{1I} = \omega_{1S}$, where $\omega_{1I} = \gamma_I B_{1I}$ and $\omega_{1S} = \gamma_S B_{1S}$, the I and S spin amplitudes. When S is a quadrupolar nucleus, the Hartmann–Hahn condition is replaced by the more general condition that ω_{1I} should be equal to ω_{nut}, the nutation frequency of the S-spin transition that is being cross-polarized to. As seen in the previous section and 5.2.2, the nutation behaviour of a quadrupolar spin depends on the relative magnitudes of the quadrupolar splitting, ω_Q (Equation (5.10)) and ω_{1S}. As we saw in the last section, there are two limiting cases. When $\omega_{1S} >> \omega_Q$, the nutation frequency is not affected by the quadrupole coupling at all, so that

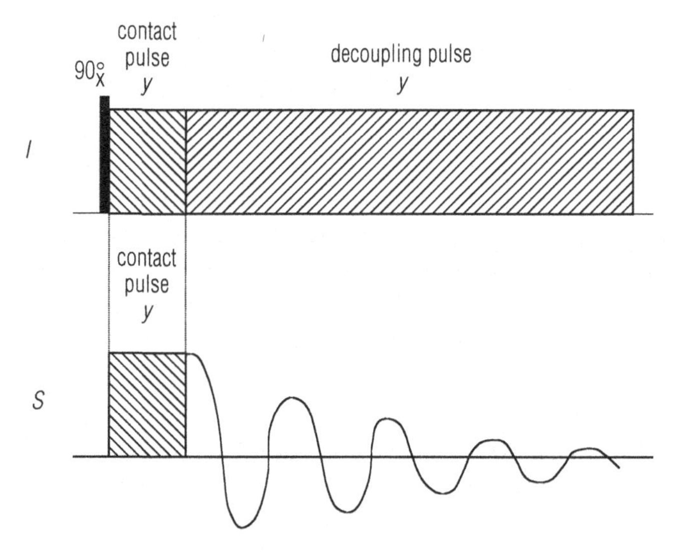

Fig. 5.19 The cross-polarization pulse sequence. This pulse sequence was introduced in Chapter 2 for spin-$\frac{1}{2}$ nuclei. Here, I is a spin-$\frac{1}{2}$ nucleus and S a half-integer quadrupolar spin. Polarization is transferred from the spin-$\frac{1}{2}$ to one or more S spin transitions, depending on the size of the quadrupole coupling relative to the pulse amplitude – see text for details.

the nutation frequency is just the usual $\omega_{nut} = \omega_{1S}$, as for any uncoupled spin system. When $\omega_{1S} << \omega_Q$, only a single transition is irradiated, that one which is on resonance. Primarily, we are interested in the central transition, for which the nutation frequency under these selective excitation conditions is $\omega_{nut} = (S + \frac{1}{2})\omega_{1S}$. Thus, for these two limiting cases, the appropriate matching conditions are:

$$\begin{aligned} \omega_{1S} >> \omega_Q: \quad & \omega_{1I} = \omega_{1S} \\ \omega_{1S} << \omega_Q: \quad & \omega_{1I} = \left(S + \frac{1}{2}\right)\omega_{1S} \end{aligned} \tag{5.50}$$

In the intermediate regime, $\omega_{1S} \sim \omega_Q$, the Hartmann–Hahn condition is not well defined. For a start, in this regime the nutation spectrum of the S spin has several frequencies, as can be seen by looking at the nutation spectra in this regime in Fig. 5.18.

As well as cross-polarizing to the central transition of half-integer quadrupolar nuclei, it is also possible to cross-polarize to symmetric $+m \rightarrow -m$ multiple quantum transitions [51]. This has aroused interest recently in connection with the MQMAS experiment. Equation (5.36) gives the nutation frequency for these multiple-quantum transitions, which can then be matched with ω_{1I} in the selective excitation limit, $\omega_{1S} << \omega_Q$, to cross-polarize to the particular multiple-quantum transition. A particularly good example of the use of cross-polarization involving quadrupolar nuclei is in the two-dimensional HETCOR (HETeronuclear CORrelation) experiment shown in Fig. 5.20(a) [52]. A HETCOR spectrum shows spatial correlations between two heteronuclear spins. In the simplest version of this type of experiment, transverse magnetization is generated for one nuclear species I (^{23}Na in the case shown in Fig. 5.20) and allowed to evolve at its characteristic frequency in t_1. The remaining magnetization is then transferred to the other nuclear species S (^{31}P in Fig. 5.20) in a cross-polarization step, and an FID collected in t_2. Fourier transformation in t_1 and t_2 results in a two-dimensional frequency spectrum with cross-peaks between I and S spin signals indicating that cross-polarization took place between the corresponding spins, and therefore that these spins are close in space. Figure 5.20(b) shows another HETCOR spectrum of the same sample, but this time the experiment has been arranged so that a high-resolution MQMAS spectrum of the I spin (^{23}Na) appears in the ν_1 dimension of the spectrum, rather than the standard magic-angle spinning spectrum of Fig. 5.20(a). This is done by using the usual MQMAS pulse sequence (Fig. 5.11) on the I spin (^{23}Na), then at the point of the MQMAS echo formation (at kt_1 after the mulitple-quantum $\rightarrow$ single-quantum transfer pulse) cross-polarization is used to transfer the I spin magnetization to the S spin (^{31}P). An S spin FID is then recorded in t_2 as in the simple HETCOR experiment.

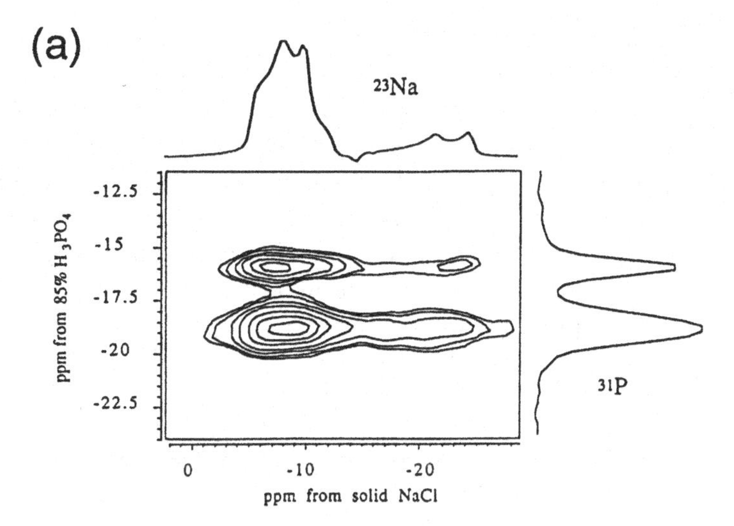

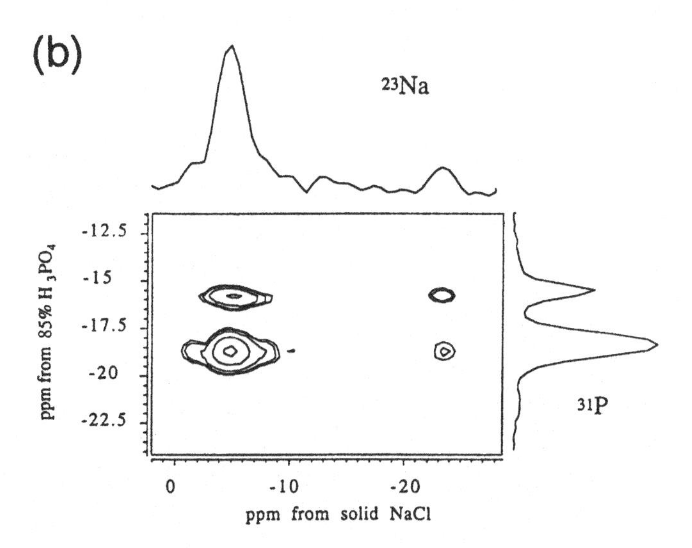

Fig. 5.20 A ^{23}Na ($I = \frac{3}{2}$) – ^{31}P ($I = \frac{1}{2}$) HETCOR (heteronuclear correlation) spectra of $Na_3P_3O_9$ [52]. (a) The HETCOR spectrum of $Na_3P_3O_9$ obtained with a 'standard' magic-angle spinning spectrum in the ^{23}Na dimension. (b) A similar spectrum, but this time the experiment has been arranged so that a MQMAS ^{23}Na spectrum appears in the ^{23}Na dimension.

Under static conditions, i.e. no sample spinning, cross-polarization to the central or symmetric multiple-quantum transition of a half-integer quadrupolar S spin is relatively straightforward. It only suffers the same difficulties as the cross-polarization between two spin-$\frac{1}{2}$ nuclei, with one exception. Only a fraction of the I-spin magnetization (a half for spin-$\frac{3}{2}$ under selective excitation conditions for the central transition, for instance) participates in the polarization transfer process, the remaining I-spin magnetization being associated with spin levels outside the central transition. In contrast, in cross-polarization between spin-$\frac{1}{2}$ nuclei, all the I-spin magnetization is involved. However, there are key differences under magic-angle spinning, which principally relate to differences in spin locking a spin-$\frac{1}{2}$ and a quadrupolar spin system. As discussed in Section 2.5, during the contact pulses on both spins, both I and S spin systems are spin locked via their respective pulses and it is while in this situation that magnetization is transferred between the spins. The conditions for spin locking a quadrupolar spin system are discussed below.

Spin locking in half-integer quadrupolar spin systems under magic-angle spinning [53]

Spin locking was discussed briefly in Section 2.5. To elaborate, a spin system is considered to be spin locked if it can be described as a collection of populated eigenstates of the hamiltonian governing it. Such a system necessarily has a constant component of magnetization (in some direction) associated with it; this is what is required in spin locking We shall consider whether or not this is the case for conditions of slow, intermediate and fast spinning rates relative to ω_Q.

The discussion which follows is in terms of eigenstates of the spin system and their populations rather than in terms of the density operator, as it is easier to define a spin-locked state in this way. We start therefore with the eigenstates of the system during a (spin-lock) rf pulse under static, i.e. no spinning, conditions.

The eigenstates in question are simply the eigenstates of a rotating frame hamiltonian[N3] incorporating a term to describe the effects of the rf field, a quadrupolar term and a term to take account of frequency offsets, i.e. difference between the rotating frame frequency, ω_{rf}, and the Larmor frequency of the spin, ω_0. Quadrupolar effects need only be considered to first order for this discussion. Thus the hamiltonian in question is

$$\begin{aligned}\hat{H}_{\text{pulse}} &= \hat{H}_{\text{rf}} + \hat{H}_Q^{(1)} + \hat{H}_\Delta \\ &= \omega_{1S}\hat{S}_y + \omega_Q(3\hat{S}_z^2 - \hat{S}^2) + (\omega_0 - \omega_{\text{rf}})\,\hat{S}_z \qquad (5.51)\end{aligned}$$

for a y-pulse. The eigenvalues of this hamiltonian (calculated numerically) are shown in Fig. 5.21 as a function of ω_Q/ω_{1S}. The nature of the eigenstates these correspond to at extreme values of ω_Q/ω_{1S} is also indicated in this diagram. This diagram simply shows that the nature of the eigenstate and its energy depends on ω_Q, which in turn depends on the molecular/crystallite orientation within the applied field $\mathbf{B}_0$.

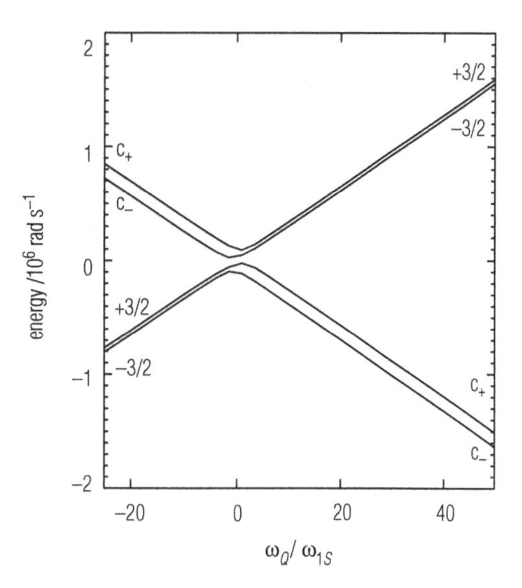

Fig. 5.21 The eigenstates of a spin-$\frac{3}{2}$ nucleus during an rf pulse. ω_Q is the quadrupole splitting defined in Equation 5.10. The eigenstates for the extreme values of ω_Q/ω_{1S} are indicated on the plot, with $c_{\pm} = \frac{1}{\sqrt{2}}\left(\left|+\frac{1}{2}\right\rangle \pm \left|-\frac{1}{2}\right\rangle\right)$. The other parameters used in the calculation are the quadrupole-coupling constant, 2 MHz, asymmetry, 0, and isotropic chemical shift, 20 ppm.

Now consider the effects of magic-angle spinning. Magic-angle spinning introduces time dependence into the system. Clearly, as the sample rotates, each crystallite experiences a change in orientation and so ω_Q for each crystallite changes. However, when the rate of spinning is very slow ($\omega_{1S}^2/\omega_Q\omega_R$ >> 1, where ω_R is the spinning rate), the change in spin state is slow enough to be adiabatic, i.e. the system remains in thermal equilibrium throughout the rotor cycle. Under these conditions, the eigenstates of the system at any point in the rotor cycle are simply those that would occur if the sample were static in that particular orientation. In other words, the eigenvalues and states of Fig. 5.21 apply. The system simply moves slowly along each of the lines in Fig. 5.21 as ω_Q changes during the spinning, with eigenstates changing smoothly between the extremes indicated. Because the system is constantly in thermal equilibrium, the eigenstates are populated according to a Boltzmann distribution. Thus, we can say that under conditions of slow speed spinning, the system can be spin locked; it is well described by a set of $2S + 1$ eigenstates with corresponding populations.

In the case of very rapid spinning, ($\omega_{1S}^2/\omega_Q\omega_R << 1$), the quadrupole interaction is being averaged rapidly relative to its size. We need only then consider the average quadrupole interaction over a rotor cycle (zero in fact if only first-order coupling is considered), which is a constant. In other words, the eigenstates of the system are effectively time independent and so, therefore, are their populations. Hence, we can also say that under conditions of fast spinning, the system can be spin locked.

However, in the case of intermediate spinning rates ($\omega_{IS}^2/\omega_Q\omega_R \approx 1$), the eigenstates of the spin system are rapidly time dependent. The ensemble average spin state for the collection of spin systems in the sample can be described as a statistical linear combination of the eigenstates, but the combination coefficients here do not correspond to eigenstate populations. This point was discussed in Chapter 1 in Section 1.2 when examining the effect of an rf pulse on the state of a collection of isolated spin-$\frac{1}{2}$ nuclei. Therefore, the system cannot be described as spin locked. The time dependence of the system means that there is no constant magnetization component in any direction.

Thus cross-polarization to the central transition of a half-integer quadrupolar spin can be conducted under slow rate (adiabatic) or very fast spinning, but not in the intermediate regime; in the latter, polarization formed in one part of the rotor cycle decays rapidly in another part.

In the adiabatic regime, it can be shown that all the *I*-spin magnetization is involved in the polarization transfer process, in contrast to the static (non-spinning) case. Moreover, *S*-spin signal can arise from direct polarization under these conditions. This can be prevented by alternating the phase of the initial *I*-spin 90° pulse along with phase alternation of the *S*-spin signal detection, while keeping the spin-locking/contact pulses at the same phase.

Finally, it is interesting to enquire whether magic-angle spinning produces a sideband-modulated Hartmann–Hahn match as it does for spin-$\frac{1}{2}$–spin-$\frac{1}{2}$ cross-polarization. In particular, in the pure spin-$\frac{1}{2}$ case under rapid magic-angle spinning, the Hartmann–Hahn match fails and is instead shifted by $\pm\omega_R$, $\pm 2\omega_R$ (see Section 2.5). The reason for this failure is that the net polarization transfer depends on the time integral of the *I–S* dipolar-coupling interaction, which vanishes under rapid magic-angle spinning (or for time periods which are integral numbers of rotor periods). However, when cross-polarizing to a quadrupolar nucleus, the polarization transfer also depends on the strength of quadrupole coupling, which, like the dipolar coupling, varies over the rotor period, but in a different manner. The net time integral of the dipolar interactions multiplied by the quadrupole coupling factor is no longer zero under magic-angle spinning. Thus, the Hartmann–Hahn match survives under magic-angle spinning, even if it is rapid compared with the dipolar-coupling strength, providing the spinning conditions are adiabatic with respect to the quadrupole coupling.

Notes

1. The same result is obtained by moving to a rotating frame of reference (one rotating about the laboratory *z*-axis at the Larmor frequency of the spin) and using average hamiltonian theory to remove the resulting time dependence. We need to use such a frame of reference

eventually to understand the outcome of NMR experiments as this is the effective frame in which we observe the spins in an experiment. We perform this frame transformation in Box 5.1. The perturbation theory used here implicitly assumes that we are in the laboratory frame. It is used here rather than the rotating frame/average hamiltonian approach as it arrives at an appropriate answer more quickly.

2. Providing the $\hat{H}_0$ is the dominant term in the hamiltonian.
3. The rotating frame here rotates about $\mathbf{B}_0$ at frequency ω_{rf}, the frequency of the spin lock rf irradiation.

References

1. Slichter, C.P. (1990) *Principles of Magnetic Resonance.* Springer-Verlag, Berlin.
2. Abragam, A. (1983) *The Principles of Nuclear Magnetism*, Chapter VII. Clarendon Press, Oxford.
3. Cohen, M.H. & Reif, F. (1957) *Solid-State Physics*, 5, 321.
4. Brink, D.M. & Satchler, G.R. (1994) *Angular Momentum.* Clarendon Press, Oxford.
5. Amoureux, J-P. (1993) *Solid-State NMR*, 2, 83.
6. Fenzke, D., Freude, D., Fröhlich, T. & Haase, J. (1984) *Chem. Phys. Lett.*, **111**, 171.
7. Man, P.P., Klinowski, J., Trokiner, A., Zanni, H. & Papon, P. (1988) *Chem. Phys. Lett.*, **151**, 143.
8. Nielsen, N. Chr., Bildsøe, H. & Jakobsen, H.J. (1992) *Chem. Phys. Lett.*, **191**, 205.
9. Poplett, I.J.F. & Smith, M.E. (1998) *Solid-State NMR*, **11**, 211.
10. Wi, S., Frydman, V. & Frydman, L. (2001) *J. Chem. Phys.*, **114**, 8511.
11. Samoson, A. (1985) *Chem. Phys. Lett.*, **119**, 29.
12. Jaeger, C., Mueller-Warmuth, W., Mundus, C. & van Wuellen, L. (1992) *J. Non-Cryst. Solids.*, **149**, 209.
13. Samoson, A., Lippmaa, E. & Pines, A. (1988) *Molec. Phys.*, **65**, 1013.
14. Chmelka, B.F., Mueller, K.T., Pines, A., Stebbins, J., Wu, Y. & Zwanziger, J.W. (1989) *Nature*, **339**, 42.
15. Bull, L.M., Cheetham, A.K., Anupold, T., Reinhold, A., Samason, A., Sauex, J., Bussemer, B., Lee, Y., Gann, S., Shore, J., Pines, A. & Dupree, R. (1998) *J. Am. Chem. Soc.*, **120**, 3510.
16. Mueller, K.T., Sun, B.Q., Chingas, G.C., Zwanziger, J.W., Terao, T. & Pines, A. (1990) *J. Magn. Reson.*, **86**, 470.
17. Grandinetti, P.J., Baltisberger, J.H., Farnan, I., Stebbins, J.F., Werner, U. & Pines, A. (1995) *J. Phys. Chem.*, **99**, 12341.
18. Frydman, L. & Harwood, J.S. (1995) *J. Am. Chem. Soc.*, **117**, 5367.
19. Medek, A., Harwood, J.S. & Frydman, L. (1995) *J. Am. Chem. Soc.*, **117**, 12779.
20. Duer, M.J. & Stourton, C. (1997) *J. Magn. Reson.*, **124**, 189.
21. Vega, S. & Naor, Y. (1981) *J. Chem. Phys.*, **75**, 75.
22. Fernandez, C. & Amoureux, J-P. (1995) *Chem. Phys. Lett.*, **242**, 449.
23. Wu, G., Rovnyak, D., Sun, B. & Griffin, R.G. (1996) *Chem. Phys. Lett.*, **249**, 210.
24. Massiot, D., Touzo, B., Trumeau, D., Coutures, J.P., Virlet, J., Florian, P. & Grandinetti, P.J. (1996) *Solid-State NMR*, **6**, 73.
25. Amoureux, J-P., Fernandez, C. & Frydman, L. (1996) *Chem. Phys. Lett.*, **259**, 347.
26. Amoureux, J-P., Fernandez, C. & Steuernagel, S. (1996) *J. Magn. Reson.*, **123**, 116.
27. Massiot, D. (1996) *J. Magn. Reson. A*, **122**, 240.
28. Wu, G., Rovnyak, D. & Griffin, R.G. (1996) *J. Am. Chem. Soc.*, **118**, 9326.
29. Ding, S. & McDowell, C.A. (1997) *Chem. Phys. Lett.*, **270**, 81.
30. Brown, S.P. & Wimperis, S. (1997) *J. Magn. Reson.*, **124**, 279.
31. Hanaya, M. & Harris, R.K. (1997) *J. Phys. Chem. A*, **101**, 6903.
32. Marinelli, L., Medek, A. & Frydman, L. (1998) *J. Magn. Reson.*, **132**, 88.

33. Ding, S. & McDowell, C.A. (1998) *J. Magn. Reson.*, **135**, 61.
34. Amoureux, J-P. & Fernandez, C. (1998) *Solid-State NMR*, **10**, 211.
35. Madhu, P.K., Goldbourt, A., Frydman, L. & Vega, S. (1999) *Chem. Phys. Lett.*, 307, 41.
36. Marinelli, L. & Frydman, L. (1997) *Chem. Phys. Lett.*, **275**, 188.
37. Duer, M.J. (1997) *Chem. Phys. Lett.*, **277**, 167; Friedrich, U., Schnell, I., Brown, S.P., Lupulesc, A., Demco, D.E. & Spiess, H.W. (1998) *Molec. Phys.*, **95**, 1209.
38. Gan, Z. (2000) *J. Am. Chem. Soc.*, **122**, 3242.
39. Ashbrook, S.E. & Wimperis, S. (2002) *J. Magn. Reson.*, **56**, 269.
40. Gan, Z. (2001) *J. Chem. Phys.*, **114**, 10845.
41. Pike, K.J., Ashbrook, S.E. & Wimperis, S. (2001) *Chem. Phys. Lett.*, **345**, 400.
42. Grandinetti, P.J., Baltisberger, J.H., Llor, A., Lee, Y.K., Werner, U., Eastman, M.A. & Pines, A. (1993) *J. Magn. Reson. A*, **103**, 72.
43. Fernandez, C., Lang, D.P., Amoureux, J.-P. & Pruski, M. (1998) *J. Am. Chem. Soc.*, **120**, 2672; Pruski, M., Bailly, A., Lang, D.P., Amoureux, J.-P. & Fernandez, C. (1999) *Chem. Phys. Lett.*, 307, 35.
44. Fyfe, C.A., Mueller, K.T., Grondey, H. & Wongmoon, K.C. (1993) *J. Phys. Chem.*, **97**, 13484.
45. Jarvie, T.P., Wenslow, R.M. & Mueller, K.T. (1995) *J. Am. Chem. Soc.*, **117**, 570.
46. Larsen, F.H., Jakobsen, H.J., Ellis, P.D. & Nielsen, N. Chr. (1998) *Molec. Phys.*, **95**, 1185.
47. Fernandez, C., Delevoye, L., Amoureux, J.-P., Lang, D.P. & Pruski, M. (1997) *J. Am. Chem. Soc.*, **119**, 6858.
48. Samoson, A. & Lippmaa, E. (1983) *Chem. Phys. Lett.*, **100**, 205.
49. Samoson, A. & Lippmaa, E. (1983) *Phys. Rev. B*, **28**, 6567.
50. Kentgens, A.P.M., Lemmens, J.J.M., Geurts, F.M.M. & Veeman, W.S. (1987) *J. Magn. Reson.*, **71**, 62.
51. Ashbrook, S.E. & Wimperis, S. (2000) *Molec. Phys.*, **98**, 1.
52. Wang, S.H., De Paul, S.M. & Bull, L.M. (1997) *J. Magn. Reson.*, **125**, 364.
53. Vega, A. (1992) *Solid-State NMR*, **1**, 17.

NMR Techniques for Studying Molecular Motion in Solids 6

6.1 Introduction

There is a huge interest in molecular motions in solids. In part, this has arisen because of the realization that many bulk material properties are dependent on the flexibility and degrees of freedom of the underlying molecules. For instance, the flexibility of a bulk polymer ultimately rests with the flexibility of the constituent molecules. The ability of a material to withstand stress depends upon the molecular degrees of freedom which can absorb the energy of the stress imposed on the material. Most solid–solid phase transformations, including glass transitions of polymers, are accompanied by the onset (or quenching) of some molecular motion, and understanding how these motions arise can lead to an understanding of the occurrence of the phase transition itself. A more general reason for studying molecular motion in solids is that any observed motion is governed by the intermolecular potential which exists in the solid. Thus by studying the geometry and rate of the motions, as a function of temperature for instance, we can experimentally probe the intermolecular potential. This is crucial if we are to understand the structure and properties of any material.

NMR is an excellent method for studying dynamics of molecules in solids; all nuclear spin interactions are in general anisotropic, i.e. they depend on the molecular orientation within the applied magnetic field of the NMR experiment. Thus, as illustrated in Fig. 6.1, a change of molecular orientation is accompanied in general by a change in strength of the chemical shielding, any dipole–dipole coupling and, for $I > \frac{1}{2}$, the quadrupole coupling.

Molecular motions in solids are incoherent processes, and as such are best described by *autocorrelation functions*. If $f(t)$ describes the time-dependent position (orientation) of a molecule in a sample of many such molecules, the autocorrelation function, $G(\tau)$ is defined by

(a)

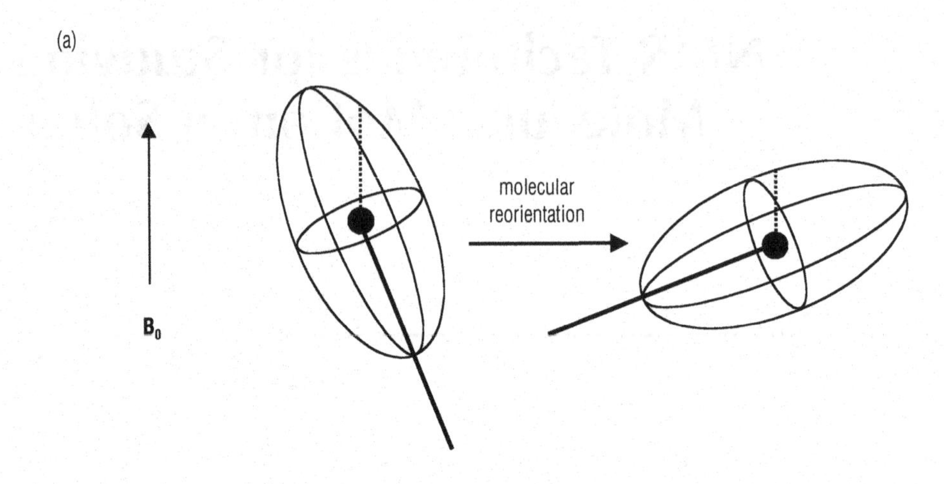

(b)

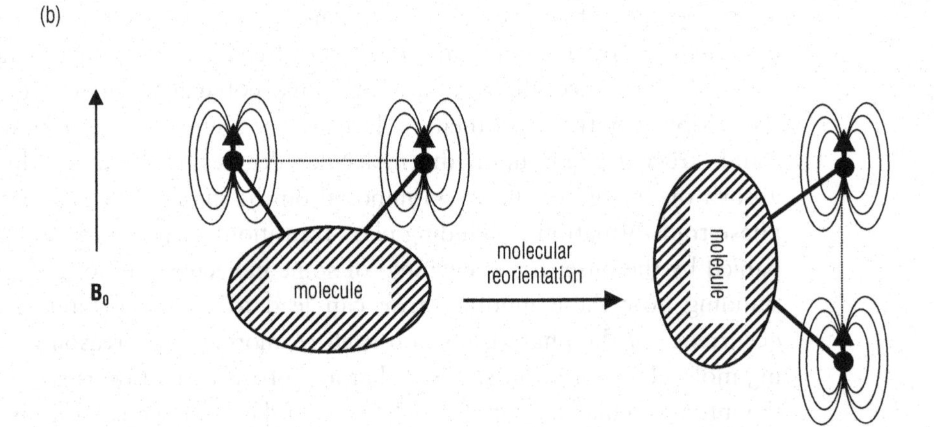

Fig. 6.1 The schematic illustration of effect of molecular reorientation on (a) chemical shift and (b) dipole–dipole coupling between two spins; the nuclei in (a) and (b) are represented by black dots. In (a), the shielding tensor is represented by an ellipsoid, whose principal axes represent the principal axis frame of the shielding tensor; the radius of the ellipsoid (dashed line) in the direction of the applied magnetic field ($\mathbf{B}_0$) is proportional to the value of the chemical shift for that molecular orientation. In (b), the strength of the dipole–dipole coupling depends on the orientation of the internuclear axis (dashed line) with respect to the applied magnetic field $\mathbf{B}_0$. The nuclear magnetic dipoles (represented by bold arrows) are orientated by the applied field. This means that the field each nuclear magnetic dipole presents to the other depends on their relative positions with respect to each other and $\mathbf{B}_0$.

$$G(\tau) = \overline{f(t)\, f(t + \tau)} \tag{6.1}$$

where the bar indicates the ensemble average over all the molecules in the sample. $G(\tau)$ is a measure of what proportion of the molecules in the sample have the same position at time $t + \tau$ as they did at time t. If most molecules have moved during the intervening period τ, and moved in many different directions, then there will be a whole variety of different values for $f(t + \tau)$ describing all the new positions of the moved molecules. The sum of $f(t)\,f(t + \tau)$ over all the molecules in the sample then gives many cancellations between the values for different molecules, resulting in a small value for $G(\tau)$. If no molecules have moved in τ, then $f(t) = f(t + \tau)$ for all molecules, and so $G(\tau)$ is large. Clearly, $G(\tau)$ will decay with increasing τ; as the gap τ between measurement of molecular positions increases, more of the molecules in the sample will have moved, resulting in decreased values for $G(\tau)$. Often the decay is assumed to be exponential, i.e.

$$G(\tau) = \exp\left(-\frac{|\tau|}{\tau_c}\right) \tag{6.2}$$

where τ_c is the *correlation time* for the molecular motion. We use the concept of a correlation time to describe molecular motion extensively throughout this chapter. The correlation time essentially provides a monitor for the speed of the motion; the smaller τ_c, the shorter the typical time between changes of molecular position.

Very slow motions ($\tau_c > 10^{-3}$ s) may be studied via two-dimensional (or higher) exchange methods. In such techniques, the strength of a particular nuclear spin interaction is monitored during the t_1 period of a two-dimensional experiment. A mixing period then follows during which molecular reorientation may occur. Finally, the new strength of the nuclear spin interaction, resulting from the change of molecular orientation/chemical site, is recorded in t_2. The final two-dimensional spectrum then correlates the strengths of the interaction during t_1 and t_2 and, from this, the angular reorientation involved in the motion can be inferred. Repeating the experiment for different mixing times allows the correlation time for the motion to be determined. These experiments are discussed in Section 6.4.

Motions with τ_c^{-1} of the order of the nuclear spin interaction anisotropy can be assessed via lineshape analysis. Here, the powder lineshape (for a powder sample) resulting from a specific nuclear spin interaction is analysed to reveal details of the molecular motion. The powder lineshapes are sensitive to motions with τ_c^{-1} of the order of the width of the powder pattern, i.e. the anisotropy of the nuclear spin interaction which causes the powder lineshape. These are generally motions with correlation times of 10^{-3} to 10^{-4} s for chemical shift and dipolar interactions, smaller for quadrupolar interactions. The specific *dynamic range*, i.e. the motional correlation times the lineshapes are sensitive to, will depend on the nucleus and its environment, and the interaction being observed. For τ_c^{-1} which are much less than the nuclear spin anisotropy, the powder pattern remains unaltered from a normal powder pattern for a static nucleus. Once the τ_c^{-1} for the motion is much greater (approximately a factor of 50 greater) than the nuclear spin interaction anisotropy, the motionally-averaged powder pattern lineshape reaches a *fast motion limit*; further decreases in the correlation time have no effect on the lineshape. Lineshape analysis for studying molecular motion is dealt with in Section 6.2.

Motions with lower correlation times (10^{-6} to 10^{-9} s) which are out of the dynamic range for lineshape analysis can be examined by spin-lattice relaxation time studies. Spin-lattice relaxation (as other relaxation processes) relies on fluctuations in nuclear spin interactions induced by molecular motion. Thus in cases where relaxation is dominated by one particular nuclear spin interaction, the spin-lattice relaxation times can be calculated for different motions and compared with experimentally derived values to reveal motional details. Other relaxation processes can also be used to study molecular motion, as discussed in Section 6.3.

In general, quadrupolar nuclei are rarely used in molecular motion studies. There are several reasons for this. For integer-spin quadrupolar nuclei, the powder pattern linewidths are of the order of the quadrupole-coupling constant, which can be of the order of MHz. Clearly, such broad lines are very difficult to observe at all, let alone use in detailed studies of molecular motion. For half-integer quadrupolar nuclei, the central transition ($+\frac{1}{2} \rightarrow -\frac{1}{2}$) is unaffected by quadrupole coupling to first order (see Chapter 5), and so is (usually) relatively easily observed. However, there is still the problem of resolving signals from different sites to overcome. The exception to all this is ^{2}H ($I = 1$) which has been and will probably continue to be used very extensively in molecular motion studies. This nucleus has a relatively small quadrupole moment and has quadrupole-coupling constants in the region of 140–220 kHz in most organic compounds, for instance. This means that its powder patterns are relatively easily observed and, moreover, are sensitive to molecular motions with correlation times in the range 10^{-4}–10^{-6} s. Its usage does not stop with lineshape studies, however. This nucleus has also been extensively used in relaxation time experiments and multidimensional exchange experiments for studying molecular motion. Accordingly, this nucleus is afforded its own section to discuss its use in molecular motion studies (Section 6.6).

6.2 Powder lineshape analysis

The orientation dependence of each nuclear spin interaction (see Chapter 1) means that, for powder samples, the NMR spectrum of a given nucleus consists of a broad powder pattern for each distinct chemical site for that nucleus. As discussed in Chapter 3, the powder pattern can be considered as being made up of an infinite number of sharp lines, one from each different molecular orientation present in the sample, the frequency of each line being determined by the molecular orientation itself. The lines from different orientations all overlap and result in the observed broad (but not featureless) line. Any molecular motion which changes a molecule's orientation, changes the spectral frequency associated with a nucleus in that molecule; the resonance line for that nucleus now moves to some other part of the powder pattern. If the motion has a τ_c^{-1} similar to the width of the powder pattern, then coalesence occurs between the lines corresponding to the different molecular orientations which arise during the course of the motion. This in turn causes distinctive distortions of the powder patterns, the distortions being dependent on both the correlation time and geometry of the molecular motion. As outlined below, powder pattern lineshapes can be simulated for likely models of the molecular motion and compared with experiment to reveal details of the molecular dynamics.

6.2.1 *Simulating powder pattern lineshapes*

In order to simulate any observable NMR spectrum, we must calculate the time evolution of the total transverse magnetization associated with the spins of interest; it is this which gives rise to the FID observed in the experiment. More precisely, the FID is proportional to the function $M^+(t)$ which describes the time evolution of the net transverse magnetization.

For the particular case of calculating a powder pattern lineshape under conditions of molecular motion, some kind of model must be assumed for the motion. It has become common practice to consider the motion as some kind of Markov process, that is, as a process involving exchange or hopping between N discrete sites, with the time taken to hop between sites being infinitesimally small compared with the residence time in each site. With this assumption, it is then straightforward to describe the time evolution of the transverse magnetization under the motional process as follows. In the *absence* of molecular motion, the time evolution of the (complex) transverse magnetization is simply given by

$$\frac{\mathrm{d}M^+(\theta, \phi; t)}{\mathrm{d}t} = M^+(\theta, \phi; t)(i\omega(\theta, \phi) + T_2^{-1}) \tag{6.3}$$

which has the solution:

$$M^+(\theta, \phi; t) = M_0^+(\theta, \phi)\exp(i\omega(\theta, \phi)t + T_2^{-1}t) \tag{6.4}$$

where θ and ϕ describe the molecular orientation in the applied field $\mathbf{B}_0$. More precisely, θ and ϕ are the polar angles describing the orientation of the applied field $\mathbf{B}_0$ in a molecule-fixed axis frame (see Fig. 6.2). T_2 is the transverse relaxation time for the spin. $\omega(\theta, \phi)$ is the resonance frequency of the spin in molecular orientation (θ, ϕ). $M_0^+(\theta, \phi)$ is the initial transverse

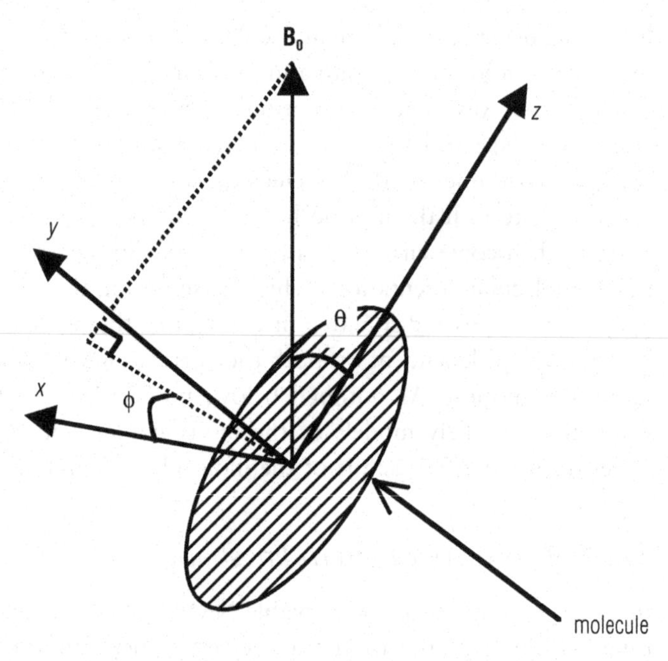

Fig. 6.2 Definition of the polar angles θ and ϕ which define the molecular orientation with respect to $\mathbf{B}_0$. The reference frame shown is a molecule-fixed frame.

magnetization associated with the particular molecular orientation. This is determined in general by the pulse sequence which generated the transverse magnetization in the experiment. $M_0^+(\theta, \phi)$ can most readily be calculated by calculating the density matrix through the pulse sequence (see Chapter 1) and then identifying the component or components of the density matrix which correspond to transverse magnetization at the end of the pulse sequence. If the transverse magnetization is generated by a single hard 90° pulse (i.e. a pulse whose amplitude, ω_1, is large compared with the nuclear spin interaction giving rise to the powder pattern), then the initial transverse magnetization $M_0^+(\theta, \phi)$ is the same for all molecular orientations, and may be nominally set to one. The time evolution of the net transverse magnetization for the whole powder sample, $M^+(t)$, is simply found from Equation (6.4) by integrating over all possible molecular orientations described by θ and ϕ:

$$\begin{aligned} M^+(t) &= \frac{1}{8\pi^2}\int_0^{2\pi}\int_0^{\pi} M^+(\theta, \phi; t)\sin\theta \, d\theta \, d\phi \\ &= \frac{1}{8\pi^2}\int_0^{2\pi}\int_0^{\pi} M_0^+(\theta, \phi)\exp(i\omega(\theta, \phi)t + T_2^{-1}t)\sin\theta \, d\theta \, d\phi \end{aligned} \tag{6.5}$$

To take account of molecular reorientations using a Markov model, Equation (6.3) is simply modified to [1]:

$$\frac{d\mathbf{M}^+(\theta, \phi; t)}{dt} = \mathbf{M}^+(\theta, \phi; t)(i\boldsymbol{\omega}(\theta, \phi) + T_2^{-1} + \boldsymbol{\Pi}) \tag{6.6}$$

$\mathbf{M}^+(\theta, \phi; t)$ is now an N-dimensional vector, each component being the complex transverse magnetization (i.e. $M_x + iM_y$) from one of the N sites involved in the motional process. $\boldsymbol{\omega}$ is an $N \times N$ diagonal matrix whose elements are the resonance frequencies associated with the N sites for a crystallite orientation (θ, ϕ). The matrix $\boldsymbol{\Pi}$ describes the exchange of magnetization between the N sites as a result of the molecular hopping process; it is also an $N \times N$ matrix whose elements Π_{ij} are given by

$$\Pi_{ij} = \tau_{c,ij}^{-1} p_j \quad \text{and} \quad \Pi_{ii} = -\sum_{j(\neq i)}^{N-1} \Pi_{ij} \tag{6.7}$$

where $\tau_{c,ij}^{-1}$ is the inverse of the correlation time for hopping from site j to site i and p_j is the population of site j. The solution to Equation (6.6) is analogous to Equation (6.4). The net transverse magnetization summed over all the molecular orientations in the powder sample is then

$$\begin{aligned}\mathbf{M}^+(t) &= \frac{1}{8\pi^2}\int_0^{2\pi}\int_0^{\pi} \mathbf{M}_0^+(\theta, \phi)\exp(i\boldsymbol{\omega}(\theta, \phi)t + T_2^{-1}t + \boldsymbol{\Pi})\sin\theta \, d\theta \, d\phi \\ &= \frac{1}{8\pi^2}\int_0^{2\pi}\int_0^{\pi} \mathbf{M}_0^+(\theta, \phi)\mathbf{L}(\theta, \phi; t)\sin\theta \, d\theta \, d\phi\end{aligned} \tag{6.8}$$

where the propagator $\mathbf{L}(\theta, \phi; t)$ is

$$\mathbf{L}(\theta, \phi; t) = \exp(i\boldsymbol{\omega}(\theta, \phi)t + T_2^{-1}t + \boldsymbol{\Pi}) \tag{6.9}$$

The propagator $\mathbf{L}(\theta, \phi; t)$ is calculated by diagonalizing the matrix ($i\omega(\theta, \phi)t + T_2^{-1} + \boldsymbol{\Pi}$) to find its eigenvectors, $\mathbf{V}$, and eigenvalues, $\mathbf{A}$. The exponential in Equation (6.9) is then found from

$$\mathbf{L}(\theta, \phi; t) = \mathbf{V}^{-1}\exp(\mathbf{A})\mathbf{V} \tag{6.10}$$

As the eigenvalue matrix $\mathbf{A}$ is diagonal, $\exp(\mathbf{A})$ is also diagonal with elements $\exp(A_{ii})$.

Some comments on the calculation of the elements of ω, i.e. the resonance frequencies of the N sites for a molecular orientation (θ, ϕ), may be useful at this point. We consider the specific case of the chemical shift anisotropy powder pattern, but the same principles apply to all nuclear spin interactions. As shown in Chapter 3, the chemical shift contribution to the spectral frequency for a given molecular orientation is given by

$$\omega_{cs}(\theta, \phi) = -\omega_0 \mathbf{b}_0^{PAF} \boldsymbol{\sigma}^{PAF} \mathbf{b}_0^{PAF} \tag{6.11}$$

where σ^{PAF} is the shielding tensor in its principal axis frame (*PAF*). The shielding tensor principal axis frame can be taken as the molecule-fixed frame which is used to describe molecular orientation. $\mathbf{b}_0^{PAF}$ is the unit vector in the direction of $\mathbf{B}_0$ in the shielding tensor principal axis frame for the particular molecular orientation, $\mathbf{B}_0$ having an orientation in this frame given by the spherical polar angles (θ, ϕ) (see Fig. 6.2). Thus, $\mathbf{b}_0^{PAF}$ is given by

$$\mathbf{b}_0^{PAF} = (\sin\theta\cos\phi, \sin\theta\sin\phi, \cos\theta) \tag{6.12}$$

After a hop changing the molecular orientation by the Euler angles $(\alpha^{hop}, \beta^{hop}, \gamma^{hop})$ (see Box 1.2, Chapter 1, for definition of Euler angles), the new shielding tensor, expressed in the original principal axis frame is given by

$$\boldsymbol{\sigma}_{new}^{PAF} = \mathbf{R}(\alpha^{hop}, \beta^{hop}, \gamma^{hop})\boldsymbol{\sigma}\mathbf{R}^{-1}(\alpha^{hop}, \beta^{hop}, \gamma^{hop}) \tag{6.13}$$

where $\mathbf{R}(\alpha^{hop}, \beta^{hop}, \gamma^{hop})$ is the rotation matrix (see Box B1.2 in Chapter 1 for discussion of rotations) describing the rotation of an object through the Euler angles $(\alpha^{hop}, \beta^{hop}, \gamma^{hop})$. The new chemical shift contribution to the spectral frequency after the hop is

$$\omega_{cs,new} = -\omega_0\mathbf{b}_0^{PAF}\boldsymbol{\sigma}_{new}^{PAF}\mathbf{b}_0^{PAF} \tag{6.14}$$

Throughout this discussion, we have referred all quantities to the shielding tensor principal axis frame in the initial (before hop) molecular orientation. The principal axis frame is a molecule-fixed frame of reference, and so moves when the molecule hops. Often, however, it is more convenient to express all the shielding tensors associated with the N sites involved in the motional process with respect to some frame of reference fixed in the crystallite that the molecule is in, and so does not move when the molecule hops. We will label this frame 'cryst' (Fig. 6.3). The shielding tensor for each site i expressed in this frame is simply

$$\boldsymbol{\sigma}_i^{cryst} = \mathbf{R}^{-1}(\alpha_i, \beta_i, \gamma_i)\boldsymbol{\sigma}_i^{PAF}\mathbf{R}(\alpha_i, \beta_i, \gamma_i) \tag{6.15}$$

where the Euler angles α_i, β_i, γ_i describe the rotation of the shielding tensor PAF in site i into the crystallite frame. σ_i^{PAF} is the shielding tensor in the principal axis frame for site i. The chemical shift contribution to the spectral frequency for each site is then

$$\omega_{cs,i}(\theta, \phi) = -\omega_0\mathbf{b}_0^{cryst}\boldsymbol{\sigma}_i^{cryst}\mathbf{b}_0^{cryst} \tag{6.16}$$

where $\mathbf{b}_0^{cryst}$ is given by an equivalent expression to that in Equation (6.12), but where θ and ϕ now describe the orientation of $\mathbf{B}_0$ in the *crystallite* frame. If the molecular motion is a rotation about a given axis, then it is sensible to make the z-axis of the crystallite frame coincide with the motion rotation axis. Then the Euler angles α_i, β_i correspond to the polar angles describing the orientation of the motion rotation axis in the (PAF, i) frame, while the third angle γ_i gives the angle of rotation about the molecular rotation axis. So, for instance, for an N site rotation about an axis inclined at an

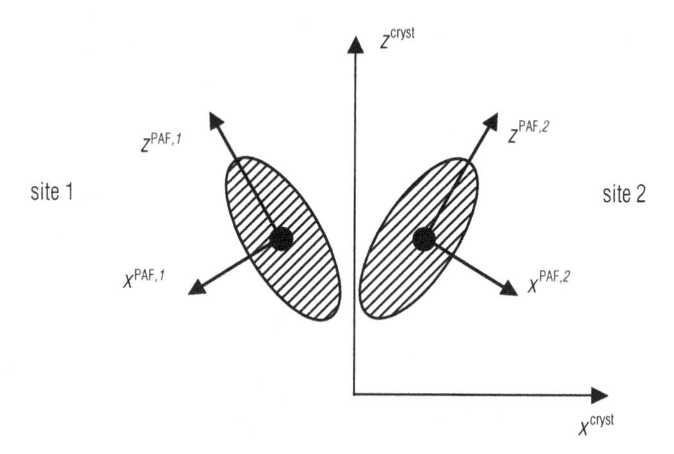

Fig. 6.3 Illustrating the crystallite-fixed frame of reference. The hashed shape represents a molecule (or part of) in which there is a nuclear spin (black dot). The molecule is involved in a motional process which means that it can exist in one of the two orientations of site shown. The shielding tensor for the nucleus in each site has a principal axis frame labelled by PAF, i where i denotes the particular site of the molecule.

angle Θ to all the PAF z-axes, the Euler angles defining the N sites are given by $\alpha_i = (i - 1)\ 360/N$; $\beta_i = \Theta$; $\gamma_i = 360 - (i - 1)\ 360/N$ (Fig. 6.4).

The final FID is proportional to the sum of the elements in the N-dimensional vector $\mathbf{M}^{+}(t)$ of Equation (6.8), each element being proportional to the FID at time t resulting from each of the N sites. Setting the proportionality constant between the FID, $F(t)$ and the time evolution of the transverse magnetization to be unity, the FID under conditions of a Markov motional process is given by

$$F(t) = \sum_{i}^{N} \mathbf{M}_i^{+}(t) \tag{6.17}$$

where the sum is over the N sites involved in the motional process. In the limit where the transverse magnetization is produced by an ideally hard 90° pulse, as discussed above, Equations (6.8) and (6.17) combine to give

$$\begin{aligned} F(t) &= \frac{1}{8\pi^2} \int_0^{2\pi} \int_0^{\pi} \sum_{i}^{N} \sum_{j}^{N} \{p_i L(t)_{ij}\} \sin\theta \, d\theta \, d\phi \\ &= \frac{1}{8\pi^2} \int_0^{2\pi} \int_0^{\pi} \{\mathbf{p} \cdot \mathbf{L}(t) \cdot \mathbf{1}\} \sin\theta \, d\theta \, d\phi \end{aligned} \tag{6.18}$$

where $\mathbf{p}$ is an N-dimensional vector containing the populations p_i of each of the N sites in the motional process. θ and ϕ are the polar angles describing the orientation of $\mathbf{B}_0$ in whichever crystallite-fixed frame has been used as the frame of reference for calculating the N site spectral frequencies.

Some examples of chemical shift anisotropy powder lineshapes calculated using Equation (6.18) for a selection of different motional models are shown

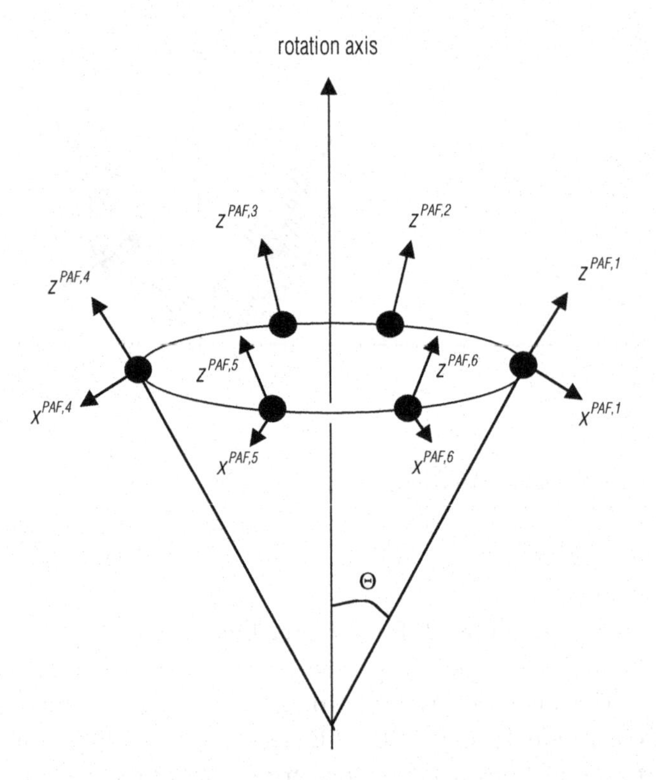

Fig. 6.4 Illustrating the relative orientations of the shielding tensor principal axis frames for the sites (in this case, six) involved in rotational motion about the axis indicated. If the rotation axis is taken as being the *z*-axis of the crystallite fixed frame of reference, the Euler angles describing the relative orientations of the principal axis frames to the crystallite-fixed frame are those given in the text.

in Fig. 6.5. Also given with this figure are the $\mathbf{\Pi}$ matrices used in the calculations.

Lineshapes in the limit of fast motion

As intimated previously, once the rate of molecular reorientation is significantly faster than the width of the powder pattern, the motionally averaged powder pattern does not change further with any further increases in the reorientation rate. It is particularly simple to calculate the fast-speed limit powder pattern within the limits of a Markov model for the molecular motion. The powder lineshape in this limit is governed by a motionally-averaged interaction tensor which is simply the average interaction tensor over the N sites involved in the motional process. The simplest way to calculate this is to express the interaction tensors for each site with respect to a single crystallite-fixed axis frame, using Equation (6.15) above. The motionally-averaged interaction tensor in the crystallite frame is then just

$$\boldsymbol{\sigma}_{\text{avg}}^{\text{cryst}} = \sum_{i}^{N} p_i \boldsymbol{\sigma}_i^{\text{cryst}} \tag{6.19}$$

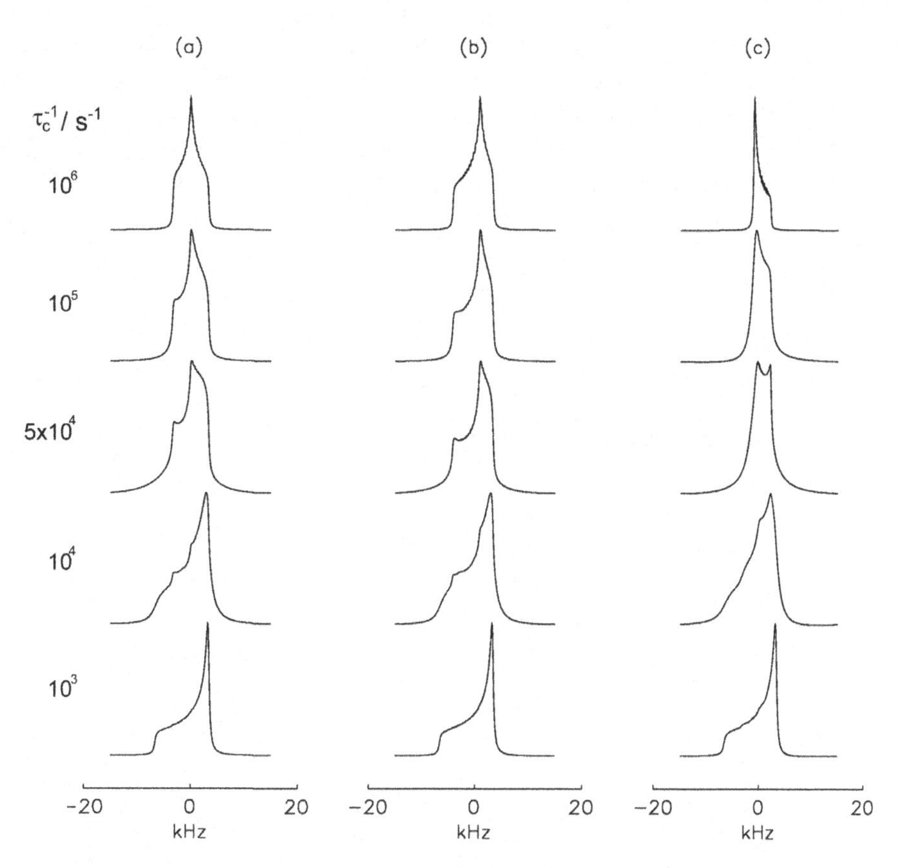

Fig. 6.5 Some chemical shift anisotropy lineshapes under conditions of molecular motion. Three different models of molecular motion are considered: (a) two-site hopping, chemical shift tensor principal z-axis reorientates by 109.5°, (b) two-site hopping, chemical shift tensor principal z-axis reorientates by 120° and (c) three-site hopping about a rotation axis orientated at 70.5° to the chemical shift tensor principal z-axis in each site. In all cases, the chemical shift tensor is axially symmetric and the populations of each site are equal. The τ_c^{-1} (Ω) for each case are given with the spectra. The $\mathbf{\Pi}$ matrices used in the calculations are

$$\begin{pmatrix} -\frac{1}{2}\Omega & \frac{1}{2}\Omega \\ \frac{1}{2}\Omega & -\frac{1}{2}\Omega \end{pmatrix} \text{ for models (a) and (b) and } \begin{pmatrix} -\frac{2}{3}\Omega & \frac{1}{3}\Omega & \frac{1}{3}\Omega \\ \frac{1}{3}\Omega & -\frac{2}{3}\Omega & \frac{1}{3}\Omega \\ \frac{1}{3}\Omega & \frac{1}{3}\Omega & -\frac{2}{3}\Omega \end{pmatrix} \text{ for model (c).}$$

Note that the lineshapes show the greatest change with hopping rate when τ_c^{-1} is of the order of the chemical shift anisotropy (10 kHz for these simulations).

where p_i is the population of each site i and σ_i^{cryst} is the interaction tensor for site i expressed in the crystallite-fixed frame. If the resulting interaction tensor is not diagonal in the chosen crystallite axis frame, it can always be diagonalized, yielding an effective principal axis frame for the motionally-averaged tensor and effective principal values, from which the effective (or motionally-averaged) interaction anisotropy and asymmetry may be determined.

A particularly simple case is that of axial reorientation about a single axis inclined at an angle Θ to the principal z-axis of the interaction tensor in the static molecule (see Fig. 6.4). This results in an axially-symmetric effective interaction tensor in the fast motion limit, with the unique axis of the effective tensor lying along the molecular rotation axis. The effective anisotropy, Δ_{eff} is given by

$$|\Delta_{eff}| = \Delta_0 \cdot \frac{1}{2}(3\cos^2\Theta - 1) \qquad (6.20)$$

where Δ_0 is the interaction anisotropy in the static molecule.

Limitations of lineshape analysis

As with any kind of analysis which relies on fitting simulated to experimental data, one must always be aware that there are potentially several 'fits', i.e. different motional models which for some motional rates give equally good fits to the experimental data. In the absence of any other data, all these fits must be considered equally possible descriptions of the actual motional process. Often, it is possible to acquire other data, i.e. use a different nucleus or intrinsic knowledge of the nature of the molecular system to remove such ambiguities. However, it is still essential to search for all possible fits, and not to be content with the first one found.

In more complex systems, the motion may be a composite motion, for instance involving rotations about several different axes with different correlation times simultaneously. In inhomogeneous systems, or glassy systems, such as polymers above the glass transition temperature, there is likely to be a distribution of correlation times and reorientation amplitudes. In either of these cases, it is likely that the lineshape is rather featureless over a wide temperature range. Under these circumstances, it may not be possible to extract unambiguous information on the motional components in the system from one-dimensional lineshape analysis; although all the information about the motion is contained in the lineshape, the information about different motional components is not resolved.

One of the problems with lineshape analysis for studying molecular motion is the model dependency of the analysis. There may well be cases where a Markov model does not describe the motional process well. Markov processes, in effect, assume the motion is a hopping between very sharp and very deep potential wells (the 'sites'). Hopping between broad, shallow potential wells goes against the main assumption of a Markov process, that the time taken for hopping is small compared with the residence time in a given site. Furthermore, broad potential wells allow diffusion within a well, so that each site is ill defined. The diffusion process itself is not well described by a Markov model. This factor should always be borne in mind when analysing lineshapes with a Markov model.

The parameters arising from a Markov model, namely the relative angular orientation of the sites involved in the dynamics process, the correlation time for hopping between sites and the populations of those sites, are not always easy to interpret at a molecular level. To be more specific, the site orientation information that arises from a lineshape analysis using a Markov model is the relative angular orientation of the nuclear spin interaction principal axis frame in the *N* different sites involved in the motion. This information is only useful, however, if the orientation of the principal axis frame is known relative to some molecular frame, as only then can the *molecular* motion (as opposed to the *principle axis frame* motion) be revealed.

Finally, it is important to understand that the vast majority of (if not all) molecular motion processes that we are likely to study by NMR are *incoherent* processes. Thus, when we refer to, for instance, a three-site hopping process about a particular axis, we do not mean that there is a coherent rotation about the axis akin to magic-angle spinning. Such a process would give rise to sidebands in the NMR spectrum. Rather, we mean a more random process, where the correlation time describes the typical time between intersite hops.

6.2.2 Resolving powder patterns

For simple materials, where there is only one (or a few) chemical sites, powder patterns can be easily measured on static, i.e. non-spinning, samples in simple one-dimensional experiments. However, many materials of interest are complex, with many different chemical sites for a given nuclear species. In a static experiment in such cases, powder patterns from different sites overlap and the concomitant lack of resolution prohibits analysis of the lineshapes. Consequently, much effort in recent years has been directed towards resolving powder patterns from different chemical sites, for the purposes of studying molecular motion.

Magic-angle spinning

One of the simplest methods is to use slow-rate magic-angle spinning. Magic-angle spinning has the effect of averaging second-rank terms in the nuclear spin hamiltonian to zero, and so removing the effects of chemical shift anisotropy, etc., from the NMR spectrum. This in itself would then remove the useful information on molecular reorientations which is contained in the anisotropic parts of the spectrum. However, under slow-rate spinning (spinning rate less than the powder pattern linewidth), spinning sidebands appear in the spectrum. For inhomogeneous nuclear spin interactions (chemical shift, heteronuclear dipole–dipole coupling and quadrupole coupling) the sidebands are sharp, and thus sideband patterns from different chemical sites are relatively easily resolved. The important point, however, is that the intensities of spinning sidebands are dependent on the anisotropic parts of the nuclear spin interaction and thus on any motional process in the sample; the linewidth of spinning sidebands is also affected by molecular motion. Thus, simulation of magic-angle spinning sideband patterns for particular models of molecular motion can yield information on molecular motions in much the same way as for static powder patterns. One interesting consequence of magic-angle spinning is that it affects the dynamic range of the experiment, i.e. what molecular motion correlation times the sideband pattern is sensitive to. This is because the width of the sideband pattern which results from magic-angle spinning depends on the rate of spinning; faster spinning reduces the width of the sideband pattern, i.e. fewer sidebands. The motions that spinning sideband patterns are sensitive to are those of the order of the width of the sideband pattern, and thus are determined in part by the sample spinning rate, as well as the nuclear spin interaction anisotropy.

The simulation of magic-angle spinning sideband patterns, resulting from chemical shift anisotropy for instance, is straightforward, if at times computationally intensive. The process is similar to that given above for the static, or non-spinning, case. However, under magic-angle spinning, the spectral frequency ω_i for each site i and crystallite orientation is now periodically time dependent, due to the time-dependent sample reorientation within the magnetic field, $\mathbf{B}_0$. This is discussed in detail in Section 2.2, for the example of chemical shift anisotropy. To recap the results from that section, the time-dependent chemical shift frequency for a site i involved in the motional process is given by

$$\omega_{\mathrm{cs},i}(t) = -\omega_0 \mathbf{b}_0^{\mathrm{R}} \boldsymbol{\sigma}_i^{\mathrm{R}} \mathbf{b}_0^{\mathrm{R}} \tag{6.21}$$

where $\mathbf{b}_0^{\mathrm{R}}$ is the unit vector in the direction of $\mathbf{B}_0$ in a rotor-fixed axis frame R. In turn, $\mathbf{b}_0^{\mathrm{R}}$ is given by

$$\mathbf{b}_0^{\mathrm{R}} = (\sin\theta_{\mathrm{R}} \cos\omega_{\mathrm{R}}t, \sin\theta_{\mathrm{R}} \sin\omega_{\mathrm{R}}t, \cos\theta_{\mathrm{R}}) \tag{6.22}$$

where θ_{R} is the sample spinning angle with respect to $\mathbf{B}_0$ and ω_{R} is the spinning speed in radians s^{-1}. The crystallite-orientation dependence of this resonance frequency will become apparent shortly. $\boldsymbol{\sigma}_i^{\mathrm{R}}$ is the shielding tensor expressed in the rotor axis frame for a spin in site i. In turn, $\boldsymbol{\sigma}_i^{\mathrm{R}}$ can be found from the shielding tensor expressed in a crystallite-fixed frame (cryst) by:

$$\boldsymbol{\sigma}_i^{\mathrm{R}} = \mathbf{R}^{-1}(\psi, \theta, \phi)\boldsymbol{\sigma}_i^{\mathrm{cryst}}\mathbf{R}(\psi, \theta, \phi) \tag{6.23}$$

where the Euler angles (ψ, θ, ϕ) express the rotation of the crystallite frame into the rotor frame, and so describe the crystallite orientation (in the rotor axis frame). The reference frames involved in the calculation of the resonance frequency under magic-angle spinning are illustrated in Fig. 6.6.

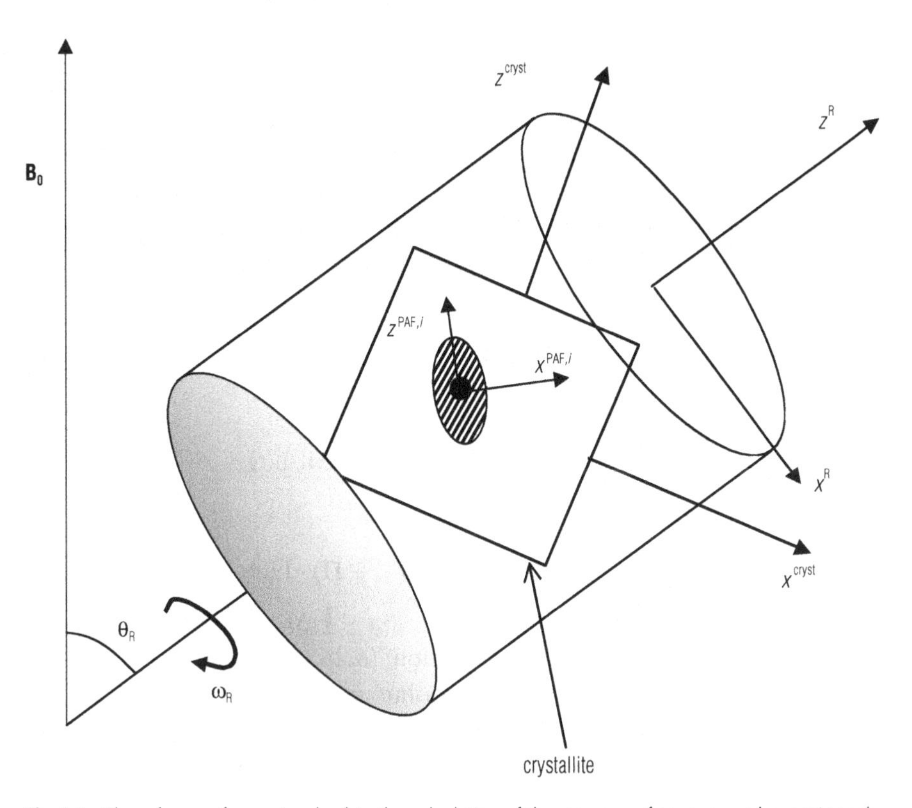

Fig. 6.6 The reference frames involved in the calculation of the resonance frequency under magic-angle spinning. One possible molecular orientation or site (labelled *i* in the text) within a single crystallite is shown; the molecule is represented by the hashed shape. In a motional process, the molecule changes orientation between *N* different sites. The shielding tensor for a nucleus (black dot) in the molecule has a principal axis frame labelled ($x^{\mathrm{PAF},\,i}$, $y^{\mathrm{PAF},\,i}$, $z^{\mathrm{PAF},\,i}$). Its orientation relative to a crystallite-fixed frame (x^{cryst}, y^{cryst}, z^{cryst}) is given by the Euler angles (α_i, β_i, γ_i). These Euler angles are clearly determined by the choice of crystallite-fixed frame, which can be any convenient frame. The orientation of the crystallite-fixed frame relative to a rotor-fixed frame (x^{R}, y^{R}, z^{R}) is given by the Euler angles (ψ, θ, ϕ); these angles vary with crystallite orientation, and so are summed over when calculating spectra for powder samples. The rotor-fixed frame conventionally has its *z*-axis along the rotor spinning axis. Finally, the orientation of the rotor frame relative to the laboratory frame *z*-axis ($\mathbf{B}_0$) is given by the Euler angles ($-\omega_{\mathrm{R}}t$, θ_{R}, 0), where θ_{R} is the spinning angle (54.74° for magic-angle spinning) and ω_{R} is the spinning rate (rad s^{-1}).

Note that in this magic-angle spinning case, three angles are required to describe the crystallite orientation, as compared to two (θ and ϕ) in the static, non-spinning case. This reduces to two (θ, ϕ) in the case of a shielding tensor with axial symmetry. Equation (6.15) above has already described how to determine $\sigma_i^{\mathrm{cryst}}$ from the shielding tensor in its principal axis frame, so we can determine the required chemical shift frequencies for the calculation of the time evolution of the transverse magnetization under magic-angle spinning conditions.

However, there are further considerations. The analytic integration of Equation (6.6) to provide a solution for the time evolution of the transverse magnetization under molecular motion performed previously is only valid when the matrix of spectral frequencies for each site $\omega(\theta, \phi)$ is time independent. When $\omega(\theta, \phi)$ is time dependent, as it is under magic-angle spinning, the matrix $(i\omega(t) + T_2 + \Pi)$ does not commute with itself at different times t, so there is no analytical solution of the differential equation equivalent to Equation (6.6) in this case. Instead, the integration must be performed numerically [2]. If we divide one rotor period up into n equally spaced periods of $\Delta t = \tau_R/n$, we can say that the solution of

$$\frac{d\mathbf{M}^+(t)}{dt} = \mathbf{M}^+(t)(i\omega(t) + T_2^{-1} + \Pi) \tag{6.24}$$

is

$$\mathbf{M}^+(t) = \mathbf{M}_0^+\mathbf{L}(t) \tag{6.25}$$

where $\mathbf{L}(t)$ is found iteratively from

$$\mathbf{L}(m\Delta t) = \exp(i\omega(t_m) + T_2^{-1} + \Pi) \cdot \mathbf{L}((m-1)\Delta t) \qquad \mathbf{L}(0) = \mathbf{1} \tag{6.26}$$

where t_m is the time point $t_m = (m - \frac{1}{2})\Delta t$ and m is an integer. The exponential of matrices in Equation (6.26) is found using Equation (6.10). Clearly, the number of time points n considered in this procedure needs to be adjusted until convergence is established. Once $\mathbf{L}(t)$ has been estimated for one rotor period, the values at all subsequent times may be derived from

$$\mathbf{L}(t + M\tau_R) = \mathbf{L}(\tau_R)^M \mathbf{L}(t) \tag{6.27}$$

Some examples of sideband patterns simulated for a selection of different motional models are shown in Fig. 6.7.

Two-dimensional techniques

The other general way of resolving powder patterns from different chemical sites is to generate multidimensional NMR spectra in which the desired powder patterns (or magic-angle spinning sideband patterns) are resolved in one dimension, separated according to (for instance) isotropic chemical shift in another dimension. Methods for resolving chemical shift anisotropy powder patterns for motional studies are discussed below. Those for ^{2}H quadrupolar powder patterns are discussed in Section 6.5.

A great many techniques exist for separating chemical shift anisotropy powder patterns. Three of the most used are discussed in Chapter 3 and are the methods due to Tycko *et al.* [3], variable-angle correlation spectroscopy (VACSY) [4] and the magic-angle turning experiment [5]. The first two of these have been adapted for molecular motion studies. However, the magic-angle turning experiment generates the isotropic dimension of the two-dimensional experiment by employing very slow speed spinning (<100 Hz), where it can be assumed that the rotor, and hence the nuclear spins, are

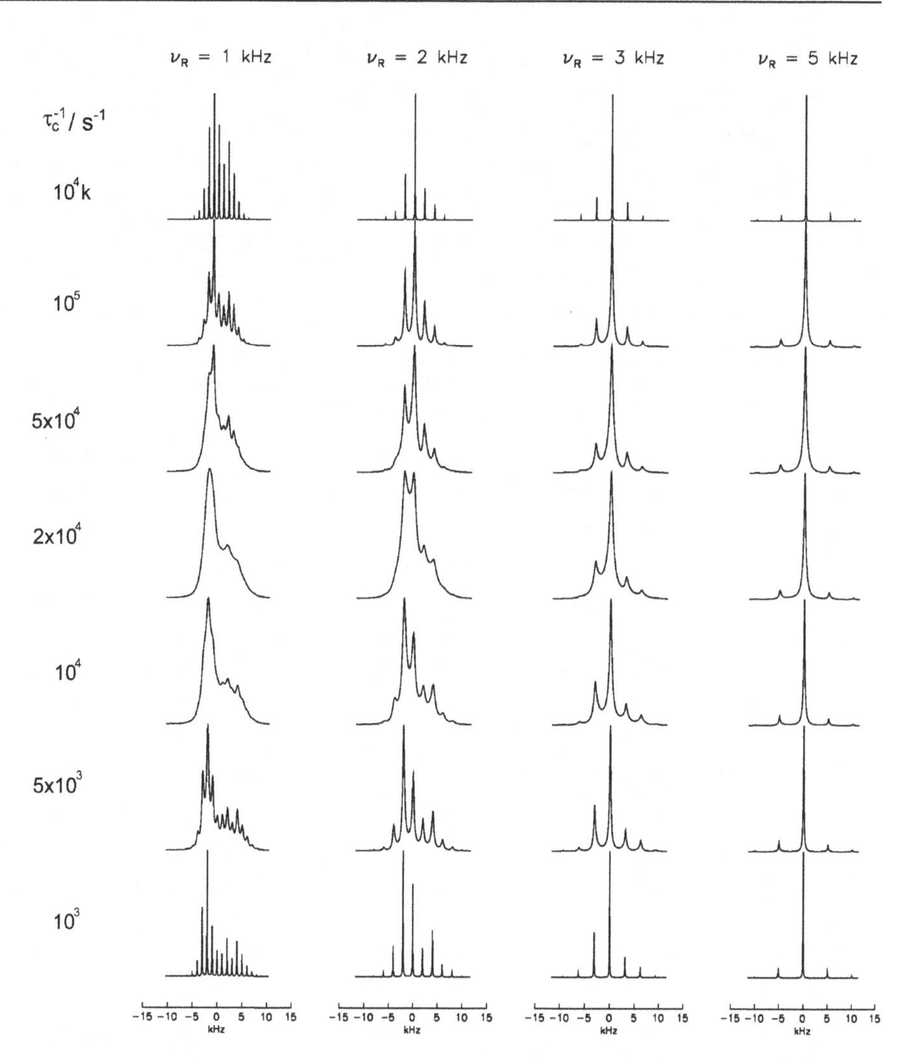

Fig. 6.7 Magic-angle sideband patterns arising from chemical shift anisotropy under conditions of molecular motion for different correlation times of motion and different spinning rates. The motional model considered is a two-site hopping, with the shielding tensor principal *z*-axis reorientating by 120° and equal populations of the two sites. The shielding tensor is axial with anisotropy (Δ_{cs}) of 10 kHz. For each spinning rate (ν_R), the sideband pattern is calculated for several different motional correlation times τ_c, given in the figure as τ_c^{-1}. It should be noted that the degree to which the pattern is distorted by the motion, i.e. the sensitivity of the pattern to the motion, depends upon the spinning rate, as well as τ_c^{-1}/Δ_{cs}.

effectively stationary for periods of $t_1/3$. Any molecular reorientation during t_1 would invalidate this assumption and lead to a much-broadened spectrum in the corresponding f_1 dimension of the final two-dimensional spectrum.

The VACSY experiment is a two-dimensional experiment, where FIDs are recorded in successive experiments in which the spinning angle is varied. Subsequent processing of the two-dimensional dataset produces a two-dimensional experiment with an isotropic spectrum in one dimension (f_2) and chemical shift powder patterns in the other (f_1); for further details see

Section 3.3.2. These powder patterns reflect motional processes in a similar manner to a normal, one-dimensional experiment, although the motionally-averaged lineshapes are different from those expected from a one-dimensional spectrum of a chemical shift anisotropy powder pattern [6]. They can, however, still be simulated to reveal the details of the molecular dynamics. Moreover, the dynamic range of the experiment is different from that of the normal, one-dimensional chemical shift anisotropy powder pattern measurement; in particular, the VACSY experiment is sensitive to slower motions than the one-dimensional experiment. To understand this, it is important to remember that powder patterns are sensitive to motions which have τ_c^{-1} of the order of the interaction anisotropy. When spinning the sample about an axis inclined at an angle θ_R to the applied magnetic field, the chemical shift anisotropy is averaged to an effective value of $\frac{1}{2}(3\cos^2\theta_R - 1)$ times the true chemical shift anisotropy, i.e. it is averaged to a smaller value in general (zero if θ_R is the magic angle). In the VACSY experiment, the sample is sequentially spun at different angles, and for every spinning angle (except $\theta = 0$), the effective chemical shift anisotropy is reduced from its true value. Thus, at all spinning angles, the spectrum recorded is sensitive to motions with longer correlation times than is the normal one-dimensional powder pattern.

The chemical shift anisotropy recoupling scheme of Tycko *et al.* [3] is particularly useful for studying molecular motions in complex solids, as its dynamic range can be 'tuned'. In Tycko's experiment, the chemical shift anisotropy is removed with magic-angle spinning, but then reintroduced during the t_1 period by series of $(2n + 2)$, $n = 0, 1, 2, 3, \ldots$ 180° pulses per rotor period; see Section 3.3.1 for further details. The effective chemical shift anisotropy which acts during t_1 is scaled from the true chemical shift anisotropy by an amount which depends on the exact timing of the pulses, with scalings of 0.4 to near zero being possible. The scaled chemical shift powder patterns, which appear in the ω_1 dimension of the final two-dimensional spectrum, are sensitive to molecular motions with τ_c^{-1} of the order of the *scaled* chemical shift anisotropy. Thus the experimenter can choose the dynamic range of this experiment to suit the particular motional processes in their sample at the temperature of interest. In the limit of ideal, hard 180° pulses, the motionally-averaged lineshapes in f_1 are identical to those which would arise from a one-dimensional static experiment, but where the chemical shift anisotropy is scaled by the particular factor in the two-dimensional experiment. Some examples of simulated f_1 chemical shift anisotropy powder patterns arising from this experiment are shown in Fig. 6.8, for dimethylsulphone (DMS) for different chemical shift anisotropy scaling factors. Note that the lineshape begins to show distortion due to molecular reorientation when the rate of motion is of the order of the scaled

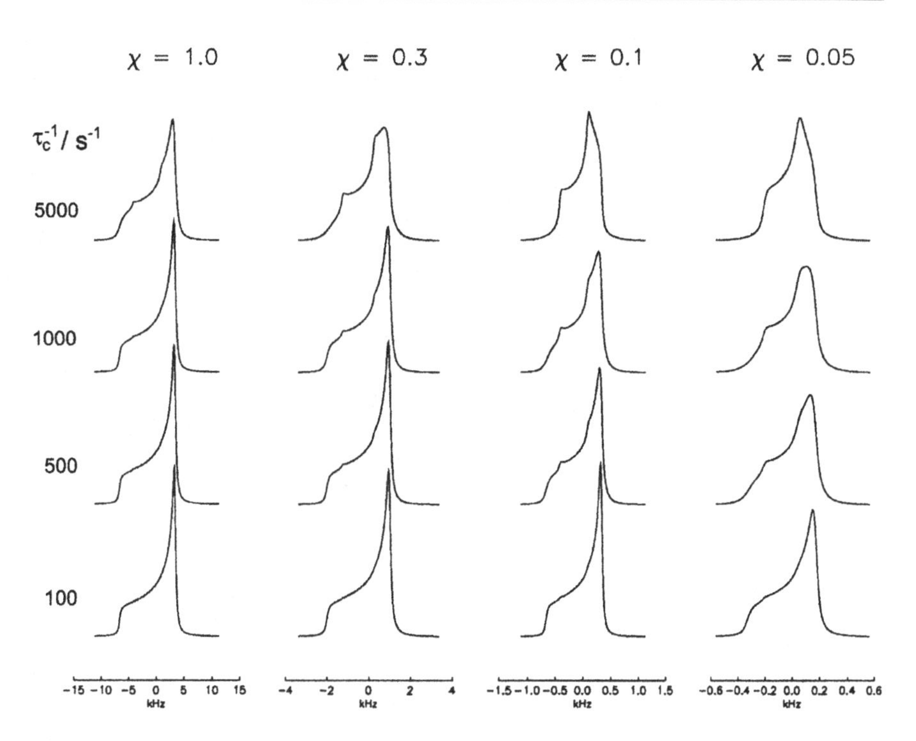

Fig. 6.8 Some simulated f_1 lineshapes arising from the chemical shift anisotropy recoupling experiment due to Tycko *et al.* [3] under conditions of molecular motion. This experiment scales the chemical shift anisotropy by a factor χ, and the resulting f_1 powder lineshapes are sensitive to motions with τ_c^{-1} of the order of the *scaled* chemical shift anisotropy. Thus the experiment is excellent for examining slow motions, as the scaling factor can be chosen by the experimenter (see text for details). Note, we use f_1 here to denote a spectral frequency axis in units of Hz, i.e. $\omega_1/2\pi$.

chemical shift anisotropy. The lineshape reaches the fast motion limit once the rate of motion is significantly larger than the scaled chemical shift anisotropy.

6.2.3 Using homonuclear dipolar-coupling lineshapes – the WISE experiment

^{1}H–^{1}H dipolar-coupling lineshapes in organic solids are frequently used as a qualitative monitor of motion in a sample. In such samples, the abundance of ^{1}H spins creates a strongly coupled spin network, which gives rise to a ^{1}H NMR spectrum which is simply a broad, gaussian line, usually of the order of several tens of kHz wide. As highlighted in Section 2.2.5, slow rate magic-angle spinning (significantly slower than the ^{1}H linewidth) has little effect on this and very rapid spinning (>30 kHz in general) is needed to produce a high-resolution spectrum. Intermediate spinning rates create sideband patterns consisting of rather broad sideband lines which are poorly resolved from one another so that the net ^{1}H lineshape is still only slightly

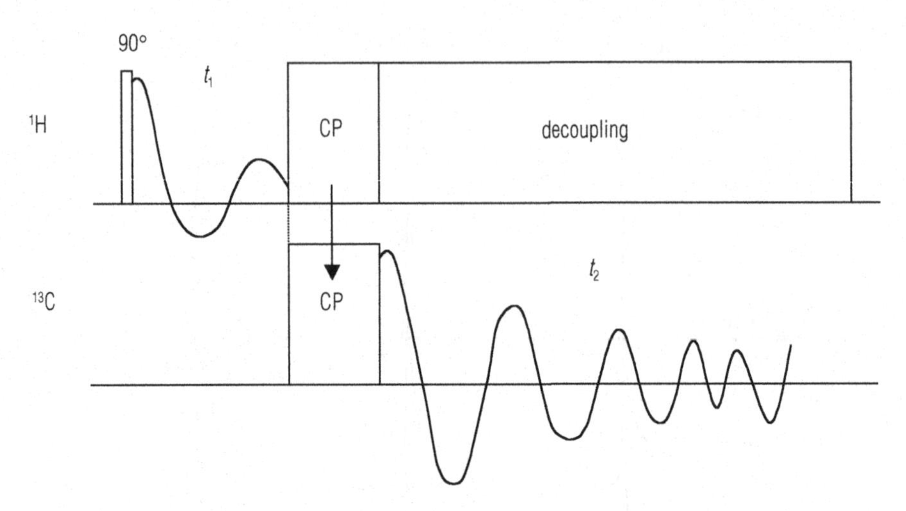

Fig. 6.9 The pulse sequence for the WISE experiment [7] which separates ^{1}H wideline spectra according to the isotropic chemical shift of a bonded heteronucleus, in this case ^{13}C. The cross-polarization step is kept short (50–100 μs) to ensure that ^{1}H magnetization is transferred only to the closest ^{13}C.

different from the static ^{1}H lineshape. However, any molecular motion which averages or partially averages the dipolar coupling on the NMR timescale, i.e. motion with τ_c^{-1} of the order of the ^{1}H linewidth or greater, can significantly reduce the ^{1}H linewidth in static spectra and much sharper sidebands are produced in magic-angle spinning spectra. Of course the problem is that in most organic samples, there are several (often many!) different ^{1}H sites and, inevitably, the broad ^{1}H lines from each overlap. Accordingly, the two-dimensional WISE (WIdeline SEparation) technique [7] has become very popular; here the ^{1}H lineshapes are separated according to the isotropic chemical shifts of the ^{13}C they are bonded to. This is achieved by the pulse sequence shown in Fig. 6.9. Essentially, transverse ^{1}H magnetization is created by an initial ^{1}H 90° pulse. The ^{1}H magnetization is then allowed to evolve during t_1; that remaining at the end of the t_1 period is then transferred to the ^{13}C spins via cross-polarization (see Section 2.5). Finally, the ^{13}C FID is recorded in t_2. The whole experiment is conducted under magic-angle spinning in order to achieve high resolution in the ^{13}C dimension of the experiment. The cross-polarization step is kept very short (50–100 μs) in order that ^{1}H magnetization is only transferred to those ^{13}C closest to the respective ^{1}H spins, i.e. those directly bonded to ^{1}H in general. A TOSS sequence (see Section 2.2.3) may be applied prior to acquisition in t_2 to suppress spinning sidebands in the ^{13}C dimension.

There is no need to record two quadrature datasets in order to obtain pure absorption lineshapes in this experiment. In most samples, the ^{1}H lineshape can be assumed to be symmetric, providing the chemical shift range of the ^{1}H spins in the sample is small. In this case, if the ^{1}H spins are set on resonance, the whole ^{1}H spectrum is then symmetric about $\omega = 0$, where ω is the offset frequency. Thus, we can record a single two-dimensional dataset and achieve a pure absorption phase two-dimensional frequency spectrum by Fourier transforming with respect to t_2, phasing the resulting ω_2 spectrum as necessary, then zeroing the imaginary part and finally Fourier transforming with respect to t_1. Zeroing the imaginary part before the final Fourier transformation allows pure absorption two-dimensional lineshapes to be obtained. It also, of course, has the effect of producing a (^{1}H) spectrum in f_1 which is necessarily symmetric about $\omega_1 = 0$, but since this is the form we expect in any case for the ^{1}H spectrum in this dimension, this is not a problem. The ω_1 lineshapes of ^{1}H spins which are slightly off resonance can usually still be monitored for signs of molecular motion; the offset from $\omega_1 = 0$ and the symmetrizing in the ^{1}H dimension does not obscure the intrinsic ^{1}H linewidth in general.

It should be remembered that the cross-polarization efficiency of mobile sites is in general somewhat less than for static sites. Thus, the ^{13}C (and therefore ^{1}H) intensities in a WISE spectrum will not in general be quantitative, with the signals from mobile sites being rather less intense than others. Despite this, however, WISE is a very useful and easy-to-implement experiment, for the qualitative detection of motion with correlation times less than about 10^{-4}–10^{-5} s in solids.

6.3 Relaxation time studies

Relaxation time measurements have long been used to characterize molecular motions in solids. All nuclear spin relaxation processes are mediated by fluctuating nuclear spin interactions, with the fluctuations (generally) arising from molecular motion. Relaxation phenomena are the subject of several whole books; indeed, the topic cannot be dealt with properly in much less space. A very brief description of the theoretical basis for understanding relaxation processes is given here simply for completeness and in order to indicate how relaxation time studies may be used to assess molecular motion in solids. For a more detailed description, the reader is referred to reference [8].

Relaxation is the change in time of the density matrix describing the spin system as the system moves back towards equilibrium from some non-equilibrium state imposed, for instance by a sequence of radiofrequency pulses. Longitudinal relaxation, for example, restores the populations of the Zeeman wavefunctions for a collection of identical spin systems to the equilibrium Boltzmann distribution. It thus affects the diagonal elements of the density matrix, which describe the populations of the spin system Zeeman wavefunctions. The rate of change of population of any given spin level can be written in terms of the transition rate to that level from all others and vice versa in combination with the populations of those spin levels, in the usual manner for any kinetic process. In turn, the transition rate between two levels i and j is given by the golden rule of quantum mechanics:

$$W_{ij} = \frac{1}{\hbar^2}\int_{-\infty}^{+\infty}\langle \mathrm{Tr}\{\mathbf{H}^*(t+\tau)\}\mathbf{H}^*(t)\rangle\, d\tau \tag{6.28}$$

where $\mathbf{H}^*(t)$ is the hamiltonian matrix in the Zeeman rotating frame (see Box 2.1 in Chapter 2 for a description of the rotating frame), describing the fluctuating nuclear spin interaction acting on the spin system. In other words, the time dependence of $\mathbf{H}^*(t)$ arises from molecular motions and the intrinsic anisotropy of the nuclear spin interaction. The hamiltonian operator describing the spin system in the Zeeman rotating frame is simply

$$\hat{H}^* = \exp(-i\hat{H}_0 t)\hat{H}\exp(+i\hat{H}_0 t) - \hat{H}_0 \tag{6.29}$$

where $\hat{H}_0$ is the Zeeman hamiltonian operator and $\hat{H}$ is the laboratory frame hamiltonian operator describing the spin system. Equation (6.28) for the transition rate involves the ensemble average over the sample (angle brackets), and the *trace*[N1] of the product of hamiltonian matrices at different times t and $t + \tau$. This latter term is a sum with components of the form $H^*_{ij}(t + \tau)H^*_{ji}(t)$, for all i and j, where H^*_{ij} and H^*_{ji} are hamiltonian matrix elements and i and j refer to elements of the basis set for hamiltonian matrices. To gain some insight into this sum, we restrict ourselves to consideration of relaxation through dipolar coupling and write the dipolar hamiltonian as a sum of spherical tensor operators:

$$\hat{H}_{\mathrm{dd}}(t) = \sum_{q=-2}^{+2}(-1)^q\Lambda^{IS}_{2-q}(t)\hat{T}^{IS}_{2q} \tag{6.30}$$

where the $\hat{T}^{IS}_{2q}$ are spherical tensor operators appropriate for dipolar coupling between two spins (I and S) (see Box 4.2 in Chapter 4); a term $\hat{T}^{IS}_{2q}$ represents a q-quantum term in the dipolar hamiltonian, which links spin levels whose z-quantum numbers differ by q. The $\Lambda^{IS}_{2q}(t)$ are components of the dipolar-coupling tensor expressed in irreducible tensor form and expressed in the laboratory frame (as in Box 4.2 in Chapter 4). The time dependence of the hamiltonian due to molecular motion is expressed through the time-varying dipolar-coupling tensor, which changes as the molecule changes orientation through angular motion with respect to the applied magnetic field in the NMR experiment. The total hamiltonian for the spin system is the sum of the Zeeman term for the spin system and the dipolar term. In the Zeeman rotating frame, this hamiltonian becomes

$$\begin{aligned}
\hat{H}^* &= \exp(-i\hat{H}_0 t)\hat{H}\exp(+i\hat{H}_0 t) - \hat{H}_0 \\
&= \exp(-i\hat{H}_0 t)(\hat{H}_0 + \hat{H}_{\mathrm{dd}})\exp(+i\hat{H}_0 t) - \hat{H}_0 \\
&= \sum_{q=-2}^{+2} (-1)^q \Lambda_{2-q}^{IS}(t)\exp(-i\hat{H}_0 t)\hat{T}_{2q}^{IS}\exp(+i\hat{H}_0 t) \\
&= \sum_{q=-2}^{+2} (-1)^q \Lambda_{2-q}^{IS}(t)\hat{T}_{2q}^{IS}\exp(i\Delta\omega_q)
\end{aligned} \tag{6.31}$$

The terms $\Delta\omega_q$ are the differences in the eigenvalues of $\hat{H}_0$ for eigenfunctions of $\hat{H}_0$ which differ in z-quantum number by q; thus $\Delta\omega_q$ is simply $q\omega_0$, where ω_0 is the Larmor frequency. When $\hat{H}^*$ is expressed in this form, it is clear that $\hat{H}^*$ only has non-zero matrix elements between Zeeman spin levels (or product spin levels) whose z-quantum numbers differ by q. The non-zero matrix elements of $\hat{H}^*$ then depend only on the corresponding $\Lambda_{2q}^{\mathrm{IS}}(t)$ and the matrix elements of the $\hat{T}_{2q}$ operators, which are just numbers.[N2] Thus, the integral in Equation (6.28) can be rewritten as a sum of integrals labelled $J_q(\Delta\omega_q)$ where

$$J_q(\Delta\omega_q) = \int_0^{\infty} \exp(-i\Delta\omega_q t)C_q(t)\,\mathrm{d}t \tag{6.32}$$

in which q is restricted to 0, 1, 2, i.e. $|q|$, and $C_q(t)$ is a correlation function describing the time dependence of the nuclear spin interaction, i.e. the molecular motion. The function $J_q(\Delta\omega_q)$ is a *spectral density* function. The spectral density is a measure of the amplitude of the q-quantum component of the nuclear spin interaction, in this case the dipolar coupling, oscillating at frequency $\Delta\omega_q = q\omega_0$ as a result of molecular motion.

The correlation function $C_q(t)$ is given by

$$C_q(t) = \langle \Lambda_{2q}^{IS}(0)\Lambda_{2q}^{IS*}(t)\rangle - |\langle \Lambda_{2q}^{IS}(0)\rangle|^2 \tag{6.33}$$

All relaxation processes can ultimately be described as some linear combination of spectral density functions, $J_q(\Delta\omega_q)$. We have only explicitly considered longitudinal relaxation processes via dipolar coupling here, but a similar case can be made for transverse relaxation, relaxation processes in the rotating frame and cross-relaxation processes and for other nuclear spin interactions. The spectral densities involved are, in each case, $J_q(\Delta\omega_\mathrm{q})$ where $\Delta\omega_q$ is the frequency of the q-quantum transition involved in the relaxation process under the particular nuclear spin interaction, whatever it may be. In transverse relaxation processes, the transitions involved include zero-quantum transitions and their respective frequencies then appear in the relevant spectral density. In rotating frame relaxation, the transitions involved are those in the rotating frame rotating about the laboratory frame z-axis at a rate ω_{rf}, the frequency of the rf irradiation applied to the spins, and so the corresponding transition frequencies are the nutation frequencies of the relevant nuclei. Hence, rotating frame relaxation processes are sensitive to motions with τ_c^{-1} in the region of the nutation frequency of the particular nuclear spin under a spin-locking pulse, generally tens to a few hundred kHz. Longitudinal relaxation processes are sensitive to molecular motions with τ_c^{-1} of the order of $q\omega_0$, i.e. the Larmor frequency (generally tens to hundreds of MHz). In contrast, transverse relaxation processes are sensitive to very low frequency motions, of the order of zero-quantum transition frequencies. Thus, a very wide range of motional frequencies may be studied by choosing different relaxation processes to monitor the motion.

In analysing relaxation data, the dominant nuclear spin interaction effecting relaxation must be known and it must exceed the effects of other interactions by at least an order of magnitude, otherwise the data become extremely complex to interpret. Accordingly, relaxation time studies are often applied to ^2H, where the dominant relaxation mechanism is nearly always through quadrupole coupling. In other cases, nuclei with ^1H bonded to them often have a dominant mechanism involving dipolar coupling with the ^1H, due to the particularly large magnetic moment of ^1H (providing that the chemical shift anisotropy associated with the nucleus is small).

Relaxation data are analysed by calculating correlation functions for the nuclear spin interactions acting on the observed spin and likely models of the molecular motion. The correlation functions are then used to calculate the relevant relaxation time. Comparison between experimental and calculated values can then lead to a description of the motion in the system. It should be noted, however, that this type of semi-classical analysis is only valid when the correlation time for the motion is much smaller than the relevant relaxation time. Furthermore, for any motions other than fairly simple ones, the analysis is likely to lead to ambiguities with several motional models calculating similar relaxation times.

6.4 Exchange experiments

Exchange experiments are invaluable for studying slow molecular motions (with correlation times of the order of milliseconds or slower) in solids, and accordingly have seen many applications in polymers, for instance. The essential concept of a two-dimensional exchange experiment is straightforward and is illustrated in Fig. 6.10. In this section we deal only with their application to spin-$\frac{1}{2}$ systems; application to spin-1 is dealt with in Section 6.5.4.

Transverse magnetization is created by an initial 90° pulse or cross-polarization step and allowed to evolve during the period t_1 under its characteristic frequency ω_1. This characteristic frequency arises from the nuclear

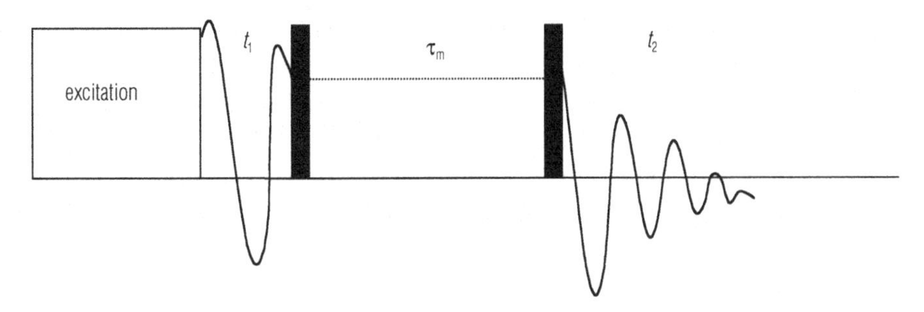

Fig. 6.10 The basic form of a two-dimensional exchange experiment to study molecular motion or chemical exchange; all pulses (black) are 90° pulses. The experiment correlates changes of molecular orientation between the t_1 and t_2 periods. Transverse magnetization is initially excited (via a 90° pulse or cross-polarization, labelled simply 'excitation' in the figure) and allowed to evolve at its characteristic frequency, ω_1, during t_1. The characteristic frequency in static solid samples is dependent on molecular orientation, as the spin interactions acting on the observed spin are anisotropic. At the end of t_1, a second pulse restores the magnetization to z (parallel to $\mathbf{B_0}$), where it is stored for a mixing time τ_m, during which molecular reorientation may occur. Finally, another pulse reconverts the magnetization into observable transverse magnetization whose evolution, this time with offset frequency ω_2, is then recorded in an FID. Subsequent processing of the resulting two-dimensional dataset produces a correlation spectrum which shows how the observed spin's offset frequency changed from ω_1 in t_1 to ω_2 in t_2. Since molecular orientation can be directly correlated with offset frequency, the two-dimensional spectrum represents a map describing the molecular reorientation between t_1 and t_2, i.e. during τ_m.

spin interaction which operates during t_1. At the end of t_1, the magnetization is stored along z (the direction of applied magnetic field in the NMR experiment) for a period τ_m, the mixing time, during which molecular dynamical processes may occur. Finally, the magnetization is returned to the transverse plane, where it again evolves under its characteristic frequency, this time ω_2, the evolution being recorded as a FID. If dipolar decoupling is required, e.g. in ^{13}C exchange spectra for organic compounds, it is applied during t_1 and t_2 only. Dipolar decoupling is switched off during the mixing time τ_m to encourage dephasing of unwanted coherences (see Section 6.4.1 below).

Appropriate processing of the resulting two-dimensional time-domain datasets yields a two-dimensional frequency correlation spectrum, correlating the characteristic frequency the spin had during t_1 with that which it subsequently had in t_2. If molecular reorientation or site exchange has occurred during the mixing time (which happens if the correlation time for the motion $\tau_c < \tau_m$), the offset frequency after the mixing time, ω_2, is different from the initial offset frequency before molecular reorientation/exchange, ω_1, and so the two-dimensional frequency spectrum contains off-diagonal intensity at (ω_1, ω_2). If there is no exchange during the mixing time, $\omega_1 = \omega_2$ ($\tau_c >> \tau_m$), spectral intensity appears only along the diagonal of the two-dimensional frequency spectrum. Analysis of the resulting spectrum in terms of molecular motion clearly relies on the offset frequencies ω_1 and ω_2 being constant during the t_1 and t_2 periods respectively, i.e. on there being no molecular motion during these periods. It is for this reason that exchange experiments are only suitable for studying slow molecular motions. By assessing the exchange intensity as a function of mixing time (with $\tau_m >> t_1, t_2$), the correlation time for the motion can be determined; the pattern of exchange intensity in the two-dimensional frequency spectrum allows the geometry of the motion to be determined [9]. A key feature here is that motional models are *not* required to extract this latter information in contrast with lineshape analyses and relaxation time studies.

For the most part, exchange experiments in the solid state use either chemical shift anisotropy (for spin-$\frac{1}{2}$) or quadrupole coupling (for spin $> \frac{1}{2}$) under static conditions, i.e. no sample spinning, to generate offset frequencies ω_1 and ω_2 which depend on molecular orientation. The projections onto the two spectral frequency axes are then the corresponding powder patterns resulting from the particular anisotropic spin interaction. The resulting two-dimensional spectrum is, in effect, a correlation map between the molecular orientations in t_1 and t_2. Although chemical shift anisotropy and quadrupole coupling are most commonly used in exchange experiments, any anisotropic nuclear spin interaction can be employed. In one example, Schmidt-Rohr and colleagues used the dipolar coupling between isolated pairs of ^{13}C spins in high-density polyethylene to label the frequencies in t_1 and t_2 [10].

High-resolution two-dimensional exchange experiments can be performed in the solid state in analogous fashion to solution-state exchange experiments by conducting the experiments (for spin-$\frac{1}{2}$) under rapid magic-angle spinning. These experiments of course monitor only chemical exchange, where the site exchange is accompanied by a change of isotropic chemical shift. Such experiments are generally much simpler to perform and analyse than the static experiments but, obviously, their field of application is much smaller. They employ the same basic pulse sequence (Fig. 6.10), with the proviso that the mixing time, τ_m, is an integral number of rotor periods. When this condition is met, the phase of the transverse magnetization at the end of the t_1 period is the same as that at the beginning of t_2; this allows pure absorption lineshapes to be achieved after appropriate processing of the data (see the following section).

Three- and higher-dimensional exchange spectra can be recorded by simple extensions of the basic two-dimensional pulse sequence (Fig. 6.11). The resulting multidimensional frequency spectra then correlate the molecular orientation at three (or more for higher-dimensional spectra) points in time. This allows assessment of the degree to which different motions or molecular jumps are correlated in time.

6.4.1 Achieving pure absorption lineshapes in exchange spectra

There are two basic methods of achieving pure absorption lineshapes in two-dimensional exchange spectra.

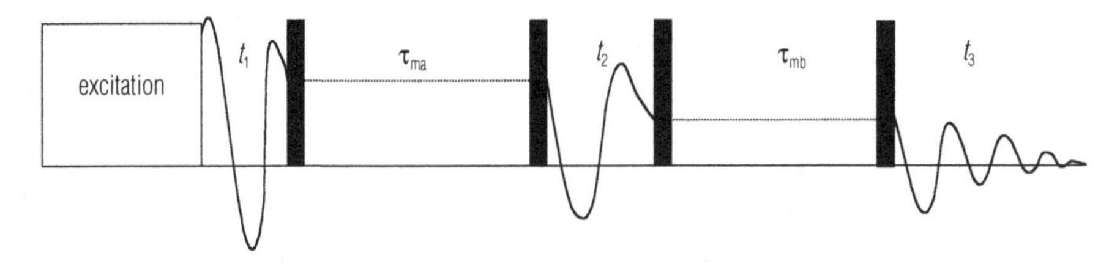

Fig. 6.11 The basic pulse sequence for a three-dimensional exchange experiment; such an experiment correlates changes of molecular orientation between three time periods, t_1, t_2, and t_3, separated by mixing times τ_{ma} and τ_{mb}. All pulses (black) are 90° pulses. The operation of the sequence is similar to that for the two-dimensional exchange experiment (Fig. 6.10), but with the addition of a second mixing time and additional evolution period. Higher-order exchange experiments can be performed by simply adding further mixing and evolution periods.

1. *Quadrature detection in* t_1. Two two-dimensional datasets are recorded, one in which the x-component of the t_1 transverse magnetization is measured (indirectly) and one in which the y-component of the t_1 magnetization is measured; x and y here refer to axes in the transverse plane of the rotating frame. In practice, this is done as follows. The initial (t_1) magnetization is excited for both datasets using, say, a 90°_x pulse. Then, for dataset 1, a 90°_{-x} storage pulse at the end of t_1 flips the $-y$-component of the t_1 transverse magnetization to z for the mixing time τ_m. This $-y$-component of the t_1 magnetization is given by $M_0 \cos \omega_1 t_1$, where M_0 is the initial transverse magnetization (along $-y$) at the start of the t_1 period, produced by the 90°_x pulse. $\omega_1 t_1$ is the angle this magnetization has precessed through (about $\mathbf{B}_0/z$) after time t_1. The x-component of the t_1 transverse magnetization is unaffected by the 90°_{-x} pulse and so remains in the transverse plane and dephases during the mixing time.[N3] The final signal recorded in t_2 is then $M_0 \cos \omega_1 t_1 \exp(i\omega_2 t_2)$. For dataset 2, a 90°_y storage pulse at the end of t_1 stores the $-x$-component of the t_1 transverse magnetization ($= M_0 \sin \omega_1 t_1$) along z for the mixing period, while the y-component now dephases. The final signal recorded in t_2 is then $M_0 \sin \omega_1 t_1 \exp(i\omega_2 t_2)$. The two datasets are then processed according to the recipe in Fig. 1.21.
2. *Off-resonance detection.* One two-dimensional dataset is recorded off resonance and processed as for the WISE experiment in Section 6.2.3. The resulting two-dimensional spectrum is symmetric about $\omega_1 = 0$, as shown in Fig. 6.12. Clearly, this method is much simpler than the quadrature detection in t_1 method, but exciting broad, off-resonance powder patterns uniformly may not be possible. In practice, then, this method is only suitable for relatively small chemical shift anisotropies of a few kHz.

In order to obtain undistorted, pure phase spectra, is it most important that the spectral data is acquired from t_1, $t_2 = 0$. However, receiver deadtime problems can lead to truncation of the data in t_2. Slightly less of a problem

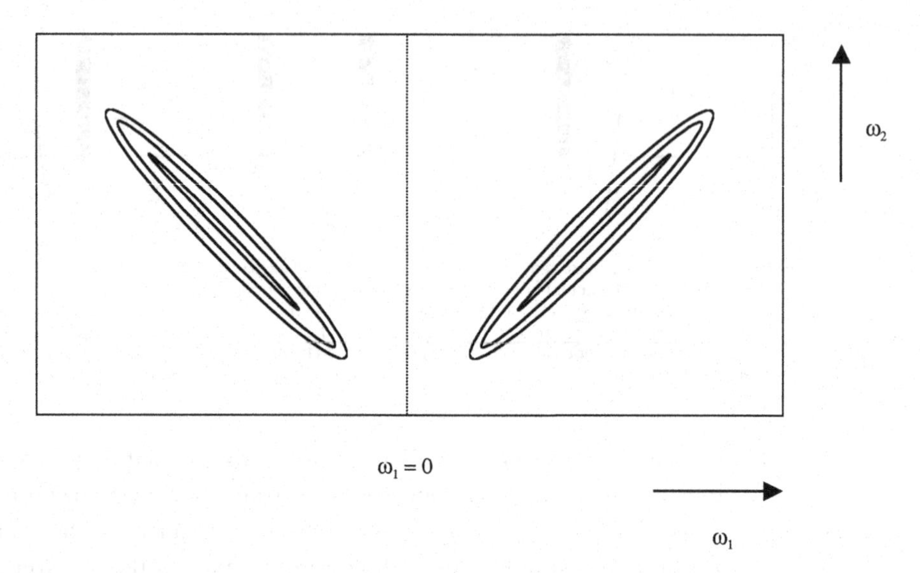

Fig. 6.12 The form of the two-dimensional exchange spectrum which arises if the data are recorded off resonance in a single experiment. The spectrum is necessarily symmetric about $\omega_1 = 0$, so that either half of the spectrum contains all the data.

is that finite pulse widths truncate the t_1 data also. Both of these features prevent spectra with t_1 and t_2 truly equal to zero from being recorded. These truncations lead to frequency-dependent phase distortions in the respective spectral dimensions. In principle, these could be corrected by first-order phase corrections, providing that the loss of information arising from the truncation is not too great. In practice, for the wider powder lineshapes such corrections are not effective. Instead, datasets from t_1, $t_2 = 0$ are produced by interpolating the experimentally-obtained signal and extrapolating back to zero [11]. This, however, is not a simple procedure and requires some experience if undistorted spectra are to be obtained. Alternatively, Hahn spin echoes (τ–180°–τ) can be used prior to the t_2 and/or t_1 periods to prevent the truncation of the data in the first place (Fig. 6.13). Then, the acquisition of the FID can be set to begin exactly at $t_2 = 0$ and, likewise, the pulse which defines the beginning of the mixing period can be set exactly on top of the first echo to acquire a $t_1 = 0$ spectrum. This greatly simplifies data processing. Such steps are essential if pure absorption three-dimensional exchange spectra are to be obtained.

6.4.2 *Interpreting two-dimensional exchange spectra*

Magic-angle spinning exchange spectra are straightforward to interpret. Off-diagonal peaks correspond to magnetization exchange between the corresponding signals in each dimension, i.e. an off-diagonal peak at (ω_1, ω_2)

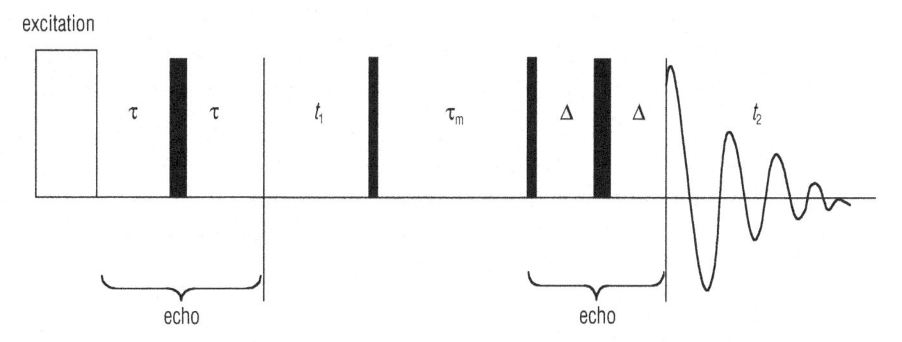

Fig. 6.13 The use of Hahn spin echoes in the recording of two- (and higher-) dimensional exchange spectra. Wide pulses (black) are 180° pulses; thin ones (also black) are 90° pulses. Truncation of the time-domain data in either t_1 or t_2 leads to gross spectral distortions in the final two-dimensional frequency spectrum in cases where broad powder lineshapes are expected in the corresponding frequency dimensions. Such truncation is usually inevitable due to finite length pulses in the case of t_1 and due to the receiver deadtime in the case of t_2. Using a τ–180°–τ (or Δ–180°–Δ) Hahn echo sequence prior to each t_1 and t_2 allows the respective time-domain datasets to be recorded from the true signal maximum and so prevents spectral distortion.

means that magnetization from the signal at offset ω_1 in the one-dimensional spectrum of the compound exchanges with that at ω_2 during the mixing time. Whether this exchange corresponds to a chemical exchange process or a spin diffusion process then has to be determined. This can be done by recording exchange spectra as a function of temperature; spin diffusion processes are independent of temperature, while chemical exchange processes vary with temperature. However, there may be elements of both processes occurring and this is more difficult to extract.

In two-dimensional static exchange spectra, the intensity at point (ω_1, ω_2) in the two-dimensional frequency spectrum is proportional to the probability that a spin had an offset frequency ω_1 (due to its particular orientation) during t_1 and reorientated such that its offset frequency was ω_2 during t_2. In this manner, model-independent information can be obtained from exchange spectra.

Static two-dimensional exchange spectra can be simulated relatively simply. In the following discussion, we restrict ourselves to the case of discrete N site-hopping motional processes, as defined in Section 6.2 on powder lineshape analysis. If, during t_1, a site has offset frequency ω_1 and then undergoes a reorientation during the mixing time τ_m, the new offset frequency ω_2 of the site in t_2 can be found using Equation (6.14) in Section 6.2.1.

The time-domain signal resulting from the exchange process between sites with frequencies ω_1 and ω_2 is then [11]

$$
\begin{aligned}
s(t_1, t_2; \tau_m) &= \langle \exp(i\omega_1 t_1)\exp(i\omega_2 t_2)\rangle \\
&= \langle \mathbf{1}\cdot\exp(i\boldsymbol{\omega}_1 t_1)\exp(\boldsymbol{\Pi}\tau_m)\exp(i\boldsymbol{\omega}_2 t_2)\mathbf{p}\cdot\mathbf{1}\rangle
\end{aligned} \tag{6.34}
$$

where $\mathbf{\Pi}$ is the kinetic matrix described in Section 6.2.1, $\mathbf{p}$ is an N-dimensional vector whose elements are the populations of each of the N sites and $\boldsymbol{\omega}_i$ is a diagonal $N \times N$ matrix whose elements are the offset frequencies during the time period t_i of the N sites involved in the hopping process, for a given crystallite orientation in the sample, exactly as described for one-dimensional lineshape analysis. The angle brackets in Equation (6.34) denote 'ensemble average', i.e. that the expression within the brackets should be summed over all possible crystallite orientations. The exponentials of matrices can be evaluated using Equation (6.10) after finding the eigenvectors and eigenvalues of the particular matrix.

Further analysis shows that for axially symmetric interaction tensors, angular reorientation by an angle Θ gives rise to an elliptical ridge in the two-dimensional frequency spectrum which results from Fourier transformation of Equation (6.34). The angle Θ is related to the major and minor axes of the ellipse, a and b, by $\tan\Theta = b/a$. An example is shown in Section 6.5. In heterogeneous samples, such as polymers, there is usually a distribution of molecular reorientational angles; two-dimensional exchange patterns can be simulated for different distributions and compared with experiment to reveal this information [12].

6.5 ^{2}H NMR

^{2}H is a spin-1 nucleus with a relatively small electric quadrupole moment ($Q = 2.8 \times 10^{-31}\,\text{m}^2$) which gives rise to quadrupole-coupling constants, χ, in the range 140–220 kHz in organic compounds, for instance. Powder NMR spectra of static samples consist of doublet patterns (Fig. 6.14), the doublet arising from the two possible spin transitions: $+1 \leftrightarrow 0$ and $0 \leftrightarrow -1$. These are often called *Pake patterns*; their horns are split by $\frac{3}{4}\chi$,

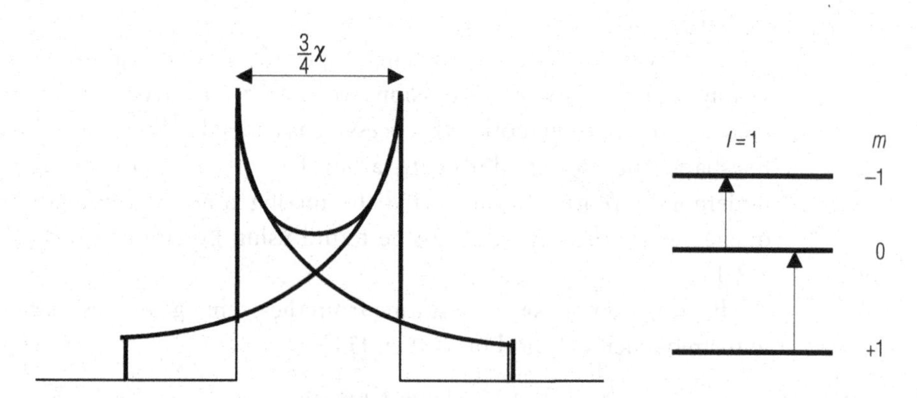

Fig. 6.14 The form of a ^{2}H ($I = 1$) quadrupole powder pattern. The doublet nature of the pattern is due to there being two allowed spin transitions.

i.e. 105–165 kHz. This moderate width of powder pattern (as compared with other quadrupolar nuclei) makes this nucleus relatively easy to deal with experimentally. Moreover, the powder pattern lineshapes are sensitive to molecular motions with correlation times of the order 10^{-4}–10^{-6} Hz, which coincidentally corresponds to a range of motional correlation times often found in solids in the temperature range accessible by most commercial NMR spectrometers, i.e. −150°C–250°C. Because of this, ^{2}H has been extensively used in motional studies, primarily via lineshape analyses, but also through relaxation time measurements and exchange experiments.

^{2}H NMR is also widely used on partially-ordered materials, such as liquid crystals, to measure order parameters.

6.5.1 Measuring ^{2}H NMR spectra

Quadrupole echo experiment

The width of ^{2}H powder patterns for static samples necessitates the use of echo techniques to record undistorted powder patterns, rather than simple pulse-acquire experiments. As discussed in Section 2.6, the receiver deadtime which must be left after a pulse means that the initial part of the FID in a pulse-acquire experiment is not recorded. These lost data constitute a significant part of the total FID for broad powder patterns, which have correspondingly short FIDs, and lead to severely distorted frequency spectra on Fourier transformation. Accordingly, ^{2}H powder patterns are generally recorded with a *quadrupole echo* or *solid echo* pulse sequence: 90°_x–τ_1–90°_y–τ_2–acquire. The delays τ_1 and τ_2 are approximately equal but, in practice, τ_2 is adjusted so that the data acquisition begins exactly at the echo maximum. τ_2 (and therefore τ_1) should be at least as long as the receiver deadtime.

There are several practical points to be considered to obtain good results with the quadrupole echo pulse sequence. As already stated, data acquisition must begin exactly at the echo maximum. Finding the echo maximum requires a bit of patience. First, the receiver phase needs to be adjusted so that all the FID intensity appears in the real part of the FID, in other words, so that the frequency spectrum that would arise after Fourier transformation has pure absorption phase. If this is not done, intensity is distributed between the real and imaginary parts of the FID. It is then very difficult to assess the *amplitude* of the FID at any point in time, as the amplitude is the square root of the sum of the squares of the real and imaginary parts. It is therefore very difficult to find the echo maximum (point of maximum amplitude of the FID). If the intensity is all in the real part of the FID, then the magnitude of the real part of the FID is synonymous with the FID amplitude at any point in time.

Having done this, an apparent echo maximum will reveal itself. However, it is important to remember that the FID signal seen on the computer screen is in fact digitized, so only measured at discrete points, despite the continuous line that is drawn between the points in most NMR spectrometer software. Thus it is quite possible (indeed, highly likely) that the true echo maximum actually falls between two recorded FID points. There are several ways of dealing with this.

One is to begin recording the FID well before the apparent echo maximum, i.e. $\tau_2 < \tau_1$. The resulting FID signal is then interpolated between the digitized points in the region of the apparent echo maximum to find the true echo maximum. A new time domain series is then generated starting from the true echo maximum, via interpolation between the recorded FID points as necessary.

Alternatively, one can change the time points at which FID is recorded, until the true echo maximum is found. This is usually done in practice by keeping τ_2 at a constant value which is much less than τ_1 and varying τ_1 in small amounts in successive experiments. The dwell time between recorded FID points is kept constant during this process. Comparing the echo FIDs between the successive experiments will then reveal the true echo maximum. The τ_1 value which produces an FID with the true echo maximum at one of the FID sampling points is then used to record the final FID with the required level of signal averaging.

Which method is used is a matter of personal choice. The former method is quicker on spectrometer time, but requires more lengthy processing; the latter uses more spectrometer time, but subsequent processing is straightforward. Where the echo FID is recorded so that the echo maximum appears some way into the FID, the time domain dataset is left-shifted by the appropriate amount prior to Fourier transformation.

Clearly, ^{2}H powder patterns should be symmetric about their isotropic chemical shifts. Experimentally, however, ^{2}H powder patterns often appear distinctly asymmetric. This is due to a variety of factors, each of which should be attended to as far as possible before the final spectrum is recorded. First, finite pulse widths do not give uniform excitation over the whole powder pattern. In order to get symmetric excitation about the isotropic chemical shift of the ^{2}H powder pattern, it is necessary that the centre of the powder pattern be on resonance. If it is not, it is unlikely that a symmetric powder pattern can ever be recorded. Second, the NMR probe response may not be symmetric. This particular feature is governed by the probe electronics and can be altered to some extent on most modern probeheads. Once the ^{2}H powder pattern has been set on resonance, the probe response can be examined. If the powder pattern is asymmetric at this point in the experiment set-up, it is probably the fault of the probe response. The match setting on the probe should be adjusted in small steps until a symmetric ^2H powder pattern is obtained.

When recording ^2H powder patterns with a quadrupole echo pulse sequence to study molecular motion, it is often useful to record powder patterns for several different echo delay times, τ_1 (and corresponding τ_2). The variation in echo intensity with τ_1 is determined by the transverse relaxation rate (as characterized by T_2). It is the anisotropic T_2 arising from molecular motion which causes the distortion of the ^2H powder lineshapes. The variation in echo intensity with τ_1, in combination with the ^2H powder lineshapes (also as a function of τ_1), thus gives useful information on the correlation time and geometry of the molecular motional process. Some simulated ^2H powder patterns as a function of τ_1 are shown in Fig. 6.15 for a two-site-hopping molecular motion.

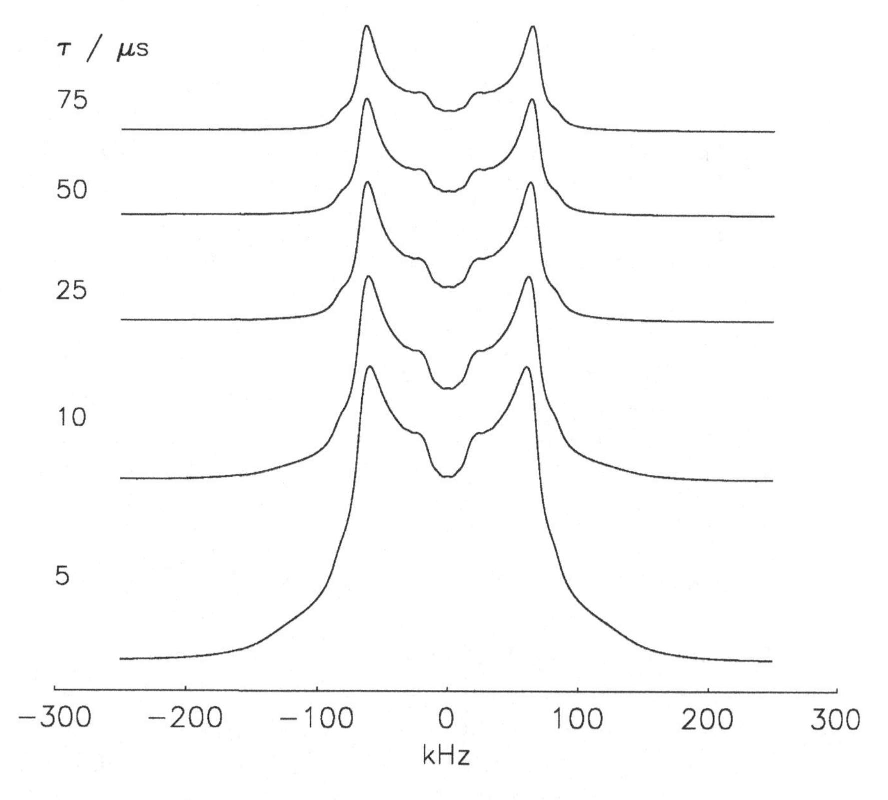

Fig. 6.15 Simulated ^2H quadrupole echo spectra for the pulse sequence 90°–τ_1–90°–τ_2–acquire, for different echo delays, $\tau = \tau_1 = \tau_2$. The molecular motion used in the simulations is a two-site hopping which reorientates the ^2H quadrupole coupling tensor z-axis by 120°. The motional correlations time is $\tau_c = 10^{-5}$ s for each spectrum.

Magic-angle spinning

In principle, magic-angle spinning can remove the effects of first-order quadrupolar linebroadening completely. In practice, this would require spinning at rates much faster than the ^2H quadrupolar powder pattern width. Such spinning rates are unlikely ever to be achievable. At the achievable spinning rates (up to 50 kHz), the ^2H powder pattern breaks up into a series of sharp spinning sidebands. This can be a distinct advantage, as the spectral intensity is then concentrated at discrete points in the frequency spectrum, rather than being smeared out over a wide frequency range in a powder pattern, so the signal-to-noise ratio is much improved.

In principle, ^2H magic-angle spinning spectra can be recorded with simple pulse-acquire sequences, rather than echo sequences, as the decay rate of the magic-angle spinning FID is significantly slower than for the static experiment. However, it is still the case that the spectral width required to record the spinning sideband pattern is usually substantial, and so the corresponding dwell time in the FID is small. Thus even a relatively small receiver deadtime before acquisition can correspond to several FID points, resulting in baseline distortions in the Fourier-transformed frequency spectrum. Hence, better quality spectra are often obtained by using a quadrupole echo pulse sequence even under conditions of magic-angle spinning. The same principles apply in setting up the experiment as in the static case with the added proviso that the echo delay should be an integral number of rotor periods so that rotational echoes are refocused at the start of the FID.

As discussed previously in this chapter, spinning sideband patterns monitor different motional regimes to static powder patterns, the particular range that they are sensitive to depending on the spinning rate as well as the ^2H quadrupole coupling constant. Thus recording ^2H spinning sideband patterns at different spinning rates can give extra information on the motional process. Some simulated ^2H spinning sideband patterns for different spinning rates and a two-site hopping molecular motion are shown in Fig. 6.16.

Quadrupolar Carr–Meiboom–Purcell–Gill pulse sequence

This pulse sequence is shown in Fig. 6.17 [13]. It consists of a standard quadrupole echo pulse sequence, after which the echo decay is recorded (step 1). Following this is a series of refocusing 90° pulses, with a complete echo being recorded after each (step 2). Finally, any remaining signal is allowed to decay and is again recorded (step 3). The complete time-domain series resulting from the outputs of steps 1, 2 and 3 strung together is then Fourier transformed. This experiment has the effect of splitting the quadrupole echo powder pattern into a manifold of spin-echo sidebands separated by $1/\tau_a$ (see Fig. 6.17 for a definition of τ_a). This in itself generates a

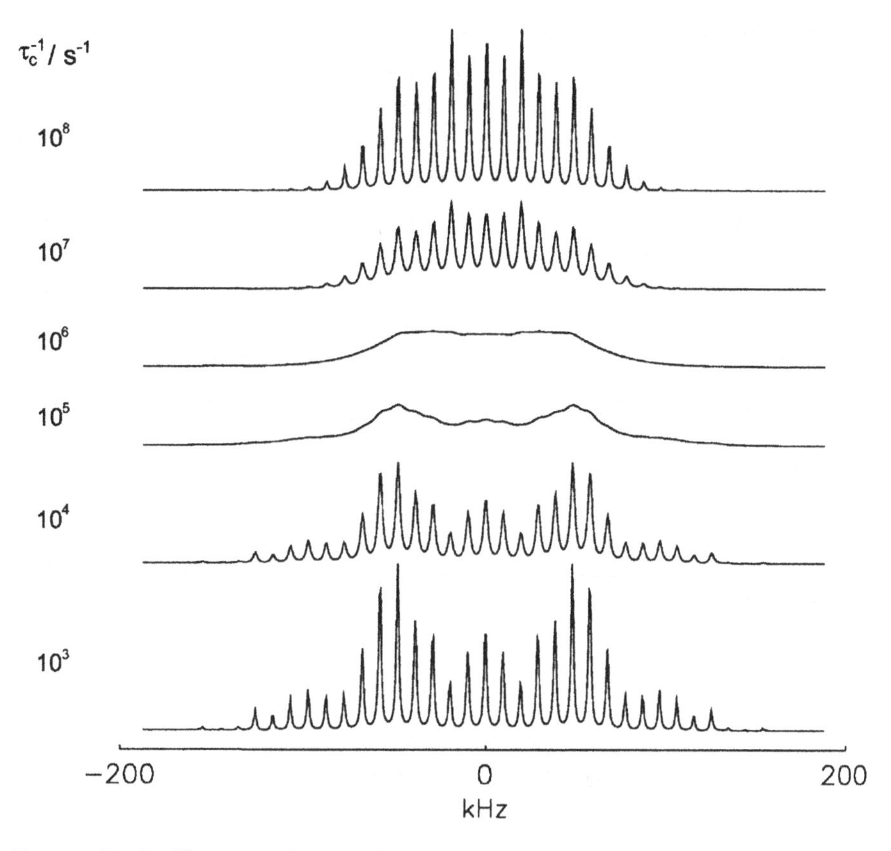

Fig. 6.16 Simulated ^{2}H magic-angle spinning spectra for different hopping rates. The molecular motion used in the simulations is a two-site hopping which reorientates the ^{2}H quadrupole coupling tensor z-axis by 120°. The correlation times for the motions are given as τ_c^{-1} with the spectra.

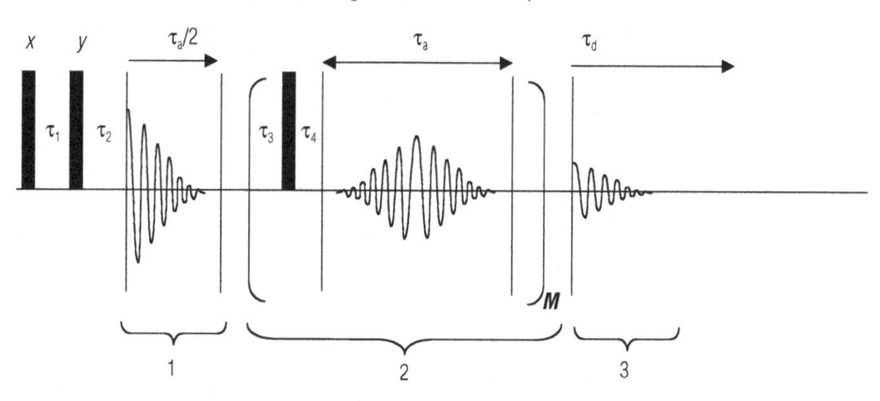

Fig. 6.17 The Quadrupolar Carr–Purcell–Meiboom–Gill (QCMPG) pulse sequence [13] for recording ^{2}H NMR spectra. Use of this sequence gives ^{2}H spectra which are sensitive to molecular motions over a much wider frequency range than static or magic-angle spinning ^{2}H spectra. The final FID from the experiment is the results from steps 1, 2 and 3 strung together in a single time-domain series; the FIDs are collected between the vertical line in the diagram for periods $\tau_a/2$, τ_a and τ_d respectively in steps 1, 2 and 3. All pulses are 90° pulses with the phases given in the diagram. The echo delays τ_1 and τ_2 are approximately equal, with τ_2 being adjusted so that the acquisition in step 1 begins exactly at the echo maximum. Step 2 is repeated M times and consists of M refocusing 90° pulses with collection of the resulting echo FID after each. τ_3 and τ_4 are short delays designed to protect the receiver from the 90° pulses in step 2.

sensitivity enhancement of an order of magnitude. Moreover, both the envelope of the spin-echo sidebands and their individual lineshapes contain information on molecular motions. It is once again obtained by simulating the spectrum for particular motional models, but the initial study [14] shows that the dynamic range of the experiment is at least two orders of magnitude larger than the conventional quadrupole echo experiment. The full dynamic range is 10^2–10^8 Hz, i.e. it is sensitive to motions which are both slower and faster than those which the quadrupole echo experiment can monitor. Sideband shapes tend to monitor the lower end of this range, while the overall sideband envelope is sensitive to higher frequency motions. Some example spectra recorded using this technique are shown in Fig. 6.18.

6.5.2 ^{2}H lineshape simulations

^{2}H powder patterns and spinning sideband patterns under conditions of molecular motion can be calculated in exactly the same way as for spin-$\frac{1}{2}$

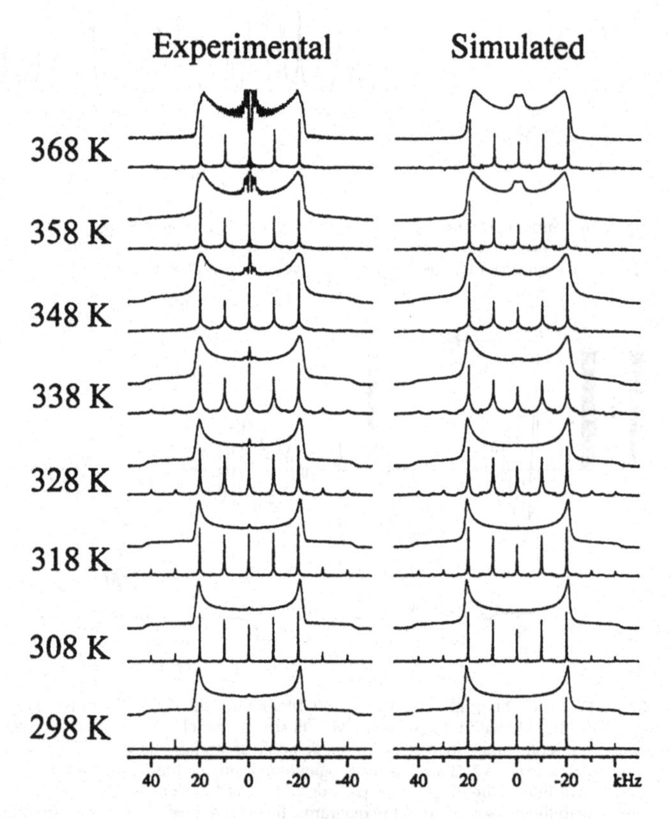

Fig. 6.18 ^{2}H spectra of deuterated dimethylsulphone (DMS) recorded with the QCMPG pulse sequence of Fig. 6.17 [14] and with the quadrupole echo pulse sequence for comparison. The top spectrum in each pair is the quadrupole echo spectrum, the lower one the QCMPG one.

chemical shift anisotropy powder patterns as detailed in Section 6.2.1, except that the shielding tensor needs to be replaced with the electric field gradient tensor ($e\mathbf{q}$) and the chemical shift frequency by the quadrupolar frequency, ω_Q. The quadrupolar frequency is determined by

$$\omega_Q = \frac{eQ}{4I(2I-1)\hbar} \mathbf{b}_0^{\mathrm{PAF}} e\mathbf{q}^{\mathrm{PAF}} \mathbf{b}_0^{\mathrm{PAF}} \tag{6.35}$$

where $\mathbf{b}_0^{\mathrm{PAF}}$ is the unit vector in the direction of $\mathbf{B}_0$ in the electric field gradient tensor principal axis frame (PAF) for the particular molecular orientation, in complete analogy with the chemical shift frequency. The electric field gradient tensor principal components are easily derived from the quadrupole coupling constant and asymmetry, and the fact that this tensor is traceless. The quadrupole-coupling constant χ is given by $e^2 q_{zz}^{\mathrm{PAF}} Q/\hbar$, where q_{zz}^{PAF} is the z-principal value of the electric field gradient tensor and the asymmetry by $\eta_Q = (q_{yy}^{\mathrm{PAF}} - q_{xx}^{\mathrm{PAF}})/q_{zz}^{\mathrm{PAF}}$. This plus the fact that $q_{xx}^{\mathrm{PAF}} + q_{yy}^{\mathrm{PAF}} + q_{zz}^{\mathrm{PAF}} = 0$ allows us to write

$$\begin{aligned} eq_{zz}^{\mathrm{PAF}} &= \left(\frac{\chi}{eQ}\right)\hbar \\ eq_{xx}^{\mathrm{PAF}} &= -\frac{1}{2} q_{zz}^{\mathrm{PAF}} (1+\eta_Q) \\ eq_{yy}^{\mathrm{PAF}} &= -\frac{1}{2} q_{zz}^{\mathrm{PAF}} (1-\eta_Q) \end{aligned} \tag{6.36}$$

Using this and the procedure in Section 6.2.1 we can then calculate the static powder pattern or spinning sideband pattern for one of the ^{2}H spin transitions. The powder pattern for the other transition is just the mirror image of this about $\omega = 0$. The sum of the powder patterns for the two transitions is then the final ^{2}H powder/spinning sideband pattern.

6.5.3 Relaxation time studies

The anisotropy (i.e. dependence on molecular orientation) in the relaxation of Zeeman order (as described by the parameter T_{1Z}) and quadrupolar order (as described by the parameter T_{1Q}) has been used with good effect to study fast molecular motions in solids [15–17]. A spin system with *quadrupolar order* is described by a density operator proportional to $3\hat{I}_z^2 - I(I+1)$ or $\hat{T}_{20}^{Q}$ in terms of spherical tensor operators (see Appendix B), while a spin system with *Zeeman order* is described by a density operator proportional to $\hat{I}_z$. For example, the spin state produced by the initial 180° pulse acting on equilibrium magnetization in an inversion recovery experiment is described by a density operator proportional to $-\hat{I}_z$; thus following the decay of this state (or equivalently, recovery of the equilibrium state) enables T_{1Z} to be determined. Quadrupolar order is produced by a 90°_x–τ–45°_y pulse sequence.

T_{1Z} anisotropy for ^{2}H has been used in a two-dimensional experiment [18] where the full T_{1Z} anisotropy is displayed in one dimension of the experiment (T_2 anisotropy was similarly dealt with too [18]). In this early work, the full two-dimensional spectrum was simulated according to given motional models and motional correlation times, and compared with experiment. However, the fitting of a two-dimensional contour plot or surface plot is notoriously difficult and perhaps it is for this reason that the technique has never been taken up in this form by other spectroscopists. In more recent studies of ^{2}H T_{1Z} and T_{1Q} anisotropy, the approach has been slightly different. The procedure in the analysis of the relaxation rate anisotropy is to calculate the spectral density functions (Equation (6.32)) for given motional models and given molecular jump correlation times, as in the earlier study [18, 19]. From the spectral densities, it is then straightforward to calculate the partially-relaxed ^{2}H spectral lineshapes for the particular inversion-recovery delays used in the experiment. A good fit of the partially-relaxed lineshapes then implies the motional model and jump correlation time are possible descriptions of the true motion in the sample. The effect of molecular motion on T_{1Z} and T_{1Q} is in general different, so it is useful to measure both parameters in order to characterize the motion as fully as possible.

6.5.4 ^{2}H exchange experiments

^{2}H two-dimensional exchange experiments follow much the same format as those for spin-$\frac{1}{2}$ nuclei, except that the storage pulse and reconversion pulses either side of the mixing time τ_m are no longer 90° pulses. The pulse sequence usually used for ^{2}H is shown in Fig. 6.19. Pure absorption spectra are produced by recording two datasets so that there is quadrature detection in t_1; the off-resonance method sometimes used for spin-$\frac{1}{2}$ is not appropriate for ^{2}H because of the difficulty of getting uniform and symmetric excitation of the ^{2}H powder pattern off resonance. Both datasets are recorded using a 90°_y pulse as the initial excitation pulse (pulse 1 in Fig. 6.19). Dataset 1 is recorded using a 54.7°_{-x} pulse as the storage pulse (pulse 2), while dataset 2 is recorded using a 54.7°_{-y} storage pulse. It can be shown that the signals arising from the two experiments are then [9]

$$\begin{aligned} &1: \quad M_0 \sin\varphi_1 \sin\varphi_2 \sin\varphi_3 \cos\omega_1 t_1 \cdot \cos\omega_2 t_2 \\ &2: \quad M_0 \frac{3}{4} \sin\varphi_1 \sin 2\varphi_2 \sin 2\varphi_3 \sin\omega_1 t_1 \cdot \sin\omega_2 t_2 \end{aligned} \tag{6.37}$$

where the φ_i are the flip angles of the various pulses in the sequence. The reason for the 54.7° pulse flip angles now becomes clear; for φ_2 and φ_3 set

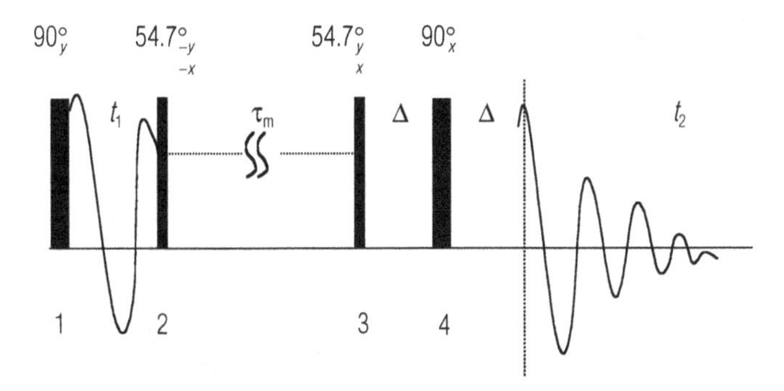

Fig. 6.19 The form of the two-dimensional exchange experiment for ^{2}H (I = 1). It differs from that for the experiment for spin-$\frac{1}{2}$ nuclei (Fig. 6.10) only in the nutation angles of the rf pulses employed. The pulses are numbered as they are discussed in the text. The initial 90°_y pulse excites transverse magnetization which evolves according to its characteristic frequency, ω_1, during t_1. Pulse 2 transfers any transverse magnetization remaining after t_1 to z for storage during the mixing time, τ_m, and pulse 3 then reconverts the magnetization back to transverse magnetization which evolves during t_2. The final Δ–90°_x–Δ sequence is an echo sequence, which allows ^{2}H powder lineshapes to be recorded in t_2 without distortion, which might arise from truncation of the dataset otherwise. Two datasets are recorded so that pure absorption lineshapes are obtained in the final two-dimensional frequency spectrum. For the first dataset, pulse 2 has phase $-y$ and pulse 3, phase y. For the second dataset, these pulses have phase $-x$ and x respectively.

to 54.7°, the scaling factors for the two datasets become equal. This considerably simplifies the data processing. The two datasets are processed in the same way as for the spin-$\frac{1}{2}$ experiment (Fig. 1.21). One point which should be borne in mind is that the two datasets may require scaling relative to each other, as a result of relaxation processes occurring during the mixing time τ_m. The density operator governing the system during τ_m for experiment 1 is proportional to $\hat{I}_z$, and so the system is subject to spin-lattice relaxation as described by the parameter T_{1Z}. However, the density operator during τ_m in experiment 2 is proportional to $(3\hat{I}_z^2 - I(I+1))$ or equivalently $\hat{T}^Q_{20}$, which corresponds to quadrupolar order, whose relaxation is governed by T_{1Q}. In general, T_{1Z} and T_{1Q} are different, and this can lead to the datasets resulting from the two experiments having different intrinsic amplitudes.

The relatively large widths of ^{2}H powder patterns mean that t_1 and t_2 data truncation of the sort described in Section 6.4.1 for spin-$\frac{1}{2}$ systems is nearly always a problem in ^{2}H exchange experiments. Using quadrupolar echoes (–τ–90°–τ–) before both the t_1 and t_2 periods (see Fig. 6.13 and the discussion in Section 6.4.1, but with the Hahn echo replaced with a quadrupole echo) is therefore very advantageous in producing undistorted spectra; an echo step before t_2 is almost essential in ^{2}H exchange experiments. Three-dimensional and reduced four-dimensional ^{2}H exchange experiments have been performed. Details of these can be found in reference [11].

A typical two-dimensional ^{2}H exchange spectrum is shown in Fig. 6.20 [9].

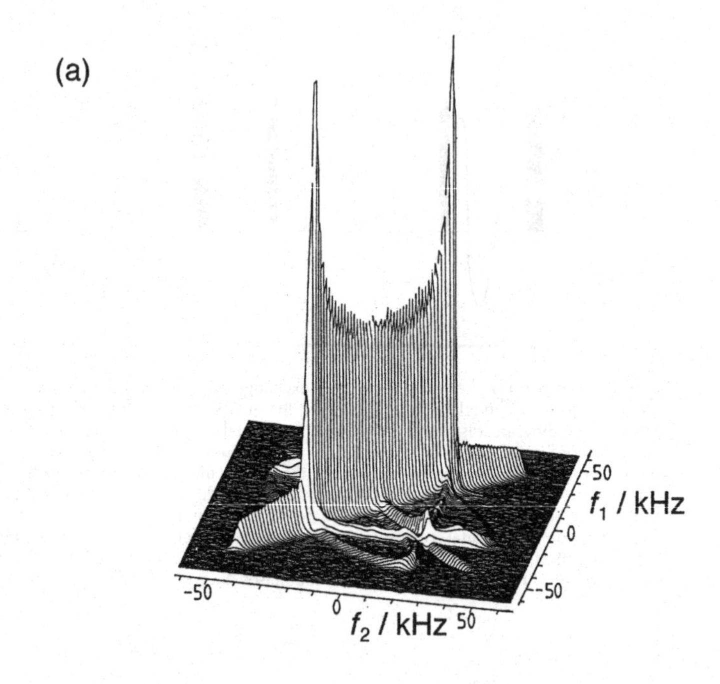

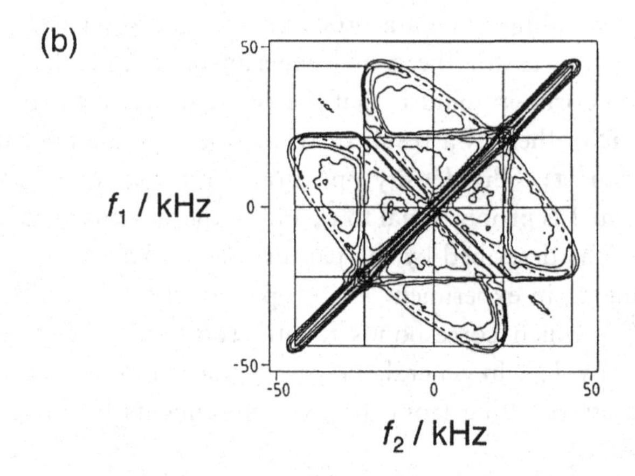

Fig. 6.20 A two-dimensional ^{2}H exchange spectrum for a static sample of DMS [9], (a) surface plot and (b) contour plot. As in the spin-$\frac{1}{2}$ case, the geometry of the elliptical ridges seen in the spectrum yields the angle of molecular reorientation occurring during the mixing period of the experiment.

6.5.5 *Resolving ^{2}H powder patterns*

Resolution of ^{2}H powder pattern lineshapes can be a problem in uniformly, or multiply-labelled samples. In many cases, ^{2}H labelling of specific sites is near impossible, especially in more complex molecules, or in naturally-occurring samples. In other cases, it would be excessively time consuming and expensive to specifically label all sites of interest in a molecule, sepa-

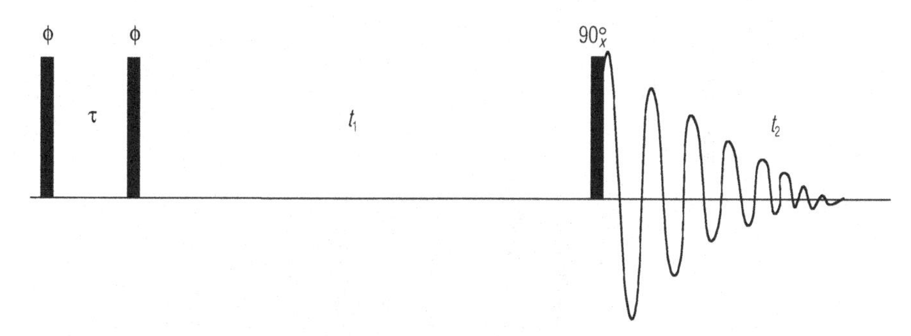

Fig. 6.21 The pulse sequence for the double-quantum ^{2}H experiment [20]. This two-dimensional experiment separates ^{2}H spinning sideband patterns (or alternatively, static-like ^{2}H quadrupole powder patterns) according to the ^{2}H double quantum chemical shift offset, so improving the resolution over a single-quantum experiment. In addition, the double-quantum transition frequency has no contribution from quadrupole coupling (to first order) so the double-quantum spectrum is not complicated by spinning sidebands. ^{2}H double-quantum coherence is excited with the initial 90°_ϕ–τ–90°_ϕ part of the pulse sequence and allowed to evolve in t_1 (double-quantum coherence is selected in t_1 by phase cycling the two 90° excitation pulses). A further 90°_x pulse then transforms the remaining double-quantum coherence into observable single-quantum coherence whose evolution is recorded in a FID in t_2. The entire experiment is conducted under magic-angle spinning. Details of molecular motion are then extracted from the separated ^{2}H spinning sideband patterns by simulation. The τ delay is of the order 10 μs. The t_1 period is rotor-synchronized so that the rotor phase at the start of t_2 is the same for all t_1 datasets. Without this, pure absorption two-dimensional spectra are impossible to produce.

rately in successive samples. Clearly, some method is needed to resolve ^{2}H signals from different sites. Magic-angle spinning can be useful, although the chemical shift range of ^{2}H is very small, so often MAS is not sufficient to resolve different sites. Two possibilities have been proposed for separating ^{2}H lineshapes in two-dimensional experiments. In the first, ^{2}H spinning sideband patterns (or static-like powder patterns) are separated according to the double-quantum ^{2}H chemical shift (Fig. 6.21) [20].

This has the effect of doubling the frequency gap between ^{2}H signals in the double-quantum dimension over what would occur in a single-quantum spectrum, and so improves the resolution over that in a normal one-dimensional MAS spectrum.

In the second, static-type ^{2}H powder patterns are separated according to the ^{13}C chemical shift of the ^{13}C nucleus the ^{2}H is bonded to [21]. The pulse sequence is shown in Fig. 6.22.

Initial ^{13}C transverse magnetization is generated by cross-polarization from ^{1}H. The entire experiment is conducted under magic-angle spinning, which of course averages the ^{13}C–^{2}H dipolar coupling to zero. It is reintroduced by a series of rotor-synchronized 180° pulses applied to the ^{13}C spins; $(2N - 2)$ 180° pulses are applied in N rotor periods. Thus multiple-quantum coherences involving the ^{13}C and ^{2}H spins can now be excited via the agency of the ^{13}C–^{2}H dipolar coupling. During the first two rotor periods of the pulse sequence shown in Fig. 6.22, zero- and double-quantum coherences between ^{13}C and ^{2}H spins are excited. In the t_1 period, transverse ^{2}H mag-

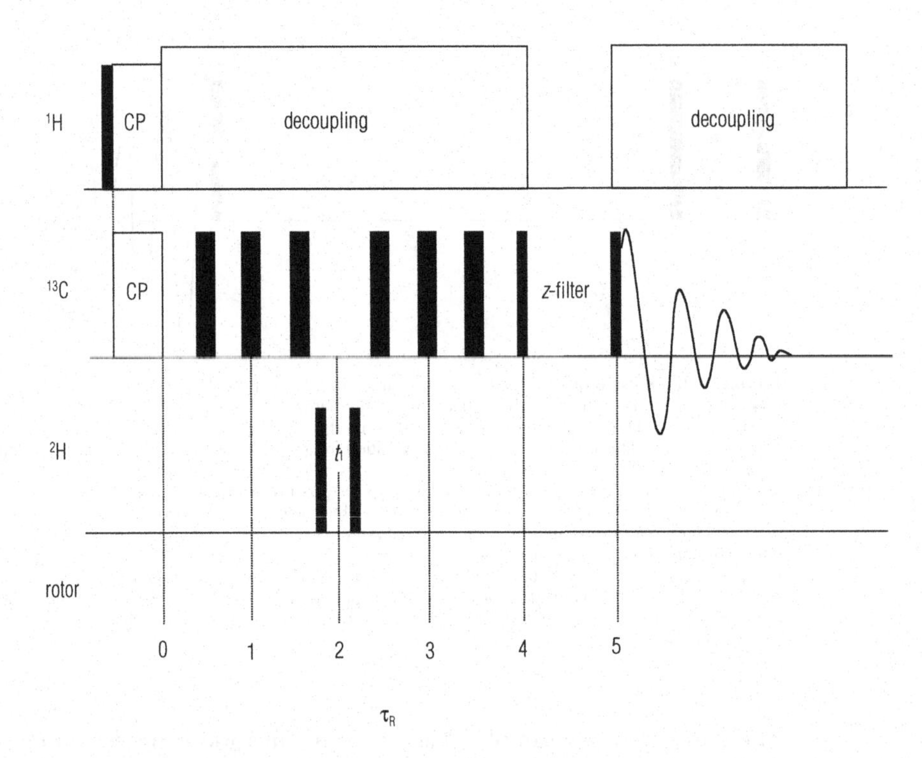

Fig. 6.22 The pulse sequence for a two-dimensional experiment to separate ^{2}H quadrupolar powder patterns according to the isotropic chemical shift of the ^{13}C spin the ^{2}H is bonded to [21] – see text for details.

netization is excited and allowed to evolve; in doing so it modulates the zero- and double-quantum coherences, and this is reflected in the final FID recorded in t_2. During the next two rotor cycles, the multiple-quantum coherences are reconverted. A *z*-filter preceding the detection period ensures a purely absorptive spectrum. Only one data set needs to be recorded, as it may be assumed that the f_1 spectrum (^{2}H dimension) is symmetric about $\omega_1 = 0$, providing of course that ^{2}H is set on resonance.

Notes

1. The trace of a matrix is the sum of its diagonal elements.
2. If we are interested in the effect of the dipolar hamiltonian on the NMR spectrum, as in Chapter 4 for instance, we would only be interested in the first-order average hamiltonian formed from Equation (6.31). Thus, only the $q = 0$ term is found to have any relevance to the appearance of the NMR *spectrum*. That is not the case here with relaxation where the process depends on the detailed time variation of the hamiltonian. Thus, terms in the rotating frame dipolar hamiltonian which would be unimportant for the spectrum, i.e. those with $q = \pm1, \pm2$, are important in relaxation.
3. If the T_2 relaxation time for the spin under study is very long, the *x*-magnetization may not fully dephase during τ_m, in which case phase cycling needs to be used to remove this component. This can be done by alternating the phase of the storage pulse and receiver together in successive scans.

References

1. Abragam, A. (1961) *Principles of Magnetic Resonance*, Chapter X. Clarendon Press, Oxford.
2. Duer, M.J. & Levitt, M.H. (1992) *Solid-State NMR*, **1**, 211.
3. Tycko, R, Dabbagh, G. & Mirau, P.A. (1989) *J. Magn. Reson. A*, **85**, 265.
4. Frydman, L., Chingas, G.C., Lee, Y.K., Grandinetti, P.J., Eastman, M.A., Barrall, G.A. & Pines, A. (1992) *J. Chem. Phys.*, **97**, 4800.
5. Gan, Z. (1992) *J. Am. Chem. Soc.*, **114**, 8307.
6. Frydman, L., Vallabhaneni, S., Lee, Y.K. & Emsley, L. (1994) *J. Chem. Phys.*, **101**, 111.
7. Schmidt-Rohr, K., Clauss, J. & Spiess, H.W. (1992) *Macromol.*, **25**, 3273.
8. Spiess, H.W. (1978) *NMR: Basic Principles and Progress*, **15**, 55.
9. Schmidt, C., Blümich, B. & Spiess, H.W. (1988) *J. Magn. Reson.*, **79**, 269.
10. Hu, W.-G., Boeffel, C. & Schmidt-Rohr, K. (1999) *Macromol.*, **32**, 1611.
11. Schmidt-Rohr, K. & Spiess, H.W. (1994) *Multidimensional Solid-State NMR and Polymers*. Academic Press, London.
12. Wefing, S., Kaufmann, S. & Spiess, H.W. (1988) *J. Chem. Phys.*, **89**, 1234.
13. Larsen, F.H., Jakobsen, H.J., Ellis, P.D. & Nielsen, N.C. (1997) *J. Phys. Chem. A*, **101**, 8597.
14. Larsen, F.H., Jakobsen, H.J., Ellis, P.D. & Nielsen, N.C. (1998) *Chem. Phys. Lett.*, **292**, 467.
15. Hoatson, G.L. & Vold, R.L. (1994) *NMR: Basic Principles and Progress*, **32**.
16. Vold, R.L., Hoatson, G.L. & Tse, T.Y. (1996) *Chem. Phys. Lett.*, **263**, 271.
17. Hoatson, G.L., Vold, R.L. & Tse, T.Y. (1994) *J. Chem. Phys.*, **100**, 4756.
18. Scheicher, A., Müller, K. & Kothe, G. (1990) *J. Chem. Phys.*, **92**, 6432.
19. Torchia, D.A. & Szabo, A. (1991) *J. Magn. Reson.*, **94**, 152.
20. Duer, M.J. & Stourton, E.C. (1997) *J. Magn. Reson.*, **129**, 44.
21. Sandström, D., Hong, M. & Schmidt-Rohr, K. (1999) *Chem. Phys. Lett.*, **300**, 213.

Appendix A NMR Properties of Commonly Observed Nuclei

Nucleus	Spin[a]	Natural abundance %	Magnetogyric ratio ($\gamma/10^6$) rad s^{-1} T^{-1}	Quadrupole moment[b] ($10^{28}Q$)/m^2	Frequency at 9.4 T ($\omega_0/2\pi$)/MHz
^{1}H	$\frac{1}{2}$	99.99	267.522		400.000
^{2}H	1	0.015	41.066	2.8×10^{-3}	61.404
^{6}Li	1	7.4	39.371	-8×10^{-4}	58.868
^{7}Li	$\frac{3}{2}$	92.6	103.975	-4×10^{-2}	155.464
^{11}B	$\frac{3}{2}$	80.1	85.847	4×10^{-2}	128.356
^{13}C	$\frac{1}{2}$	1.1	67.283		100.58
^{14}N	1	99.6	19.338	1×10^{-2}	28.912
^{15}N	$\frac{1}{2}$	0.37	−27.126		40.544
^{17}O	$\frac{5}{2}$	0.04	−36.281	-2.6×10^{-2}	54.244
^{19}F	$\frac{1}{2}$	100	251.815		376.376
^{23}Na	$\frac{3}{2}$	100	70.808	0.1	105.864
^{27}Al	$\frac{5}{2}$	100	69.763	0.15	104.308
^{29}Si	$\frac{1}{2}$	4.7	−53.190		79.468
^{31}P	$\frac{1}{2}$	100	108.394		161.92
^{79}Br	$\frac{3}{2}$	50.5	67.228	0.37	100.52
^{81}Br	$\frac{3}{2}$	49.5	72.468	0.31	108.356
^{107}Ag	$\frac{1}{2}$	51.8	−10.889		16.188
^{109}Ag	$\frac{1}{2}$	48.2	−12.518		18.612
^{111}Cd	$\frac{1}{2}$	12.75	−56.926		84.862
^{113}Cd	$\frac{1}{2}$	12.3	−59.550		88.772
^{117}Sn	$\frac{1}{2}$	7.6	−95.78		142.528
^{119}Sn	$\frac{1}{2}$	8.6	−100.21		149.163
^{129}Xe	$\frac{1}{2}$	24.4	−74.521		111.299
^{195}Pt	$\frac{1}{2}$	33.8	57.68		85.656
^{207}Pb	$\frac{1}{2}$	22.1	55.805		83.730

[a] Ground state spin.

[b] For many nuclei, there are several different values of nuclear quadrupole moments reported in the literature; the values given here are intended only to indicate the relative size of the quadrupole moments.

Appendix B The General Form of a Spin Interaction Hamiltonian in Terms of Spherical Tensors and Spherical Tensor Operators

We stated in Chapter 1 that any spin interaction hamiltonian describing an interaction A can be expressed in the form

$$\hat{H}_A = -\gamma \hat{\mathbf{I}} \cdot \mathbf{A}_{\mathrm{loc}} \cdot \hat{\mathbf{J}} \tag{B.1}$$

where $\mathbf{A}_{\mathrm{loc}}$ is a second-rank Cartesian tensor which describes the strength and orientation dependence of the local spin interaction and $\hat{\mathbf{J}}$ is a Cartesian vector operator whose exact nature depends on the particular spin interaction (see Section 1.4); $\hat{\mathbf{I}}$ is the usual Cartesian nuclear spin operator.

Expanding the scalar products in (B.1) and incorporating the factor of $-\gamma$ into $\mathbf{A}_{\mathrm{loc}}$, we have:

$$\begin{aligned} \hat{H}_A &= \sum_{\alpha,\beta=x,y,z} \hat{I}_\alpha A^{\mathrm{loc}}_{\alpha\beta} \hat{J}_\beta \\ &= \sum_{\alpha,\beta=x,y,z} A^{\mathrm{loc}}_{\alpha\beta} \hat{I}_\alpha \hat{J}_\beta \end{aligned} \tag{B.2}$$

Then collecting together the operators which act on the spin coordinates in Equation (B.2), i.e. the products $\hat{I}_\alpha \hat{J}_\beta$, we can rewrite Equation (B.2) as:

$$\hat{H}_A = \sum_{\alpha,\beta=x,y,z} A^{\mathrm{loc}}_{\alpha\beta} \hat{R}_{\alpha\beta} \tag{B.3}$$

where $\hat{R}_{\alpha\beta} = \hat{I}_\alpha \hat{J}_\beta$ are the elements of a second-rank Cartesian operator $\hat{\mathbf{R}}$.

Now the nine components $\hat{R}_{\alpha\beta}$, $\alpha, \beta = x, y, z$ of a second-rank Cartesian tensor can be conveniently decomposed into the sum of:

1. a scalar $\hat{T}^{(0)}$

$$\hat{T}^{(0)} = \frac{1}{3} \sum_\alpha \hat{R}_{\alpha\alpha} \tag{B.4}$$

2. an antisymmetric tensor $\hat{\mathbf{T}}^{(1)}$ with three components given by

$$\hat{T}^{(1)}_{\alpha\beta} = \frac{1}{2}(\hat{R}_{\alpha\beta} - \hat{R}_{\beta\alpha}) \qquad \text{(B.5)}$$

3. a symmetric tensor $\hat{\mathbf{T}}^{(2)}$ with zero trace and five components given by

$$\hat{T}^{(2)}_{\alpha\beta} = \frac{1}{2}(\hat{R}_{\alpha\beta} + \hat{R}_{\beta\alpha}) - \frac{1}{3}\hat{R}_{\gamma\gamma}\delta_{\alpha\beta} \qquad \text{(B.6)}$$

such that an element of the Cartesian tensor is given by

$$\hat{R}_{\alpha\beta} = \delta_{\alpha\beta}\hat{T}^{(0)} + \hat{T}^{(1)}_{\alpha\beta} + \hat{T}^{(2)}_{\alpha\beta} \qquad \text{(B.7)}$$

The relevance of this decomposition is the particular transformation properties under rotations of the components $\hat{T}^{(0)}$, $\hat{\mathbf{T}}^{(1)}$, $\hat{\mathbf{T}}^{(2)}$.

$\hat{T}^{(0)}$ is a scalar and so is invariant under rotations. In other words, it transforms under rotations as an *s*-orbital would or a spherical harmonic of rank 0.

The three components of $\hat{\mathbf{T}}^{(1)}$ form a vector and just transform among themselves under rotations or, in other words, under a rotation, a component of $\hat{\mathbf{T}}^{(1)}$ transforms in general into a linear combination of the three components of $\hat{\mathbf{T}}^{(1)}$. Thus it behaves just like a set of *p*-orbitals under rotations.

Finally, the five components of $\hat{\mathbf{T}}^{(2)}$ also transform among themselves under rotations and thus behave like a set of *d*-orbitals.

It is now not difficult to see that the tensors $\hat{T}^{(0)}$, $\hat{\mathbf{T}}^{(1)}$, $\hat{\mathbf{T}}^{(2)}$ are representations of the spherical point group (equivalently termed the full rotation group), just like atomic orbitals are; any rotation transforms a component of the tensor into a linear combination of the tensor components.

Our aim is to rewrite the interaction hamiltonian of Equation (B.1) in terms of just such tensors. We define irreducible spherical tensors through their transformation properties under a rotation. An irreducible spherical tensor of rank k transforms like the $\Gamma^{(k)}$ representation of the spherical point group, and so it has $2k + 1$ components. In order to provide an irreducible representation of the $\Gamma^{(k)}$ representation, a component of the irreducible spherical tensor T_{kq} (k defines the *rank* of the tensor, i.e. the irreducible representation that the tensor transforms as, and q defines the component or *order* of the tensor) must transform under a rotation according to

$$\hat{R}(\alpha,\beta,\gamma)T_{kq}\hat{R}^{-1}(\alpha,\beta,\gamma) = \sum_{q'} T_{kq'}D^{k}_{q'q}(\alpha,\beta,\gamma) \qquad \text{(B.8)}$$

where $\hat{R}(\alpha, \beta, \gamma)$ is the rotation operator which rotates an object by the Euler angles (α, β, γ) within the axis frame the object is defined in (see Box 1.2 for further details of rotations) and the $D^{k}_{q'q}(\alpha, \beta, \gamma)$ are elements of Wigner rotation matrices (see Box 4.2 for their definition). Equation (B.8) describes the *active* rotation of the T_{kq} component within the axis frame it is defined in. An equivalent definition of the T_{kq} is that they must satisfy

$$\begin{aligned} [\hat{J}_z, T_{kq}] &= qT_{kq} \\ [\hat{J}_\pm, T_{kq}] &= \sqrt{k(k+1) - q(q \pm 1)}T_{kq\pm 1} \end{aligned} \tag{B.9}$$

where $\hat{J}_z$ is the operator for rotation about z of the frame in which T_{kq} is defined and the $\hat{J}_\pm$ are defined by

$$\hat{J}_\pm = \hat{J}_x \pm i\hat{J}_y \tag{B.10}$$

where $\hat{J}_x$ and $\hat{J}_y$ are operators for rotation about x and y respectively.

The Cartesian tensor operator, $\hat{\mathbf{R}}$, with elements $\hat{R}_{\alpha\beta} = \hat{I}_\alpha \hat{J}_\beta$, does not satisfy these criteria, although it is clear that linear combinations of its components do, since $\hat{\mathbf{R}}$ can be written in terms of $\hat{T}^{(0)}$, $\hat{\mathbf{T}}^{(1)}$, $\hat{\mathbf{T}}^{(2)}$ (Equation (B.7)) which do transform as representations of the spherical point group (although not satisfying the criteria provided by Equations (B.8) and (B.9), so $\hat{T}^{(0)}$, $\hat{\mathbf{T}}^{(1)}$, $\hat{\mathbf{T}}^{(2)}$ do not transform as *irreducible* representations of the spherical point group).

To find *irreducible* tensor components T_{kq} in terms of linear combinations of the components of a second rank Cartesian tensor **R** (we treat a general Cartesian tensor here and do not restrict ourselves to Cartesian tensor operators), we begin by noting that a second-rank Cartesian tensor transforms under rotations like a set of products $u_\alpha v_\beta$, $\alpha, \beta = x, y, z$, where u_α and v_β are simply components of two general Cartesian vectors **u** and **v** respectively. It is straightforward to find the irreducible spherical tensor components of a Cartesian vector, i.e. the linear combinations of $\{u_\alpha\}$ (or v_β) which obey the commutation relations of Equation (B.9):

$$\begin{aligned} u_{11} &= -\sqrt{\frac{1}{2}}(u_x + iu_y) \\ u_{10} &= u_z \\ u_{1-1} &= \sqrt{\frac{1}{2}}(u_x - iu_y) \end{aligned} \tag{B.11}$$

The irreducible spherical tensor components of the products $u_\alpha v_\beta$ are then formed from products of u_{1q} and v_{1q} as defined in Equation (B.11). The product of two irreducible spherical tensors $\mathbf{u}^{(k)}$ and $\mathbf{v}^{(k')}$ is itself a linear combination of irreducible spherical tensor components $\{w_{KQ}\}$, with K running from $k + k'$ to $|k - k'|$ and $Q = q + q'$ where q and q' denote the possible components of $\mathbf{u}^{(k)}$ and $\mathbf{v}^{(k')}$ [1]. In more detail, the w_{KQ} component of this linear combination is given by

$$w_{KQ} = \sum_{q,q'} \langle kk'qq'|KQ\rangle u_{kq} v_{k'q'} \tag{B.12}$$

The daunting looking symbol $\langle kk'qq'|KQ\rangle$ is just a Clebsch–Gordan coefficient (sometimes known as a Wigner coefficient) [1]. These coefficients are related to *Wigner-3j* symbols, denoted $\begin{pmatrix} K & k & k' \\ Q & q & q' \end{pmatrix}$, which are tabulated in many places:

$$\begin{pmatrix} K & k & k' \\ Q & q & q' \end{pmatrix} = \frac{(-1)^{K-k-q'}}{\sqrt{2k'+1}} \langle kk'qq'|KQ\rangle \tag{B.13}$$

Thus the T_{00} irreducible spherical tensor component of the set of products $\{u_\alpha v_\beta\}$ is given by

$$\begin{aligned} T_{00}^{(u,v)} &= \sum_{q,q'} \langle 11qq'|00\rangle u_{1q} v_{1q'} \\ &= -\sqrt{\frac{1}{3}}(u_{11}v_{1-1} - u_{10}v_{10} + u_{1-1}v_{11}) \\ &= -\sqrt{\frac{1}{3}} \sum_\alpha u_\alpha v_\alpha \end{aligned} \tag{B.14}$$

Since the set of products $\{u_\alpha v_\beta\}$ transform like the components of a second-rank Cartesian tensor $\{R_{\alpha\beta}\}$, we can replace the product $u_\alpha v_\beta$ in Equation (B.14) with $R_{\alpha\alpha}$, giving us

$$T_{00} = -\sqrt{\frac{1}{3}} \sum_\alpha R_{\alpha\alpha} \tag{B.15}$$

To find T_{10}, we use

$$\begin{aligned} T_{10}^{(u,v)} &= \sum_{q,q'} \langle 11qq'|10\rangle u_{1q} v_{1q'} \\ &= \sqrt{\frac{1}{2}}(u_{11}v_{1-1} - u_{1-1}v_{11}) \\ &= \sqrt{\frac{1}{2}} i(u_x v_y - u_y v_x) \end{aligned} \tag{B.16}$$

which enables us to write

$$T_{10} = \sqrt{\frac{1}{2}} i(R_{xy} - R_{yx}) \tag{B.17}$$

The $T_{1\pm1}$ are then found by application of the second of Equation (B.9) to T_{10} in Equation (B.17):

$$T_{1\pm1}^{(u,v)} = \frac{1}{\sqrt{1(1+1)}}[\hat{J}_{\pm}, T_{10}]$$
$$= \mp\frac{1}{2}i(u_y v_z - u_z v_y \pm i(u_z v_x - u_x v_z)) \tag{B.18}$$

which gives us

$$T_{1\pm1} = \mp\frac{1}{2}i(R_{yz} - R_{zy} \pm i(R_{zx} - R_{xz})) \tag{B.19}$$

The components T_{2q} are found in a similar manner to be:

$$T_{20} = \sqrt{\frac{1}{6}}(2R_{zz} - R_{xx} - R_{yy})$$
$$T_{2\pm1} = \mp\frac{1}{2}(R_{xz} + R_{zx} \pm i(R_{yz} + R_{zy})) \tag{B.20}$$
$$T_{2\pm2} = \frac{1}{2}(R_{xx} - R_{yy} \pm i(R_{xy} + R_{yx}))$$

We now return to our equation for $\hat{H}_A$:

$$\hat{H}_A = \sum_{\alpha,\beta=x,y,z} A_{\alpha\beta}^{\text{loc}} \hat{R}_{\alpha\beta} \tag{B.21}$$

It is clear from this equation that $\hat{H}_A$ is a scalar product of two second-rank tensors (it must be a *scalar* product, as $\hat{H}_A$ is itself a scalar, energy being a scalar quantity), $\mathbf{A}^{\text{loc}}$ and $\hat{\mathbf{R}}$. As we saw at the beginning of this discussion, we can replace each of these tensors by a sum of zeroth-, first- and second-rank irreducible spherical tensors, i.e.

$$\mathbf{A}^{\text{loc}} \rightarrow A_{\text{loc}}^{(0)} + \mathbf{A}_{\text{loc}}^{(1)} + \mathbf{A}_{\text{loc}}^{(2)}$$
$$\hat{\mathbf{R}} \rightarrow \hat{R}_{\text{loc}}^{(0)} + \hat{\mathbf{R}}^{(1)} + \hat{\mathbf{R}}^{(2)} \tag{B.22}$$

The scalar product of two irreducible spherical tensors $\mathbf{R}^{(k)}$ and $\mathbf{S}^{(k)}$ is defined as

$$\hat{\mathbf{R}}^{(k)} \cdot \mathbf{S}^{(k)} = \sum_q (-1)^q R_{kq} S_{kq} \tag{B.23}$$

Replacing the second-rank Cartesian tensors of Equation (B.21) with the equivalent irreducible spherical tensors, we obtain;

$$\hat{H}_A = \sum_{k=0}^{2} \sum_{q=-k}^{+k} (-1)^q \Lambda_{k-q}^{A} \hat{T}_{kq}^{A} \tag{B.24}$$

It is worth noting that when the interaction Cartesian tensor $\mathbf{A}^{\text{loc}}$ is traceless and symmetric, as it is for dipolar coupling and quadrupole coupling, all the terms Λ_{1q}^{A} are zero, as is the isotropic term Λ_{00}^{A}. Thus, only the $\Lambda_{k-q}^{A}\hat{T}_{kq}^{A}$ terms remain in Equation (B.24), i.e. the summation over k reduces to the single term $k = 2$. This is not the case, however, for the chemical shielding interaction in which the Λ_{1q}^{cs} terms (the antisymmetric part of the Cartesian shielding tensor σ) are non-zero, as discussed in Section 3.1.2. However, the $k = 1$ terms in Equation (B.24) are unimportant even for chemical shielding when the Zeeman term dominates the total hamiltonian describing the spin system.

The irreducible spherical tensor spin operators $\hat{T}_{kq}^{A}$ and irreducible spherical tensors Λ_{kq}^{A} describing the various possible spin interactions A can be derived from Equations (B.2) and (B.15)–(B.20) and are given below:

Chemical shift for spin I

$$\hat{H}_{cs} = \gamma\hat{\mathbf{I}}\cdot\boldsymbol{\sigma}\cdot\mathbf{B}_0$$

$$T_{00}^{cs} = -\sqrt{\frac{1}{3}}\sum_{\alpha=x,y,z}\hat{I}_{\alpha}B_{0\alpha}$$

$$T_{10}^{cs} = i\sqrt{\frac{1}{2}}(\hat{I}_xB_{0y} - \hat{I}_yB_{0x})$$

$$T_{1\pm1}^{cs} = -\frac{1}{2}(\hat{I}_zB_{0x} - \hat{I}_xB_{0z} \pm i(\hat{I}_zB_{0y} - \hat{I}_yB_{0z}))$$

$$\hat{T}_{20}^{cs} = \sqrt{\frac{1}{6}}(2\hat{I}_zB_{0z} - \hat{I}_xB_{0x} - \hat{I}_yB_{0y})$$

$$\hat{T}_{2\pm1}^{cs} = \mp\frac{1}{2}(\hat{I}_xB_{0z} + \hat{I}_zB_{0x} \pm i(\hat{I}_yB_{0z} + \hat{I}_zB_{0y}))$$

$$\hat{T}_{2\pm2}^{cs} = \frac{1}{2}(\hat{I}_xB_{0x} - \hat{I}_yB_{0y} \pm i(\hat{I}_xB_{0y} + \hat{I}_yB_{0x}))$$

The irreducible spherical tensor components of the shielding tensor in its principal axis frame (PAF) are:

$$\Lambda_{00}^{cs} = -\gamma\sqrt{\frac{1}{3}}(\sigma_{xx}^{\mathrm{PAF}} + \sigma_{yy}^{\mathrm{PAF}} + \sigma_{zz}^{\mathrm{PAF}})$$

$$\Lambda_{10}^{cs} = \Lambda_{1\pm1}^{cs} = 0$$

$$\Lambda_{20}^{cs} = \gamma\sqrt{\frac{1}{6}}(2\sigma_{zz}^{\mathrm{PAF}} - \sigma_{xx}^{\mathrm{PAF}} - \sigma_{yy}^{\mathrm{PAF}})$$

$$\Lambda_{2\pm1}^{cs} = 0$$

$$\Lambda_{2\pm2}^{cs} = \frac{1}{2}\gamma(\sigma_{xx}^{\mathrm{PAF}} - \sigma_{yy}^{\mathrm{PAF}})$$

Dipolar coupling between spins I and S

$$\hat{H}_{dd} = -2\hat{\mathbf{I}}\cdot\mathbf{D}\cdot\hat{\mathbf{S}}$$

$$\hat{T}_{00}^{IS} = -\frac{2}{\sqrt{3}}\left(\hat{I}_z\hat{S}_z + \frac{1}{2}(\hat{I}_+\hat{S}_- + \hat{I}_-\hat{S}_+)\right)$$

$$\hat{T}_{10}^{IS} = \frac{1}{\sqrt{2}}(\hat{I}_-\hat{S}_+ - \hat{I}_+\hat{S}_-) \quad \hat{T}_{1\pm1}^{IS} = (-\hat{I}_\pm\hat{S}_z + \hat{I}_z\hat{S}_\pm)$$

$$\hat{T}_{20}^{IS} = \sqrt{\frac{1}{6}}(3\hat{I}_z\hat{S}_z - \hat{\mathbf{I}}\cdot\hat{\mathbf{S}}) \quad \hat{T}_{2\pm1}^{IS} = \mp\frac{1}{2}(\hat{I}_\pm\hat{S}_z + \hat{I}_z\hat{S}_\pm) \quad \hat{T}_{2\pm2}^{IS} = \frac{1}{2}\hat{I}_\pm\hat{S}_\pm$$

The irreducible spherical tensor components of the dipole coupling tensor in its principal axis frame (PAF) are:

$$\begin{aligned}
\Lambda_{00}^{IS} &= 0 \\
\Lambda_{10}^{IS} &= \Lambda_{1+1}^{IS} = 0 \\
\Lambda_{20}^{IS} &= -\sqrt{6}D_{zz}^{PAF} = -\sqrt{6}d \\
\Lambda_{2\pm1}^{IS} &= 0 \\
\Lambda_{2\pm2}^{IS} &= 0
\end{aligned}$$

where d is the dipolar coupling constant.

Quadrupole coupling on spin I

$$\hat{H}_Q = \frac{eQ}{2I(2I-1)\hbar}\hat{\mathbf{I}}\cdot\mathbf{V}\cdot\hat{\mathbf{I}}$$

$$\hat{T}_{00}^{Q} = -\frac{1}{\sqrt{2}}$$

$$\hat{T}_{10}^{Q} = \sqrt{2}\hat{I}_z \qquad \hat{T}_{1\pm1}^{Q} = \mp\sqrt{\frac{1}{2}}(\hat{I}_x \pm i\hat{I}_y)$$

$$\hat{T}_{20}^{Q} = \sqrt{\frac{1}{6}}(3\hat{I}_z^2 - \hat{\mathbf{I}}\cdot\hat{\mathbf{I}}) \qquad \hat{T}_{2\pm1}^{Q} = \mp(\hat{I}_\pm\hat{I}_z + \hat{I}_z\hat{I}_\pm) \qquad \hat{T}_{2\pm2}^{Q} = \frac{1}{2}\hat{I}_\pm^2$$

The irreducible spherical tensor components of the quadrupole coupling tensor in its principal axis frame (PAF) are:

$$\begin{aligned}
\Lambda_{00}^{Q} &= 0 \\
\Lambda_{10}^{Q} &= \Lambda_{1\pm1}^{Q} = 0 \\
\Lambda_{20}^{Q} &= \frac{\sqrt{6}\chi_{zz}^{\mathrm{PAF}}}{4I(2I-1)} = \frac{\sqrt{6}e^2qQ}{4I(2I-1)\hbar} \\
\Lambda_{2\pm1}^{IS} &= 0 \\
\Lambda_{2\pm2}^{IS} &= \frac{e^2qQ}{4I(2I-1)\hbar}\eta_Q
\end{aligned}$$

where χ is the quadrupole coupling constant and η_Q is the asymmetry.

References

1. Brink, D.M. & Satchler, G.R. (1994) *Angular Momentum*, Clarendon Press, Oxford.

Index